00034549
WNRE LIBRARY, PINAWA, MANITOBA.

D1527229

CONSTITUTIVE EQUATIONS IN PLASTICITY

CONSTITUTIVE EQUATIONS IN PLASTICITY

Edited by Ali S. Argon

The MIT Press
Cambridge, Massachusetts, and London, England

PUBLISHER'S NOTE
This format is intended to reduce the cost of publishing certain works in book form and to shorten the gap between editorial preparation and final publication. The time and expense of detailed editing and composition in print have been avoided by photographing the text of this book directly from the author's transcript.

Printed in the United States of America

Library of Congress Cataloging in Publication Data

Constitutive equations in plasticity.

Includes indexes.
1. Deformations (Mechanics) 2. Dislocations in crystals. 3. Plasticity. 4. Boundary value problems.
I. Argon, Ali S.
TA417.6.C66 620.1'1233 75-22409
ISBN 0-262-01042-9

PREFACE

Modern technology has increasingly many applications in which complex boundary value problems of inelastic deformation must be solved with high precision to assure reliable and safe performance of structural components subjected to complex histories in service. Often important strain-dependent and time-dependent micro-structural changes occur during the operation of the components which must be taken into account. This requires knowledge of accurate constitutive relations based directly on the micro-structural parameters which govern the principal mechanisms of inelastic deformation. Only in this way can the changes in the behavior of structural materials be modelled to account for the effects of complex service histories. Unfortunately most of the currently used constitutive relations for inelastic behavior are still of a highly empirical nature lacking even the refinement of proper dimensionless normalization of the important parameters. Although they are generally useful in the range of the parameters where actual measurements were made, they are often quite inadequate for extrapolation and lack the proper mechanistic structure to account for the effects of micro-structural alterations. This state of affairs is a result of poor communications between workers in different areas. The materials scientist is usually preoccupied with observing and measuring phenomena in pure form on materials of little engineering utility under conditions of stress far removed from real applications, and is often unwilling to generalize his fundamental findings into a form useful to a practical stress analyst. The applied mechanician on the other hand has gone to great lengths of developing constitutive relations of all types for all kinds of imaginable (and unimaginable) behavior, having abstract symmetry properties and formulated in all encompassing generalized, compact notations. However, seldom is there an attempt made to connect these formulations to the behavior of real materials and prescribe some fundamental experiments necessary to flesh out the seemingly powerful skeletal structure. Meanwhile the stress analyst-engineer, often untrained either in materials science or in deciphering the all-powerful notation of the applied mechanician to extract a potentially useful formulation, is required to come up with working hardware.

This book is an attempt to bring together a group of researchers, in both applied mechanics and materials science, who have recognized the need for a fundamental micro-structural approach in constructing constitutive equations. The object of the undertaking was not so much to come up with a catalog of forms of equations which can be used directly by the practicing engineer but rather to indicate the dimensions of the required task and furnish detailed descriptions of the required ingredients for representing the various complex behavior patterns and thereby stimulate both the applied mechanician and materials scientist. Accordingly, the first five chapters of the book strive to furnish the proper micro-structural foundations for constitutive relations for inelastic behavior from the point of view of applied mechanics and the internal defect structure which governs this behavior. The following eight chapters describe as accurately as possible the changes in the defect structure upon both monotonic and cyclic straining in both pure and particle strengthened metals, and in a temperature range from $0^{o}K$ to near the melting point. The remaining two chapters give detailed examples of the construction and use of specific constitutive relations for nuclear reactor applications which is a prime area requiring highly reliable constitutive relations.

I am grateful for the cooperation of all the contributors to this book who have done their best to come up with a certain amount of internal integration. The collection of their individual expertise, of course, deserves the major credit. In producing this book I have obtained the generous financial help of the Department of Mechanical Engineering at M.I.T. for which I am grateful to Professor Ascher H. Shapiro. I am grateful in addition to Margaret Underdown who did all the typing expertly, and finally to my wife Xenia and daughter Alice who furnished key help in preparing the author and subject indices.

Cambridge, Massachusetts
February, 1975

Ali S. Argon

CONTRIBUTORS

A. S. Argon
Massachusetts Institute of Technology, Cambridge, Massachusetts, U.S.A.

M. F. Ashby
University Engineering Department, Cambridge, England

L. M. Brown
Cavendish Laboratory, Cambridge, England

G. Y. Chin
Bell Laboratories, Murray Hill, New Jersey, U.S.A.

H. J. Frost
University Engineering Department, Cambridge, England

J. H. Gittus
United Kingdom Atomic Energy Authority, Reactor Fuel Element Laboratories, Springfields, Salwick, Preston, England

J. C. Grosskreutz
Black and Veatch Consulting Engineers, Kansas City, Missouri, U.S.A.

E. W. Hart
General Electric Corporation Research and Development Laboratory, Schenectady, New York, U.S.A.

B. Ilschner
Institut für Werkstoffwissenschaften I der Universität Erlangen-Nürnberg, Federal Republic of Germany

V. Z. Jankus
Argonne National Laboratory, Argonne, Illinois, U.S.A.

U. F. Kocks
Argonne National Laboratory, Argonne, Illinois, U.S.A.

C. Y. Li
Cornell University, Ithaca, New York, U.S.A.

V. K. Lindroos
Helsinki University of Technology, Otaniemi, Finland

H. M. Miekk-oja
Helsinki University of Technology, Otaniemi, Finland

H. Mughrabi
Max-Planck-Institut für Metallforschung, Stuttgart, Federal Republic of Germany

P. Neumann
Max-Planck-Institut für Eisenforschung, Düsseldorf, Federal Republic of Germany

R. B. Poeppel
Argonne National Laboratory, Argonne, Illinois, U.S.A.

J. R. Rice
Brown University, Providence, Rhode Island, U.S.A.

W. M. Stobbs
Cavendish Laboratory, Cambridge, England

R. W. Weeks
Argonne National Laboratory, Argonne, Illinois, U.S.A.

G. L. Wire
Cornell University, Ithaca, New York, U.S.A.

H. Yamada
Cornell University, Ithaca, New York, U.S.A.

J. Zarka
Ecole Polytechnique, Paris, France

CONTENTS

ABSTRACT. *The physical-microstructural basis of constitutive relations for plastic deformation involving dislocation motion is reviewed and the more detailed developments appearing in the following chapters is put in perspective. Areas where recent advances have occurred are pointed out as well as areas where further developments are urgently needed.*

1. PHYSICAL BASIS OF CONSTITUTIVE EQUATIONS FOR INELASTIC DEFORMATION

A. S. Argon

1.1 INTRODUCTION

The connection between externally measurable parameters such as stress, strain rate, temperature, and some internal parameters such as hardness, which govern the inelastic behavior of solids are called constitutive relations. Modern technology has increasingly many applications in which complex boundary value problems of inelastic deformation must be solved with high precision requiring the use of constitutive relations. Developments in numerical techniques for solution of boundary value problems have now made unnecessary the earlier reliance on convenient but inexact constitutive relations and have focused on the need for more accurate relations. The inelastic behavior of crystalline solids is accomplished almost exclusively by mobile structural imperfections such as dislocations and vacancies, which are altered not only by the deformation history but also by thermal motion. It is almost mandatory that accurate constitutive relations be based on a physical and mechanistic foundation. In this chapter we will describe a general philosophy of approach to constitutive relations based on such a mechanistic point of view, making use of basic thermodynamics and absolute reaction rate theory. The approach given will be only in skeletal outline since many of the specific aspects will be discussed in greater detail in the following chapters in this book.

We will begin first by giving a general, semi-quantitative, description of how macroscopic constitutive relations could be synthesized starting with a detailed description of the defect structure of the crystal grains making up a polycrystalline aggregate and a description of the mechanisms of inelastic deformation in them. This multi-tier summation problem will serve principally as a guide to establish proper forms and functional relationships for the external variables which could then be used in phenomenological approaches as a skeleton to be fleshed-out by a minimum number of experimental measurements, and where extrapolation and interpolation can then be performed with greater reliability. Accordingly, we will then describe such phenomenological approaches inspired by the mechanistic approach. In many areas both the mechanistic understanding and the phenomenological description still remain incomplete such as in the case of anelastic behavior and the Bauschinger effect; while in some other areas much of the mechanistic developments seem to have missed the mark entirely - notably in the case of the mechanism of diffusion controlled power-law creep. Hence our presentation here as well as the various contributions in this book, taken collectively will fall short of furnishing a complete guide to the construction of constitutive relations but will go a long way toward elucidating the physical-structural bases of such relations to hopefully stimulate further detailed developments.

1.2 BASIC INGREDIENTS OF CONSTITUTIVE RELATIONS

1.2.1 *Mechanisms of Inelastic Deformation*

Inelastic deformation in crystalline matter can occur by one or a combination of several mechanisms: twinning or martensitic shear transformation, glide and climb of dislocations, or by diffusional transport of matter often accompanied by grain boundary sliding. Twinning and martensitic shear transformations are largely temperature independent. This is a result of the relatively large free energies involved in bringing about a transformation nucleus to its saddle point configuration. Hence, these shear transformations are activated by relatively large critical shear stresses that are found, at least locally, in work hardened materials (Christian, 1969). Although in some instances twinning and martensitic shears can contribute substantial amounts of deformation, they are generally far less prevalent than slip. We will consider these processes here as a special case of rate independent slip, albeit of a very coarse nature. The process of deformation by diffusional transport of matter is relatively straight forward and comparatively well understood. The structural parameter of any major importance in this mode of deformation is the grain size. The process is dominant only at very high temperatures near the melting point or in very fine grained material. It has been fully discussed by Nabarro (1948) and Herring (1950), and has been recently elucidated further by Raj and Ashby (1971). Its proper relationship to other mechanisms is discussed in Chapter 4 by Ashby and Frost. We will not consider it further. By and large the vast majority of inelastic deformation of crystalline matter occurs by the glide and climb motion of dislocations. Our discussion will center on these mechanisms.

1.2.2 *Dislocation Glide at Low Temperatures*

In the low temperature range where diffusional processes are too slow, the motion of dislocations in structural alloys is obstructed by either an inherent lattice resistance or by more or less localized obstacles which pin down dislocations. The flexibility of dislocations permits them to overcome such obstacles locally in a substantially uncoupled manner (Kocks, 1966, 1967, Argon, 1972). The local free energy changes in unpinning are often small enough to permit significant thermal overcoming of obstacles. The details of this process, which has been a subject of both extensive and intensive study over the past decade, have recently been evaluated and formalized by Kocks, Argon, and Ashby (1975). We will find it useful now to start with a short account of part of their development relating the basic local resistances on each dislocation element to the macroscopic flow stress.

1.2.3 *The Resistances to Glide in Individual Grains at Low Temperature*

If the entire obstacle structure in the interior of a crystal were given, and the specific mechanical interaction of these obstacles with dislocation elements were known, it would be possible to construct a contour of <u>element glide resistance</u>

τ_{ELEM} , describing the resistance to the motion of an infinitesimal segment of a dislocation. A visualization of such a contour for the motion of a dislocation in a particular direction in a given plane is shown in fig. 1.1. When a long dislocation is placed in such a topography of element resistances under an applied shear stress σ_1 each segment would come to equilibrium at a different point, by bowing out into an undulated configuration, subject to the additional constraints of self stresses resulting from the local curvatures. Under a constant applied stress producing a constant driving force,

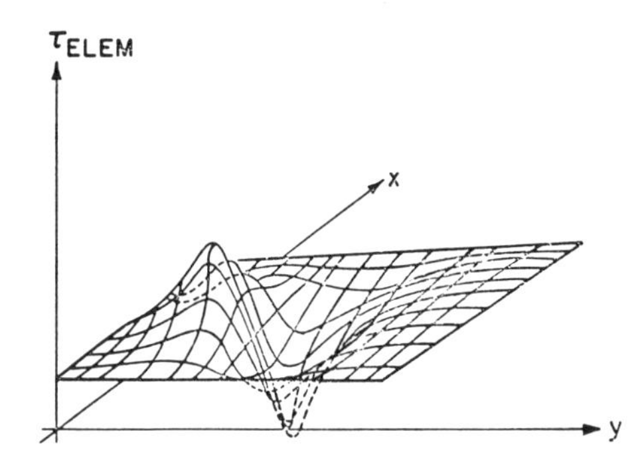

Fig. 1.1 A possible element glide resistance diagram for a dislocation parallel to the x direction and advancing in the y direction (from Kocks, et al., 1975, courtesy Pergamon Press).

the dislocation will reach a final shape for which the local resistive force on each element will also be constant. This will be achieved by the dislocation by bowing into regions of lower element glide resistance and exactly compensating for the deficiency of element glide resistance in such areas by curvature induced self stress. Thus, the bowing of the dislocation produces the first level of smoothing of the local element resistances into a uniform <u>line glide resistance</u> τ_{LINE} . As the applied stress is increased, it is initially possible to balance the driving force on the dislocation along its entire length with a line glide resistance. In this range the external behavior is elastic, albeit with a very slightly reduced shear modulus due to the reversible bowing of the dislocations. As the stress is increased, however, the line glide resistance in certain portions of the slip plane becomes insufficient and dislocations overcome one or more successive maxima of local line glide resistance and advance into regions of higher line glide resistance, shown schematically in fig. 1.2. This represents the local dislocation motion at $0^{o}K$, producing the first increments of inelastic deformation. In certain situations of heterogeneous line glide resistance where the successive maxima are high and far apart, groups of dislocations can form in front of the highest maxima and thereby amplify the applied stress on such regions roughly in direct proportion to the number of dislocations in the group. This grouping of dislocations in the slip plane in front of regions of large line glide resistance permits penetration of slip through such regions at stresses less than what would be necessary for a single dislocation alone. This represents the second level of smoothing in which the largest peaks in the line glide resistance of a slip plane are reduced to define a plane glide resistance τ_{PLANE} . Once the applied stress is

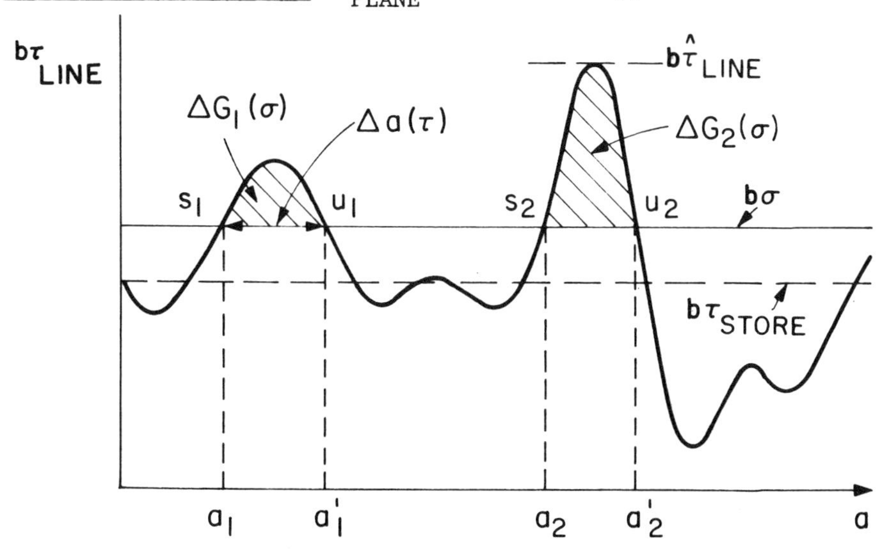

Fig. 1.2 A possible line glide resistance contour as a function of area traversed by a dislocation in a series of equilibrium shapes.

large enough to permit the overcoming of these peaks by at least some of the dislocations in the highest regions of line glide resistance long range dislocation motion can occur on any slip plane in isolation. Since moving dislocations in parallel planes interact with each other, slip at any one time, under a given applied stress, can develop only on planes with a spacing larger than a certain minimum value. This requires that the stress be increased somewhat above the level of the maximum plane glide resistance to enforce a long range cross flow of mobile dislocations in the crystal on a set of parallel planes. At a temperature above 0°K but below that where diffusion becomes prevalent, the line glide resistance and the plane glide resistance will be reduced and become time dependent because of the thermally activated penetration of dislocation segments through at least some of the slip plane obstacles. Figure 1.2 shows the line glide resistance a dislocation experiences in a succession of equilibrium configurations sweeping out increasing areas of the slip plane. At an applied stress σ less than the maximum line glide resistance $\hat{\tau}_2$ the dislocation would become pinned first at obstacle 1. If the dislocation can be carried over from its stable equilibrium position s_1 to its unstable equilibrium position u_1 it could proceed further to obstacle 2. This becomes possible when the required free energy of activation

$$\Delta G_1 = b\int_{a_1}^{a_1} \tau(a)\, da = b\int_{\sigma}^{\hat{\tau}_1} \Delta a(\tau)\, d\tau \qquad (1.1)$$

can be obtained from the pool of thermal energy. The availability of this required free energy is not only governed by the Boltzmann statistics of fluctuations of energy in the crystal but is also subject to a general restriction imposed by the second law of thermodynamics. The latter states that random thermal energy cannot be converted into free energy in cyclic or steady state processes. Hence,

$$\sigma > \tau_{STORE} = \lim_{A \to \infty} \frac{1}{A} \int_0^A \tau(a)\, da \qquad (1.2)$$

is a lower limit on stress for thermally activated penetration of a thermally stable obstacle structure. For many discrete obstacles considerable thermal penetration is possible resulting in a temperature (and strain rate) dependent flow stress. For other obstacles thermal penetration may not be possible because either $\sigma < \tau_{STORE}$, or it may be very improbable because the activation free energy ΔG may become very large as σ drops under $\hat{\tau}_{LINE}$ even by a small amount. In these instances the dislocation can only move large distances at stresses $\sigma \approx \hat{\tau}_{LINE}$ without any significant reduction with increasing temperature.

The actual kinetics of the motion of dislocations in a slip plane depends on the specific details of the interaction of the dislocation line with the current obstacles. These interactions have been discussed in detail by Kocks, Argon and Ashby (1975), and the interested reader is referred to this work. For our purposes here it will suffice to extract out of this work a general expression for the strain rate which can be written as

$$\dot{\gamma} = \dot{\gamma}_G \exp(-\Delta G(\sigma, \hat{\tau})/kT \qquad (1.3)$$

where

$$\dot{\gamma}_G = (b\,a_o)(\rho_m(\sigma))(\nu_G) \qquad (1.4)$$

is the pre-exponential factor: it gives the increment of strain (per unit volume) per single release event, the time average number ρ_m of released dislocation segments per unit volume and the frequency factor of the single release event. Of these terms both a_o, the area swept out by a released dislocation segment, and ρ_m are dependent on the applied stress, - with the best estimate for the combined stress dependence being a power function with an exponent of about 4. (For a theory of the stress dependence of the pre-exponential factor see Appendix 1.) The frequency factor ν_G depends to some extent on the stiffness of the interaction of the dislocation with the obstacle (Granato, et al., 1964): it can be taken, however, for most discrete obstacles as $10^{-2}\nu_D$ where ν_D is the Debye frequency.

The free energy of activation in eqn. (1.3) has been given to be merely dependent on the applied stress and an appropriate moment of the distributed glide resistances - here the choice being the peak value of the plane glide resistance, as would be suggested by eqn. (1.1). In reality $\hat{\tau}$ must be interpreted as the peak plane glide resistance of a generalized obstacle which is mechanically equivalent to the sum total effects of the slip plane obstacles which are in reality distributed both in space and in strength. The reader must be alerted here to the sweeping nature of this statement, which indicates that the net mechanical effect of the complex obstacle structure, once averaged over all the possible interactions it can have with moving dislocations, can be represented by a single parameter. Both theoretical studies of the thermally activated motion of dislocations through discrete obstacles (for an assessment of recent developments see Kocks, et al., 1975) as well as phenomenological studies of

Hart, et al., of low temperature plastic deformation, discussed in Chapter 5, indicate the general validity of this statement. In particular Hart, et al. find that all of their stress relaxation experiments at low temperatures give a one parameter family of curves relating $\ln \sigma$ to $\ln \dot{\varepsilon}$. It is readily shown that this behavior is consistent with a form

$$\dot{\gamma} = \dot{\gamma}_o \left(\frac{\sigma}{\mu}\right)^m \exp[-\Delta\Gamma(\sigma/\hat{\tau})] \qquad (1.5)$$

where $\Delta\Gamma(\sigma/\hat{\tau})$ stands for $\frac{\Delta G(\sigma/\hat{\tau})}{kT}$. This makes the equation resemble the required form given by eqn. (1.4)*. Comparison of eqn. (1.5) with (1.3) and (1.4) shows that part of the pre-exponential factor reflects the stress dependence of the steady-state density of mobile segments of dislocations which was referred to above as that part of the dislocation population that organizes itself into dynamic groups to level down some of the extreme differences in the line glide resistance. When the stress is changed in magnitude these currently mobile dislocations will have to adjust to new requirements by increasing or decreasing in density during a transient phase of deformation. When the deformation direction is reversed, this dynamic mobile dislocation structure built up to smooth the obstacle contour in the forward direction can be decomposed without having to penetrate many obstacles and thereby produce premature and easy reverse plastic deformation of an amount that must be proportional to the magnitude of the pre-exponential factor. This is the well known Bauschinger effect. When the deformation direction is merely shifted away from the forward direction instead of complete reversal, these dislocations still become largely available, giving rise to the lateral softening effects studied by Edwards and Washburn (1954) and to the well known corners in the yield loci. The rather remarkable shift property reported in stress relaxation experiments by Hart, et al. in Chapter 5 which gives rise to eqn. (1.5) suggests that these transients too may be rather reproducible, following identical histories, at least for the portions immediately preceding the changes of direction of deformation. The transients in a stress relaxation experiment performed during monotonic loading will, of course, be rather

*In Chapter 3 Kocks has given another possible formulation that is consistent with the stress relaxation behavior reported by Hart, et al. in Chapter 5.

different from a transient that occurs in a deformation reversal experiment or in cyclic reversal experiments. Although models can be made to represent these transients and anelastic deformations (see Chapter 14 for some examples), accurate representations will require detailed models related to the specific microstructure. Such detail is particularly necessary in the case of anelastic effects in particle strengthened materials where plastic deformation leaves strongly polarized dislocation debris around particles to build up an orderly arrangement of internal stress that tends to produce rather orderly back deformation to undo the debris. This so-called perfect memory behavior is discussed by Brown and Stobbs in Chapter 10.

1.2.4 *Plastic Shear Resistance of Polycrystals at Low Temperature*

In polycrystalline samples compatibility of deformation among the individual grains must be assured by enforcing slip on other than the most favorably oriented slip systems. This produces several additional considerations. First, for each grain based on its given orientation relative to the principal axes of overall deformation a set of five independent slip systems capable of producing a shape change matching the externally imposed shape change and involving the lowest possible expanditure of plastic work is selected. Taylor (1938) has shown that an average over all possible orientations of grains increases the shear deformation resistance by about 80%* - considering no interaction of the five slip systems (see also Rice in Chapter 2 and Chin in Chapter 11). Since the dislocations of these five systems act as obstacles to each other, however, the flow stress of the polycrystal would be somewhat higher.

This picture holds in general for fine grained polycrystals where all parts of any particular grain deform by one set of slip systems and there is little effect of grain size. In coarse grained polycrystals, however, the compatibility between grains can be maintained by inhomogeneous accommodating deformations restricted to thin layers around grain boundaries.

---

*The so-called Taylor factor of 3.1 relates the shear deformation resistance of a single crystal to the tensile deformation resistance of a polycrystal. The increase discussed above relates the shear resistance of the single crystal to the shear resistance of the polycrystal.

This permits selection of the best set of slip planes near every boundary and should therefore result in some reduction of the flow stress in coarse grained polycrystals from that in fine grained polycrystals (see also Thompson, et al., 1973).

The required activity of the several slip systems in each grain to assure compatible deformation necessitates the intersection of dislocations of such slip systems. Thus, in material in which the intrinsic slip plane obstacles are precipitate particles or solute atoms the dislocation intersection process will contribute another effective slip plane obstacle offering different glide resistance. Thus, an unavoidable superposition problem arises in the deformation of polycrystalline materials in which the particular strain rate sensitivity of the flow stress may become grain size dependent because of the potentially different composition of accommodation slip in grains of different size. Such problems remain largely unexplored.

1.2.5 *Intragranular Deformation at Elevated Temperature*

As mentioned in Section 1.2.1, deformation by combined Nabarro-Herring creep and grain boundary sliding at elevated temperatures is now well understood. The mechanism of trans-crystalline creep deformation (so-called power-law creep) at intermediate stresses where dislocation motion is involved, however, is one of the least understood of all processes, in spite of the fact that a great wealth of information exists describing the microstructural changes that accompany this deformation (see e.g. Lindroos and Miekk-oja in Chapter 8). The process has been well described phenomenologically (see Ashby and Frost in Chapter 4) and it has been widely appreciated that it consists of an interplay between hardening and recovery reaching a balance when a steady state gets established (Bailey, 1926, Orowan, 1946). Notwithstanding these observations and some qualitative understanding, nearly all of the theoretical effort in modelling high temperature creep over a decade has centered around the mechanism of screw dislocations dragging jogs - in spite of the fact that the mechanism is basically unstable and offers no understanding whatever of the microstructure. It is now belatedly becoming clear that the mechanism of intracrystalline creep is locked in the establishment of the reversible dislocation subgrain structure and cannot be understood without it.

Recent experiments have shown that in pure metals and ionic compounds the final balance gives rise to a dislocation subgrain structure (Clauer, et al., 1970a,b on molybdenum, Hasegawa, et al., 1970, 1971 on copper, Hüther and Reppich, 1973, on MgO, Poirier, 1972 on NaCl, Cropper and Pask, 1973 on LiF) which relates "uniquely" to the applied stress. Exell and Warrington (1972) have observed that once established, the sub-grains continue to alter as their boundaries migrate to produce strain. Thus small sub-grains disappear and large ones grow. As the average steady state size of sub-grains does not change with time, such coarsening sub-grains must undergo refinement by production of new boundaries in them, probably as a result of increased deformation and strain hardening followed by polygonization and other forms of recovery not yet fully understood.

On the other hand, it has become clear that in solid solution alloys (e.g. Horiuchi, 1972) no sub-grain structure develops, but steady state is achieved with a relatively unclustered density of dislocations that again relates rather uniquely to the applied stress. This indicates that recovery in solid solution alloys cannot proceed to as high a degree as in pure metals, but levels off at an earlier level of constrained equilibrium. This points out that an important missing ingredient is a detailed "process" theory of recovery which goes beyond a statement of thermodynamic dis-equilibrium for a starting state but can be used in the understanding of the decomposition of a work hardened structure into various levels of constrained equilibrium. The theory of Holt (1970) is in this respect quite inadequate.

The painstaking studies of Lindroos and Miekk-oja, discussed in Chapter 8, and that of Hasegawa, et al. (1971) are an important ingredient for any detailed understanding of the mechanism of creep, but an internally consistent detailed mechanism has yet to be worked out. This problem is of course closely related mechanistically to the process of structure changes by strain hardening touched on in the next section.

1.2.6 *Strain Hardening*

Although considerable advances have been made in the description of the rate of plastic deformation or creep for a given internal obstacle state (internal parameters), as we already discussed in Section 1.2.3 above, comparatively little has been done definitively to describe the structural alterations occurring in plastic straining of pure metals where the process is essentially statistical. To be sure much has been written on the subject of strain hardening in the past. Nearly all of such developments, however, have been deficient by being either semi-empirical, using experimental slip line data for assuring internal consistency (Seeger, 1956), or by making various reasonable sounding but nevertheless arbitrary assumptions, such as similitude of obstacle structures (Kuhlmann-Wilsdorf, 1962, 1966) or type of secondary slip (Hirsch, 1964) etc. The several attempts at statistical theories for structure change have been too specialized (e.g. Kocks, 1966, Argon, 1969) and have dealt with only portions of the stress-strain curve, such as Stage II and Stage I without being able to map out the entire stress-strain curve for a given orientation. It is against this background that the process theory of Zarka discussed in Chapter 9 appears as a novel departure. Yet the ingredients of this theory are simple. It starts out with an initial dislocation network, makes segments of dislocations move on the appropriate slip planes according to some assumed glide resistance law which has no critical importance in the storage process, it merely regulates time. The dislocation segments are stopped and stored according to an assumed law (acquisition of a critical density of jogs). Adjustments are made for the stress fields of the stored segments and for orientation changes, then new dislocation segments are released to repeat the process over a second step, and so on. The earlier theories of strain hardening showed that different individual details of the glide resistance and dislocation storage criteria often lead to similar results. In keeping with this fact, several weak criteria in the theory of Zarka are of little importance. Its strength lies in a reasonably careful accounting of dislocation activity on all possible slip planes in incremental steps. Not surprisingly this produces not only all the familiar features of multi-stage stress-strain curves but nets a wealth of internal detail which is beyond the level of available information on dislocation structures in the deformed state (see Mughrabi in Chapter 6), giving perhaps a new stimulus to electron microscopists.

Zarka's theory (as many of the earlier strain hardening theories also do) indicates that the rate of strain hardening depends on the current obstacle state (i.e. hardness, or glide resistance) of the material. Recent experiments by Wire, Ellis and Li (1975) demonstrate that the rate of strain hardening depends in addition on the strain rate and that the two parameters of hardness and strain rate govern the strain hardening rate as the only important state parameters. The physical meaning of this is clear. The rate of dislocation retention is governed directly by the density of obstacles and by the density of mobile dislocation segments and by not much else.

The description of structure change due to plastic deformation in particle strengthened metals has been more successful because in such materials the rate of dislocation storage is primarily governed by the necessity to accomodate displacement incompatibilities between non-deforming particles and deforming matrices - the retention of the so-called "geometrically necessary dislocations" (see e.g. Ashby, 1966, Russell and Ashby, 1970). These developments are discussed in greater detail by Brown and Stobbs in Chapter 10.

1.2.7 *Deformation under Multi-Axial Stress*

The glide process discussed in Section 1.2.3 occurs on well defined crystallographic planes at a rate governed by the shear stress on these planes. The same conclusion holds for twinning and martensitic shears. Thus, when a multi-axial state of stress is applied to a single crystal, it is the level of the resolved shear stress on the possible slip systems which governs the shear strain on these systems. It is then possible to represent states of multi-axial stress producing constant glide rate on particular slip systems as planar portions of polyhedral surfaces in stress space. In polycrystals it is natural to expect that the inter-grain constraints will round out the corners of the polyhedra and that plasticity on the whole will be governed by a critical state of deviation from a state of pure pressure or negative pressure, as proposed initially by von Mises (1913). These surfaces in stress space have been known for a long time in rate-independent plasticity as "yield surfaces" where they also serve as "potentials" which give the components of strain increments from the associated flow rule that states that these strain increments are proportional to the components of the outward normal vector to the yield surface at the point representing the flow state. How these concepts can be carried over to rate dependent deformation to define flow potentials is discussed by Rice in Chapter 2, while the basis of some of the violations of the associated flow rule are pointed out by Kocks in Chapter 3.

1.3 FORMS OF CONSTITUTIVE EQUATIONS

The previous discussion now permits us to put down some general forms for constitutive equations for plastic deformation.

Given a current state of internal obstacles and other relevant structural parameters characterizable by a set of plastic resistances $\tau^{(k)}$ (where k may stand for any of the possible slip planes in a single crystal, or any of a number of spatially located planes of easy shear in a textured polycrystal: in some instances the $\tau^{(k)}$ may be elements of a tensor), the material will deform under a given state of stress σ_{ij} and temperature T by a plastic strain rate of

$$\dot{\varepsilon}_{ij} = \dot{\varepsilon}_{ij}(\sigma_{ij}, \tau^{(k)}, T) = \lambda(\sigma_{ij}, \tau^{(k)}, T)(\frac{\partial\Phi}{\partial\sigma_{ij}})_{\dot{\lambda}} \qquad (1.6)$$

where $\dot{\lambda}$ implies an equivalent plastic strain rate referred to a reference experiment such as a tension test (where the product of the tensile flow stress with $\dot{\lambda}$ furnishes the same plastic work dissipation rate) and that the magnitude of the particular strain rate is proportional to the corresponding component of the gradient of a flow rate potential Φ evaluated at the particular equivalent flow rate $\dot{\lambda}$ given by σ_{ij}, $\tau^{(k)}$, and T . The latter condition, of course, implies the applicability of an associated flow rule. The experiments of Hart, Li and co-workers show that in many instances (most probably in polycrystalline metals where only weak deformation textures develop) only a single scalar plastic shear resistance is sufficient to describe the obstacle state and the deformation rate expression could then be written as

$$\dot{\varepsilon}_{ij} = \dot{\lambda}(s, \tau, T)(\frac{\partial\Phi}{\partial\sigma_{ij}})_{\dot{\lambda}} \qquad (1.6a)$$

where τ is the above-mentioned scalar plastic shear resistance, s the deviatoric shear stress, and $\dot{\lambda}$ the well known equivalent plastic strain rate.

This rate of flow expression in eqn. (1.6) or (1.6a) assumes that loading transients resulting from a stress reversal, a re-application of stress, or a change of direction of stress have died down and that no structural changes due to strain hardening or recovery and re-crystallization are occurring. In reality structural changes, of course, do occur incrementally and must be described by a coupled incremental change-of-state equation which can be written in its simplest form as

$$d\tau^{(k)} = d\tau^{(k)}(\sigma_{ij}, \tau^{(k)}, T) = d\tau^{(k)}(\tau^{(k)}, \dot{\varepsilon}_{ij}dt, dt\ T) \qquad (1.7)$$

where it is understood that the incremental change in the set of plastic resistances depends on the current plastic resistances $\tau^{(k)}$, the temperature T , the increment of time for recovery, and the plastic strain rate tensor. It is well worth emphasizing here that in this interpretation structural parameters other than slip plane obstacles, such as sub-grain boundaries, etc. are assumed lumped into the set of plastic resistances $\tau^{(k)}$. It is, of course, possible that the $d\tau^{(k)}$

have either a positive sign, a negative sign, or become zero when a structural balance during flow becomes established.

Once again, in weakly textured materials where the plastic shear resistance appears isotropic and can be given by a scalar, eqn. (1.7) for the change in plastic resistance takes on a degree of simplicity as

$$d\tau = d\tau(\tau\,,\ \dot{\lambda}dt\,,\ dt\,,\ T) \tag{1.7a}$$

reducing the multiplicity of plastic resistances to only two, i.e. eqn. (1.7a) and a corresponding one for the tensile plastic resistance.

Equations (1.6) and (1.7) represent a relatively ideal process in which transients are neglected. In nearly all instances, however, transients due to stress change occur and must be coped with. The most prominent of the transients is the Bauschinger effect. Such transients are either governed by the "steady state" mobile defect structure of the state preceding the stress change (for the Bauschinger effect), or by the difference between the "steady state" mobile defect structures of the two states (for cyclic stress variations). In either case therefore their total magnitude in strain is limited. Furthermore, the rates of transient deformations are often unrelated to the steady state deformation rate due to the presence of internal stresses (certainly in the case of the Bauschinger effect) or an anisotropic build-up of deformation induced obstacles. Hence the transient deformation is more in the nature of anelastic deformation. A form for such deformation with particular emphasis on the Bauschinger effect is

$$a_{ij} = a_{ij}(\tau^{(k)}\,,\ T\,,\ (\sigma_{ij} - \sigma^{o}_{ij})) \tag{1.8}$$

where a_{ij} gives the anelastic transient strain tensor and σ^{o}_{ij} refers to the flow stress just preceding a change to σ_{ij}. Often in the case of the Bauschinger effect the anelastic strain which is relatively rate independent, can be represented as a power function of the stress difference during the first change of the stress (Deak, 1962). Repeated reversals or change of direction of stress, or cycling entirely within the flow potential surface tends to increase the exponent in the power function representation of the anelastic strain while decreasing the proportionality coefficient with a net decrease in the total anelastic strain. In all cases the total magnitude of the anelastic strain obtainable at complete reversal of stress is of the order of magnitude of the pre-exponential factor divided by the frequency factor of the preceding "steady state" plastic deformation rate.

Primarily because of the difficulty in precisely describing the anelastic strains, integration of eqn. (1.6) along arbitrary deformation paths leads to discrepancies in strain in the form of additive constants.

In strongly textured materials where there are several plastic resistances the more general forms of eqns. (1.6) and (1.7) must be used. In this instance there is a multiplicity of equations for both eqns. (1.6) and (1.7).

The forms discussed above pointed out only the significant variables in the constitutive relations without giving specific forms of relations. Such specific forms can in many instances be given as summation problems if the types and distribution of the obstacles are known. Several such cases have been discussed by Kocks et al. (1975). Similarly, the change of state equations for strongly textured materials or single crystals can be modelled, albeit laboriously, by an approach discussed by Zarka in Chapter 9.

The currently most reliable approach, however, is to obtain the constitutive connections between variables by a limited number of well-chosen experiments, such as, for example, those discussed by Hart, Li and co-workers. The results of such experiments should then be compared with the forms arising from mechanistic models. In this way of model-inspired phenomenology it should be possible to come up with physically sound constitutive relations which can stand the test of interpolation and extrapolation. The reader will find many such examples in the following chapters in this book.

ACKNOWLEDGEMENTS

This work was supported by the Department of Mechanical Engineering at M. I. T. I have benefited from discussions with nearly all the authors of this book, but particularly from discussions with Dr. U. F. Kocks. In addition, I am grateful to Professor S. Takeuchi for pointing out to me some recent key work in the field of creep.

REFERENCES

Argon, A. S., (1969) in Physics of Strength and Plasticity, edited by A. S. Argon (Cambridge, Mass.: M.I.T. Press), p. 217.

Argon, A. S., (1972) Phil. Mag., 25, 1053.

Ashby, M. F., (1966) Phil. Mag., 14, 1157.

Bailey, R. W., (1926) J. Inst. Met., 35, 27.

Christian, J. W., (1969) in Physics of Strength and Plasticity, edited by A. S. Argon (Cambridge, Mass.: M.I.T. Press), p. 85.

Clauer, A. H., Wilcox, B. A. and Hirth, J. P., (1970) Acta Met., 18, 367; Ibid., 18, 381.

Cropper, D. R. and Pask, J. A., (1973) Phil. Mag., 27, 1105.

Deak, G. I., (1962) "A Study on the Causes of the Bauschinger Effect", Sc.D. Thesis (Cambridge, Mass.: M.I.T.).

Edwards, E. H. and Washburn, J., (1954) J. Metals, 200, 1239.

Exell, S. F. and Warrington, D. H., (1972) Phil. Mag., 26, 1121.

Friedel, J., (1956) Les Dislocations (Paris: Gauthier-Villars).

Granato, A. V., Lücke, K., Schlipf, J. and Teutonico, L. J., (1964) J. Appl. Phys., 35, 2732.

Hasegawa, T., Hasegawa, R. and Karashima, S., (1970) Trans. Japan Inst. Met., 11, 101.

Hasegawa, T., Karashima, S. and Hasegawa, R., (1971) Metal. Trans., 2, 1449.

Herring, C., (1950) J. Appl. Phys., 21, 437.

Hirsch, P. B., (1964) in Dislocations in Solids, Disc. of the Faraday Society, No. 38 (London: The Faraday Society), p. 111.

Hirsch, P. B. and Humphreys, F. J., (1970) Proc. Roy. Soc. (London) A318, 45.

Holt, D. L., (1970) J. Appl. Phys., 41, 3197.

Horiuchi, R. and Otsuka, M., (1972) Trans. Japan Inst. Met., 13, 284.

Hüther, W., and Reppich, B., (1973) Phil. Mag., 28, 363.

Kocks, U. F., (1966) Phil. Mag., 13, 541; (1967) Canad. J. Phys., 45, 737.

Kocks, U. F., Argon, A. S. and Ashby, M. F., (1975) Thermodynamics and Kinetics of Slip (Progress in Materials Science) edited by B. Chalmers, et al., (Oxford: Pergamon Press) vol. 19.

Kuhlmann-Wilsdorf, D., (1962) Trans. A.I.M.E., 224, 1047.

Kuhlmann-Wilsdorf, D., (1968) in Work Hardening, edited by J. P. Hirth and J. Weertman (New York: Gordon and Breach) p. 97.

von Mises, R., (1913) Goettinger Nachr., Math.-phys. Klasse, p. 582.

Nabarro, R. F. N., (1948) in Report of a Conference on the Strength of Solids (London: Physical Soc.), p. 75.

Orowan, E., (1946) J. West of Scotland Iron and Steel Inst., 54, 45.

Poirier, J. P., (1972) Phil. Mag., 26, 713.

Raj, R. and Ashby, M. F., (1971) Met. Trans., 2, 113.

Russell, K. C. and Ashby, M. F., (1970) Acta Met., 18, 891.

Seeger, A., (1956) in Dislocations and Mechanical Properties of Crystals, edited by J. C. Fisher et al., (New York: J. Wiley), p. 243.

Taylor, G. I., (1938) J. Inst. Metals, 62, 307

Thompson, A. W., Baskes, M. I. and Flanagan, W. F., (1973) Acta Met., 21, 1017.

Wire, G. L., Ellis, F. V. and Li, C. Y., (1975) Acta Met., in the press.

Appendix I. Stress Dependence of the Pre-Exponential Factor

The pre-exponential factor appearing in eqn. (1.4) can be expressed for jerky glide in a somewhat more appropriate form as (see also Kocks, 1966)

$$\dot{\gamma}_G = \frac{b}{d} \nu_G \tag{A1.1}$$

where d is the spacing of currently active slip bands. This comes about by considering that there is only one dislocation in a volume of $a_o d$ which sweeps out an area a_o during one activated event. The distance d between active slip planes is governed by the necessity of passing dislocations of opposite type by each other in adjacent slip planes in a region which is undergoing homogeneous strain where no average lattice curvature need develop. If the plane glide resistance were completely absent (i.e. $\tau_{PLANE} = 0$) then the stress σ_m required to just move dislocations by each other on slip planes a distance d apart is

$$\sigma_m = \alpha \frac{\mu b}{d} \tag{A1.2}$$

where $\alpha = 1/8\pi(1-\nu)$ for edge dislocations and $1/4\pi$ for screw dislocations.

When the plane glide resistance is not zero, but thermally activated overcoming of discrete obstacles is possible, only a portion of $\sigma - \sigma_m$ of the applied stress is available for forcing the dislocations through the thermally penetrable obstacles. This simple picture is proper when the mean obstacle spacing λ along the dislocation line in the slip plane is much smaller than the active slip plane spacing d so that the plane glide resistance can be considered as a property of a plane without any significant fringe effect in parallel planes a distance d away.* Hence the activation free enthalpy $\Delta G'$ becomes, according to eqn. (1.1)

---

*In some instances where d is very small this assumption may not be possible, requiring a more sophisticated analysis balancing the plane glide resistance against the inter-plane glide resistance. For such an analysis see Hirsch and Humphreys (1970).

$$\Delta G' = b\int_{\sigma-\sigma_m}^{\hat{\tau}} \Delta a(\tau)\, d\tau = b\int_{\sigma}^{\hat{\tau}} \Delta a(\tau)\, d\tau + b\int_{\sigma-\sigma_m}^{\sigma} \Delta a(\tau)\, d\tau \tag{A1.3}$$

and

$$b\int_{\sigma-\sigma_m}^{\sigma} \Delta a(\tau)\, d\tau \approx b\, \sigma_m \Delta a(\sigma - \sigma_m/2) \simeq b\, \sigma_m \Delta a(\sigma) \tag{A1.4}$$

This gives for the strain rate,

$$\dot{\gamma} = \frac{\nu_G \sigma_m}{\alpha\, \mu} \exp\left(- \frac{(\Delta G(\sigma,\, \hat{\tau}) + b\, \sigma_m \Delta a(\sigma))}{kT}\right) \tag{A1.5}$$

where $\Delta G(\sigma\,, \hat{\tau})$ is the activation free enthalpy which would be expected if the slip plane had been in isolation.

It is reasonable to expect that the slip plane spacing (or its interaction stress σ_m) adjust itself to maximize the strain rate, i.e.,

$$\frac{d\dot{\gamma}}{d\sigma_m} = 0 \tag{A1.6}$$

governs the effective slip line spacing

$$\frac{d\dot{\gamma}}{d\sigma_m} = \frac{\nu_G}{\alpha\mu} \exp\left(- \frac{(\Delta G + b\, \sigma_m \Delta a(\sigma))}{kT}\right)\left[\, 1 - \frac{b\, \sigma_m \Delta a(\sigma)}{kT}\, \right] = 0 \tag{A1.7}$$

giving

$$\sigma_m = \frac{kT}{b\Delta a(\sigma)} \tag{A1.8}$$

and

$$\dot{\gamma} = \frac{\nu_G}{\alpha\, e} \frac{kT}{\mu b \Delta a(\sigma)} \exp\left(- \frac{\Delta G(\sigma,\, \hat{\tau})}{kT}\right) \tag{A1.9}$$

In jerky glide, however, the activation area $\Delta a(\sigma)$ is

$$\Delta a(\sigma) = \lambda(\sigma)\, \Delta x \tag{A1.10}$$

where λ , the distance between obstacles along the dislocation line, is given by the well known Friedel relation (see Friedel, 1956)

$$\lambda = \ell \left(\frac{2\mathcal{E}}{\sigma b \ell}\right)^{1/3} \tag{A1.11}$$

and the activation distance Δx of the obstacle can be taken as a linear one without much loss of generality,

$$\Delta x = w\left(1 - \frac{K}{\hat{K}}\right) \tag{A1.12}$$

where ℓ is the mean obstacle spacing in the slip plane, $\mathcal{E}$ is the line energy of the dislocation (considered to be equal to the line tension), w the effective obstacle size, $\hat{K}$ the maximum resistive force of the obstacle and $K = \sigma b \lambda$ the force exerted on the obstacle by the applied stress. Substitution of the last two equations in eqn. (A1.10) gives

$$\frac{\Delta a}{w\ell} = \left(\frac{2\mathcal{E}}{\sigma b \ell}\right)^{1/3}\left(1 - \frac{(\sigma b \ell / 2\mathcal{E})^{2/3}}{(\hat{K}/2\ \ )}\right) \tag{A1.13}$$

where $\hat{K}/2\mathcal{E}$ is the normalized obstacle strength. Equation (A1.13) gives the sought stress dependence of activation area. Figure A1.1 gives a plot of the normalized activation area as a function of the normalized stress for five different levels of obstacle strength which are in the range of what might be appropriate for forest dislocations. The vertical lines are asymptotic levels of normalized stress where the activation area goes to zero. These curves show that at a temperature above $0^{o}K$ where the flow stress is about 0.7-0.8 of the value at $0^{o}K$ a power function match to the $\Delta a(\sigma)$ curve would be

$$\frac{\Delta a}{w\ell} \simeq \beta \left(\frac{2\mathcal{E}}{\sigma b \ell}\right)^{m} \tag{A1.14}$$

with an exponent $m \simeq 4$. This exponent would increase at lower temperatures and decrease at higher temperatures. Hence, the phenomenological form of the pre-exponential factor would be

$$\dot{\gamma}_G = \frac{\nu_G kT}{\alpha e w \ell b \mu \beta}\left(\frac{\sigma b \ell}{2\mathcal{E}}\right)^{m} \tag{A1.15}$$

Equation (A1.15) would appear to be the basis of the pre-exponential factor in the phenomenological form of the shear strain rate equation (eqn. 1.5) found by Hart, Li and their co-workers (see Chapter 5).

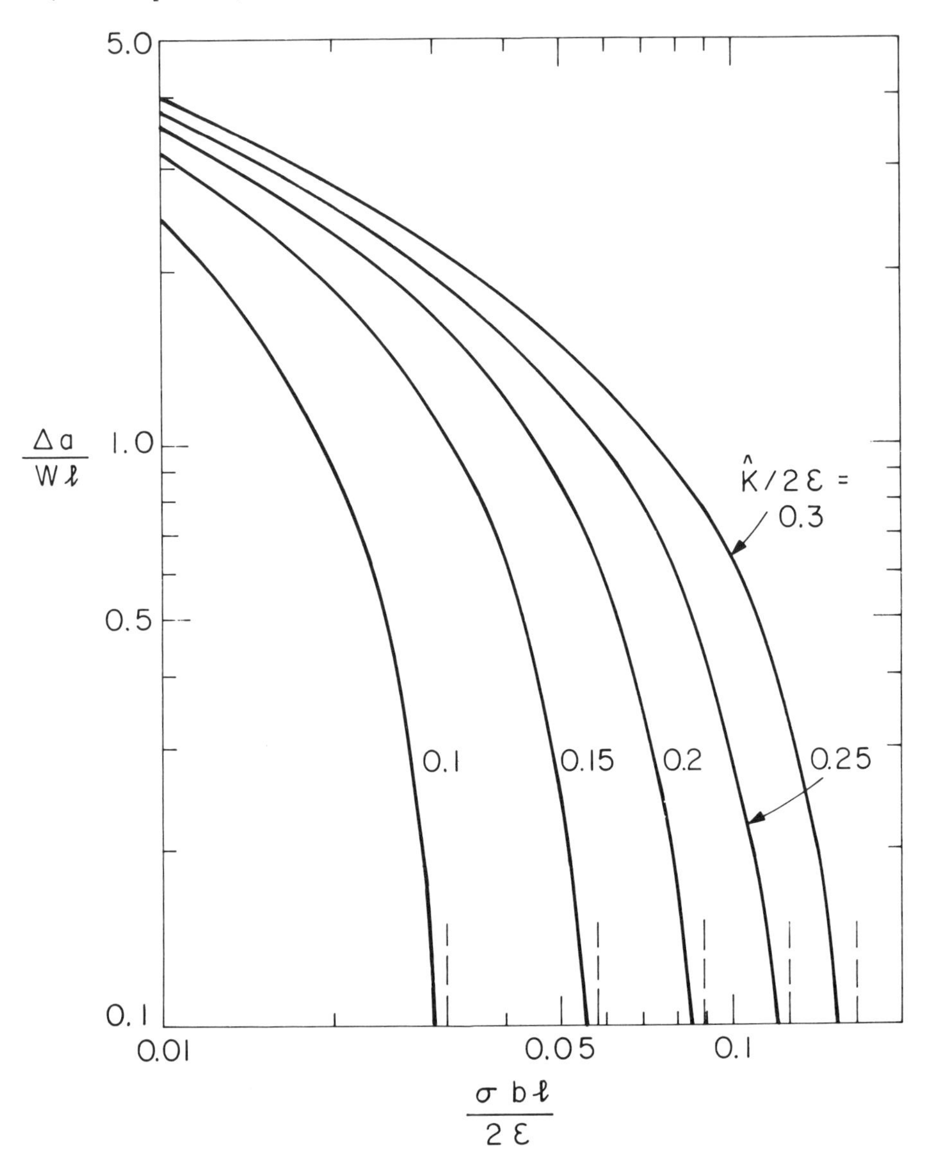

Fig. A1.1 The stress dependence of the activation area for a dislocation segment overcoming a simple discrete obstacle.

ABSTRACT. *This chapter presents the foundations in continuum mechanics and irreversible thermodynamics for constitutive relations governing plasticity. Special emphasis is given to the connection of macroscopic formulations to deformation mechanisms as operative on the microscale. Indeed, much of the chapter is organized around an internal variable framework by which inelastic structural rearrangements of a representative material sample can be related to its macroscopic deformation. This is illustrated most extensively for deformation by slip, but other mechanisms such as diffusion and phase changes are discussed as well. An extended synopsis is given in the summary section.*

2. CONTINUUM MECHANICS AND THERMODYNAMICS OF PLASTICITY IN RELATION TO MICROSCALE DEFORMATION MECHANISMS

J.R. Rice

2.1 INTRODUCTION

The aim of this chapter is to discuss macroscopic constitutive relations in metal plasticity, with special reference to the form and framework for these relations as set by underlying microscale mechanisms of deformation. Subsequent chapters (e.g., those by Kocks and Ashby) will study the detailed kinetic laws of separate deformation mechanisms. Here, after a review of what can be said on constitutive laws from a purely macroscopic or continuum standpoint, a general thermodynamic formalism is developed by which structural rearrangements of material elements on the microscale, by slip, diffusion, and the like, can be related to corresponding increments of macroscopic plastic strain. This framework is already sufficient for ascertaining certain broad structural features of macroscopic constitutive laws, for example a plastic "normality" structure which follows from only the most essential features of a broad class of microscale relations for the kinetics of processes of structural rearrangement. Specifically, this entails that the rate of some scalar rearrangement process be stress state dependent only through the thermodynamic force conjugate to the extent of that rearrangement.

The general framework set forth for the micro-macro transition leaves unresolved the detailed steps by which one averages, for example, over all the possible slip systems within a given grain and over all grains in a polycrystalline material sample to arrive finally at specific functional forms for constitutive relations. There are, however, approximate means for doing this, involving models of the "self consistent" type for polycrystals, and work in that area is reviewed. Other chapters (e.g., by Zarka and Chin) provide detailed studies of such averaging procedures.

Such models result in predictions of behavior which do reflect, at least approximately, some of the very real, complex features of plastic response when general states of combined stress and non-proportional stressing paths are examined.

Still, the polycrystal models involve great complexity of analysis with moderately large computer programs and large storage requirements to evaluate stress-strain relations for even relatively simple processes of macroscopically homogeneous deformation. Hence they do not seem to be feasible as input to, say, large scale finite element analyses of inhomogeneously stressed structural or machine elements.

Thus, in addition to the desire for physical and mathematical rigor in formulating constitutive laws from microscale processes, there is a compelling desire for simplicity of description in terms of a comparatively small number of averaged microstructural parameters or, equivalently, in terms of parameters that are defined from relatively simple functionals of prior deformation history. This brings one back to the context of continuum descriptions but in the effort, microscale information can be gainfully utilized if only in a suggestive rather than a rigorously derived manner. Indeed, the chapter closes with some suggestions of the forms that might be employed for such relations. The same approach is, of course, discussed in several other chapters of the book (e.g., those by Ilschner, Gittus, and Weeks and Poeppel).

The discussion in Sections 2.2 to 2.7 is intended for arbitrary amounts of strain, although it is frequently specialized, where noted, to small strains or, more commonly, to small elastic distortions from a finitely deformed plastic state. This finite deformation context, or at least the latter variant with small elastic distortions, is essential to any full discussion not just because many practical problems involve large plastic strains. Even in circumstances of small strain, by comparison to unity, the distinctions that arise in a rigorous finite strain analysis between different measures of stress and stress rate become important, e.g. in the case of a time-independent material, whenever the rates of increase of stresses with strains have magnitudes that are comparable to those of the stresses themselves. Indeed, it seems objectionable in a general thermo-mechanical theory to contemplate other than arbitrarily large plastic distortions, for the current shapes of many metallic objects are obtained by the finite strains of, say, rolling or forging or extrusion from some prior shape in the solid state.

2.2 CONTINUUM REPRESENTATIONS OF INELASTIC CONSTITUTIVE LAWS

We admit the concept of a macroscopic stress tensor T in a continuum and suppose that alterations of stress at a point of the continuum are determined solely by the history, experienced at that point in processes emanating from a standard reference state, of the temperature θ and deformation gradient F. The latter is defined by $F = \partial x/\partial X$ or, on a cartesian frame, $F_{ij} = \partial x_i/\partial X_j$ where X_1, X_2, X_3 and x_1, x_2, x_3 are coordinates of material points before and after deformation, respectively, referred to fixed background axes. Hence it is elementary to write that the stress tensor T at time t is expressible by a relation of the form

$$T(t) = \mathscr{F}[F(s), \theta(s) \; ; \; 0 \leq s \leq t] \tag{2.1}$$

where the notation indicates that T is a functional of all prior values of F and θ, including the current values. Here T is the Cauchy (or "true") stress. We shall later introduce stress tensors σ with work conjugate properties relative to rotation invariant strain measures.

Now, F may be written as RU where U, with $U_{ij} = U_{ji}$, denotes a pure deformation and R denotes a rigid rotation, hence satisfying $R^T = R^{-1}$ and $\det(R) = 1$ where R^{-1} is the matrix inverse and R^T the transpose of R. If the deformation history $F(s)$ is altered by an additional rotation history $R'(s)$ in eqn. (2.1), it is evident that the resulting stress at time t is changed only insofar as it is further rotated by $R'(t)$. Hence, with $R'(s) = R^{-1}(s)$, eqn. (2.1) becomes

$$R^T(t)T(t)R(t) = \mathscr{F}[U(s), \theta(s) \; ; \; 0 \leq s \leq t] \tag{2.2}$$

so that the stress tensor, when rotated backward by the inverse to R, depends only on the history of pure deformation (and temperature) at the material point.

There are many possible definitions of a strain tensor. Except where noted otherwise, we shall consistently use ε to denote some rotation invariant, or objective, strain tensor, by which is meant that ε has the same principle axes as U and that the three principle values ε_α and U_α, α = I to III, of each are related by a monotonic scalar function $h(...)$, such that

$$\varepsilon_\alpha = h(U_\alpha) \ , \quad \alpha = \mathrm{I, II, III} \ , \tag{2.3}$$

where $h(1) = 0$, $h'(1) = 1$.

The latter conditions assure that ε agrees with the infinitesimal strain tensor for small stretches and rotations; each different monotonic function h, vanishing at unity with unit slope, defines a different finite strain measure.

To each strain measure there corresponds a symmetric work conjugate stress tensor σ. This is defined so that, if material elements are instantaneously given virtual velocities $\dot{x}$ to which there correspond strain rates $\dot{\varepsilon}$, then

$$\sigma_{ij}\dot{\varepsilon}_{ij} = \text{rate of stress working per unit volume of reference state} = \det(F)T_{ij}\partial\dot{x}_j/\partial x_i \tag{2.4}$$

By using $\partial\dot{x}/\partial x = \dot{F}F^{-1}$, $F = RU$, and noting that T is symmetric, this becomes

$$\sigma_{ij}\dot{\varepsilon}_{ij} = \det(U)U^{-1}_{ij}(R^T TR)_{jk}\dot{U}_{ki} \tag{2.5}$$

Further, since ε is expressible, at least in principle, as a function of U and conversely $U = U(\varepsilon)$, where a symmetrical dependence on ε and ε^T is adopted, this defines σ as

$$\sigma_{mn} = \det(U)U^{-1}_{ij}(R^T TR)_{jk}\partial U_{ki}/\partial\varepsilon_{mn} \tag{2.6}$$

Since only $R^T TR$ and functions of U are involved, we can evidently write eqn. (2.2) as

$$\sigma(t) = \mathcal{G}[\varepsilon(s), \theta(s) \ ; \ 0 \le s \le t] \ . \tag{2.7}$$

This is a _memory functional_ representation of a constitutive law. Later we shall specialize it to elastic-plastic materials, which can exhibit a purely elastic response to alterations of ε and θ when these take place rapidly or, in the time-independent material idealization, are appropriately directed. This means that $\sigma(t)$ is a direct function of $\varepsilon(t)$ and $\theta(t)$, of a form that can be regarded as having a memory functional dependence on those portions of history at which other than elastic response occurred. Alternatively, the dependence on history can be replaced by a dependence on what it has produced, namely, the current pattern of structural arrangement, on the microscale, of material elements. The latter suggests an _internal variable_ representation in the constitutive law. Such forms have been introduced for viscoelastic deformation by Eckart (1948), Meixner (1953), and Biot (1954), and have been studied in some generality by Coleman and Gurtin (1967). In these studies the internal variables are taken to be macroscopic parameters, but the appropriate variables and their rate laws are often not readily identifiable. By contrast, the internal variable formulation in Sections 2.5 to 2.8 takes the variables to characterize specific, local structural rearrangements at sites throughout a representative sample of material, the emphasis being on a rigorous formulation of the rate laws of the sample at that level, with the required averaging to form the macroscopic constitutive relations being taken up separately.

2.3 SOME STRESS AND STRAIN TENSORS AND STRESS RATES

The simplest material strain tensor to calculate for general deformations is the Green strain, and its conjugate is the symmetric Piola-Kirchhoff stress (e.g., Truesdell & Toupin, 1960). These are given by

$$\varepsilon = \frac{1}{2}(U^2 - I) = \frac{1}{2}(F^T F - I)$$

$$\sigma = \det(U)U^{-1}(R^T TR)U^{-1} = \det(F)F^{-1}TF^{-1T} . \tag{2.8}$$

However, in metal plasticity we are almost always concerned with small elastic distortions from a nearby, unloaded state, even though that state itself may be finitely removed from some prior reference configuration. Hence it is convenient to have a stress measure which can be interpreted, apart from the effect of small elastic distortions, as a force per unit area of that unloaded state. The stress measure $R^T\text{T}R$ would be ideal to that purpose, although it is not a conjugate stress. We may think of this as a rotation-invariant "true" stress, having components that are always properly referenced to material elements for use in constitutive relations. There is a conjugate stress measure which has many of the properties of this rotation-invariant true stress, equaling or differing negligibly from the volume ratio, $\det(U)$, times it in some important circumstances. This is the σ associated with logarithmic strain, the latter being defined on principal axes as the logarithm of the corresponding stretch ratios,

$$\varepsilon_\alpha = \log U_\alpha \quad , \quad \alpha = \text{I, II, III} \quad , \tag{2.9}$$

and for any axes orientation, $\varepsilon = \log U$ and $U = \exp \varepsilon$ where $\log U$ and $\exp \varepsilon$ are understood as infinite matrix power series in $U - I$ and ε , respectively. Thus, following an analysis by Hill (1968), one may show from eqn. (2.6) that σ has the following properties:

(i) $\sigma = \det(U)R^T\text{T}R$ precisely when principal axes of deformation and rotation-invariant true stress coincide,

(ii) The normal components of σ along the principal directions of deformation coincide always with the corresponding normal components of $\det(U)R^T\text{T}R$, and

(iii) σ differs from $\det(U)R^T\text{T}R$ by terms that are of quadratic order in ε .

In particular, when the cartesian background frame is aligned with the principal directions of ε , one computes that

$$\sigma_{ij} = \det(U)(R^T\text{T}R)_{ij} \; [1 - \frac{1}{6}(\varepsilon_{ii} - \varepsilon_{jj})^2 + \ldots] \; . \tag{2.10}$$

(no sum here on repeated indices)

Even at the finite strain state $\varepsilon_{11} = .25$, $\varepsilon_{22} = -.25$, with all other $\varepsilon_{ij} = 0$, the maximum difference between components of σ and of $\det(U)R^T\text{T}R$ is approximately 4% , and occurs for σ_{12} . Further, since plastic flow is nearly volume preserving, the volume ratio $\det(U)$ differs from unity only by terms of the size of elastic distortions. Hence, $\sigma \approx R^T TR$ to moderately large strains. A more consequential result of (iii) above and of eqn. (2.10) is that

$$\frac{d\sigma}{dt} = \frac{d}{dt}[\det(U)R^T TR] \qquad (2.11)$$

precisely when the reference state for strain measurement is chosen to coincide, instantaneously, with the current state, apart from an arbitrary rotation R (i.e., when $U = I$ and, hence, $\varepsilon = 0$ instantaneously). Further, for $\varepsilon \neq 0$, eqn. (2.10) assures that the difference between the two rates in eqn. (2.11) is of order σ times ε times $d\varepsilon/dt$ and this would, for example, typically be negligible in comparison to $d\sigma/dt$ if ε was a strain of elastic order. This is an unusual feature. In general, the difference between two stress rates is of order σ times $d\varepsilon/dt$ (Hill, 1968), and hence is not negligible in comparison to $d\sigma/dt$ if the latter were, for example, computed well into the plastic range, where the slope of a stress-strain diagram has a magnitude comparable to that of the stress itself. This is readily demonstrated by writing the analogous formula to eqn. (2.11) in terms of the Green strain and symmetric Piola-Kirchhoff stress of eqn. (2.8). When the reference state is likewise chosen to correspond instantaneously with the current state, except for an arbitrary rotation R,

$$\frac{d\sigma}{dt} + \frac{d\varepsilon}{dt}\sigma + \sigma\frac{d\varepsilon}{dt} = \frac{d}{dt}[\det(U)R^T TR] , \qquad (2.12)$$

although in the current state itself, $\det(U) = 1$ and $\sigma = R^T TR$, just as for the stress σ conjugate to logarithmic strain.

As Hill (1968) has shown, the derivative on the right sides of eqns. (2.11,12) coincides with the Jaumann or co-rotational derivative of Kirchhoff stress [= det(U)T] when the reference state corresponds instantaneously with the current state. If it corresponds except for an arbitrary rotation R, this result is generalized to

$$\frac{d}{dt}[\det(U)R^T TR] = R^T\left\{\frac{\mathscr{D}}{\mathscr{D}t}[\det(U)T]\right\}R , \qquad (2.13)$$

where $\mathscr{D}/\mathscr{D}t$ denotes the Jaumann rate, defined as the ordinary time rate of components of the tensor involved, when these are referred to a coordinate frame that is instantaneously coincident with the background frame but spinning relative to it with an angular velocity given by the antisymmetric part of $\dot{F}F^{-1}$ $(= \partial\dot{x}/\partial x)$.

The utility of these results is as follows. For the type of elastic-plastic constitutive law introduced subsequently, it is possible to associate an instantaneously unloaded state, having deformation gradient F^p, with any state encountered in a process. The rotation R^p is not uniquely determined, and must be specified by, say, making F^p a pure deformation or by fixing certain material or, as appropriate, lattice or fiber directions relative to the background frame, or by some other means conducive to simplicity of constitutive description of the deformation from F^p to F. In most cases of practical interest the deformation $F^p \rightarrow F$ involves (elastic) strains that are minute fractions of unity, but possibly large rotations.

The procedure, then, is to write constitutive relations relative to some fixed reference state that is appropriately chosen for the instant at hand, so as to coincide instantaneously with the unloaded state (i.e. $F^p = I$ instantaneously). These relations will, in general, be most simply expressed in terms of the rotation-invariant true stress $R^T TR$, where R is now the rotation from the instantaneous reference state to the current state. But, with the assumption of small elastic dimension changes, both $R^T TR$ and its rate coincide, from what has been said, with those of the stress σ conjugate to logarithmic strain. Thus one may identify $R^T TR$ as a proper conjugate stress σ for the constitutive rate relations of the following sections and, further, directly calculate its time derivative in terms of the Jaumann rate in eqn. (2.13). The strain rate $d\varepsilon/dt$ can then be identified, for small elastic dimension changes, as $R^T DR$ where D is the instantaneous rate of deformation tensor (i.e. the symmetric part of $\dot{F}F^{-1}$) and R is again the rotation from the instantaneous reference state to the current state. Further, within the small elastic strain assumption, it is permissible to replace R by F, or by $F(F^p)^{-1}$ when both F and F^p are measured from some distant reference state, in the expression for $d\varepsilon/dt$ and on the right in eqn. (2.13). One may also set $\det(U) = 1$ in eqn. (2.13) with impunity. The practical result of all this is that one need finally involve only F^p, F, $\mathscr{D}T/\mathscr{D}t$ and D in the constitutive rate law; there is no need to directly calculate R and U. Similar simplifications of constitutive representations at finite deformation have been discussed by Willis (1969).

2.4 ELASTIC-PLASTIC MATERIALS AND THERMODYNAMIC CONSIDERATIONS

To further specialize eqn. (2.7) it is assumed that the materials under consideration may, under appropriate conditions, exhibit a purely elastic response to stress or temperature alterations at any stage of their deformation history. This response is regarded as due to lattice stretching, in the absence of any structural rearrangements of constituent elements of the material by plastic processes such as slip, twinning, diffusion, or the like. It is taken to be fully reversible. Such elastic response is actually elicited only in the limit of fast alterations of σ and θ in the case of time-dependent materials, for which instantaneous σ, θ alterations cause an alteration only in the plastic contribution to strain rate, but not to strain itself. This is often called "instantaneous elasticity". Within the time-independent idealization of plasticity, when the concept of a yield surface in stress space bounding an elastic range is introduced, such elastic response is elicited from a point of elastic-plastic deformation only when the σ, θ alterations are directed into the current elastic range. Of course, it does not matter that any finite range of purely elastic response exists, from the standpoint of a micro-mechanical model of the material, in that one can always define the purely elastic response as the deformation that would have resulted in a given case had the lattice stretching not been accompanied by plastic processes of structural rearrangement.

The further discussion in this section follows the work of Hill and Rice (1973) in showing how the existence, on thermodynamic grounds, of a work potential for the elastic response sets a concise structure for elastic-plastic constitutive laws in conjugate variables. We let H denote, symbolically, the current plastic state of the material, in the sense that variations of σ and θ at fixed H necessarily induce a purely elastic response. From the standpoint of the discussion leading to eqn. (2.7), H may be viewed as denoting some functional taken over those portions of the prior deformation history during which inelastic response occurred. Alternatively, from the standpoint of a mechanical micro-model of the material, H denotes, symbolically, the current pattern of microstructural arrangement of constituent elements of the material.

With either viewpoint, we may evidently specialize eqn. (2.7) to

$$\sigma = \sigma(\varepsilon, \theta, H) \quad , \tag{2.14}$$

where the notation means that at fixed H , the current stress is a direct function, with form dependent on the plastic state, of the current strain ε and temperature θ. Alterations of state at fixed H are evidently reversible and hence must be in accord with the existence of thermodynamic potentials. Letting η denote entropy and u the internal energy, both per unit volume of the adopted reference state,

$$\sigma_{ij}d\varepsilon_{ij} + \theta d\eta = du \quad \text{at } \underline{\text{fixed}} \ H \tag{2.15}$$

It proves more convenient to employ the Helmholtz free energy,

$$\phi = u - \theta\eta = \phi(\varepsilon, \theta, H) \ , \tag{2.16}$$

and its dual on ε, to be introduced shortly, so that

$$\sigma_{ij}d\varepsilon_{ij} - \eta d\theta = d\phi(\varepsilon, \theta, H) \quad \text{for fixed } H \tag{2.17}$$

Hence the stress-strain relations (2.14) have the structure

$$\sigma_{ij} = \partial\phi(\varepsilon, \theta, H)/\partial\varepsilon_{ij} \ , \tag{2.18}$$

when ϕ is symmetrized in the components of ε. Also, $\eta = -\partial\phi/\partial\theta$. It is important to realize that eqn. (2.18) applies throughout some program of inelastic deformation, the potential ϕ being taken at each instant as that appropriate to purely elastic response at the plastic state H prevailing at that instant. Equation (2.18) has been obtained alternatively from the Coleman-Noll thermodynamic formalism (Coleman and Noll, 1963; Coleman, 1964) for materials exhibiting instantaneous elasticity (Bowen and Wang, 1966), and this would include many representations of time-dependent plasticity. It was also derived by adapting the same formalism to the time-independent idealization by Green and Naghdi (1965). Indeed, in at least the present case, for which constitutive relations of the type eqn. (2.14) apply, the Coleman-Noll formalism and the classical thermodynamic approach adopted in this chapter lead to identical results (Kestin and Rice, 1970; Rivlin, 1970).

The dual potential to ϕ is

$$\psi = \sigma_{ij}\varepsilon_{ij} - \phi = \psi(\sigma, \theta, H) \ , \tag{2.19}$$

and in terms of it, when ψ is symmetrized in components of σ,

$$\varepsilon_{ij} = \partial\psi(\sigma, \theta, H)/\partial\sigma_{ij} \ . \tag{2.20}$$

Provided that eqn. (2.19) can be defined, either directly or by some process of analytic continuation, for the current H when $\sigma = 0$ and $\theta = \theta_o$ (the reference temperature) the strain that remains can be called the plastic strain associated with the plastic state H. That is, if $\varepsilon = \varepsilon(\sigma, \theta, H)$, then

$$\varepsilon^p = \varepsilon(0, \theta_o, H) . \tag{2.21}$$

When dealing with increments of constitutive functions of σ, or ε, θ, and H, the prefix d^p will denote the "plastic" part of the change in that function, defined as the change in the function when H is changed to $H+dH$ but σ or ε, respectively, and θ are given the same values. In particular

$$d^p\phi = \phi(\varepsilon, \theta, H+dH) - \phi(\varepsilon, \theta, H) , \tag{2.22}$$

$$d^p\psi = \psi(\sigma, \theta, H+dH) - \psi(\sigma, \theta, H) , \tag{2.23}$$

where, for example, $d^p\psi$ is to be regarded as a function of σ θ, H and whatever differential parameters comprise dH. Further, one may show that $d^p\psi$ and $-d^p\phi$ are numerically equal when evaluated at corresponding values of σ and ε. The plastic part of a strain increment is

$$d^p\varepsilon = \varepsilon(\sigma, \theta, H+dH) - \varepsilon(\sigma, \theta, H) , \tag{2.24}$$

and since ε at each of the two plastic states is given by eqn. (2.20), by differentiating eqn. (2.23) one has for $d^p\varepsilon$ (and analogously for $d^p\sigma$)

$$d^p\varepsilon_{ij} = \partial(d^p\psi)/\partial\sigma_{ij} , \qquad d^p\sigma_{ij} = \partial(d^p\phi)/\partial\varepsilon_{ij} . \tag{2.25}$$

Thus the plastic variations in the potentials are themselves potentials for the plastic variations in stress and strain. A full strain increment includes the parts due to variations in σ and θ as well as H:

$$d\varepsilon_{ij} = M_{ijk\ell}d\sigma_{k\ell} + \alpha_{ij}d\theta + d^p\varepsilon_{ij}, \tag{2.26}$$

where the instantaneous compliances $M = \partial^2\psi/\partial\sigma\partial\sigma$ (so that M is symmetric on interchange of ij and $k\ell$) and thermal expansivities $\alpha = \partial^2\psi/\partial\sigma\partial\theta$.

It is simplest to see the relation of $d^p\varepsilon$ to $d\varepsilon^p$ by assuming that M is independent of σ (as appropriate for small lattice stretches) and considering isothermal deformation at θ_o . Then

$$\varepsilon = M\sigma + \varepsilon^p \quad \text{whereas} \quad d\varepsilon = Md\sigma + d^p\varepsilon . \tag{2.27}$$

Thus

$$d^p\varepsilon - d\varepsilon^p = (d^pM)\sigma , \tag{2.28}$$

and the two agree precisely only when the moduli are unaltered by changes in H . This is certainly an exceptional circumstance for moduli based on conjugate variables, even if moduli based on a "true" stress measure, say R^TTR , are essentially unaltered by deformation. On the other hand, it seems plausible that d^pM is of the same order as $Md\varepsilon^p$, which means that the right side of eqn. (2.28) is of the order $M\sigma$ times $d\varepsilon^p$. But $M\sigma$ is the strain due to elastic lattice stretching and, so long as this is a minute fraction of unity, we have $d^p\varepsilon = d\varepsilon^p$ to a suitable approximation.

Implicit to the framework is the assumption that thermodynamic potentials have meaning during processes of inelastic deformation. This can be given justification within the internal variable framework of Sections 2.5 to 2.7, where processes necessary to reversibly separate, shift, and rejoin lattice planes, or to reversibly accomplish other microstructural rearrangements can be considered in principle for purposes of finding the change in potentials owing to changes in H . Otherwise, the existence of the potentials is simply postulated (as in the Coleman-Noll formalism) although the resulting equations of the theory do not allow their complete operational determination, even in principle. For example, a function varying arbitrarily with H , but not depending on ε or σ and θ , could be added with impunity to ϕ or ψ , since only their derivatives on ε or σ are involved in, say, eqns. (2.18), (2.20), and (2.25).

For general irreversible processes (i.e., those in which H changes), let Q be the heat supply to a material element, per unit of reference volume, so that the first law reads

$$\sigma_{ij} d\varepsilon_{ij}/dt + Q = du/dt \tag{2.29}$$

It will suffice to assume that all of Q results by radiant heating at spatially uniform temperatures over an element, so that the heat flux times $\partial\theta/\partial x$ does not complicate the entropy production inequality. This is permissible because $\partial\theta/\partial x$ is assumed to not enter the constitutive relations (2.7,14,18). Hence

$$d\eta/dt \geq Q/\theta . \tag{2.30}$$

By substituting the inequality into eqn. (2.29) and recovering ϕ from eqn. (2.16), this becomes

$$\sigma_{ij} d\varepsilon_{ij}/dt - \eta d\theta/dt \geq d\phi/dt . \tag{2.31}$$

But $d\phi$ may be split into $d^p\phi$ plus another part which, by eqn. (2.17), cancels the left side, to leave as the second law requirement,

$$\frac{d^p\phi}{dt} \equiv \frac{\phi(\varepsilon,\ \theta,\ H+dH) - \phi(\varepsilon,\ \theta,\ H)}{dt} \leq 0 \ , \tag{2.32}$$

where dH is the variation in plastic state occurring during time dt. This attains a very clear meaning when written subsequently in terms of internal variables.

We shall later see that various microscale mechanisms for plastic deformation lead to macroscopic constitutive laws that are in accord with the following "normality" structure. Specifically, for time-dependent plastic flow, this entails that a scalar function $\Omega = \Omega(\sigma,\ \theta,\ H)$ exist at each H so that the plastic portion of the instantaneous strain rate is

$$d^p\varepsilon_{ij}/dt = \frac{\partial\Omega(\sigma,\ \theta,\ H)}{\partial\sigma_{ij}} \tag{2.33}$$

Further, in the time-independent idealization, for which an H dependent yield surface

$$F(\sigma,\ \theta,\ H) = 0 \tag{2.34}$$

can be assumed to exist in stress-temperature space, the plastic portion of the strain increment is given by

$$d^p\varepsilon_{ij} = d\Lambda \frac{\partial F(\sigma, \theta, H)}{\partial \sigma_{ij}} \tag{2.35}$$

when the yield surface is smooth at the considered point. The scalar $d\Lambda$ is homogeneous of degree one in $d\sigma$, $d\theta$ and of a sign such that $d^p\varepsilon$ aligns with the outer normal in σ space, at the current θ, from the elastic to elastic-plastic domain. Alternatively, if the yield surface is not smooth at the current stress state, but has instead limiting segments $F_\alpha = 0$, $\alpha = 1, 2, \ldots, m$, then

$$d^p\varepsilon_{ij} = \sum_\alpha d\Lambda_\alpha \frac{\partial F_\alpha(\sigma, \theta, H)}{\partial \sigma_{ij}} \tag{2.36}$$

so that $d^p\varepsilon$ lies within the cone of limiting outer normals.

Hill and Rice (1973) were able to prove, on the basis of the existence of the potentials ϕ, ψ and $d^p\phi, d^p\psi$ as in eqns. (2.18, 20,25), that:

(i) If the above normality structure applies for any one choice of conjugate stress and strain measures and choice of reference state, then it necessarily applies for <u>every</u> choice of conjugate variables and reference state; and

(ii) If a composite material is made up of subelements that can be modelled as continua in which the above normality structure applies to the local stress-strain relations, then the same normality structure is necessarily transmitted to the overall stress-strain relations of the composite, when these are phrased in work-conjugate variables.

The last remark is of particular importance for polycrystalline aggregates, to the extent that these can be modelled as simple composites of single crystal continua. It is then well known that when elastic distortions of the crystal are small, and Schmid's law or its time-dependent generalization applies to describe the plastic response of the crystal, that the normality structure is identically satisfied for the stress-strain relations of each crystal. Hence it is necessarily satisfied by the macroscopic stress-strain relations for an aggregate of such crystals. This is a case for which the general framework developed here reveals a key structural feature of a constitutive law, even though very little can be said on the detailed form of the constitutive law for the polycrystal. There is an extensive literature on the approximate calculation of polycrystal properties from those of single crystals; this is discussed in Section 2.8. The works of Havner (1973), Hill and Rice (1972), and Rice (1971) may be consulted for rigorous formulations of crystal constitutive laws in a finite strain context.

Two separate quasi-thermodynamic postulates (Drucker, 1951, 1960; Il'yushin, 1961) have been proposed in the time-independent plasticity literature, and lead to the normality structure of eqns. (2.35,36). That by Il'yushin postulates that non-negative net work is done in enforcing an isothermal cycle of straining that begins and ends at the same arbitrary state. This is a separate postulate rather than a consequence of the second law because restoration of ε and θ does not fully restore the state - H has changed. Drucker's postulate deals with a material element under an arbitrary prestress and postulates that the net work done by an external agency in a cycle of adding and removing some additional set of loadings is non-negative. The postulate is not independent of the adopted stress measure and hence cannot be unambiguously interpreted in general. However, if limited to cycles involving only an infinitesimal accumulation of plastic straining, it is invariant to stress measure and leads to the normality structure of eqns. (2.35,36) in conjugate variables.

2.5 GENERAL FRAMEWORK FOR TRANSITION FROM MICROSCALE PROCESSES TO MACROSCOPIC STRAIN

This section presents a general framework by which microstructural deformation mechanisms are related to macroscopic plastic straining. The formalism is wide enough to include crystalline slip, which will be the mechanism of primary concern here, and also phase transformations, twinning, diffusional transport, etc. For maximum generality, a collection of "internal variables" are introduced to describe the local, microstructural rearrangements of a material sample by such mechanisms. The approach followed is due to Rice (1971); it is related to, and provides a unified setting for similar general studies by Havner (1969), Hill (1967), Kestin and Rice (1970), Lin (1968), Mandel (1966), Rice (1970), and Zarka (1972).

Consider a representative macroscopic sample of material, having volume V in an unloaded reference state. This is subjected to boundary loadings causing macroscopically homogeneous deformation; σ_{ij} and ε_{ij} are the conjugate macroscopic stresses and strains thus induced, and these are supposed to satisfy

$$V\sigma_{ij}d\varepsilon_{ij} = \text{work increment of boundary loadings} \tag{2.37}$$

Here, for simplicity, results are given as appropriate to isothermal behavior; θ is not explicitly listed as a variable in constitutive functions except when it proves fruitful to indicate the manner of generalization to the non-isothermal regime. The latter regime is fully discussed in the work by Rice (1971). Given that we now focus on a definite macroscopic sample of material, σ and ε have exact specifications in terms of averages of local surface stress and displacement fields over the boundary of the body. These can be converted to volume averages involving the local, inhomogeneous stress and strain fields. A full discussion is given by Hill (1967) and, in conjugate variables for finite strain, by Rice (1971) and Hill (1972).

The material sample may deform by: (1) elastic stretching of lattice bonds, and (ii) local microstructural rearrangements of its constituent elements by slip, etc. Following the notation of the last section, we let H denote symbolically the current pattern of microstructural rearrangement; this pattern is due to the prior distory of inelastic deformation experienced by the sample, and H may equally be thought of as representing this history. The free energy Φ of the sample depends on ε and H. When isothermal processes at fixed H are considered, we have, as in eqns. (2.16,17)

$$V\sigma_{ij}d\varepsilon_{ij} = [d\Phi]_{H\text{ constant}}, \quad \text{and} \quad \sigma_{ij} = \frac{1}{V}\frac{\partial\Phi(\varepsilon,H)}{\partial\varepsilon_{ij}} \tag{2.38}$$

Alternatively, if we introduce the dual potential

$$\Psi = V\sigma_{ij}\varepsilon_{ij} - \Phi \quad , \quad \text{then} \quad \varepsilon_{ij} = \frac{1}{V}\frac{\partial\Psi(\sigma,H)}{\partial\sigma_{ij}} \tag{2.39}$$

Evidently, we can identify $\Phi/V = \phi$, $\Psi/V = \psi$ and use the formulae of the last section.

Consider two neighboring patterns of microstructural rearrangement denoted by H, $H+dH$. We suppose that a set of incremental internal variables $d\xi_1$, $d\xi_2$, ..., $d\xi_n$ characterize the specific local rearrangements, which are represented collectively by dH , at sites throughout the sample. The required number of such variables increases in proportion to the size of the sample. Indeed, we shall see that an essentially infinite number of continuous variables defined piecewise throughout the sample is required for, say, a description of crystalline slip within this framework (Rice, 1971), but the structure of the theory is made evident more simply in terms of the discrete $d\xi$'s and there is no loss of generality. In analogy with the definition of the thermodynamic "force" on a dislocation line or crack front, we define a set of forces $f_\alpha = f_\alpha(\varepsilon,H)$ conjugate to the variables by

$$\Sigma\, f_\alpha(\varepsilon,H)d\xi_\alpha \;=\; -\,[\Phi(\varepsilon,H+dH) - \Phi(\varepsilon,H)] \equiv -\,d^p\Phi\ . \tag{2.40}$$

The analogous expression in terms of the dual potential is

$$\Sigma\, f_\alpha(\sigma,H)d\xi_\alpha \;=\; [\Psi(\sigma,H+dH) - \Psi(\sigma,H)] \equiv d^p\Psi\ . \tag{2.41}$$

The intent of these relations is to define the f's from a continuum model at the microscale level. Specifically, atomic scale fluctuations with position of the energy of a configuration are regarded as being averaged out, although the size of these fluctuations could well be important to determining the kinetic relations satisfied by the rates of the rearrangements, as e.g. in thermal activation models. In this regard, the f's differ from the force defined by Kocks (Chapter 3), who includes these short range fluctuations as well. In the case of a dislocation moving in a perfect lattice, the present f's would include the effects of the applied and other long range stress fields acting on the slip plane, but would not include the periodic Peierls stress due to the lattice itself, whereas Kocks' force definition would include both. The next section identifies various internal variables and their conjugate forces.

On rewriting eqns. (2.24,25) for the plastic part of a strain increment, namely

$$d^p\varepsilon_{ij} \;=\; \varepsilon_{ij}(\sigma,H+dH) - \varepsilon_{ij}(\sigma,H) \;=\; \frac{\partial}{\partial\sigma_{ij}}\left(\frac{1}{V}\,d^p\Psi\right)\ , \tag{2.42}$$

and then expressing $d^p\Psi$ in terms of the conjugate forces by eqn. (2.41), we obtain the following fundamental relation between a plastic increment of macroscopic strain and the corresponding extents $d\xi$ of microstructural rearrangements:

$$d^p\varepsilon_{ij} \;=\; \frac{1}{V}\ \Sigma\ \frac{\partial f_\alpha(\sigma,H)}{\partial\sigma_{ij}}\,d\xi_\alpha\ . \tag{2.43}$$

The summation extends over all the individual sites in the sample where rearrangements take place and, in view of the factor $1/V$, provides a volume average over the sample. The relation plays a key role in establishing a normality structure to constitutive laws that is valid for a wide range of kinetic mechanisms for the rearrangements. When the elastic stretching of the lattice is small and suitably linearized in its stress dependence over the range of interest,

$$\varepsilon_{ij} = \varepsilon^p_{ij}(H) + M_{ijk\ell}(H)\sigma_{k\ell} . \tag{2.44}$$

Ψ may then be determined by integrating eqn. $(2.39)_2$ on σ, noting from eqn. $(2.39)_1$ that $\Psi = -\Phi$ when $\sigma = 0$; the "locked-in" free energy at $\sigma = 0$ will be denoted $\Phi^o = \Phi^o(H)$. Thus

$$\Psi = -\Phi^o(H) + V\sigma_{ij}\varepsilon^p_{ij}(H) + \frac{1}{2} V\sigma_{ij}M_{ijk\ell}(H)\sigma_{k\ell} , \tag{2.45}$$

and by inserting this into the definition given by eqn. (2.41),

$$\Sigma\, f_\alpha(\sigma,H)d\xi_\alpha = -d\Phi^o(H) + V\sigma_{ij}d\varepsilon^p_{ij}(H)$$
$$+ \frac{1}{2} V\sigma_{ij}\sigma_{k\ell}dM_{ijk\ell}(H) . \tag{2.46}$$

Since this must hold for arbitrary $d\xi$'s , with corresponding dH's on the right side, it is seen that in general the f's are <u>quadratic</u> functions of stress. However, as discussed in connection with eqns. (2.27,28), elastic moduli when phrased in terms of, say, the stress measure R^TTR and strain increment measure $U^{-1}dU + dU\,U^{-1}$ are very little affected by processes such as dislocation motion. This means that dM based on some conjugate measure will be of the order $Md\varepsilon^p$, and hence that the third term in eqn. (2.46) is of the order $M\sigma$ times the second. This means that it is negligible and hence each f is linear in stress, taking the form

$$f_\alpha(\sigma,H) = f^o_\alpha(H) + \sigma_{ij}f^1_{\alpha,ij}(H) . \tag{2.47}$$

It is seen that f^o_α is associated with the "locked-in" free energy whereas, to the order of the approximation,

$$d^p\varepsilon = d\varepsilon^p = \frac{1}{V}\Sigma f^1_{\alpha\ ,ij}(H)d\xi_\alpha \ . \qquad (2.48)$$

The situation is otherwise when the mechanism of inelastic deformation is due to a phase change or, as a simple limiting case, due to the stable growth of elastic-brittle Griffith microcracks. If the material sample contains no initial stresses, then its shape will be fully recovered on unloading and $\varepsilon^p = 0$. In this case all of $d^p\varepsilon$ arises from the σdM term of eqn. (2.28), where dM is the change in compliance due to the crack growth. Hence in this case the forces f , conjugate to increments $d\xi$ of crack extension, contain no linear term but only the quadratic term in eqn. (2.46).

For non-isothermal behavior, θ is simply entered as another canonical variable in Φ and Ψ , as in Section 2.4, and in the f's ; eqn. (2.43) remains valid with $f_\alpha = f_\alpha(\sigma,\theta,H)$. Also eqn. (2.44) and, under the conditions cited, eqns. (2.47,48) remain valid if re-interpreted so that all H-dependent terms now depend on θ and H . In particular, ε^p is then re-interpreted as the strain when $\sigma = 0$ but θ and H remain at their current values, which differs from eqn. (2.21) unless $\theta = \theta_o$.

Now, eqn. (2.46) defined the f's in terms of macroscopic stresses and the associated plastic changes in macroscopic quantities such as Φ^o, ε^p and M . However, for each type of mechanism, f can be expressed in terms of the <u>local</u> field of stress and lattice strain prevailing near the site of the associated rearrangement. Some examples follow.

2.6 EXAMPLES OF INCREMENTAL INTERNAL VARIABLES AND CONJUGATE FORCES

2.6.1 *Crystalline slip*

Suppose that the transition between H and $H+dH$ can be desribed as due to incremental glide motions of the dislocations in a metal, where the dislocations are regarded as line defects and dn is a continuous variable along each dislocation loop, denoting the local advance of the line normal to itself in its slip plane. Then in eqns. (2.40,41)

$$\Sigma f_\alpha d\xi_\alpha \rightarrow \int_L [q\ dn]dL \quad , \tag{2.49}$$

where q is the force per unit length of dislocation line and L denotes an integration along all lines in the material sample. With this representation eqn. (2.43) becomes

$$d^p\varepsilon_{ij} = \frac{1}{V}\int_L [\frac{\partial q(\sigma,H)}{\partial\sigma_{ij}}\ dn]dL \ . \tag{2.50}$$

Here the notation means that q is a function of the macroscopic stress and of the entire current pattern of dislocations within the sample. It is known that within the linear elastic model the force on a dislocation contains a term τb, where τ is the local shear stress in the slip direction at a point along the dislocation line, acting in addition to the self stress of that segment of the line itself; b is the Burgers vector. Thus

$$q = q_o(H) + \bar{\tau}b \tag{2.51}$$

where $\bar{\tau}$ is a homogeneous linear function of σ, giving the shear stress at the dislocation site in the slip direction as would be induced elastically by applying macroscopic stresses σ to the material sample while the dislocations are held in position (i.e., at fixed H). The term q_o represents contributions from the self stress and from all other sources of internal stress within the material sample. Note that in addition to containing geometric orientation factors, $\bar{\tau}$ also varies from point to point within the sample because the elastic application of σ induces a highly non-uniform stress field on the scale of, say, the grains and inclusions making up the material sample. Thus we obtain, in correspondence to eqn. (2.48) of the general formalism, an expression given by Rice (1970),

$$d^p\varepsilon_{ij} = \frac{1}{V}\int_L [\frac{\partial\bar{\tau}}{\partial\sigma_{ij}}\ b\ dn]dL \ . \tag{2.52}$$

The terms dn and dL of eqns. (2.49,50) must be measured in units of, say, lattice spacing at finite lattice strain, to remain invariant under elastic distortion (Rice, 1971).

More generally, we shall average out the individual dislocations, specifying instead the local amounts of shear $d\gamma^{(1)}, d\gamma^{(2)}$, ... on the operative slip systems of the crystalline subelement encompassing any considered point of the material sample. If eqn. (2.52) were applied to a single crystal under macroscopically homogeneous deformation we would have

$$d\gamma^{(k)} = \frac{1}{V} \int_{L^{(k)}} [b^{(k)} dn] dL = \rho^{(k)} b^{(k)} \langle dn \rangle^{(k)} \tag{2.53}$$

where now the integral extends only over the dislocations on system (k) , and we express the result in terms of dislocation density $\rho = L/V$, b , and average advance $\langle dn \rangle$ on that system. The same interpretation is adopted locally within the heterogeneous material sample in that, e.g., we consider the ρ's and γ's to be defined locally throughout each grain of a polycrystal. Thus in eqns. (2.40,41)

$$\Sigma\, f_\alpha d\xi_\alpha \rightarrow \int_V [\Sigma\, \tau^{(k)} d\gamma^{(k)}] dV \quad , \tag{2.54}$$

where this defines thermodynamic "stresses" conjugate to the $d\gamma$'s , and from eqn. (2.43)

$$d^P\varepsilon_{ij} = \frac{1}{V} \int_V [\Sigma \frac{\partial \tau^{(k)}(\sigma,H)}{\partial \sigma_{ij}} d\gamma^{(k)}] dV \ . \tag{2.55}$$

Here, from eqns. (2.51-53), it is evident that the τ's will have the form

$$\tau^{(k)} = \tau_o^{(k)}(H) + \bar{\tau}^{(k)} \tag{2.56}$$

where, again when the lattice elasticity can be treated as linear, $\bar{\tau}^{(k)}$ is the homogeneous linear function of σ giving the shear stress that would be elastically induced on the slip plane in the slip direction by σ . The term $\tau_o^{(k)}$ includes the shear stress induced by misfits and other sources of residual stress, and also a part accounting for energy that is "locked-in" at the discrete dislocation level.

An alternate, direct mechanical identification of the thermodynamically defined τ's is made by calculating the total reversible work, or change in Φ, done in the following steps, which are carried out under fixed overall deformation of the sample. These are assumed to take place under circumstances of small elastic stretches with moduli, referred to crystallographic axes, being unaffected by plastic shears.

(i) Cut free an element δV, applying to it and the walls of the cavity thus created surface tractions appropriate to the local stress field s_{ij} at that point; no work is involved.

(ii) Remove s_{ij} from δV; the work of elastic unloading is done.

(iii) Move the dislocations as appropriate to accomplish the considered increment $d\gamma$ of plastic shear; $\delta V\, d^p\phi_o$ is the energy change, where $\phi_o(H)$ is the density of locked-in energy at the discrete dislocation level.

(iv) Elastically reload the body to the shape it had in (i); this regains the work of (ii) except that there is a deficit $-s_s d\gamma$ times δV (where s_s is the resolved shear component of s_{ij}) which arises because the required elastic reloading strain is less by the shear $d\gamma$, plus a term of order $(d\gamma)^2$ which arises because s_{ij} is now altered by some amount ds_{ij}.

(v) Put the element back into the cavity, removing the unwanted layer of body force arising from ds_{ij} of (iv); this causes displacements of its own order and hence the net work of this step is of order $(d\gamma)^2$

Thus, the net change in Φ, which must equal $-(\tau d\gamma)\delta V$ from eqns. (2.40,54), is, by summing the contributions of steps (iii), (ii) and (iv)

$$d^p\Phi = -\tau d\gamma\ \delta V = d^p\phi_o \delta V - s_s d\gamma\ \delta V\ . \tag{2.57}$$

If we divide s_s into a part due to residual stresses, day s_s^{res}, plus the part $\bar{\tau}$ that is induced elastically by σ this becomes

$$\tau \;=\; -d^p\phi_o(H)/d\gamma + s_s^{res} + \bar{\tau} \;, \tag{2.58}$$

in conformity with the discussion following eqn. (2.56). Rice (1971) and Hill and Rice (1972) have discussed the identification of τ when no simplifying assumptions are made as to the size of lattice stretches or the effect of slip on moduli.

Often eqn. (2.54) will be insufficient to completely represent the H change in Φ ; e.g., during annealing, dislocations may annihilate one another without creating $d\gamma$'s. Put another way, increments in dislocation density ρ on the various slip systems cannot, especially at elevated temperature, be considered universally related to the corresponding $d\gamma$'s. Thus eqns. (2.54,55) may be generalized to

$$\Sigma\, f_\alpha d\xi_\alpha \;\rightarrow\; \int_V [\Sigma\; \tau^{(k)} d\gamma^{(k)} + A^{(k)} d\rho^{(k)}]dV \tag{2.59}$$

$$d^p\varepsilon_{ij} \;=\; \frac{1}{V}\int_V [\Sigma\, \frac{\partial \tau^{(k)}}{\partial \sigma_{ij}}\; d\gamma^{(k)} + \frac{\partial A^{(k)}}{\partial \sigma_{ij}}\; d\rho^{(k)}]dV \;, \tag{2.60}$$

where $A^{(k)}$ are affinities conjugate to the dislocation densities on the various slip systems. A simpler generalization would involve just one ρ , the sum of those for all systems.

But alterations in ρ when the γ's are fixed can create no strain, unless there is an effect of ρ on elastic moduli [and even then the resulting effect is second order in σ by eqn. (2.46)]. Thus from eqn. (2.60) $\partial A^{(k)}/\partial\sigma = 0$ and $A^{(k)} = A^{(k)}(H)$. The simplest assumption is that the locked-in energy ϕ^o depends only on the ρ's , and not on the γ's . Then, through the type of argument leading to eqns. (2.57,58), one has

$$A^{(k)} \;=\; -\partial\phi^o/\partial\rho^{(k)} \quad , \quad \tau^{(k)} \;=\; s_s^{res} + \bar{\tau} \;. \tag{2.61}$$

2.6.2 *Diffusion*

For simplicity, consider a single, foreign diffusing substance which can move through the lattice of the crystallites in the sample and along its internal surfaces A , where A includes grain interfaces, boundaries of cavities, etc. Let the structural rearrangement of the material sample be characterized by the increase dn in concentration of the substance per unit volume of V and dN per unit area of interface A . Then the local chemical potential μ of the substance, at points throughout the sample, is defined by writing the change in Helmoltz free energy at fixed temperature and overall strain of the sample as

$$d^{P}\Phi = \int_V [\mu dn]dV + \int_A [\mu dN]dA - \int_S [\mu dN]dS \tag{2.62}$$

for arbitrary alterations dn, dN in composition. In the last integral over the external surface S of the material sample, dN represents the amount of substance that has exited. Hence one can make the identity

$$\Sigma f_\alpha d\xi_\alpha \rightarrow - \int_V [\mu dn]dV - \int_A [\mu dN]dA + \int_S [\mu dN]dS \tag{2.63}$$

However, as is well known, quantities such as dn/dt and dN/dt cannot be written in rate equations of a local kind. Hence the $d\xi$'s are identified with increments of diffusive flux. These characterize structural rearrangements by the local amount dq_i $(i=1,2,3)$ of material crossing unit area in the x_i direction within V , and the amount dQ_i $(i=1,2)$ crossing unit length in the z_i direction on A , where z_1, z_2 is an orthogonal cartesian coordinate system locally tangent to A . Further, for mass balance

$$dn = - \partial(dq_i)/\partial x_i \quad \text{in } V \tag{2.64}$$

and, in the simple form when A is locally flat,

$$dN = -\partial(dQ_i)/\partial z_i + (\nu_i dq_i)^+ + (\nu_i dq_i)^- \quad \text{on } A \ , \tag{2.65}$$

where the last two terms represent fluxes from the two sides of A , ν being the local outward normal to the portion of V supplying the dq (the divergence term is more complicated when A is curved). Thus, applying the divergence theorem in eqn. (2.62), taking $dN = \nu_i dq_i$ on S , assuming that incoming and outgoing fluxes dQ balance at all intersections of internal surfaces, and assuming that μ is continuous and piecewise differentiable, there results

$$\Sigma f_\alpha d\xi_\alpha \rightarrow \int_V [-(\partial\mu/\partial x_i)dq_i]dV + \int_A [-(\partial\mu/\partial z_i)dQ_i]dA \quad , \tag{2.66}$$

and this final result is valid for locally curved A as well.

The dependence of μ on the local stress field within the sample can be ascertained directly from eqn. (2.62). Suppose that addition of an amount dN of the substance to a grain boundary thickens it by an amount kdN where $k=k(H)$, which means that k depends on the current arrangement of the material sample and concentration level at the place where dN is added. Then the work done by inserting the matter in presence of the local stress field contains the term $-s_n kdN$ where s_n is the normal stress acting on the interface. Hence

$$\mu = \mu_b(H) - s_n k \tag{2.67}$$

Similarly, if the addition of dn to an element of volume causes the shape change strain $\beta_{ij}dn$ then, by arguments analogous to that for the continuum slip model,

$$\mu = \mu_v(H) - s_{ij}\beta_{ij} \tag{2.68}$$

when elastic lattice distortions are small and moduli negligibly affected, where s_{ij} is the local stress field at the site within the material sample. As for τ , s_n and s_{ij} can be split into portions $\bar{s}_n$ and $\bar{s}_{ij}$ which are the parts elastically induced by the macroscopic stress, and therefore homogeneous linear functions of σ with position-dependent coefficients, in general, plus portions due to residual and misfit stresses. Thus the force terms $-\partial\mu/\partial x_i$ and $-\partial\mu/\partial z_i$ of eqn. (2.66) can be thought of as functions of σ and H , linear in the former for the present circumstances, and the relation analogous to eqn. (2.43) can be written. In the special case when the above linear forms are used,

$$d^P\varepsilon_{ij} = \frac{1}{V}\int_V [\frac{\partial}{\partial x_\ell}(\beta_{qr}\frac{\partial\bar{s}_{qr}}{\partial\sigma_{ij}})dq_\ell]dV + \frac{1}{V}\int_A [\frac{\partial}{\partial z_\ell}(k\frac{\partial\bar{s}_n}{\partial\sigma_{ij}})dQ_\ell]dA \tag{2.69}$$

Here, in the differentiation on σ , $\bar{s}_n$ and $\bar{s}_{ij}$ are regarded as functions of σ and H , whereas the differentiations on x and z are, of course, total spatial derivatives in the usual sense. It is reasonably straightforward to generalize this discussion to multi-species diffusion, including self diffusion, and hence to provide a framework for inelastic deformation by diffusional creep.

2.6.3 *Phase changes*

Here we consider diffusionless phase changes, examples being martensitic transformations and twinning, in which a form α of a solid is converted to form β at an interface. Let A be the locus of all internal $\alpha - \beta$ interfaces in the material sample. The structural rearrangement in this case is characterized by the amount dz of normal advance of the interface into the α phase, where dz is a function of position on A . The thermodynamic force per unit area of the interface is denoted by p , so that eqns. (2.40,41) become

$$\Sigma f_\alpha d\xi_\alpha \rightarrow \int_A [p dz] dA \; . \tag{2.70}$$

Thus eqn. (2.43) gives

$$d^P\varepsilon_{ij} = \frac{1}{V} \int_A [\frac{\partial p(\sigma,H)}{\partial \sigma_{ij}} dz] dA \; . \tag{2.71}$$

Eshelby (1970) has given the explicit formula which relates p to the local field at the interface. This can be obtained by (i) cutting out the volume of α involved in the advance dz over an element of A and applying tractions in accord with the local stress field acting there, (ii) unloading, (iii) transforming the α to β , (iv) applying stresses to regain the initial shape, (v) reinserting and removing the unwanted layer of surface force. Eshelby's result is

$$p = (\phi)_\alpha - (\phi)_\beta - T_i \left[(\frac{\partial u_i}{\partial n})_\alpha - (\frac{\partial u_i}{\partial n})_\beta \right] \; , \tag{2.72}$$

where the ϕ's are local free energy densities on the two sides of the interface, T_i is the surface traction vector, and the terms $\partial u_i/\partial n$ are the normal spatial derivatives of the displacement vector on the two sides. Of course, the base levels for the ϕ's are not arbitrary since β obtains from α or vice-versa.

When the transformation corresponds to the growth of a void by cutting away material at a traction free surface, this reduces to the formula derived by Rice and Drucker (1966) with the ϕ difference being the strain energy density at the void surface. If the interface is considered to have a surface free energy γ , Eshelby's formula must be given an additional term of the form $-\gamma(\kappa_1 + \kappa_2)$, where the κ's are principal curvatures of the interface.

2.6.4 *Griffith cracks*

The formalism can also be applied to a material sample containing some distribution of Griffith cracks and this forms a final example. Indeed, in rock and some other brittle materials microcrack growth is an important mechanism of inelastic deformation. Let the locus of all crack fronts be denoted by L and let $d\ell$ be a function of position along L describing the amount of local advance of the cracks, and hence constituting the structural rearrangements. The advances considered here will be such that the surfaces of cracking have continuously turning tangent planes, without abrupt forking or branching. If F denotes the thermodynamic crack extension force per unit length along L , then eqns. (2.40,41) become

$$\Sigma\, f_\alpha d\xi_\alpha \rightarrow \int_L [F\, d\ell]dL\ . \qquad (2.73)$$

The model conventionally adopted, following Griffith, computes the change in free energy $d^p\Phi$ in quasistatic crack advance as the sum of the change in that part of Φ representing elastic deformation (i.e., in the "strain energy") plus that in the part of Φ representing surface energy. Thus

$$F = G - 2\gamma \qquad (2.74)$$

where G is the elastic energy release rate as introduced by Irwin (1957) and γ is the surface free energy. Thus, when the local crack tip energy release rate is expressed in terms of the macroscopic stress σ on the sample, there results from eqn. (2.43)

$$d^p\varepsilon = \frac{1}{V}\int_L \left[\frac{\partial G(\sigma,H)}{\partial \sigma_{ij}}\, d\ell\right] dL \,. \tag{2.75}$$

In the absence of residual stress, ε^p will always be zero and hence, in terms of eqns. (2.27,28) all of $d^p\varepsilon$ reflects changes in overall elastic compliances under stress.

The local representation for G, within the linear elastic treatment of lattice stretching, can be given as a homogeneous quadratic function of the crack tip stress intensity factors, denoting the strength of the characteristic $r^{-1/2}$ stress singularity. Indeed, while a somewhat different notation is usually employed, these are here defined for the most general anisotropic material so that the stress vector T_i at distance r ahead of the crack, on the plane of prospective growth, is given by

$$T_i = k_i\, r^{-1/2} + \ldots \,, \qquad i=1,2,3 \tag{2.76}$$

the dots representing non-singular terms, whereas the crack opening $u_i^+ - u_i^-$ at a small distance r behind the tip is

$$u_i^+ - u_i^- = C_{ij} k_j\, r^{1/2} + \ldots \,. \tag{2.77}$$

Here C_{ij} are certain coefficients dependent on the local elastic compliances, and the stress intensity factors k_i, i=1,2,3, are linearly dependent on the applied stress σ. Thus, following Irwin's (1957) method of calculation of the work of unloading the crack surfaces,

$$G = \lim_{\Delta\ell \to 0} \frac{1}{\Delta\ell}\,\frac{1}{2}\int_0^{\Delta\ell} [k_i\, r^{-1/2} + \ldots]$$

$$[C_{ij} k_j (\Delta\ell - r)^{1/2} + \ldots] dr = \frac{\pi}{4}\, C_{ij} k_i k_j \tag{2.78}$$

One may write

$$k_i = k_i^{res} + \bar{k}_i \tag{2.79}$$

where k^{res} is induced by residual stresses and where $\bar{k}$ is a homogeneous linear function of σ, denoting the stress intensity factor induced by elastic application of σ. Also, writing $d^P\varepsilon = d\varepsilon^P + dM\sigma$ from eqn. (2.28), one finds from eqns. (2.75,78,79) that the increments of plastic strain and of overall elastic compliances due to crack advance are:

$$d\varepsilon^P_{ij} = \frac{1}{V}\int_L [\frac{\pi}{2} C_{qr}\, k^{res}_q \frac{\partial \bar{k}_r}{\partial \sigma_{ij}}\, d\ell]dL$$

$$dM_{ijk\ell} = \frac{1}{V}\int_L [\frac{\pi}{2} C_{qr} \frac{\partial \bar{k}_q}{\partial \sigma_{ij}} \frac{\partial \bar{k}_r}{\partial \sigma_{k\ell}}\, d\ell]dL \qquad (2.80)$$

The latter formula is closely connected with Irwin's (1960) relation between the stress intensity factor and load-point compliance changes for a cracked body. It also provides a general solution for the effect of cracking on overall elastic moduli, as has been considered in some particular cases by Walsh (1965). Equation (2.78) for G is generalized to nonlinear elastic behavior as the crack tip J Integral (Rice, 1968; see also Cherepanov, 1967) which takes the same form as Eshelby's (1956,1970) general formula for the force on a point or line defect in an elastic field.

2.7 KINETIC RELATIONS AND PLASTIC NORMALITY

The last two sections have outlined the general method by which macroscopic strains are related to structural rearrangements on the microscale, and some specific $d\xi$'s with their conjugate f's have been identified. The framework is completed in principle by a specification of _kinetic relations_ for the rates $d\xi/dt$ of structural rearrangement.

Of course, the assumption throughout is that such f's as we have introduced (force on dislocation or interface, chemical potential, etc.) retain meaning at states far removed from _global_ thermodynamic equilibrium for the material sample. Specifically, the relations that define the f's and relate the $d\xi$'s to $d^P\varepsilon$ are strictly valid when one considers the transition from one to another _constrained equilibrium_ state, having structural arrangements H and $H+dH$ respectively. But the structural arrangements are unconstrained during actual processes and it is assumed that eqn. (2.43) and its various specializations of the last section can be applied then as well, provided always that the f's are identified as those of the imagined state of constrained equilibrium corresponding to the current H . The same assumption is tacitly adopted in much of dislocation theory and physical metallurgy, but bears statement before proceeding to kinetic aspects.

The kinetic equations are restricted by the second law as in eqn. (2.32). Indeed, writing $d^p\Phi$ as in eqn. (2.40), this becomes

$$\Sigma\, f_\alpha d\xi_\alpha/dt \;\geq\; 0 \; . \tag{2.81}$$

In the special cases for which f_α is linear in σ as in eqn. (2.47) we can note that $\Sigma\, f^o_\alpha d\xi_\alpha = -\, d\Phi^o(H)$ and use eqn. (2.48) to convert this to

$$\sigma_{ij} d\varepsilon^p_{ij}/dt \;\geq\; d\Phi^o/dt \tag{2.82}$$

for isothermal processes, where Φ^o is the locked-in energy. Hence the macroscopic plastic work rate need not be positive, but can be negative when "locked-in" energy is being taken from the sample. This is evidently the case for a Bauschinger effect which commences during unloading in a tensile test, while the stress is still acting in tension. It is also the case for time-dependent strain recovery after reduction of the load on a specimen to a small but still tensile value.

In several instances it has been seen that the structural rearrangements can be resolved into individual, scalar processes characterized by some set (effectively infinite) of scalar variables $d\xi_\alpha$. For these cases it is natural to think of the rate $d\xi_\alpha/dt$ at which a particular rearrangement takes place as being primarily dependent on its associated thermodynamic force f_α , for a given θ and current pattern H of structural arrangement. That is, when some instantaneous rate $d\xi_\alpha/dt$ is thought of as a function of $\sigma, \theta,$ and H , the dependence on σ occurs primarily through dependence of the rate on the associated scalar force $f_\alpha = f_\alpha(\sigma,\theta,H)$. To this approximation, kinetic rate laws have the form

$$d\xi_\alpha/dt \;=\; r_\alpha(f_\alpha,\theta,H) \tag{2.83}$$

with each rate being stress dependent only via its conjugate thermodynamic force. The second law requirement will be satisfied so long as the rate function r_α has always the same sign as does f_α. The form is suggested by thermal activation models, in which case f_α represents the effect of applied stresses on biasing the pre and post-barrier energy levels. However, as a secondary effect, local stress terms not included in f could alter the size of the barrier itself, at least to the extent that lattice dimensions are affected, so that eqn. (2.83) should be thought of only as an approximation and not as a physical law.

Kinetic relations of the Schmid type that are usually taken to describe crystalline slip are, in fact, in accord with this class of rate law. Specifically, these relations entail that the rate of slip $d\gamma^{(k)}/dt$ [or average dislocation velocity $\langle dn/dt\rangle^{(k)}$] on a given slip system is dependent on the local stress state only via the resolved shear stress $s_s^{(k)}$ on that system. As seen in eqns. (2.56,58,61), the net resolved shear stress differs from the thermodynamic shear stress only by a term dependent on H (and θ in a non-isothermal analysis) so that the type of kinetic law just described for crystalline slip has the form

$$d\gamma^{(k)}/dt = \Gamma^{(k)} (\tau^{(k)}, \theta, H) \quad , \tag{2.84}$$

and is thus a special case of the general class of eqn. (2.83). This form includes, of course, such effects as direct and latent hardening against further slip, via the dependence on H. Again, some approximation is entailed. The large isotropic compression of a crystalline lattice by high pressure would alter the resolved shear stress for a given shear force only insofar as the area is changed, but sufficiently large pressures are likely to alter eqn. (2.84) in a more complicated way. Of greater interest, however, is the possibility, pointed out in private communication by F. Kocks, that when dislocations split into partials, not only is the resolved shear stress on the slip system important but so also are the resolved shears on different planes which tend to either coalesce or widen the partials. In such cases the stresses which "set-up" the deformation are not synonymous with those that drive the dislocation.

For vector processes of structural rearrangement as in diffusion, it is in general not possible to associate a given component dq_i/dt of the matter flux only with its force $-\partial\mu/\partial x_i$. Nevertheless, it is usually taken as sufficient to assume, in the absence of, say, large lattice strain effects on the intrinsic height of energy barriers, that the flux vector for some diffusing species at a point depends only on the various forces $-\partial\mu/\partial x$ as defined for all the diffusing species at that point. Hence, within the general framework, considerable interest attaches to the case for which an individual rate $d\xi_\alpha/dt$ depends, at a given θ and H, on some set of forces f, conjugate to that rate and to others present locally. The special case in which these are taken to be instantaneously linear,

$$d\xi_\alpha/dt = \Sigma L_{\alpha\beta}(\theta,H)f_\beta \quad , \tag{2.85}$$

with Onsager reciprocity $L_{\alpha\beta} = L_{\beta\alpha}$ is, in fact, just a particular case of eqn. (2.83). This is because, as Kestin and Rice (1970) have remarked, re-definition of the f's and $d\xi$'s by linear combinations, in accord with the rotation to principal axes of L in f space, reduces eqn. (2.85) to a diagonal form, in which each instantaneous rate is stress-state dependent only via its conjugate force.

It is of interest that rate laws of the class of eqn. (2.83), for which conjugate forces govern rates, lead to a remarkable normality structure for macroscopic constitutive laws (Kestin and Rice, 1970; Rice, 1970, 1971), independently of the detailed form of the rate equations. Indeed, the balance of this section is devoted to an exploration of the topic. The structure under consideration is such that a macroscopic scalar flow potential $\Omega = \Omega(\sigma,\theta,H)$ exists at each instant of the deformation such that the instantaneous plastic portion of the macroscopic strain rate is given by

$$d^p\varepsilon_{ij}/dt = \frac{\partial\Omega(\sigma,\theta,H)}{\partial\sigma_{ij}} \quad . \tag{2.86}$$

It is easiest to derive the equation by directly writing the microscale representation for Ω :

$$\Omega(\sigma,\theta,H) = \frac{1}{V}\Sigma\int_0^{f_\alpha(\sigma,\theta,H)} r_\alpha(f_\alpha,\theta,H)\,df_\alpha \quad , \tag{2.87}$$

where the integration is done at fixed θ and H, and with this eqn. (2.86) may be proven as a direct application of eqn. (2.43):

$$\frac{\partial \Omega}{\partial \sigma_{ij}} = \frac{1}{V} \Sigma \frac{\partial f_\alpha}{\partial \sigma_{ij}} r_\alpha = \frac{1}{V} \Sigma \frac{\partial f_\alpha}{\partial \sigma_{ij}} d\xi_\alpha/dt = d^P \varepsilon_{ij}/dt .$$

If we specialize these results to the crystalline slip model, with the rate law of eqn. (2.84),

$$\Omega(\sigma,\theta,H) = \frac{1}{V} \int_V \Sigma \left\{ \int_0^{\tau^{(k)}(\sigma,\theta,H)} \Gamma^{(k)}(\tau^{(k)},\theta,H) d\tau^{(k)} \right\} dV , \tag{2.88}$$

and now eqn. (2.86) may be proven directly from the corresponding special version (2.55) of eqn. (2.43). This shows also that the macroscopic flow potential is just the volume average of local flow potentials for each slip system of each individual crystallite of the material sample.

Consider the significance of the flow potential from a purely macroscopic standpoint: Now instead of requiring six separate constitutive relations for the stress and history dependence of the instantaneous plastic strain rate components, we require only one for the scalar Ω, from which the others are generated. Further, this has a geometric interpretation in a stress space having coordinate axes which are the components of σ. At each apoch in the history of deformation, a family of surfaces of the form Ω = constant exists in this space, and have the property that the instantaneous plastic strain rate has a direction 'normal' to the Ω surface through the current stress point, and a magnitude equal to the gradient between neighboring Ω surfaces. Provided that the local rates are steadily increasing functions of the conjugate forces for any given θ and H, and that conditions for the forces to be linear in σ are met, each Ω surface may be shown to be convex, in that a plane which is tangent to the surface at any point will never cross it. Indeed, if we consider two different stress states σ^A, σ^B but the same plastic state H and θ,

$$\Sigma\ (f_\alpha^A - f_\alpha^B)[(d\xi_\alpha/dt)^A - (d\xi_\alpha/dt)^B] \geq 0 \tag{2.89}$$

by the assumed monotonicity of the rate law. But by eqns. (2.47,48) which apply when the forces are linear (in which case $d\varepsilon^P/dt = d^P\varepsilon/dt$), the inequality becomes, successively

$$\Sigma\ (\sigma_{ij}^{A} - \sigma_{ij}^{B}) f_{\alpha,ij}^{1} [(d\xi_\alpha/dt)^A - (d\xi_\alpha/dt)^B] \geq 0$$

$$(\sigma_{ij}^{A} - \sigma_{ij}^{B}) [(d^p\varepsilon_{ij}/dt)^A - (d^p\varepsilon_{ij}/dt)^B] \geq 0 \qquad (2.90)$$

This, in combination with eqn. (2.86), proves the convexity of Ω surfaces (see Rice, 1970).

Often the kinetic relation of eqn. (2.84) is strongly non-linear: At any given H and θ, an essentially zero $d\gamma/dt$ results for a certain range of τ values, whereas $d\gamma/dt$ takes on very large magnitudes for values of τ only slightly beyond the limits of this range. Of course, these limits change with accumulating H. Evidently, if we consider a restricted range of deformation rates, then the resulting behavior is well described by a time-independent idealization in which the limits represent critical shear stresses for yielding a slip system, and in which the changes in the limits with H represent strain hardening.

Within this time-independent model, we may take state B in the preceding inequality to coincide with a point within the elastic domain at the current H and A to lie on the current yield surface. Thus it reduces to the classical inequality of maximum plastic work,

$$(\sigma_{ij}^{A} - \sigma_{ij}^{B})(d^p\varepsilon_{ij}/dt)^A \geq 0 \ , \qquad (2.91)$$

except that the present form, in conjugate deformation variables, is properly invariant to rigid rotations in going from A to B, under the tacit assumption of small elastic lattice stretches between the two states. This inequality is well known to lead to the time independent normality structure discussed in connection with eqns. (2.35,36) and also to require that all yield surfaces be convex. Rice (1971) and Hill and Rice (1973) have further shown, in fact, that validity of the inequality (2.89), in the time-independent theory, implies the Il'yushin inequality, and hence normality in conjugate variables, without regard to the magnitude of elastic deformations.

By directly adopting the time-independent idealization in connection with a somewhat more restricted crystalline slip model than necessitated in the present framework, Hill (1967), Mandel (1966), and Rice (1966) independently derived such results on plastic normality and convexity as just discussed. They show that the plastic strain increment $d^p\varepsilon$ has the direction of the outward normal to the current macroscopic yield surface in σ space when that surface is smooth, and a direction within the cone of limiting normals at a vertex. In fact, as Hill has emphasized, a vertex is usually to be expected, within the model, on subsequent yield surfaces at points of sustained deformation. This is because the macroscopic yield surface is the envelope of an infinite family of yield planes in σ space, each corresponding to a critical shear stress on a local slip system. These planes may translate as residual stress contributions to τ build up and as direct or latent hardening occurs, but their normals remain of fixed orientation. Every individual plane corresponding to a slip system active in the sustained deformation must pass through the current stress state, and this creates the vertex. Hill has also proposed that vertex-free large offset "yield surfaces" can be interpreted as families of plastic limit states as defined in terms of the local distribution of hardness in a polycrystalline aggregate. This is tantamount to treating the material as rigid-plastic, as in a study by Bishop and Hill (1951).

Rice (1970) has discussed the time-independent idealization in terms of a clustering of Ω surfaces outside the non-yielding domain of σ space. Also, he has shown that if the relation between $d\gamma/dt$ and τ on each slip system is continuous, then the Ω surfaces do not contain vertices except possibly when a surface corresponds to zero strain rate. This latter case arises when a non-yielding domain, in which Ω = constant, exists.

The dual potential to Ω is of some utility. Let us suppose that the relation of $d^p\varepsilon/dt$ to σ is invertible to the extent that a function (Rice, 1973)

$$\Lambda = \Lambda(d^p\varepsilon/dt,\theta,H) = \sigma_{ij}d^p\varepsilon_{ij}/dt - \Omega \tag{2.92}$$

may be defined. Then by eqn. (2.86) when θ and H are considered fixed,

$$d\Lambda = \sigma_{ij}d(d^p\varepsilon_{ij}/dt) \quad , \text{ or } \quad \sigma_{ij} = \frac{\partial\Lambda(d^p\varepsilon/dt,\theta,H)}{\partial(d^p\varepsilon_{ij}/dt)} \tag{2.93}$$

_if_ components of $d^p\varepsilon/dt$ can be varied independently. Usually they cannot be, because plastic straining is incompressible. In this case it is easy to see that the differential form of eqn. $(2.93)_1$ allows solution for the deviatoric part of σ .

By using eqns. (2.43,87) and by writing r_α for $d\xi_\alpha/dt$ and integrating by parts, the microstructural interpretation of Λ is

$$\Lambda = \frac{1}{V}\Sigma\left[\int_0^{r_\alpha} f_\alpha(r_\alpha,\theta,H)dr_\alpha - (f_\alpha - \sigma_{ij}\frac{\partial f_\alpha}{\partial \sigma_{ij}})r_\alpha\right], \qquad (2.94)$$

where, for purposes of the integration at fixed θ,H, the kinetic law given by eqn. (2.83) is supposed to have been inverted to obtain f in terms of r. Of course, when f is linear in σ this becomes

$$\Lambda = \frac{1}{V}\Sigma\left[\int_0^{r_\alpha} f_\alpha(r_\alpha,\theta,H)dr_\alpha - f_\alpha^o(\theta,H)r_\alpha\right], \qquad (2.95)$$

where f_α^o is now the value of f_α when $\sigma = 0$. This would, for example, take the form

$$\Lambda = \frac{1}{V}\int_V \Sigma\left[\int_0^{\eta^{(k)}} \tau^{(k)}(\eta^{(k)},\theta,H)d\eta^{(k)} - \tau_o^{(k)}(\theta,H)\eta^{(k)}\right]dV \qquad (2.96)$$

for the slip model which averages out the individual dislocations, where $\eta^{(k)}$ is written for $d\gamma^{(k)}/dt$, and where the rate law of eqn. (2.84) is supposed to have been inverted in the integrand.

For the time-independent idealization, f_α will have a definite value (namely, the current yield value) if the associated r_α or $d\xi_\alpha/dt$ is non-zero. Hence each of the integrals in eqn. (2.94) amounts in a term $f_\alpha r_\alpha$, and so

$$\Lambda = \frac{1}{V}\Sigma\,\sigma_{ij}\frac{\partial f_\alpha}{\partial \sigma_{ij}} r_\alpha = \sigma_{ij}\, d^p\varepsilon_{ij}/dt \qquad (2.97)$$

by eqn. (2.43). Thus eqn. (2.93) reproduces a known result for time-independent materials satisfying the normality rule: that components of σ are derivatives of the rate of plastic working with respect to corresponding components of $d^p\varepsilon/dt$. This identification of Λ could also be developed directly from eqn. (2.92) in the rate-insensitive limit.

2.8 THE AVERAGING PROBLEM; POLYCRYSTAL MODELS

The general framework must be completed by some procedure of averaging a given set of rate relations over all the local sites of rearrangement within a material sample, to arrive finally at specific macroscopic representations of constitutive laws. Here we shall examine some approaches to the averaging problem for the plastic behavior of polycrystalline aggregates deforming by slip, on the assumption that kinetic relations of the kind (2.84) are given, a priori, from dislocation dynamics considerations and/or experiment for the operative slip systems within the individual crystals of the aggregate. Of course, these relations are not known precisely in general and very simple forms have been employed in the studies under review. Nevertheless, they do presumably show the manner in which constraints of neighboring grains and induced residual stresses affect the macroscopic constitutive behavior of polycrystals. All the discussion of this section is carried on within a small displacement gradient approximation, for which distinctions between stress and deformation measures of different kinds and their rates are ignored. Further, elastic moduli are taken to be uninfluenced by slip.

Local stress and strain fields within individual crystalline elements of the aggregate are denoted by s and e. The plastic strain is given in terms of the local shears γ by

$$e^p_{ij} = \Sigma\, \mu^{(k)}_{ij} \gamma^{(k)} , \qquad (2.98)$$

where the summation extends over all operative slip systems of the element and where

$$\mu_{ij} = \frac{1}{2} (n_i m_j + n_j m_i) , \qquad (2.99)$$

with n and m being unit vectors describing the slip plane normal and slip direction for a given system. The mechanical shearing stress acting on a given system is

$$\tau^{(k)} = s_{ij} \mu^{(k)}_{ij} ; \qquad (2.100)$$

this differs from the thermodynamic shear stress of eqn. (2.54) only in that the latter contains an additional part accounting for energy which would remain stored in the dislocation substructure even if the local stress were reduced to zero ($s \to 0$).

Hence τ as we now use it includes the long range residual shear stress as well as that induced elastically by σ , and called $\bar{\tau}$ earlier.

Taylor (1938) and Bishop and Hill (1951) considered a single phase polycrystal and neglected elastic strains (rigid-plastic model), further supposing that each individual grain sustains the macroscopic strain ε^p of the aggregate. While their considerations were for time-independent behavior only, we can in fact consider the general time-dependent case, presuming that by inversion of eqn. (2.84), rate laws are given in the form (isothermal for simplicity)

$$\tau^{(k)} = \tau^{(k)}(\eta^{(k)},H) \quad , \quad \text{where} \quad \eta^{(k)} = d\gamma^{(k)}/dt \qquad (2.101)$$

The procedure is to directly calculate the potential Λ from eqn. (2.96) as

$$\Lambda = \frac{1}{V}\int_V \Sigma\left[\int_o \tau^{(k)}(\eta^{(k)},H)d\eta^{(k)}\right] dV \quad . \qquad (2.102)$$

Now the term with $\tau_o^{(k)}$ of eqn. (2.96) has seemingly disappeared. This is because its part representing stored energy in the dislocation substructure has already been incorporated, due to the discussion following eqn. (2.100), and because the remaining long-range residual stress part does no net work on the $d\gamma$'s by the principle of virtual work, which can here be applied because elastic strains are neglected and hence the γ's give the total strain and correspond to a compatible deformation field.

To calculate each $\eta^{(k)}$ of an individual grain so that Λ may be computed, one recognizes that these are to be constrained by the approximation that each grain sustains the same strain. Hence

$$\Sigma\, \mu_{ij}^{(k)}\eta^{(k)} = d\varepsilon_{ij}^p/dt \qquad (2.103)$$

for each. We must further choose the η's so that the associated set of τ's as computed from eqn. (2.101) are, in fact, derivable from a local stress field s by eqn. (2.100). The correct η's are given by minimizing the bracketed terms in eqn. (2.102) subject to the constraint of eqn. (2.103), for by the method of Lagrange multipliers, this is equivalent to

$$\delta\left\{\left[\Sigma\int_{o}^{\eta^{(k)}}\tau^{(k)}(\eta^{(k)},H)d\eta^{(k)}\right]-\lambda_{ij}\left[\Sigma\ \mu_{ij}^{(k)}\eta^{(k)}\right]\right\}=0\ ,$$

$$\text{or}\quad\left\{\tau^{(k)}(\eta^{(k)},H)-\lambda_{ij}\mu_{ij}^{(k)}\right\}\delta\eta^{(k)}=0\ , \tag{2.104}$$

where λ_{ij} are the multipliers. Evidently, the equation is solved when

$$\tau^{(k)}=\lambda_{ij}\mu_{ij}^{(k)}\ , \tag{2.105}$$

which is the same as saying that the τ's are derivable from a stress field. In fact, $\lambda = s$.

By performing this constrained minimization, the bracketed term of eqn. (2.102) is determined as a function of $d\varepsilon^{p}/dt$ for each grain orientation. The remaining volume integral means that Λ is given by the average of this function over all grain orientations, and σ is computed from eqn. (2.93). The time-independent version of this general approach is exactly that employed by Bishop and Hill (1951). The net result is that Taylor orientation factors have been determined showing, for example, that the flow stress of an fcc polycrystal loaded in simple tension is approximately 3 times the corresponding shear strength on its $(1\ 1\ 1)[1\ \bar{1}\ 0]$ slip systems (assumed equal for all). Lin (1957) has further extended this approach to the elastic-plastic case by assuming that the total strain is constant in each grain; this allows an estimate of the entire stress-strain curve. Of course, the constraint that each grain deforms the same makes it impossible for stress equilibrium to hold and also causes an overestimate of the resistance to flow.

Batdorf and Budiansky (1949) have proposed a slip theory of plasticity which, if reinterpreted in the present context, can be seen as complementing the above approach by assuming that each individual grain carries the same stress, equal to σ. Hence eqn. (2.100) becomes

$$\tau^{(k)}=\sigma_{ij}\mu_{ij}^{(k)} \tag{2.106}$$

and with this together with rate laws of the type (2.84), phrased in terms of the mechanical shear stress, one may directly calculate Ω from eqn. (2.88) as the average over all orientations of the flow potential which an individual grain would have, if subjected to the stress σ . The corresponding plastic strain rate is then given by eqn. (2.86). Of course, this approach does not satisfy displacement continuity between adjacent grains, nor can it account for development of residual stresses which tend to build up preferentially on the systems of greatest slipping; hence it underestimates the resistance to flow.

Recently Clough and Simmons (1973) have proposed an approach to rate-dependent flow which, on examination, may be seen as amounting to the formalism outlined above with a rate law for which $d\gamma/dt$ varies as a hyperbolic sine of τ , with no effect of H on the relation. The original Batdorf-Budiansky application was to rate-independent slip, with hardening of active systems, but no latent hardening or reverse hardening. This led to a pronounced vertex at the current load point. Also, for any stress path which continusouly activated every slip system, once initially activated, the total strain was seen to depend only on the stress -- i.e., 'deformation theory' applied (Budiansky, 1959).

Lin and Ito (1965,1966) analyzed by methods of three dimensional elasticity the behavior of a polycrystalline model of $4 \times 4 \times 4$ square blocks, each containing one permissible set of slip planes with three equally spaced slip directions. Orientations were chosen to simulate a macroscopically isotropic polycrystal. They showed that a vertex formed at the current load point when a zero offset strain definition of yield was adopted, but they also showed that this vertex became a rounded bulge when a small but finite offset definition was used.

The bulk of work on predicting elastic-plastic behavior of polycrystals has been based on the self-consistent model of Kröner (1961) and Budiansky and Wu (1962). It considers s and e to take on constant values within each grain. Apart from any constitutive connection between the two, these are related to σ and ε by the same formulae that would apply if the grain were a homogeneous spherical inclusion imbedded in an infinite homogeneous matrix, having the overall elastic properties of the aggregate, and carrying the remotely uniform fields σ and ε . Thus, if these overall properties are isotropic with shear modulus G and Poisson ratio ν ,

$$s_{ij} = \sigma_{ij} - \frac{3-5\nu}{4-5\nu} G\delta_{ij}(e_{kk}-\varepsilon_{kk}) - \frac{7-5\nu}{4-5\nu} G\,(e_{ij}-\varepsilon_{ij}) \ . \qquad (2.107)$$

Here, ε,σ are volume averages of e,s ,

$$\varepsilon_{ij} = \frac{1}{V}\int_V e_{ij}\,dV \quad , \quad \sigma_{ij} = \frac{1}{V}\int_V s_{ij}\,dV \tag{2.108}$$

and the first of these will imply the second by eqn. (2.107).

Now, in the special case when each grain is idealized as being elastically isotropic with the same constants ν and G eqn. (2.107) may be re-written solely in terms of plastic strain as

$$s_{ij} = \sigma_{ij} - \frac{2}{15}\,\frac{7-5\nu}{1-\nu}\,G\,(e^p_{ij}-\varepsilon^p_{ij}) \quad , \tag{2.109}$$

and in this case, although not generally (Rice, 1970; Hill, 1971), ε^p is the volume average of e^p . Hence, using eqns. (2.98,100) the shear stress associated with a slip system in a given grain, having the orientation parameter $\mu^{(k)}$, is

$$\tau^{(k)} = \sigma_{ij}\mu^{(k)}_{ij} - \frac{2}{15}\,\frac{7-5\nu}{1-\nu}\,G\,\Big[\Sigma\mu^{(\ell)}_{ij}\mu^{(k)}_{ij}\gamma^{(\ell)}$$

$$- \frac{1}{V}\int \Sigma\,\mu^{(\ell)'}_{ij}\mu^{(k)}_{ij}\gamma^{(\ell)'}dV'\Big] \quad , \tag{2.110}$$

where the first sum, on (ℓ) , extends over all slip systems of the same grain and the second, on $(\ell)'$, extends over all systems of every grain as it is encountered in the volume integral (or orientation average); the primes distinguish those variable quantities in the integration. This gives an explicit representation for the long range residual stress, as a linear function of all the γ's in all the grains. The procedure is then to solve the kinetic relations for the γ's , given a history of σ variation, and to thereby compute ε^p .

Hutchinson (1964) has applied this procedure to time-independent calculations, both without hardening and with Taylor hardening, for fcc and bcc polycrystals. His results include the calculation of Bauschinger effects and of the response to proportional loading under combined stress. Bui (1970) has adopted the model to compute subsequent yield surfaces and shows a clear vertex formation. Kocks (1970) has given an extensive general survey of work with the self-consistent model and with the Taylor model in predicting the yield behavior of polycrystals, including experimental comparisons. Brown (1970a) and Zarka (1972) have considered time dependent behavior; the former has adopted a power law relation between $d\gamma/dt$ and τ for fcc polycrystals, and has computed the surfaces of constant flow potential in tension-torsion stress space for various deformation histories. These seem, to a fair approximation, to show kinematic translation without much shape change. Brown (1970b) has also attempted direct experimental measurement of Ω surfaces. These, for an aluminum alloy at elevated temperature, seem to show the same pronounced anisotropy as did the lower temperature yield surface for tension-torsion specimens of the same material.

Hill (1965) has suggested a more elaborate self-consistent model which is intended to take account of directional weaknesses developing with continuing deformation in a time-independent plastic framework; the corresponding generalization for time-dependence is, however, unclear. Recently Hutchinson (1970) has given an extensive review, contrasting the Hill model with that of Kroner-Budiansky-Wu. The latter gives limit states which agree with the Taylor model, and are thus overestimates, whereas the Hill model seems to give lower values. Hutchinson also calculates the plastic moduli governing increments of shear after tensile loading. These are considerably nearer to the predictions of 'deformation' theory than to those of a 'flow' theory with a smooth yield surface, although the theory itself is, of course, of the flow type.

Similar averaging procedures, to obtain macroscopic constitutive laws, could presumably be carried out for other of the internal variable and conjugate force sets of Section 2.6. The point which must be achieved, in general, is the development of an equation analogous to (2.110), which expresses the forces f_α in terms of the macroscopic stress σ and the various microstructure parameters whose increments are measured by the $d\xi$'s . These, together with a kinetic relation of, say, the type (2.83) relating $d\xi/dt$ to f enable one to write differential equations for $d\xi/dt$ in which σ enters as a forcing function. The resulting macroscopic plastic strain rate is expressed in terms of the f's and $d\xi/dt$'s by eqn. (2.43), or by eqn. (2.48) when appropriate, and this is in the form of a volume average expression. As for the self-consistent crystalline slip models, the balance of the analysis consists of carrying out the averaging over all sites and their orientations within a representative material sample to arrive finally at specific macroscopic constitutive relations.

Judging from the progress with slip models as just reviewed, this general procedure does seem to reveal many of the observed features of combined stress behavior, including Bauschinger effects, kinematic like translation of yield and flow potential surfaces, tendency toward vertex formation on small offset yield surfaces, etc. However, the computations involved in estimating the constitutive response to even simple deformation histories are quite complex and involve large storage requirements since, e.g., shears γ on all systems of crystals of all the representative orientations chosen for the calculation must be analyzed in each step. Indeed, when each $d\gamma/dt$ is expressed in terms of its associated τ, eqn. (2.110) becomes a large system of coupled, non-linear differential equations.

Thus the use of averaging procedures that involve, even with substantial approximations, a direct calculation from micro-scale models entails substantial complexity, and this would seem overwhelming if required in each increment of deformation for each element of, say, a finite element computer formulation for some structural problem involving inhomogeneous deformation. This means that any reasonably direct prediction of material response is unlikely to displace the phenomenological and less rigorously based structure-parameter models, discussed in subsequent chapters, as a basis for practical calculation. A brief discussion of this type of approach follows.

2.9 PHENOMENOLOGICAL AND MACROSCOPIC STRUCTURE-PARAMETER FORMULATIONS

There is an extensive literature on purely phenomenological approaches to constitutive laws within the time-independent plasticity idealization (see reviews by Drucker, 1956; and Naghdi, 1960) and some attempts at generalization to the time-dependent range have been made. Here we examine a simple formulation, intended for problems of time-dependent plastic flow under variable temperature and non-proportional stressing, which incorporates and generalizes, in what seems to be a physically acceptable form, notions such as kinematic and isotropic hardening as developed in the time-independent theory. We deal with initially isotropic materials.

Taking the viewpoint that the instantaneous plastic strain rate $d^p\varepsilon/dt$ is some function of σ, θ and the current plastic state H, we can define a <u>rest stress</u> tensor λ_{ij} (see Rice, 1970; Ahlquist and Nix, 1969) associated with θ and H as that for which $d^p\varepsilon/dt$ vanishes when $\sigma = \lambda$. The tensor λ is approximately interpretable as a macroscopic structure parameter, or internal variable, that measures the intensity of residual stress contributions to the forces f. As seen in eqn. (2.47) and in the slip and diffusion examples of Section 2.6, these typically have a form in which there is a term directly proportional to σ plus another term of long and short range residual stress origin, the latter being reflected by λ . Still, λ is not purely a <u>structure</u> parameter because its definition, as the stress corresponding instantaneously to a null plastic strain rate, involves the kinetic relations as well. For this reason, and also because temperature alterations can cause residual stresses in heterogeneous materials, λ will vary at least slightly with θ at a given H .

Now, λ can be taken as a measure of the anisotropy that has been induced by plastic deformation. It is known within the time-independent framework, however, that when offset strains of the order 1/2 to 2% are taken to define yield, subsequent yield surfaces in stress space are essentially isotropic, with little evidence of the pronounced anisotropy that shows on small offset yield surfaces. Thus there is need for a further <u>scalar</u> structure parameter, called ρ here, which characterizes the intrinsic resistance to flow that exists apart from anisotropic and related strain transient effects (see the chapter by Hart et al.). We may think of ρ as denoting a parameter such as the net dislocation density, or some average measure of the kind b/L where L is a distance between strong dislocation pinning points, or instead just as the flow stress (suitably averaged among directions to free the definition of anisotropy) at some fixed temperature, sufficiently low that creep effects are absent. This single parameter characterization would seem suitable so long as the deformation is not so large as to induce significant preferred orientation and texturing.

The balance of the discussion is done within the conventional small displacement gradient approximation. All that is said can, however, be taken to apply as well for large plastic deformations, within the limitation of texturing, by using the procedure outlined at the end of Section 2.3 and taking F^p there to correspond to a pure deformation. In that case σ is to be interpreted as the stress conjugate to logarithmic strain based on a reference state instantaneously coincident with the elastically unloaded state, or as $R^T TR$ to the order of the approximation, and the rates $d\sigma/dt$ and $d\varepsilon/dt$ are interpreted as discussed there in terms of $\mathscr{D}T/\mathscr{D}t$ and D .

The assumption made is that the instantaneous $d^p\varepsilon/dt$ depends only on (i) the stress difference $\sigma-\lambda$, (ii) temperature θ , and (iii) the scalar structure parameter ρ . The assumption (i) is, of course, far too simply to closely match either observed behavior or the predictions of detailed microscale models for general loading paths, particularly in the neighborhood of yield surface vertices or sharply rounded portions of flow potential surfaces. It does, however, comprise a suitably simple basis for applications to stress analysis.

It is assumed that the microscale mechanisms of deformation are such that the flow potential Ω exists and, from (i) above, it is clear that this can be stress-state dependent only through the three invariants of the stress difference $\sigma - \lambda$. If, however, the plastic response $d^p\varepsilon/dt$ is volume preserving, or conversely if the microscale forces are uninfluenced by hydrostatic stress of the levels considered, then Ω depends only on the second and third invariants of the deviatoric part of $\sigma - \lambda$. This means that any hydrostatic part of λ is without effect and λ can therefore be taken as a deviatoric tensor. It usually constitutes a suitable approximation to assume, as in the Prandtl-Reuss equations, that there is a dependence only on the second deviatoric invariant, which can be expressed as an equivalent shear stress τ_{eq}:

$$\tau_{eq} = \tau_{eq}(\sigma-\lambda) = [\tfrac{1}{2}(\sigma'_{ij}-\lambda_{ij})(\sigma'_{ij}-\lambda_{ij})]^{1/2} ,$$

$$\text{where } \sigma'_{ij} = \sigma_{ij} - \tfrac{1}{3}\delta_{ij}\sigma_{kk} . \tag{2.111}$$

Thus we arrive at the form

$$\Omega = \Omega[\tau_{eq}(\sigma-\lambda), \theta, \rho] , \tag{2.112}$$

and there follows from eqn. (2.86) the plastic strain rate

$$d^p\varepsilon_{ij}/dt = \frac{\sigma'_{ij}-\lambda_{ij}}{2\,\tau_{eq}(\sigma-\lambda)} \frac{\partial\Omega[\tau_{eq},\theta,\rho]}{\partial\tau_{eq}} . \tag{2.113}$$

Note that the explicit dependence of Ω on τ_{eq} at different states θ and ρ can be determined experimentally by examining the variation of instantaneous plastic strain rate with stress at various stages throughout a program of uniaxial tension and compression, or of pure shear deformation, provided results are fitted to the assumed symmetric dependence on $\sigma-\lambda$.

The constitutive description must be completed by a specification of equations governing alterations in the structure parameters λ and ρ. This is perhaps the most arbitrary part, and there is little hope of including all possible effects. An appealing form for ρ is

$$\frac{d\rho}{dt} = h(\rho,\theta)\left[\frac{d^p\varepsilon_{ij}}{dt}\ \frac{d^p\varepsilon_{ij}}{dt}\right]^{1/2}$$

$$+\ \beta(\rho,\theta)\ \lambda_{ij}\ \frac{d^p\varepsilon_{ij}}{dt} - r(\rho,\theta)\quad , \tag{2.114}$$

where h represents an intrinsic rate of "hardening", the same for all directions, and β represents the extent to which the hardening induced by a strain increment is biased by its direction relative to the current rest stress. In fact, the first two terms are independent of the time scale and contribute net hardening rates

$$h(\rho,\theta) + \beta(\rho,\theta)[\lambda_{ij}\ \lambda_{ij}]^{1/2}\quad ,\quad h(\rho,\theta) - \beta(\rho,\theta)[\lambda_{ij}\ \lambda_{ij}]^{1/2}$$

when $d^p\varepsilon$ is respectively co-directional and oppositely directed to λ . The minus sign in the latter form may be thought of as representing the annihilation of dislocations that have not spread widely from their sources, but rather have been blocked by obstacles in the deformation that produced λ . The last term $r(\rho,\theta)$ of eqn. (2.14) is the temperature-dependent rate of hardness recovery. Provided that λ can be ascertained, the functions h and β can be determined in principle by loading a tensile specimen to a given hardness state ρ , and then measuring the increments of ρ due to rapid increments of further loading and reverse compressive loading, both done on a time scale for which the recovery effect is negligible. Also, $r(\rho,\theta)$ is accessible either from recovery studies or from observed values of ρ and the strain rate in states of steady creep at various temperatures, provided that these correspond to ρ = constant.

It is plausible that λ be chosen codirectional with σ' in a program of proportional stressing, commencing from a state at which $\lambda = 0$. Further, λ should approach some saturation magnitude, with ongoing deformation, that increases with increasing σ', and is here supposed to take the form

$$\lambda_{ij}^{sat} = q(\rho,\theta)\ \sigma'_{ij}\quad , \tag{2.115}$$

where $0 \le q < 1$, although the saturation magnitude itself may never be attained if σ' increases indefinitely. A simple manner of describing the change in λ is by writing

$$\frac{d\lambda_{ij}}{dt} = p(\rho,\theta)\left[\lambda^{sat}_{ij} - \lambda_{ij}\right]\left[\frac{d^p\varepsilon_{ij}}{dt}\,\frac{d^p\varepsilon_{ij}}{dt}\right]^{1/2} \tag{2.116}$$

so that changes in λ always follow the direction from λ to the instantaneous saturation value associated with the current state. It is seen that p is a relaxation parameter when phrased in terms of plastic strain arc length

$$\ell = \int_0^t \left[\frac{d^p\varepsilon_{ij}}{dt}\,\frac{d^p\varepsilon_{ij}}{dt}\right]^{1/2} dt\ . \tag{2.117}$$

For example, if p and q are taken as constant during some program of creep deformation at constant stress σ, commencing from a state at $t = 0$ for which $\lambda = \lambda^o$, then

$$\lambda_{ij} = e^{-p\ell}\,\lambda^o_{ij} + (1 - e^{-p\ell})q\,\sigma'_{ij}\ , \tag{2.118}$$

and this shows the decay of prior influences as the new saturation state, appropriate to σ, is approached. Evidently, p would have to be of a magnitude so that prior memory is lost in strains ℓ of order 1/2 to 2% . As it stands, eqn. (2.116) has no time scale, but λ like ρ should be subject to recovery at high temperature and one way of incorporating this is by adding a term proportional to $-\lambda$ in eqn. (2.116).

For an application of the formulation, consider isothermal deformation and suppose that $d^p\varepsilon/dt$ in eqn. (2.112) varies from negligible to very large values as τ_{eq} is increased through a certain critical magnitude, dependent on the current plastic state. Thus the time-independent idealization is adopted and we interpret the structure parameter ρ, for convenience, as the critical stress magnitude for flow. In the notation of eqn. (2.35), the yield condition is then $F \equiv \tau_{eq}(\sigma-\lambda) - \rho = 0$, and eqn. (2.112) is now replaced by

$$d^p\varepsilon_{ij} = \frac{\sigma'_{ij}-\lambda_{ij}}{2\,\tau_{eq}(\sigma-\lambda)}\,d\Lambda = \frac{1}{\sqrt{2}}\,N_{ij}\,d\Lambda\ , \tag{2.119}$$

where N is the indicated unit tensor ($N_{ij}N_{ij} = 1$) giving the direction of stressing relative to λ. We solve for $d\Lambda$ in the standard way (e.g. Naghdi, 1960), writing $dF = 0$ during a plastic deformation process and expressing $d\rho$ and $d\lambda$ as in eqns. (2.114,116), ignoring the recovery term in the former. This results in

$$d^p\varepsilon_{ij} = \frac{1}{E_t} N_{ij} N_{k\ell} d\sigma_{k\ell} , \quad \text{for } N_{k\ell}d\sigma_{k\ell} \geq 0 , \tag{2.120}$$

where the total or overall hardening rate E_t is given by

$$E_t = \sqrt{2} (h + p q \rho) - [- \sqrt{2} \beta + p(1-q)]N_{ij} \lambda_{ij} , \tag{2.121}$$

and where explicit relations for the change in ρ and λ during a deformation process with $Nd\sigma \geq 0$ are

$$d\rho = \frac{1}{E_t} [h + \beta N_{ij} \lambda_{ij}]N_{k\ell} d\sigma_{k\ell} \tag{2.122}$$

$$d\lambda_{ij} = \frac{p}{E_t} [\sqrt{2} q \rho N_{ij} - (1-q)\lambda_{ij}]N_{k\ell} d\sigma_{k\ell} \tag{2.123}$$

Examining eqn. (2.121) for E_t , we see that a result of the formulation is anisotropy of the overall hardening, depending on the direction N of the stress difference $\sigma-\lambda$ at flow relative to the direction of λ . Indeed, unless the bias parameter β in eqns. (2.114,122) is large by comparison to unity, the dominant term in the latter part of the expression for E_t is that containing p , since from the discussion following eqn. (2.118), this would have to be very much larger than unity. Thus, when the rest stress λ is near its saturation level [i.e. the bracketed term in eqn. (2.123) vanishes], the anisotropy is most pronounced, with the greatest differences being between the E_t for continued stressing in the direction of λ and that for reversed stressing. The former is of order h , whereas the latter is of order $h + 2 p q \rho$.

It is also of interest to note that the anisotropic effects represented by λ are indeed transient, and that large amounts of deformation under a fixed stressing direction result ultimately in strain increments that become normal to an isotropic hardening "yield" surface. This surface is of the form $\tau_{eq}(\sigma)$ = constant, and has an apparent hardening rate of order h in all directions, when saturation conditions have been achieved. However, it is not an actual yield surface, but rather an envelope of individual yield surfaces of the kind $\tau_{eq}(\sigma-\lambda)$ = constant, each being generated by a different loading history and each exhibiting a pronounced anisotropy of hardening.

There are many issues to be further explored here, concerning both the general formulation and the assignment of specific forms to the functions involved. An interesting question of the former kind is the following: Given an initially isotropic material in which the plastic state is assumed to be fully characterized by a scalar ρ and second order tensor λ, what is the most general possible class of flow and structure parameter equations? The present efforts have generated a member of the class, but others are possible that could, for example, model substantial shape distortions in flow potential and yield surfaces.

2.10 SUMMARY

This chapter has presented the basis in continuum mechanics and thermodynamics for constitutive descriptions of plasticity, with special attention being given to the connection between macroscopic formulations and deformation mechanisms as operative on the microscale. Conjugate stress and strain measures have been introduced in Sections 2.2 and 2.3, and procedures outlined there for materially objective formulations of elastic-plastic constitutive equations at large or small deformations. Further, the thermodynamic framework for inelastic constitutive laws, as set by the existence of potentials governing purely elastic response, has been reviewed in Section 2.4 and second law restrictions have been stated.

Section 2.5 has presented the general procedure by which structural rearrangements, on the microscale, of the elements of a representative material sample can be related to its macroscopic plastic deformation. This involves the thermodynamic forces f conjugate to the extents $d\xi$ of the rearrangements, and each f is shown to be a "plastic potential" for the macroscopic strain induced by its associated $d\xi$ [eqn. (2.43)]. The procedure is applied in Section 2.6 to inelastic deformation arising from crystalline slip, diffusion, phase changes, and micro-cracking, and the appropriate $d\xi$'s and f's are identified in each case.

In Section 2.7 it is remarked that kinetic relations for crystalline slip in accord with a Schmid resolved shear stress dependence, and also for linear diffusion with Onsager coefficient symmetry, fall into the general class for which a given rate $d\xi/dt$ is stress state dependent only through its conjugate f. All microscale kinetic laws of this class are shown to lead to a unifying normality structure in macroscopic constitutive relations, for which components of the instantaneous plastic strain rate are given by derivatives of a scalar flow potential on corresponding stress components. Associated results in terms of normality to yield surfaces are demonstrated within the time-independent idealization of crystalline slip.

The problem of averaging microscale kinetic relations over all sites within a representative material sample, to arrive at specific macroscopic constitutive descriptions, is illustrated in Section 2.8 by review of procedures for predicting the behavior of polycrystals deforming by slip. It is suggested that polycrystal models of the type considered seem capable of modelling some of the real complexities of path dependence in plastic response and of shape distortions of flow potential or yield surfaces, but the necessary computations are very extensive, even for simple deformation paths. Thus considerable interest remains in phenomenological and less rigorously based structure parameter formulations. A class of such constitutive relations is presented in Section 2.9, where the effect of prior deformation on an initially isotropic material is taken to be represented by a single scalar hardness parameter and by a rest stress tensor, the latter coinciding with the stress state at which the instantaneous plastic strain rate vanishes.

ACKNOWLEDGEMENTS

Studies in preparation of this chapter have been supported by the U. S. Atomic Energy Commission. Section 2.8 and portions of Sections 2.5 to 2.7 have been adopted with minor modifications from a paper of similar title by the writer (Rice, 1973).

2.11 REFERENCES

Ahlquist, C. N. and Nix, W. D. 1969 Scripta Met. 3, 679.

Batdorf, S. B. and Budiansky, B. 1949 NACA TN 1871.

Biot, M. A. 1954 J. Appl. Phys. 25, 1385

Bishop, J. F. W. and Hill, R. 1951 Phil. Mag. 42, 414 and 1298.

Bowen, R. and Wang, C. -C. 1966 Archiv. Rational Mech. Anal. 22, 79.

Brown, G. M. 1970a J. Mech. Phys. Solids 18, 367; 1970b J. Mech. Phys. Solids 18, 383.

Budiansky, B. 1959 J. Appl. Mech. 26, 259

Budiansky, B. and Wu, T. T. 1962 in Proc. 4th U. S. Nat. Cong. Appl. Mech., ASME, N. Y., p. 1175.

Bui, H. D. 1970 Sc. D. Thesis, Paris.

Cherepanov, G. P. 1967 Prikl. Mat. Mekh. 31, 476

Clough, R. B. and Simmons, J. A. 1973 in Rate Processes in Plastic Deformation (Li, J. C. M. and Mukherjee, A. K., eds.), Amer. Soc. Metals, Cleveland.

Coleman, B. D. 1964 Archiv. Rational Mech. Anal. 17, 1 and 230.

Coleman, B. D. and Gurtin, M. E. 1967 J. Chem. Phys. 47, 597.

Coleman, B. D. and Noll, W. 1963 Archiv. Rational Mech. Anal. 13, 167.

Drucker, D. C. 1951 in Proc. 1st U. S. Nat'l. Congr. Appl. Mech., ASME, N. Y. p. 487; 1956 in Rheology, Vol. I (Eirich, H., ed.), Academic Press, N. Y., chp. 4, p. 97; 1960 in Structural Mechanics (Goodier, J. N. and Hoff, N. J., eds.), Pergamon, N. Y., p. 407.

Eckart, C. 1948 Phys. Rev. 73, 373.

Eshelby, J. D. 1956 in Solid State Physics, Vol. III (Seitz, F. and Turnbull, D., eds.), Academic Press, N. Y.; 1970 in Inelastic Behavior of Solids (Kanninen, M. F. et al., eds.), McGraw-Hill, N. Y., p. 77.

Green, A. E. and Naghdi, P. M. 1965 Archiv. Rational Mech. Anal. 18, 251.

Havner, K. S. 1969 Int. J. Solids Structures 5, 215; 1973 J. Mech. Phys. Solids 21, 383.

Hill, R. 1965 J. Mech. Phys. Solids 13, 89; 1967 J. Mech. Phys. Solids 15, 79; 1968 J. Mech. Phys. Solids 16, 229 and 315; 1971 Prikl. Mat. Mekh. 35, 31; 1972 Proc. Roy. Soc. Lond. A326, 131.

Hill, R. and Rice, J. R. 1972 J. Mech. Phys. Solids 20, 401; 1973 SIAM J. Appl. Math. 25, 448.

Hutchinson, J. W. 1964 J. Mech. Phys. Solids 12, 11 and 25; 1970 Proc. Roy. Soc. Lond. A319, 247.

Il'yushin, A. A. 1961 Prikl. Mat. Mekh. 25, 503.

Irwin, G. R. 1957 J. Appl. Mech. 24, 361; 1960 in Structural Mechanics (Goodier, J. N. and Hoff, N. J., eds.), Pergamon, N. Y., p. 557.

Kestin, J. and Rice, J. R. 1970 in A Critical Review of Thermo-dynamics (Stuart, E. B. et al., eds.), Mono Book Corp., Baltimore, p. 275.

Kocks, U. F. 1970 Met. Trans. 1. 1121.

Kröner, E. 1961 Acta Met. 9, 155.

Lin, T. H. 1957 J. Mech. Phys. Solids 5, 143; 1968 Theory of Inelastic Structures, Wiley, N. Y., chp. 4.

Lin, T. H. and Ito, M. 1965 J. Mech. Phys. Solids 13, 103; 1966 Int. J. Engng. Sci. 4, 543.

Mandel, J. 1966 in Proc. 11th Cong. Appl. Mech. (Munich, 1964) (Görtler, H., ed.), Springer-Verlag, Berlin, p. 502.

Meixner, J. 1953 Kolloid-Z. 134, 2.

Naghdi, P. M. 1960 in Plasticity (Lee, E. H. and Symonds, P. S., eds.), Pergamon, N. Y., p. 121.

Rice, J. R. 1966 Tech. Rept. ARPA SD-86 E-31, Brown Univ., Providence; 1968 in Fracture, Vol. 2 (Liebowitz, H., ed.), Academic Press, N. Y., p. 191; 1970 J. Appl. Mech. 37, 728; 1971 J. Mech. Phys. Solids 19, 433; 1973 in Metallurgical Effects at High Strain Rates (Rohde, R. W. et al., eds.), Plenum, N. Y., p. 93.

Rice, J. R. and Drucker, D. C. 1966 Int. J. Fracture Mech. 3, 19.

Rivlin, R. S. 1970 in Inelastic Behavior of Solids (Kanninen, M. F. et al., eds.), McGraw-Hill, N. Y., p. 117.

Taylor, G. I. 1938 J. Inst. Metals 62, 307.

Truesdell, C. and Toupin, R. A. 1960 in Encyclopedia of Physics, Vol. III (Flügge, S., ed.), Springer-Verlag, Berlin.

Walsh, J. B. 1965 J. Geophys. Res. 70, 381

Willis, J. R. 1969 J. Mech. Phys. Solids 17, 359.

Zarka, J. 1972 J. Mech. Phys. Solids 20, 179.

LIST OF SYMBOLS

A	affinity conjugate to dislocation density
β	bias hardening parameter
β	local shape-change strain tensor per unit of diffusive matter addition
b	Burgers vector
det(...)	determinant of the matrix (...)
dN	increment in the amount of diffusing substance per unit of material interface
dq	incremental diffusive flux vector in volume of material sample
dQ	incremental diffusive flux vector along internal interface of material sample
dz	incremental advance of a phase interface normal to itself
$d\ell$	incremental advance in length of a Griffith crack
$d\Lambda$	scalar multiplier in plastic flow law
dn	incremental advance of dislocation line normal to itself; also used for increment in the amount of diffusing substance per unit volume
$d\gamma$	increment of shear strain on crystal slip system
$d\xi_{\alpha}$	incremental variable characterizing local structural rearrangement within material sample
ε	rotation invariant material strain tensor
e	local strain tensor within polycrystalline aggregate
E_t	total hardening rate
F	thermodynamic force on a Griffith crack

$F = \partial x/\partial X$	deformation gradient tensor
$F(\sigma, \theta, H)$	yield function
f_α	thermodynamic force conjugate to incremental structural rearrangement $d\xi_\alpha$
G	elastic energy release rate for Griffith crack
h(...)	function used to define strain measure
H	symbolic representation for current plastic state of a material (deformations at fixed H are necessarily elastic)
h	rate of isotropic strain hardening
k	local thickening coefficient for grain boundary with matter addition; also stress intensity factor for Griffith crack
ℓ	arc length measure in plastic strain space
M	tensor of incremental elastic compliances
n,m	unit vectors respectively normal to and in slip direction for crystal slip system
ρ	dislocation density; also general symbol for scalar structure parameter
p	thermodynamic force on a phase interface; also relaxation parameter for rest-stress
p as a superscript	denotes plastic part of an increment of some state variable, arising from alterations of H
Q	heat supply rate per unit reference volume

q	force per unit length of dislocation line; also parameter relating saturation value of rest-stress to current stress state.
R	rotation part of F, F=RU
r	rate of hardness recovery
s	local non-uniform stress tensor at points within material sample
s_s	resolved shear stress component of s, for crystal slip system
s_n	normal component of s on interface
$\bar{s}$	the part of local stress tensor s that is induced by elastic application of macroscopic stress tensor σ
T	Cauchy (or true) stress tensor
T as a superscript	denotes matrix transpose
t	time
$\bar{\tau}$	the part of shear stress on a local crystal slip system induced by elastic application of macroscopic stress tensor σ
τ	thermodynamic shear stress conjugate to increment $d\gamma$ of shear strain
$\tau_{eq}(\sigma)$	equivalent shear stress parameter, based on second deviatoric invariant of σ
U	pure deformation part of F, F=RU
u	internal energy per unit reference volume
V	reference volume of material sample

x_1, x_2, x_3	Cartesian coordinates of material points in deformed state
X_1, X_2, X_3,	Cartesian coordinates of material points in reference state
θ	thermodynamic temperature
σ	stress tensor conjugate to the strain tensor ε
-1 as a superscript	denotes matrix inverse
μ	chemical potential of diffusing substance; also slip plane orientation tensor for crystalline slip.
γ	surface free energy
Λ	dual potential to the flow potential
C	elastic compliance coefficients
ν	Poisson ratio for isotropic elastic materials
λ	rest stress tensor
σ'	deviator part of the tensor σ
η	entropy per unit reference volume
ϕ	Helmholtz free energy, $\phi = u-\theta\eta$
ψ	dual potential to ϕ, $\psi = \sigma_{ij}\varepsilon_{ij} - \phi$
α	tensor of incremental thermal expansion
Ω	flow potential for macroscopic plastic deformation rate
Φ	total Helmholtz free energy of material sample
Ψ	dual potential to Φ
$r_\alpha = d\xi_\alpha/dt$	rate of structural rearrangement

ABSTRACT. *The general restrictions imposed on constitutive relations for macroscopic plasticity by the micromechanism of slip are outlined. Thermodynamic considerations show that it is important to distinguish between energy-storing and dissipative slip resistance mechanisms, and that only the former must obey the "flow rule". Mechanistic considerations further suggest a delineation of different ranges of the external variables in which different behavior should be expected: at strains below or above a critical limit of the order of magnitude of the elastic strains; and at stresses below or above the mechanical threshold (i.e. the flow stress at zero temperature). These critical values are parameters of state that should appear in any constitutive relation; others are discussed. On the other hand, history parameters are studiously avoided. The meaning and importance of 'equations of state' for plasticity, and the tests of their validity, are reviewed; especially, the limitations of the stress relaxation test are discussed.*

3. CONSTITUTIVE RELATIONS FOR SLIP*

U.F. Kocks

3.1 INTRODUCTION

The basic mechanisms that control the macroscopic plastic behavior of materials occur on an atomic scale. For any known or postulated local mechanism, a fair amount of averaging must, therefore, be undertaken before the macroscopic properties can be understood. This averaging procedure can conveniently be divided into a series of steps, progressing from one scale of the 'microstructure' to the next.

Most materials of interest in plasticity are crystalline -- to be more precise, polycrystalline: they consist of "grains" in which the atoms are regularly arranged in orientations that differ from one grain to the next. Each 'volume element' of the macroscopic specimen must contain many such grains to be treated as a continuum. When the plastic properties of the individual grains are known, the macroscopic constitutive relations follow from an averaging procedure that involves pri-

*Work performed under the auspices of the U.S. Atomic Energy Commission.

marily geometry and kinematics and few physical assumptions. Some important physical assumptions are: are the grains allowed to slide over each other; and, does deformation tend to proceed from grain to grain by a nucleation-and-growth process, or does it proceed more gradually with some degree of spatial uniformity? In many practical cases, quasi-homogeneous deformation with nonsliding grain boundaries is a realistic assumption (pure or dispersion strengthened materials of the FCC or BCC lattice structures at temperatures up to at least half the melting point are good examples). For these cases, Rice has outlined, in the preceding chapter, the basis for solving some of these averaging problems. A more extensive, though less rigorous, discussion can be found in Kocks (1970a).

The constitutive behavior of individual grains follows from the physical processes that occur during plastic deformation. These may generally be of three kinds: twinning, slip, and diffusional flow. For large classes of materials, such as those of FCC or BCC lattice structures mentioned above, twinning is a less important mode of deformation than slip (although its occurrence at low temperatures may have a bearing on fracture problems). Diffusional flow is important in most materials at sufficiently high temperature; Ashby discusses some aspects of this in the following chapter. Over a temperature range from near absolute to well above half the melting point, *slip* dominates plastic deformation. In the higher-temperature ranges, it may be influenced by diffusional processes, such as dislocation climb; but the general form of the kinematic and kinetic relations is determined by dislocation glide. These constitutive relations for slip are our concern in this chapter.

The process of slip is inherently non-uniform on various scales: the atomic planes on which slip actually occurs are many atomic distances apart; within these planes, the moving dislocation lines, which cause the slip, are themselves well separated from each other; finally, they generally move one segment at a time along the length of the line. In the next three paragraphs, we summarize procedures that may be employed to determine macroscopic slip behavior, and the limitations of the averages derived, from the local interaction of the dislocation segments with physical obstacles. A more extensive and more rigorous treatment of these problems can be found in Kocks et al. (1975).

A knowledge of the physical obstacles, and of the way in which they influence macroscopic properties, forms the basis of any attempt to influence mechanical properties through materials selection and metallurgical processing. In this chapter, however, we assume the material and metallurgical history to be given, and the obstacles to have only some very general properties. We wish to derive from such general properties and the general nature of slip, what the general form of constitutive relations for slip may be: what the appropriate variables are, what kind of equations they occur in, and how many parameters appear in those equations.

3.2 LIMITING YIELD CONDITIONS, THE PLASTIC POTENTIAL

On the coarsest scale, slip consists in the rigid translation of parts of a crystal over each other on discrete planes. Both the orientation of the plane and the direction of slip are crystallographically prescribed; we shall call them the 1- and 2-direction, respectively. If one plane of area A is slipped by a displacement δa the work done by the applied stresses σ_{ij} is

$$\delta W = \sigma_{12} A \, \delta u \tag{3.1}$$

To find out what magnitude of applied stress was necessary to achieve this deformation, we consider the change in (Helmholtz) free energy δF associated with this deformation at the given temperature T and demand that, according to the Second Law,

$$\delta W \geq \delta F \qquad (\delta T = 0, \quad \delta\sigma_{ij} = 0) \tag{3.2}$$

The free energy of the specimen may have changed as a consequence of the prescribed slip, because various defects may have been produced: slip on this particular plane may have sheared second-phase particles and produced new particle-matrix interface on them; it may have sheared intersecting dislocations and produced jogs on them; or it may, by any number of mechanisms, have led to an increase in the number of vacancies in the crystal*. In all these cases, the free energy changes of the specimen may be said to be localized at various points in the slip plane, so that we can write

$$\delta F = \int_A \tau_{ELEM} \, dx \, dy \cdot \delta u \tag{3.3}$$

*In the last of these examples, an extra term $-Pv\delta n$ would have to be added to the right side of eqn. (3.1), where P is the applied pressure, v the macroscopic volume change per vacancy, and δn the number of vacancies generated.

The integrand τ_{ELEM} is, for now, not completely defined: eqn. (3.3) would describe the total free energy change even if τ_{ELEM} had additional contributions that average out to zero over the slip plane area A, i.e. if free energy increases in some elements were compensated by decreases in others.* The average of τ_{ELEM} over a large area A,

$$\tau_{STOR} \equiv \frac{1}{A} \int_A \tau_{ELEM} \, dx \, dy = \frac{\delta F}{A \, \delta u} \tag{3.4}$$

represents the net energy storage per unit area and unit displacement. Comparing eqns. (3.4) and (3.1), the Second Law (eqn. 3.2) demands that

$$\sigma_{12} \geq \tau_{STOR} \qquad \text{(thermodynamic threshold)} \tag{3.5}$$

Equation (3.5) is in the nature of a yield criterion: it specifies which component of the applied stress is relevant** and what its critical value is. If deformation were in fact to occur at this lower limit, it would have to be reversible; for then the yield criterion given by eqn. (3.5) could be written (generalized to arbitrary stresses and strain increments):

$$f(\sigma_{ij}) = \frac{1}{A \, \delta u} (\delta F - V \, \sigma_{ij} \, \delta \varepsilon_{ij}) = 0 \tag{3.6}$$

The quantity in brackets is the change in free enthalpy, δG, which is zero at given stress and temperature only for revers-

---

*Also, τ_{ELEM} in eqn. (3.3) may depend on the magnitude of δu. We shall define τ_{ELEM} more precisely in eqn. (3.10).

**In terms of a general applied stress, eqn. (3.5) would read $m^{(s)}_{ij}\sigma_{ij} > \tau^{(s)}_{STOR}$ where $m_{ij}^{(s)}$ are the generalized Schmid factors for each system s; see Kocks, (1970a).

ible changes.* In this (very unrealistic) case, the yield function $f(\sigma_{ij})$ defined in eqn. (3.6) is also the plastic potential, for then the strain increments could be derived from its gradient:

$$\delta\varepsilon_{ij} = -\frac{1}{V}\frac{\partial\ \delta G}{\partial\sigma_{ij}} = \frac{\partial f}{\partial\sigma_{ij}}\frac{A\delta u}{V} \tag{3.7}$$

This is the case treated in detail by Rice in Chapter 2 of this book.

Yield stresses in real materials are usually far in excess of any reasonable estimate of τ_{STOR}; equivalently, the energy stored during plastic deformation is usually but a small fraction of the plastic work. There is thus cause for concern that a treatment which uses this "thermodynamic threshold" as a reference state is inadequate to the problem at hand. This may be illustrated by the following example, which represents the other extreme.

Remaining with the model of homogeneous slip, i.e. the rigid translation of one half of the crystal over the other, let us now consider the effect of the discreteness of the lattice (and ignore the imperfections whose shearing caused energy storage above). Using a simple quasi-chemical model, all atomic bonds crossing the slip plane would now have to be stretched and then broken simultaneously, only to be reformed when the displacement δu has become equal to one lattice spacing b. Thus, there would be a uniform τ_{ELEM} over the entire plane, which undulates as a function of δu with a wavelength b and an amplitude that is called the 'theoretical' or 'ideal' shear strength τ_{IDEAL}. No energy is stored in this process when the displacement is a multiple of b; thus, the thermodynamic threshold τ_{STOR} equals zero. At any finite stress σ_{12}, even if it is much smaller than τ_{IDEAL}, there is a finite "thermodynamic force conjugate to slip", as used by Rice, and if one is prepared to wait forever, the crystal

---

*Note that this is true even if δF and τ_{STOR} depend on the state of stress, as the free energy of vacancies may depend on the pressure for example; for their derivatives with respect to stress would, during reversible changes, be exactly compensated by corresponding derivatives of the work terms, such as the pressure dependence of the vacancy volume.

must indeed slip. However, to achieve a strain rate in the range of any practical interest, one must apply a stress essentially equal to τ_{IDEAL} - although at this level itself, the strain rate becomes indeterminate. We define this level as the "mechanical threshold for homogeneous slip" and include, for generality, the additional stress necessary for any energy-storing mechanisms:

$$\sigma_{12} = \tau_{IDEAL} + \tau_{STOR} \quad \text{(mechanical threshold for homogeneous slip)} \qquad (3.8)$$

If it made any sense to talk about a 'plastic potential' in this limit, it would clearly <u>not</u> coincide with the yield surface; for the theoretical shear strength, being the extreme limit of nonlinear elastic behavior, is strongly influenced by other stress components, such as an applied pressure; but other strain components, such as a volume change, would remain in the elastic range while the shear in the prescribed direction increases without bound.

Whereas the "zero-rate" limit τ_{STOR} was smaller than any reasonable flow stress of interest, the "infinite-rate" limit τ_{IDEAL} is larger than any stress reached in plastic flow--of order $\mu/20$, where μ is a shear modulus. Real crystalline materials do not deform by homogeneous slip but by gradually spreading slip (of unit displacement b) over the slip plane. In this way, disruption of atomic bonding is restricted to the boundaries of the slipped areas, the dislocations.

3.3 DISLOCATION GLIDE

In homogeneous slip, the resistance to localized obstacles was averaged over the whole slip plane (eqn. 3.4) and thereby made quite low; on the other hand, the resistance to shear of the ideal lattice was high because it acted everywhere at the same time. In dislocation glide, the situation is reversed: the lattice resistance to dislocation motion is considerably smaller than to homogeneous shear (chiefly because it is now localized and subject to thermal activation); but the dislocations are exposed to the <u>local</u> values of τ_{ELEM}, which may be much higher than their average, τ_{STOR}. Note that dislocations traversing the entire slip plane (or sufficiently large areas) leave behind at least the same structure changes and energy stored as homogeneous slip; thus, the thermodynamic limit, eqn. (3.5), must of course still hold. Yet, in addition, dislocations experience fluctuations in the resistance to their motion.

As dislocations move, they sweep out small increments of area δa on the slip plane, producing an average slip displacement

$$\delta u = b \frac{\delta a}{A} \tag{3.9}$$

The local change in free energy of the crystal, which was previously introduced as an integrand in eqn. (3.3), may now be more precisely defined as

$$\tau_{ELEM} = \lim_{\delta a \to 0} \frac{\delta F}{b\ \delta a} \tag{3.10}$$

We call it the 'element glide resistance' (Kocks et al. 1975). Equivalently, its average may now be defined as

$$\tau_{STOR} = \lim_{\delta a \to \infty} \frac{\delta F}{b\ \delta a} \tag{3.11}$$

In fact, only small elements of dislocation feel the element glide resistance as it varies from place to place: the dislocation, by adapting its shape to the variations in glide resistance, effectively averages the element resistance along its line. This gives rise to a glide resistance that is larger than the area average of the element resistance (eqn. 3.11), but smaller than the local maximum, $\hat{\tau}_{ELEM}$. For example, a dislocation in front of a linear periodic array of identical obstacles of spacing L which has bowed out to its equilibrium between the obstacles, feels a line glide resistance:

$$\tau_{LINE} = \frac{1}{L} \int_0^L \tau_{ELEM}\, dx \tag{3.12}$$

If the element resistance is concentrated in only a few areas, so that the obstacles are 'discrete', one may integrate it up to give the local obstacle strength

$$K \equiv \int_0^w \tau_{ELEM}\, b\ dx \tag{3.13}$$

where w is the width of the obstacle. Thus the line glide resistance is

$$\tau_{LINE} = \frac{K}{b\,L} \tag{3.14}$$

It is determined by the strength K and the spacing L of the obstacles in the slip plane*.

The line glide resistance still varies from place to place: both the strength and the spacing of the obstacles will generally be nonuniform. Figure 3.1 shows in a schematic way how

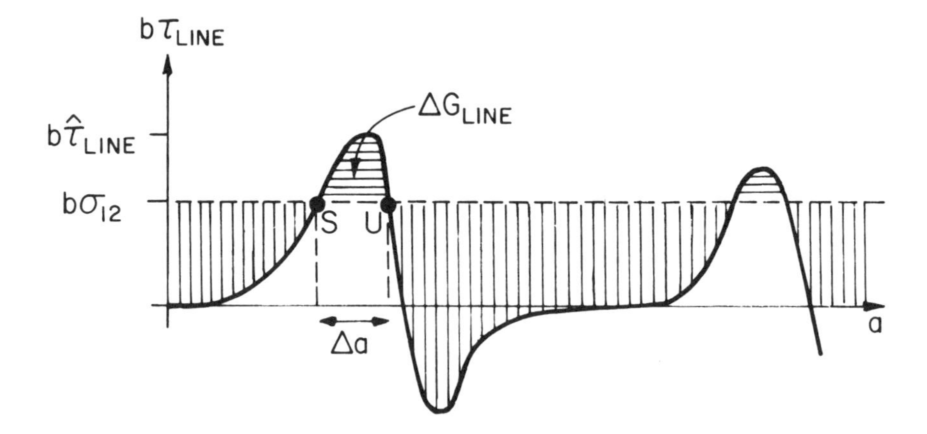

Fig. 3.1 Schematic diagram of the line glide resistance τ_{LINE} to a dislocation segment versus the area a swept out by it. Under the (resolved) applied stress σ_{12}, the driving force is positive in some areas (vertical shading), negative in others (horizontal shading). The areas shaded in the figure are the energy dissipated and the activation energy, respectively.

---

*It is an important consequence of the dislocation mechanism of plastic deformation that the glide resistance is proportional to the inverse obstacle <u>spacing</u> rather than the inverse of the <u>area</u> per obstacle, as τ_{STOR} is: thus, the latter is proportional to the concentration of obstacles, whereas the former is proportional to the square root of the concentration. This is what makes even dilute alloys strong.

τ_{LINE} may vary for a particular dislocation as a function of the area swept out by it. At any one place, the dislocation may be in static equilibrium if the applied stress σ_{12} is such that the work it would do in a small forward displacement δa (eqns. 3.1, 3.9)

$$\delta W = \sigma_{12} \; b \; \delta a \tag{3.15}$$

would be just enough to supply the free energy change

$$\delta F = \tau_{LINE} \; b \; \delta a \tag{3.16}$$

If the applied stress is constant (dotted line in fig. 3.1), there are at best a few such equilibrium positions (such as S and U in the figure). Everywhere else there is a finite driving force per unit length of dislocation

$$b \; (\sigma_{12} - \tau_{LINE}) \tag{3.17}$$

Where it is positive, the dislocation must move, under the given applied stress; the area shaded by vertical bars is the energy dissipated in this process. In the regions shaded by horizontal bars, the driving force is negative. The dislocation is arrested in front of these regions, as in the position S . If, however, an energy ΔG equal to the area shaded by horizontal bars in fig. 3.1 were made available to the dislocation through thermal activation, it could proceed to the unstable equilibrium position U and from there on again move under positive driving force. Thus, some dislocation motion is possible at all stresses, given sufficient time (and, again, provided that the applied stress is sufficient to supply any energy stored over this increment swept; i.e., that the total energy dissipated exceeds the total energy used for thermal activation).

There is an important threshold, though, when the applied stress equals the maximum line glide resistance:

$$\sigma_{12} = \hat{\tau}_{LINE} \quad \text{(mechanical threshold for dislocation segment)} \tag{3.18}$$

Above this stress, the dislocation is under positive driving stress everywhere and it must accelerate until its motion is checked by viscous forces; below it, the dislocation can proceed in the jerky manner described above with the aid of thermal activation only.

Figure 3.2 shows a schematic diagram of the kinds of kinetic

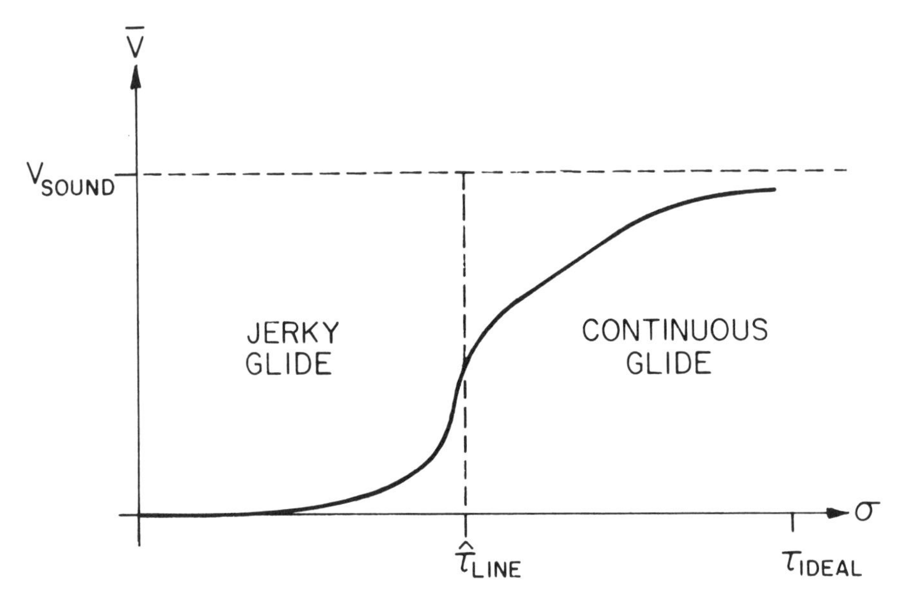

Fig. 3.2 Schematic diagram of average dislocation velocity $\bar{v}$ versus applied stress σ_{12}. Below the threshold $\hat{\tau}_{LINE}$, glide is jerky and controlled by thermal activation; above it, glide is continuous and controlled by visco-dynamic resistances.

relations to be expected in these two ranges. (Also shown is a third range in which the dislocation velocity would be limited by radiation losses, near the velocity of sound.) Above the mechanical threshold the velocity stress relation is basically linear, below it, essentially exponential. In the latter régime, the velocity is meant to be the average over many "jerks", the waiting time at each one being given by

$$\frac{1}{t_w} = \nu_o \cdot \exp - \frac{\Delta G_{LINE}(\sigma_{12}, \hat{\tau}_{LINE})}{kT} \qquad (3.19)$$

where ν_o is an 'attempt frequency' of order 10^{11} sec^{-1}, and the activation energy is, according to the description given in fig. 3.1, given by

$$\Delta G_{LINE} = \int_{\sigma_{12}}^{\hat{\tau}_{LINE}} b \, \Delta a \, d\tau_{LINE} \tag{3.20}$$

It typically depends on stress in one of the two ways shown in fig. 3.3. It goes to a well-defined limit, namely zero, for

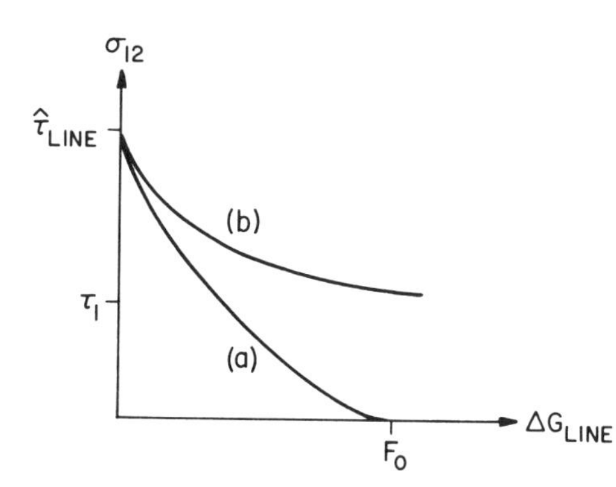

Fig. 3.3 Typical dependences of the local activation energy ΔG_{LINE} on the applied stress σ_{12} : (a) for 'short-range' obstacles; (b) for longer-range obstacles or two sets of obstacles, one long-range and one short-range.

large stresses. For small stresses, it may go to a well-defined total activation energy F_o (curve a); for other cases, it may increase without bound; and for others yet, it may reach a limiting value at a finite stress τ_1 . The latter two cases look similar in practice (curve b).

As the line glide resistance varies from place to place on each slip plane, so does ΔG_{LINE}. Even if one defined a 'typical' or 'average' obstacle pair, however, its values of $\hat{\tau}_{LINE}$ and ΔG_{LINE} at any given stress σ_{12} do not characterize slip over an entire slip plane, because of the interactions of distant parts of each dislocation with each other and of many dislocations among each other.

3.4 SHORT-RANGE SLIP AND LONG-RANGE SLIP

Figure 3.4 illustrates a typical distribution of driving forces over a (two-dimensional) slip plane. The areas in which the driving force is positive are left white, those in which it is negative are shaded. Part (a) of the figure shows the case when the stress is so high that the region in which the driv-force is positive is a continuous area throughout the slip

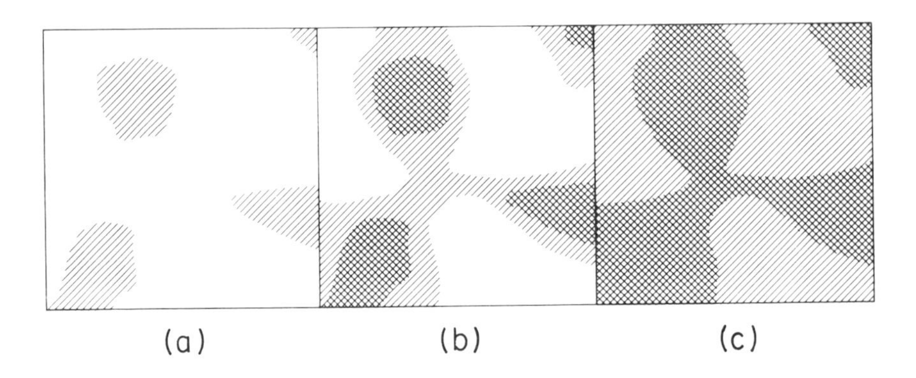

Fig. 3.4 Soft spots (white) and hard spots (shaded) in a slip plane. In (a), the soft spots are contiguous, in (b), corresponding to a lower applied stress, the hard spots are. At an even lower stress (c), thermal activation is required everywhere -- in the softer regions less, in the harder regions more.

plane. A dislocation, once emitted from any source, can thus proceed indefinitely -- although it may leave rings around the 'hard spots' which may or may not subsequently shrink. In part (b), the stress is lowered just enough to reverse the topological situation: now, the region in which the driving force is negative forms a continuous area in the slip plane, and the 'soft spots' are isolated. Dislocations that are emitted from any source in one of the soft areas may sweep out that area, but no more, under the action of the applied stress alone. If thermal activation is possible across one of the 'critical gates' that connect this soft spot with the next, indefinite slip is again possible, at a rate given by the properties of the obstacles near the critical gates. Finally, in part (c) of fig. 3.4, the stress is so low that thermal fluctuations are required everywhere to make dislocation motion possible. However, the rate at which thermally activated glide occurs is larger in the soft spots and again smaller near the critical gates. Dislocations may now be described as having an average, relatively high, velocity within the soft regions and then spending a relatively long waiting time in front of the critical gates.

It is evident that long-range slip is controlled by the properties at the critical gates, i.e. in the hardest spots the dislocation must traverse in order to proceed through the entire plane. When dislocations do not proceed through the entire slip plane, the plastic regions are isolated, and the total strain is limited to the order of magnitude of the elastic strain.

Needless to say, the individuality of dislocation is not relevant here: if one dislocation proceeds through one soft spot right to its limits and another leaves from just about the same place in the next region, this is entirely equivalent, within a macroscopic framework, to one dislocation going the whole way. Similarly, if one dislocation in a particular soft region is held up at its boundaries and further dislocations are generated within the same region until the increased stress on the first dislocation is enough to propagate it out of the region, this is entirely equivalent to the last dislocation having gone through the whole region and out of it. While the critical applied stress is changed, the total strain due to dislocation motion is still of the order of the elastic strains.

Some of these small-strain effects may be anelastic in nature, such as the rearrangement of ideal steady-state pile-ups upon a stress increase (see Hart et al. in Chapter 5); others clearly involve irreversible structure changes and are not linear in stress.

For the specific case of randomly distributed discrete obstacles, Kocks (1966, 1967) has treated the elasto-plastic region in detail. The behavior may be characterized by the size of the area in which the driving force is positive, as a function of the applied stress (fig. 3.5). It is seen that this

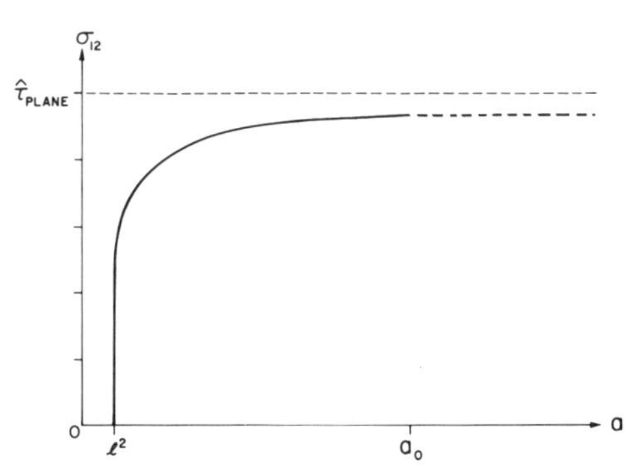

Fig. 3.5 The area swept by a segment of dislocation as a function of the applied stress, at a constant spacing ℓ of discrete obstacles. At a_0 (qualitative), the soft regions become contiguous.

area remains near ℓ^2, the area per obstacle point, over a significant range of stresses, and then very rapidly approaches an asymptote. This asymptote corresponds to the stress level, called $\hat{\tau}_{PLANE}$, at which the free area becomes continuous. The qualitative form of this area function can be derived from

Kocks's (1966) statistical theory to be

$$a = \frac{\ell^2}{1-\exp\left\{1 - \left(\frac{\sigma}{\hat{\tau}_{PLANE}}\right)^2\right\}} \tag{3.21}$$

Actually, the curve loses its meaning at the value of the average area the soft spots have when they do join up; this is called a_o in the figure. The total strain that can be achieved in this short-range slip mode is

$$\gamma_o = b\, N_m\, a_o \tag{3.22}$$

where N_m is the number of active soft areas per unit volume (or the number of mobile segments, or the number of dislocation sources, all per unit volume). The maximum of N_m corresponds to the situation where all soft spots in each slip plane are active (Kocks 1970b):

$$N_m \leq \frac{1}{d \cdot a_o} \tag{3.23}$$

(d: slip plane spacing).

The strain rate to be achieved in <u>long-range slip</u> or 'flow', as it is given by the thermal activation at the critical gates (and neglecting any time spent by the dislocation in the soft regions), is then

$$\dot{\gamma}_{FLOW}(\sigma_{12}, T) = \gamma_o \nu_o \cdot \exp\left(-\frac{\Delta G_{PLANE}}{kT}\right) \tag{3.24}$$

We shall give an operational definition of the specific value $\dot{\gamma}_{FLOW}$ of the dislocation strain rate in the next section. The activation energy ΔG_{PLANE} will be of the same form as that given for the local process in eqn. (3.20), except that it relates to the obstacles in the critical regions. Obviously, short-range slip occurring only in the soft regions will be governed by a lower value of the activation energy for the same stress; on the other hand, the regions in which the activation energy would be higher are never activated.

Equation (3.24) holds in the range (see eqn. 3.5)

$$\tau_{STOR} < \sigma_{12} < \hat{\tau}_{PLANE} \quad \text{(thermal activation)} \qquad (3.25)$$

The threshold stress

$$\sigma_{12} = \hat{\tau}_{PLANE} \quad \text{(mechanical threshold for long-range slip)} \qquad (3.26)$$

at which long-range slip becomes a mechanical necessity, plays a central role in constitutive relations for slip: it separates the region in which the kinetic laws are governed by visco-dynamic effects from that in which thermal activation is rate determining.

For the thermally activated regime, fig. 3.6a illustrates the qualitative behavior of the macroscopic flow rate as a function of applied stress, according to eqns. (3.24-25). It allows for the possibility that, in the case of long-range ob-

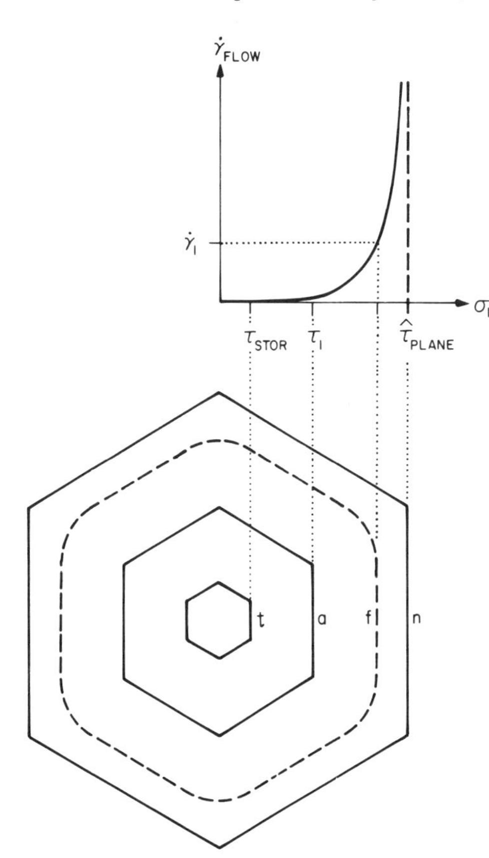

Fig. 3.6 (a) Schematic diagram of flow rate $\dot{\gamma}_{FLOW}$ versus applied stress σ_{12} in the regime controlled by thermal activation. (b) The various yield surfaces (solid) and a flow surface for a given strain rate and temperature (dashed).

stacles shown as curve (b) in fig. 3.3, there may be, in addition to τ_{STOR}, an 'athermal' lower limit for flow called τ_1. Actually, both limits are often unimportant because of the very sharp variation of the flow rate with stress as shown schematically in fig. 3.6a, which is a direct consequence of the exponential relation (3.24). It is often convenient to describe the stress dependence by the differential coefficient

$$m \equiv \left.\frac{\partial \ln \dot{\gamma}_{FLOW}}{\partial \ln \sigma}\right|_T \gg 1 \qquad (3.27)$$

It is large compared to 1 in the overwhelming number of practical cases in which slip controls plastic flow (the exceptions being at high temperatures or near τ_{STOR}).

The various limits shown in fig. 3.6a may be interpreted as a series of yield surfaces. Figure 3.6b shows a symbolic representation of a part of stress space with the kind of plastic yield surfaces that describe slip. (For a more thorough discussion, see Kocks 1970a.) The three concentric surfaces shown delineate ranges of possible behavior. For all practical applications, we are concerned with the region marked 'thermally activated flow', which lies <u>below</u> the mechanical threshold. Any stress point in this region may now lead to deformation on all slip systems, the rate of glide in each system being determined by how far below its mechanical threshold facet the stress point lies. If eqn. (3.27) is true, the rate on one slip system will far outweigh that on any other, unless the stress points directly into a corner. Thus, we may draw a surface of constant flow rate into this region which has essentially the same shape as the yield surface. (See also Rice in Chapter 2.)

One may now think of reversing this procedure and prescribing a macroscopic strain rate, such as it is done in conventional tensile tests, and asking what macroscopic stress is necessary to insure plastic deformation at this rate. This corresponds to rewriting eqn. (3.24) as an implicit equation for the flow stress σ_{FLOW} at a prescribed $\dot{\gamma}$

$$\dot{\gamma}(\sigma_{FLOW}, T) = \gamma_o \nu_o \cdot \exp\left(-\frac{\Delta G}{kT}\right) \qquad (3.28)$$

(where we have dropped the subscript on ΔG). The dashed line in fig. 3.6b is a 'flow surface':

$$\sigma_{12} = \sigma_{FLOW}(\dot{\gamma}, T) \, . \tag{3.29}$$

We will give an operational definition of σ_{FLOW} in the next section.

At a prescribed strain rate and temperature, σ_{FLOW} plays the same role as $\hat{\tau}_{PLANE}$ does at zero kelvin (i.e. in the absence of thermal fluctuations): it marks the line of separation between long-range slip and short-range slip (figs. 3.4 and 3.5). One may call it τ_{PLANE} (without the 'hat').

The value of the mechanical threshold, or of the flow stress at any given temperature and strain rate, depends on the strength of the obstacles in a nonlinear way, as opposed to the linear relation between $\hat{\tau}_{LINE}$ and the strength. This is a consequence of the statistics of interaction between dislocations of a finite line tension and discrete obstacles (Foreman and Makin 1966, Kocks 1967). For random distributions, this relation is well described by the 'Friedel formula' (1956)

$$\tau_{PLANE} = \alpha \left[\frac{K}{\mu b^2} \right]^{3/2} \cdot \frac{\mu b}{\ell} \tag{3.30}$$

where ℓ is the average obstacle spacing in the plane and α is of order 1 and depends on the (anisotropic) line tension of dislocations in the material.

This equation may be read as a relation between the mechanical threshold $\hat{\tau}_{PLANE}$ and the maximum strength $\hat{K}$ of the obstacles; or it may be read as an expression for the plane glide resistance at a given temperature and strain rate through the effective obstacle strength $K(T,\dot{\gamma})$.

3.5 FLOW STRESS, WORK HARDENING, AND CREEP

We have emphasized in the preceding sections that plastic deformation occurs, by nature of its mechanism, in a non-uniform way throughout the crystal; and that this leads to a distinction between short-range slip, which is of a transient nature and can only occur for small strains, and long-range slip, which is necessary for large deformations. In a similar manner, experiments show two stages of behavior: fig. 3.7a shows, as a schematic example, the beginning portion of stress-strain curves taken, say, in a tensile test at constant strain rate and temperature. The different curves are different in that they have different initial behavior; however, these transients

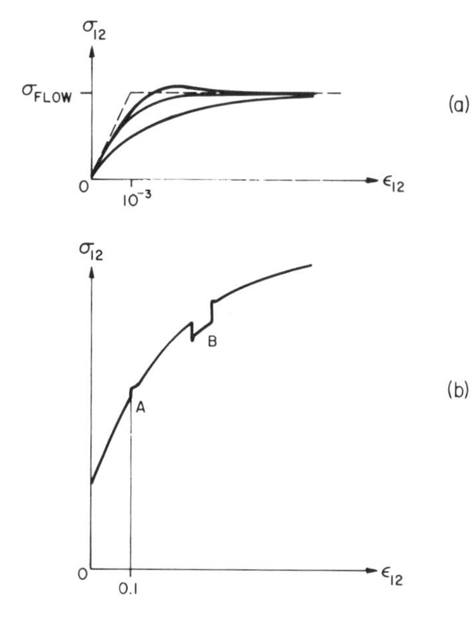

Fig. 3.7 Schematic stress-strain diagrams plotted to different strain scales. In (a), the dashed line corresponds to an elastic-ideally-plastic material, the solid lines show various transients. In (b), the test was interrupted by unloading (A) and strain-rate cycling (B). If the transients are left out, (b) shows the flow stress σ_{FLOW} as a function of strain.

extend only to strains of the order of a few times the elastic strain and the material then flows at what, on this scale, looks like a constant stress. We call it the flow stress, σ_{FLOW}.

Actually, the flow stress is not constant, as a stress-strain diagram drawn to a different strain scale shows in fig. 3.7b. Here, the elastic slope is so steep that it cannot be seen. On the other hand, there is a definite slope to this stress strain curve which starts at about 1/100 of the elastic slope and gradually decrease towards zero. Since this slope is usually so small compared to the elastic constants for all strains larger than the elastic ones, it appears to be zero in fig. 3.7a. This fortunate fact is responsible for the possibility of easily distinguishing transients on the one hand from 'work hardening' on the other*.

The transients are hard to see in fig. 3.7b. If they were left out entirely, the curve would truly be a locus of flow

*In dispersion strengthened alloys, the sharp break should occur on a diagram of σ^2 vs. γ (Kocks 1969) or of σ vs. $\sqrt{\gamma}$ (Ashby 1968).

stress versus plastic strain, rather than of applied stress versus strain. This is evident if one intersperses an unloading experiment (A) or a strain-rate change (B). In each case, upon reloading under the same conditions, the stress rises towards its previously achieved value, goes through some transient behavior and then continues along an extrapolation of the previous curve. The flow stress is thus one natural parameter of the state this crystal was left in by its former history. The value in any particular test depends on the strain rate and temperature employed in that test, as it is theoretically explained in eqn. (3.29); but for any standardized test conditions it could be a measure of the present state.

Instead of tensile tests, in which a certain strain rate is suddenly imposed and then left constant, one may, in a completely analogous way, employ creep tests, in which a certain stress is suddenly imposed and then left constant (fig. 3.8).

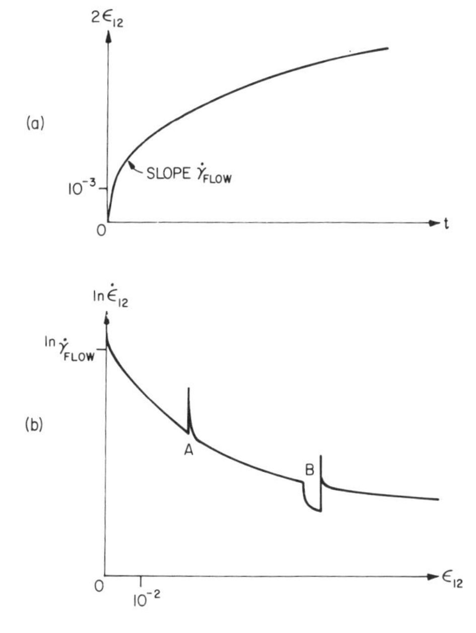

Fig. 3.8 (a) A schematic strain-time diagram in creep, plotted to a very sensitive strain scale, defining the flow rate $\dot{\gamma}_{FLOW}$ at the given stress, structure and temperature. (b) The corresponding diagram of the logarithmic strain rate versus strain, plotted to a less sensitive scale. A and B are interruptions as in fig. 3.7(b).

Upon the application of the stress there may be an instantaneous, or almost-instantaneous, strain due to elastic and anelastic components due to the lattice strain or short-range dislocation movement. Then, there is a rather sharp break to a relatively constant strain rate; it is the 'flow rate' $\dot{\gamma}_{FLOW}$

introduced before, for this particular state of the material, under the given stress and temperature. It diminishes gradually as the material hardens. The equivalent of the flow stress versus strain curve in fig. 3.7b is now the flow rate versus strain curve in fig. 3.8b (if one ignores the transients): after unloading (A) or stress cycling (B), the flow rate returns to the previous value at the same stress. In principle, $\dot{\gamma}_{FLOW}$ could thus serve as a parameter of state for any standard temperature and applied stress; in practice, it is not useful, because its values would span more orders of magnitude than can conveniently be measured - and it becomes unusable near zero temperature, whereas σ_{FLOW} approaches the important parameter $\hat{\tau}_{PLANE}$ in this limit.

Some features of the work hardening behavior of the classes of materials under discussion here are worth noting. They are evident from fig. 3.9, taken from Kocks et al. (1968), and concerning aluminum polycrystals over a wide range of temper-

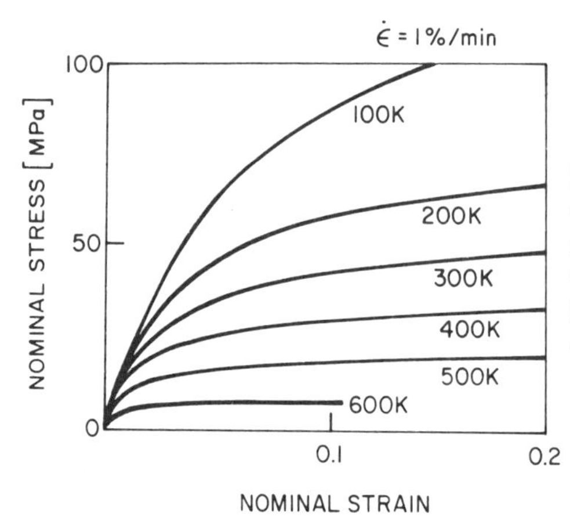

Fig. 3.9 Tensile stress versus strain diagrams for aluminum polycrystals at various temperatures, (after Kocks et al., 1968, courtesy of Gordon and Breach).

atures. There is a clear tendency at high temperatures for the flow stress to saturate at high strains; this is exactly equivalent to reaching 'steady-state creep' at constant stress. At lower temperatures, the curves appear, for all intents and purposes, to follow the same kind of behavior and seem also to extrapolate to a saturation stress even though it may never be reached. (More sophisticated methods of extrapolation, including correction for the area change, show the same extrapolation behavior.) In this sense one may regard the entire stress-strain curve as a transient into steady state, much as all stages of creep before steady-state creep have been termed transient creep. Such a procedure is appropriate when the primary interest lies in the steady-state behavior; here, we are more concerned with the work hardening régime.

At smaller strains, the low temperature curves start with a finite and relatively steep slope -- they can by no means be described as a square-root dependence of the stress on the strain. The curves from higher temperatures seem to merge with those from low temperatures within this steep range. From investigations on single crystals of FCC and BCC lattice structures, it has become customary to identify this steep range at 'stage II of work hardening', and the high-strain region as 'stage III'. While stage II is sometimes described as 'linear hardening', it is rarely truly linear; its important characteristics are its independence of temperature and strain rate and its relatively constant value (within a factor of 2) for many materials: about $\mu/300$ (or about E/100 for polycrystals). Stage III is very much more temperature dependent and goes continuously into the steady-state régime.

3.6 HISTORY, MEMORY, STATE, AND STRUCTURE

The phenomenon of work hardening is one expression of the fact that the properties of a material depend on its history. Different detailed histories may, however, give rise to the same properties: to this extent, the 'memory' of the material is limited. In continuum mechanics, it is customary to regard the 'memory' as a functional of the history (see, for example, Chapter 2 by Rice). In this chapter, we have taken the view common in Materials Science: that the memory manifests itself in the 'state' of the material, much like the 'memory' of a computer consists of a number of parameters characterizing its current state, irrespective of how this state was arrived at.

We regard it as self-evident, then, that all current properties of a material are entirely determined by its current state. Among those properties of concern to us is, for example, the current flow stress and its dependence on temperature and strain rate.* Another one of these properties is the current rate of change of the flow stress with strain: the current work hardening coefficient. It specifies how history (strain) changes the current state differentially. Inasmuch as all straining potentially changes the current state, the 'current' properties can only be obtained by back-extrapolation. This is one of the central problems in data evaluation.

---

*The fact that a rate is relevant in the property of interest should not concern us: all properties must be uniquely determined, not only equilibrium properties.

We have pointed out in the last section that the 'flow stress' at a given strain rate (fig. 3.7a -- or the 'flow rate' at a given stress, fig. 3.8a) is a property that lends itself to easy back-extrapolation, because of the abrupt change in slope of experimental stress-strain diagrams by at least an order of magnitude. Another important experimental fact is that work hardening, i.e. an irreversible change in the relevant current "state", occurs only at stresses very near the flow stress, i.e. in long-range slip; straining into the transition region does not change the flow stress for subsequent tests, except after many cycles.

These two observations make the philosophical separation of the constitutive relations into two sets operationally feasible: one describing the dependence of the current properties on the current state, the other the (differential) dependence of the current state on the current changes in history.

In metallurgical terms, the 'current state' of a material is characterized by its current structure: the density and distribution of all lattice defects. In principle, it needs an immense number of parameters to characterize a metallurgical structure. If they were all relevant to the mechanical properties of interest, not much would be gained by this description over one that specifies the precise history. It is a cardinal observation of crystal plasticity that, in fact, some mechanical properties are sufficiently well described by a very few structure parameters. Moreover, only some of these vary during a test; for example, the distribution of second-phase particles would not change at temperatures below which diffusion is widespread, but the dislocation density would.

For a physical understanding of the mechanisms of plastic flow, it is imperative that the relevant structure variables be identified: then, for any structural state that one has determined by metallographic means, one can predict the macroscopic properties; and conversely, for any desired set of macroscopic properties, one can prescribe what change in the metallurgical structure should be attempted.

Here, we have a much less ambitious aim: to characterize the current state of material to the extent that it enters the constitutive relations -- so that, for example, one or a few well chosen tests allow one to predict the current mechanical behavior under all other test and service conditions. For this, it is not necessary to characterize the state in structural terms; well chosen phenomenological parameters are adequate if not better. For example, the physical mechanism may suggest the presence of a number of constants, all of which may be structure dependent, which however occur only in a certain combination, such as $\gamma_o \nu_o$ in eqn. (3.24): a single parameter combining both is all that is phenomenologically relevant.

On the other hand, a knowledge of the basic physical mechanism may help keep the number of phenomenological parameters down; for example, a temperature function such as the exponential term in eqn. (3.24) would require the specification of many parameters if it were represented by a power series.

The total number of history-, state-, or structure-dependent parameters appearing in a constitutive equation is of crucial importance in its usefulness: the fewer parameters there are, the fewer measurements must be made to characterize the current state. When there is only one state parameter and all other quantities in the equation are either material constants or external variables measurable at every instant, a single measurement of all the external variables characterizes the 'state'; the value of any one of these variables when all the others are changed in a subsequent test can now be predicted. In this case, the equation may be called an 'equation of state'* (Hart 1970), in the same sense as it is used, say, in the ideal-gas law (where the number of molecules is the 'structure' parameter). Similarly, if a constitutive law consists of a set of n equations and there are no more than n independent state parameters, it may be a set of equations-of-state.

While such 'equation-of-state behavior' lends itself to simple experimentation and to sweeping generalizations, it must be emphasized that it is simpler only as a matter of degree: when it is not obeyed, a few more measurements must be made, such as of derivatives of the external variables with respect to each other, rather than merely of their instantaneous values. The matter of qualitative importance is that the constitutive relation contain only instantaneously measurable variables (such as T, σ, $\dot{\gamma}$) and no history variables (such as strain and time): then the parameters in it will be parameters of state, interpretable as parameters of the current structure.**

*Again, the fact that one of the instantaneously measurable variables is here a rate rather than a static quantity is immaterial in this context.

**In Rice's "internal variable" theory, (cf. Chapter 2) the distinction is not as clearly made: changes in the internal variables are synonymous with strains; instead of "structure", he uses "structural rearrangement". In real materials, changes of structure are common without changes in strain; and large strains with little attendant change in structure are the rule rather than the exception.

In the following section, we shall give an example of an equation of state derivable from the physical considerations of the preceding treatment, and discuss the conditions under which it might be expected to hold or not to hold.

3.7 EQUATIONS OF STATE AND TRANSIENTS

We have seen in the preceding that the constitutive behavior of materials may have significantly different features under different circumstances, even if slip controls the deformation; for example, short-range slip and long-range slip may be markedly different; the kinetic laws are fundamentally different for stresses above and below the mechanical threshold for long-range slip; and details of the behavior are different for rate sensitive and rate insensitive materials. In general, there will thus be a large number of parameters necessary to describe this behavior.

In special cases of practical importance, however, the situation may simplify significantly. We shall summarize below such a special case, which corresponds to probably the most common case of interest: long-range slip at stresses below the mechanical threshold, where thermal activation controls rate effects (eqn. 3.25); and, for simplicity, for the case that these rate effects are relatively mild (eqn. 3.27). We can then write a flow condition at prescribed strain rate and temperature (eqn. 3.29).

$$\sigma_{12} = \sigma_{FLOW}(\dot{\gamma},T) \tag{3.31}$$

where the dependence of the flow stress on strain rate and temperature is implicitly given in eqn. (3.28). Let us examine the number of structure parameters this relation may contain.

First, there is the mechanical threshold $\hat{\tau}_{PLANE}$, which the activation energy ΔG must contain in order to go to the correct limit in the absence of thermal fluctuations. In fig. 3.3, we have seen that ΔG may sometimes contain an additional threshold-type parameter, called τ_1 in that figure; however, it is sometimes not independent but proportional to $\hat{\tau}_{PLANE}$. In addition, ΔG must contain a total activation energy F_o which describes the behavior in the approximation of very low stresses; in all cases that have been explicitly treated, this is a parameter that depends only on the nature of the obstacles, which remains constant, not on their density, which changes; or else (Basinski et al. 1972) it is itself a function of $\hat{\tau}_{PLANE}$. Thus, F_o is not expected to be an independent parameter of state. Also in the low-stress region, there is, of course, the possibility of the existence of a thermodynamic limit τ_{STOR}. Thus, we have at least one stress parameter, the mechanical threshold, and possibly one or two more.

The preexponential terms in eqn. (3.28) represent, from a phenomenological point of view, one additional constant. It may be structure sensitive, for example by way of the number of mobile dislocation segments N_m in eqn. (3.22). However, under the conditions of long-range slip considered here, the product of this density and the average area swept will usually attain a steady-state value (reached in strains small compated to the ones considered here), which depends only on the applied stress or on τ_{PLANE} (not very strongly even at that). While there is doubtless the possibility of an independent structure parameter entering through this preexponential factor, there is also the likely possibility that this is not the case.

If $\hat{\tau}_{PLANE}$ is the only independent structure parameter, we have an equation of state of the form

$$\sigma_{12} = \sigma_{FLOW}(\dot{\gamma}, T, \hat{\tau}_{PLANE}) \tag{3.32}$$

with only a single state parameter contained in it. It is a special case of a more general equation for long-range slip below the mechanical threshold:

$$\sigma_{12} = \sigma_{FLOW}(\dot{\gamma}, T, \hat{\tau}_{PLANE}, \tau_1, \dot{\gamma}_o) \tag{3.33}$$

where the $\dot{\gamma}_o$ is used for the combination $\gamma_o \nu_o$.

If eqn. (3.32) does in fact hold, we have an easy way of replacing the single structure parameter $\hat{\tau}_{PLANE}$ by a phenomenological parameter, such as the flow stress at a <u>standard</u> (STD) temperature and strain rate; we call it

$$\tau_{FLOW} \equiv \sigma_{FLOW} (\dot{\gamma}_{STD}, T_{STD}, \hat{\tau}_{PLANE}) \tag{3.34}$$

The use of the symbol σ_{FLOW} again means that the stress is to be determined by the back-extrapolation procedure given in fig. 3.7. For any other strain rate and temperature, the flow condition is then

$$\sigma_{12} = \sigma_{FLOW}(\dot{\gamma}, T, \tau_{FLOW}) \tag{3.35}$$

Here, τ_{FLOW} is the parameter of state: it can be determined at any instant by an excursion to standard test conditions. (It is equal to the quantity τ_{PLANE} - without hat - introduced in connection with eqn. (3.30), when $\dot{\gamma}_o$ does not depend on stress.)

From a theoretical point of view, the mechanical threshold $\hat{\tau}_{PLANE}$ is a more direct measure of the state of the material. Thus, one may be tempted to suggest that $\hat{\tau}_{PLANE}$ itself be measured, i.e. the flow stress at zero temperature. From an operational point of view, this would not be wise; for it would require that the entire $\Delta G(\sigma, \hat{\tau}_{PLANE})$ relation be known from zero temperature to the range of real interest - even though the physical mechanisms may well be different in different ranges. It seems best to select a standard strain rate, and a standard temperature at the lower end of the actual range of interest, and measure τ_{FLOW} for the purposes of state characterization.

To complete the constitutive relations, the rate of change of the flow stress with strain may now be written as

$$\dot{\tau}_{FLOW} = \theta\, \dot{\gamma} \tag{3.36}$$

Note that the flow stress must always be measured (intermittently) at the standard test conditions, even though the straining may take place at a different strain rate and temperature; for the work hardening coefficient θ may itself depend on $\dot{\gamma}$ and T:

$$\theta = \theta(\dot{\gamma}, T, \tau_{FLOW}, \ldots) \tag{3.37}$$

It also may depend on the current flow stress (unless work hardening is linear) and possibly on other state parameters. The work-hardening coefficient θ as defined in eqn. (3.36) is the total derivative of the flow stress with respect to strain and includes time dependent effects such as dynamic recovery (see Kocks 1974).

If no more than one additional parameter enters eqn. (3.37), it is also an 'equation of state'. In the limit of stage II work hardening, θ is in fact a constant independent of structure (and also of strain rate and temperature). Then, the constitutive relations of this material 'exhibit equation-of-state behavior'. In the large-strain limit, i.e. at the saturation stress (or equivalently, in steady-state creep), θ is zero and $\hat{\tau}_{PLANE}$ has vanished from equation (3.32): equations of state are obviously obeyed in this regime, after the work hardening transients have been allowed to elapse.

It is well known that in the intermediate stage III of work hardening, at least one state parameter in addition to τ_{FLOW} enters the work hardening in eqn. (3.27): different thermomechanical histories that give the same flow stress frequently do not give the same initial work-hardening rate (Mecking and Lücke 1969, Alden 1973). One of the major questions of work-hardening theory is whether one additional work-hardening parameter is in fact enough; in that case equation-of-state behavior can again be expected.

The experiments of Hart and Li and their co-workers (cf. Chapter 5), which will be summarized in a different chapter in this book, do show equation-of-state behavior over a significant range of variables. In view of the foregoing, this may be interpreted as follows. If there is only one state parameter, it ought to be $\hat{\tau}_{PLANE}$. Then, the applied stress should be scaled by $\hat{\tau}_{PLANE}$; if the strain rate $\dot{\gamma}$ is scaled at all, for different histories, its scaling factor must be a function of $\hat{\tau}_{PLANE}$. The particular relation Hart and Li found can be written as

$$\dot{\gamma} = (\hat{\tau}_{PLANE})^M \cdot f(\sigma/\hat{\tau}_{PLANE}) \tag{3.38}$$

where the arbitrary function f contains the Boltzmann factor and any dependence of the pre-exponential factors in eqn. (3.24) on the applied stress (normalized). In addition, the pre-exponential factor (basically the density of mobile segments) is apparently proportional to the M-th power of the mechanical threshold, say through the steady-state slip plane spacing d (eqn. 3.23).

At a steady-state structure, the ratio $\sigma/\hat{\tau}_{PLANE}$ becomes a constant (dependent on strain rate and temperature but not on history); then, the strain rate becomes proportional to σ^M. Indeed, the experiments often showed the scaling exponent M to be the same as the high-temperature steady-state creep exponent. In this extrapolation to steady state, equation-of-state behavior must obviously be obeyed. As one backs away from the steady-state limit, it would be reasonable to expect that first one state parameter comes in, then a second, and so on. If a second state variable ever enters, equation-of- state behavior can no longer be expected. Indeed, both Hart and Li did occasionally observe deviations from this behavior, always in the range furthest removed from steady state. In this sense, their experiments merely confirm reasonable expectations -- and delineate the range where such expectations are met.

It must be emphasized that the entire discussion was based on long-range slip only or, in terms of the phenomemological description of the stress-strain curve, for strain increments in excess of γ_o, which is at most of the order of a few tenths of a per cent. In stress relaxation experiments, the total strain achieved is, in hard machines, of the same order as the elastic strain of the specimen. While this assures the desirable feature of the absence of any essential work hardening, it also means that strain is most likely restricted to isolated soft spots. The behavior in these soft spots could, in principle, be quite different from that in the critical gates. For the above discussion, we have assumed that they are correlated -- an assumption implicit in Hart's recommendation to use these stress relaxation data for the prediction of large-strain properties. In some cases, a correlation has in fact been found experimentally, while actual values of the rate sensitivity were different: they were correlated, not identical (Cheng and Kocks 1970).

In the small-strain region, eqn. (3.31) and none of the subsequent ones are expected to hold. Firstly, there is no question but that in this range, the pre-exponential term in a rate equation of the form (3.24) is severely stress dependent according to eqn. (3.22) and according to the distribution of dislocation sources. Also it very likely depends on some separate structure parameters such as the density of dislocation sources, or the number of soft spots actually active. Secondly, the activation energy ΔG_{PLANE} in eqn. (3.24) would have to be replaced by the appropriate average of local values ΔG_{LINE} which would depend very much on the total volume actually active. Most fundamentally, the strain rate in the rate equation is now not $\dot{\gamma}_{FLOW}$ (as defined by the specific extrapolation procedure of fig. 3.8) but any intermediate strain rate at any stress equal to or less than the flow stress; the entire stress-strain curve, both as it describes the elasto-plastic transition and as it describes work hardening, is now described by a single equation. This gives rise to added terms in the rate equation, which describe the strain increment achieved for purely mechanical reasons due to an increase in stress (fig. 3.5) and due to concurrent changes in obstacle density. Kocks (1970b, see also Kocks et al. 1974) has shown that the general form of the rate equation is

$$\dot{\gamma} = \gamma_o [\nu_o \exp - \frac{\overline{\Delta G_{LINE}}}{kT} + p\cdot(\dot{\sigma} - \theta\gamma)] \tag{3.39}$$

where p is an additional stress and structure dependent parameter; it may be made to contain all dislocation elastic and anelastic effects. Thus, in addition to all the problems with the rate dependence of thermal activation, we have here a clear necessity for stress <u>rate</u> effects. When the stress rate is substantial, as it usually is when the applied stress is not near the flow stress, this term must be taken into account. It may well account for the difficulties encountered in some stress relaxation experiments (Law and Beshers, 1972, see also Clark and Alden 1974).

A higher level of understanding of mechanisms will be necessary before any meaningful constitutive relation can be proposed for this transient range.

3.8 SUMMARY

Constitutive laws are discussed which relate instantaneously measurable quantities such as stress, temperature, and strain rate to each other, but involve history parameters such as strain and time only in differential form. Such laws contain parameters that characterize the 'state' or the 'structure'. From a consideration of the basic physical mechanisms underlying slip, the following general conclusions are drawn.

a) The mechanical threshold $\hat{\tau}_{PLANE}$ for long-range slip separates régimes of stress in which equilibrium of dislocations on slip planes is possible from those in which is is not. This is the most fundamental parameter that must enter all constitutive relations (unless it itself is determined by a steady-state balance at a given temperature and strain rate).

b) Mechanical thresholds are, in practice, predominantly of the 'activation' type rather than energy-storing. For this reason, plastic potential and yield surface may deviate from each other if the 'activated state' depends on the state of stress; the 'flow rule' is then violated.

c) Processes that lead to energy storage rather than dissipation during plastic deformation, provide a lower limit to stresses at which macroscopic deformation can occur, the thermodynamic threshold τ_{STOR}. It is a plastic potential as well as a yield condition.

d) Kinetic laws for plastic flow differ fundamentally above and below the mechanical threshold. Of most practical interest is the range $\tau_{STOR} < \sigma_{12} < \hat{\tau}_{PLANE}$, in which thermal activation is rate controlling. The ratio $\sigma_{12}/\hat{\tau}_{PLANE}$ frequently describes all influences of the stress sufficiently well in this range; sometimes, an 'athermal stress level' τ_1 must also be defined.

e) When the influence of stress on strain rate is very strong ($m > 10$, say), 'flow surfaces' may be defined for each particular strain rate and temperature, which are parallel to the τ_{STOR}- and $\hat{\tau}_{PLANE}$-surfaces in stress space. Their parameter τ_{FLOW}, for any standard strain rate and temperature, may be used as a parameter of state in preference to the structure parameter $\hat{\tau}_{PLANE}$, which can only be measured by an extrapolation to zero temperature. Work hardening (and recovery) may then be described by the (differential) dependence of this standard flow stress on strain and time.

f) When the flow stress is the only parameter of state entering the kinetic law, and there is no more than one additional state parameter entering the work hardening law, an 'equation of state' is obeyed for these variables. This is expected in stage II of work hardening in FCC and BCC metals and also near steady-state flow in all materials. In an intermediate strain range, deviations might occur. However, even then, a sufficient number of measurements should define current material behavior independent of the path by which it arrived at that state.

g) The heterogeneous nature of slip suggests that all of the above conclusions hold for long-range slip only. Short-range dislocation motion may occur at any stress (even below τ_{STOR}) in isolated 'soft spots'. The kinetics of this motion should depend on a number of distribution parameters rather than only the slip-plane averages τ_{PLANE}, etc. It will also show strain rate effects that depend on the stress rate and on the rate of change of all structure parameters. These transients, which would be hard to describe in any phenomenologically meaningful way, are expected to occur for strain increments of a magnitude up to a few times the elastic strains, but not beyond.

h) The use of stress relaxation experiments to establish macroscopic constitutive relations is subject to serious doubt inasmuch as they extend over very small strains only; any correlation with long-range slip properties must always be tested.

ACKNOWLEDGEMENTS

This work was performed under the auspices of the U.S. Atomic Energy Commission. I am grateful to Drs. A. S. Argon, E. W. Hart, J. R. Rice, and A. P. L. Turner for fruitful discussions.

REFERENCES

Alden, T. H., (1973) Met. Trans., 4, 1047.

Ashby, M. F., (1968) in Symposium on Oxide Dispersion Strengthening, edited by G. S. Ansell et al. (New York: Gordon and Breach), p. 143.

Basinski, Z. S., Foxall, R. A., and Pascual, R., (1972) Scripta Met., 6, 807.

Cheng, C. Y., and Kocks, U. F., (1970) Sec'd. Int'l. Conf. Strength Metals Alloys (Metals Park, Ohio: ASM), p. 199.

Clark, M. A. and Alden, T. H., (1975) in Rate Processes in Plastic Deformation of Materials, edited by J. C. M. Li and A. K. Mukherjee (Metals Park, Ohio: ASM), p.656.

Foreman, A. J. E., and Makin, M. J., (1966) Phil. Mag., 14, 911.

Friedel, J., (1956) Les Dislocation, (Gauthier-Villars).

Hart, E. W., (1970) Acta Met., 18, 599.

Kocks, U. F., (1966) Phil. Mag., 13, 541. (1967) Canad. J. Phys., 45, 737. (1969) in Physics of Strength and Plasticity, edited by A. S. Argon (Cambridge, Mass.: MIT Press), p.143. (1970a) Met. Trans., 1, 1121. (1970b) in Fundamental Aspects of Dislocation Theory, edited by J. A. Simmons et al. (Washington, D. C.: Nat. Bur. St.), Spec. Publ. 317, p.1077. (1975) in Rate Processes in Plastic Deformation of Materials, edited by J. C. M. Li and A. K. Mukherjee (Metals Park, Ohio: ASM), p.356.

Kocks, U. F., Argon, A. S., and Ashby, M. F., (1975) Prog. Mat. Sci., edited by B. Chalmers et al. (Oxford: Pergamon Press) Vol. 19, p. 1.

Kocks, U. F., Chen, H. S., Rigney, D. A., and Schaefer, R. J., (1968) in Work Hardening, edited by J. P. Hirth and J. Weertman (Gordon and Breach), p. 151.

Law, C. C., and Beshers, D. N., (1972) Scripta Met., 6, 635.

Mecking, H., and Lücke, K., (1969) Acta Met., 17, 279.

LIST OF SYMBOLS

A	area of slip plane
δa	increment of area swept out by dislocation
Δa	activation area
a_o	mean free-slip area
b	amount of Burgers vector
d	slip plane spacing
E	Young's modulus
F	(Helmholtz) free energy
F_o	total activation energy
$f(\sigma_{ij})$	yield function
G	free enthalpy (Gibbs free energy)
ΔG_{LINE}	activation free enthalpy for dislocation segment
$\Delta G_{PLANE} \equiv$	ΔG: average activation free enthalpy ("activation energy")
K	obstacle strength
k	Boltzmann's constant
L	spacing of discrete obstacles in periodic array
ℓ	average spacing of discrete obstacles on slip plane
M	empirical exponent
m	relative stress dependence of flow rate
$m_{ij}^{(s)}$	generalized Schmid factor
N_m	number of active "soft" areas ("mobile segments") per unit volume.
δn	number of vacancies generated during increment of slip

P	pressure
p	coefficient of stress rate in strain rate equation
S	symbol for stable equilibrium position of dislocation
T	temperature
t_w	average waiting time of dislocation segment for thermal activation
U	symbol for unstable equilibrium position of dislocation
δu	incremental surface displacement
v	volume change of specimen per vacancy generated
V	total volume of specimen
δW	incremental work done on specimen by applied forces
w	width parameter for discrete obstacle
α	proportionality constant: (dislocation line energy)/ $(\mu b^2/2)$
$\dot{\gamma}$	rate of plastic shear (slip)
$\dot{\gamma}_{FLOW}$	"long-range" part of $\dot{\gamma}$, back-extrapolated
$\dot{\gamma}_o$	proportionality constant in rate equation
γ_o	total shear achieved in short-range slip
ε_{ij}	total strain tensor
ε_{12}	total strain in slip plane and slip direction
θ	work-hardening rate at constant strain rate and temperature
μ	shear modulus
ν_o	"attempt frequency" for thermal activation

σ_{ij}	(applied) stress tensor
σ_{12}	stress in slip plane and slip direction (macroscopic average)
σ_{FLOW}	"flow stress" at any particular strain rate and temperature
τ_1	athermal glide resistance
τ_{ELEM}	element glide resistance
τ_{FLOW}	flow stress at a standard temperature and strain rate: a measure of the "state" of the material
τ_{IDEAL}	ideal (or "theoretical") shear strength
τ_{LINE}	line glide resistance
τ_{PLANE}	plane glide resistance ($\hat{\tau}_{PLANE}$: mechanical threshold)
τ_{STOR}	average element glide resistance

ABSTRACT. *A crystalline solid can deform plastically by a limited number of basic atomistic processes: dislocation motion, diffusion, grain boundary sliding, and twinning or a phase transformation. These combine to give at least twelve distinguishable deformation mechanisms: yielding or slip, power-law creep, Nabarro-Herring creep, Coble creep, and super-plastic flow are examples. Approximate constitutive equations exist, or are now being developed, to describe how fast a polycrystalline sample deforms when a single mechanism operates. From these, deformation mechanism diagrams can be constructed which display the fields of stress and temperature in which a given mechanism of flow is dominant, and the strain-rate which all mechanisms, acting together, produce. Properly normalized, the diagrams for the group of materials with a particular crystal structure and bond type are essentially identical: a rule of corresponding states applies to the members of the group. This permits the approximate prediction of the plastic properties of materials for which very little data is available.*

4. THE KINETICS OF INELASTIC DEFORMATION ABOVE $0^{\circ}K$

M.F. Ashby and H.J. Frost

4.1 INTRODUCTION

Although it is often convenient to think of a solid as having a well-defined yield stress, below which it does not flow, this is inaccurate. Above absolute zero, any stress will cause a polycrystalline solid to flow (although the rate at which it does so may be indetectably small on the time-scale of a laboratory experiment or an engineering application).

This is because the macroscopic flow behavior is simply the volume-average of processes occurring on the microscale: glide or climb of dislocation segments, diffusive motion of atoms or ions, relative sliding at grain or phase boundaries, and so on. All these are kinetic processes, activated by the thermal energy that the solid contains, and biased in their direction to a greater or lesser extent by an applied stress.

The macroscopic behavior reflects these microscopic kinetic processes. At high temperatures, the stress biasing may be a linear one, leading to a strain-rate which varies linearly, or as a small power, of the stress. But at lower temperatures (when less thermal energy is available) the biasing becomes non-linear, giving such a strong dependence of strain-rate on stress that the solid may appear to have a unique yield strength - though more careful experiments would show it to depend on strain-rate. This low-temperature behavior is discussed more fully by Kocks in Chapter 3. Here we consider, in a less-detailed manner, the mechanisms by which a polycrystal may flow at both low and high temperatures, and the construction of diagrams which display the distinguishable regimes of plastic flow.

The discussion throughout has its basis in microscopic models of the kinetic processes involved in plastic flow.

4.2 RATE EQUATIONS

Flow occurs when atoms move their relative positions. There are relatively few ways of doing this on the microscopic scale in a crystalline solid: the (uncorrelated) diffusive motion of atoms; the (correlated) motion of atoms when a dislocation segment moves by glide or climb; the (correlated) motion of atoms when grains or phases slide over each other - probably because of the motion of dislocations in the boundary plane; and the (correlated) atom motion associated with twinning and certain phase transformations.

The macroscopic behavior, however, is more diverse. The microscopic processes combine in various ways to give six groups of deformation mechanism, each of which can be limited in its rate in several ways. Before discussing each group in detail, it is convenient to list them all:

(1) Shear "Collapse": catastrophic flow when the ideal shear strength is reached or exceeded.

(2) Twinning and Martensitic Transformations: stress induced phase transformations that probably occur by a dislocation mechanism, when they become a sub-group of (3), below.

(3) Flow by Dislocation Glide alone: slip.

  (a) Limited by a lattice resistance

  (b) Limited by discrete obstacles

  (c) Limited by phonon or other drags.

(4) Flow by Dislocation Glide and Climb: power-law creep.

(a) Glide plus lattice-diffusion controlled climb ("High Temperature Creep")

(b) Glide plus core-diffusion controlled climb ("Low Temperature Creep").

(5) Diffusional Flow:

(a) Lattice-diffusion controlled flow ("Nabarro-Herring Creep")

(b) Boundary-diffusion controlled flow ("Coble Creep")

(c) Interface-reaction controlled flow.

(6) Coupled Mechanisms

(a) Power-law breakdown

(b) Grain-boundary sliding with accommodation by creep

(c) Superplastic flow

(d) Dynamic recrystallization.

For each mechanism a rate-equation exists. This is an equation linking shear strain rate $\dot{\gamma}$ to shear stress σ, temperature T and to structure:

$$\dot{\gamma} = \dot{\gamma}\ (\sigma, T, \text{structure})$$

or, more generally,

$$\dot{\varepsilon}_{ij} = \dot{\varepsilon}_{ij}\ (\sigma_{lm}, T, \text{structure})$$

(where $\dot{\varepsilon}_{ij}$ is the strain-rate tensor and σ_{lm} the stress tensor). Structure includes all the internal characteristics of the deforming polycrystal: the parameters describing its atomic structure; its bonding and crystal class; its defect structure, including grain-size, dislocation density and arrangement, solute or precipitate concentration; and so on. To simplify the formulation of a rate-equation, it is usual to make one of two assumptions. At high temperatures, when creep is the dominant flow mechanism, it is assumed that a steady state exists: the parameters describing the structure do not change with time, though they do readjust themselves to new, steady values when the stress or the temperature are changed. But at low temperatures, the rate at which a steady state is reached may be very slow; it is then more practical to adopt the simplifying assumption of a steady structure, which does not depend on small changes of stress, temperature or time.

We now consider rate-equations for the deformation mechanisms listed above

4.2.1 *Shear "Collapse"*

The ideal shear strength, τ_{IDEAL}, defines a stress level above which any crystal, even a perfect one, is mechanically unstable: in the terminology used to describe the mechanics of frames, the crystal structure becomes a mechanism, and "collapse" ensues. For practical purposes this can be described by the rate-equation

$$\dot{\gamma} = 0 \qquad \sigma \leq \tau_{IDEAL}$$

$$\dot{\gamma} = \infty \qquad \sigma \geq \tau_{IDEAL} \tag{4.1}$$

The quantity τ_{IDEAL} has been calculated from atomistic models (see, for example, Kelly 1966); the result is always about $\mu/20$, where μ is the shear modulus.

4.2.2 *Stress induced twinning and martensitic transformations*

Deformation dominated by twinning is commonly observed among crystals with hexagonal and lower symmetry, and - coupled with the closely-related martensitic shears - may be an important mode of deformation of crystalline polymers. It seems most probable that both proceed by the propagation of dislocations through the deforming material so, although the shear is not a lattice-invariant one, rate-equations for these mechanisms of flow are derived in the manner outlined in the next paragraphs.

4.2.3 *Dislocation Glide*

At low temperatures (less than 0.3 T_m , where T_m is the absolute melting temperature) the deformation is dominated by the glide motion of dislocations. To describe the strain-rate produced by glide we assume it to depend on the thermal activation of dislocation segments over barriers, either due to a Peierls resistance, or to discrete obstacles. The approach adopted here is a simplification of a very complex situation described more fully in Chapter 3, but the resulting equations nevertheless describe experimental observations tolerably well.

In a complete description, the strain-rate depends on how quickly the dislocations overcome the barriers to their glide and how quickly they move from one barrier to the next. Under most conditions the dislocations move relatively quickly from one barrier to the next, and the strain-rate can be adequately described in terms of the waiting time spent at barriers. At very high strain-rates the limit on the free velocity of the dislocation imposed by phonon and other drags does become important. But under less-than-shock-loading conditions and except at very low temperatures such effects play no important role.

Barriers to dislocation glide can be divided into various classes (see, for example, Kocks et al., 1975). In particular, we can distinguish between barriers that can be overcome with the aid of thermal activation and those that cannot (thermal and athermal barriers respectively). Typical examples of athermal barriers are long-range stress fields due to dislocation groups, and large incoherent precipitates; solute atoms, or a Peierls resistance typify thermal barriers. We may further distinguish different types of thermally activatable barriers. Here we will consider two of these: localized obstacles (such as forest dislocations), and a Peierls potential barrier.

In pure FCC and HCP metals the dislocations glide easily without any appreciable Peierls resistance. The flow stress is then determined by the number and strength of localized obstacles such as forest dislocations. Under these conditions the flow stress at absolute zero, $\hat{\tau}_o$, is proportional to $\mu b/\ell$, where ℓ is the obstacle spacing, and b is the Burgers vector. The constant of proportionality is complicated, depending on the strength of the obstacles and on the statistics of their distribution, but as an adequate approximation one may write $\hat{\tau}_o = \mu b/\ell$. At higher temperatures part of the energy necessary to overcome an obstacle can be supplied by thermal activation. The amount of energy required for activation has the approximate form $\Delta F\ (1 - \sigma/\hat{\tau}_o)$, where F is the total free energy required to overcome the obstacle in the absence of a stress. (This is equivalent to the assumption of a square form for the strength-distance curve of the obstacle). If we assume that the strain-rate is proportional to the frequency of activation, (and thus ignore the problems of the distribution of obstacles and the particular grouping of mobile dislocation segments) we obtain:

$$\dot{\gamma}_1 = \dot{\gamma}_o \exp\left[- \frac{\Delta F}{kT} (1 - \sigma/\hat{\tau}_o)\right] \qquad (4.2)$$

where $\dot{\gamma}_o$ is an appropriate pre-exponential term which includes the number of mobile dislocation segments. The reader is referred to Chapter 3 for a full discussion of the form of this equation.

Like most of the equations in this chapter, this one is based on a model which is physically sound, but which is insufficiently precise to predict useful values of the constants $\dot{\gamma}_o$, ΔF and $\hat{\tau}_o$. Theory gives the form of the equation; we have to resort to experimental data to obtain the physical constants which enter it. This approach of "model inspired phenomenology" is a useful one in dealing with phenomena too complicated to model exactly, and allows one to obtain a rate equation that can be extrapolated outside the range for which data is available; a purely empirical equation cannot.

The three physical constants of eqn. (4.2) have well defined meanings. The quantity $\hat{\tau}_o$ is the flow stress at 0^oK. For pure FCC or HCP metals, $\hat{\tau}_o$ describes the state of work-hardening; at the level of approximation with which we are concerned here, ℓ can be thought of as the spacing of forest dislocations. In the computations described below, we chose ℓ to give the observed saturation flow stress (a typical value of ℓ is $10^2 b$). The quantity ΔF is the total Helmholtz free energy required to cut or overcome an obstacle (in this case, a forest dislocation: roughly $\mu b^3/5$). It determines the temperature-dependence of the flow stress. Finally, $\dot{\gamma}_o$ sets the absolute magnitude of the strain-rate. Its value is less critical, since it does not enter the exponential. A value between 10^6/sec and 10^8/sec fits experimental observations well.

Dislocation motion in BCC and diamond cubic crystals, and in oxides and carbides, is more difficult. There is a resistance to motion produced by the crystal lattice itself. This Peierls resistance increases rapidly with decreasing temperature. Following Guyot and Dorn (1967) we use a Peierls potential of parabolic form, such that the activation energy required to move a dislocation has the form:

$$\Delta F = \Delta F_k \left(1 - \frac{\sigma}{\hat{\tau}_p} \right)^2$$

Here $\hat{\tau}_p$, as before, is the flow stress at absolute zero and ΔF_k is the energy of formation of a kink pair. The strain-rate is given by $\dot{\gamma}_2$ where

$$\dot{\gamma}_2 = \dot{\gamma}_p \exp \left[- \frac{\Delta F}{kT} \right] \quad (4.3)$$

and $\dot{\gamma}_p$ is an appropriate pre-exponential term. Like the equation for obstacle-controlled glide, which it closely resembles, the constants $\hat{\tau}_p$, ΔF_k , and $\dot{\gamma}_p$ are found by matching experimental data. Again, $\dot{\gamma}_p$ is typically 10^6 to 10^8/sec.

A material which exhibits a lattice resistance will - in general - also contain obstacles. The superposition of these two strengthening agents is complicated; for the purposes of this chapter it is sufficient to assume that the strain-rate due to dislocation glide is equal to the smaller of eqns. (4.2) and (4.3).

4.2.4 *Dislocation Creep*

At temperatures above one-third of the melting point there is sufficient mobility of vacancies to allow dislocations to climb as well as glide. Deformation is possible at a lower stress than would be needed for glide alone. The steady-state creep-rate for high temperatures and moderate stresses can be described by the semi-empirical equation:

$$\dot{\gamma}_3 = A' \frac{D_{eff} \mu b}{kT} \left[\frac{\sigma}{\mu} \right]^m \quad (4.4)$$

where A' and m are material constants, D_{eff} is a diffusion coefficient (usually the lattice diffusion coefficient, D_v), and the other factors are as defined above. This equation has a reasonable theoretical basis, as has been discussed by Mukherjee, et al. (1969). (We have converted it from its usual tensile stress-tensile strain-rate form into an equivalent shear-stress/shear strain-rate form by using a Von-Mises definition of equivalent shear stress and strain-rate:

$A' = A\left(\sqrt{3}\right)^m \Big/ \left(2(1+\nu)\right)^{m-1}$, where A is the constant used by Mukherjee et al. and ν is Poisson's ratio.) Dislocation climb is limited by the diffusion of vacancies to or from the dislocation, giving a linear dependence of the strain-rate on the diffusion coefficient. The power dependence of strain-rate on stress has not been conclusively explained; empirically, the value of m is usually between 3 and 7.

An additional dislocation-creep mechanism exists. At lower temperatures, transport of matter via dislocation core diffusion contributes significantly to the overall diffusive transport of matter, and may even become the dominant transport mechanism. Robinson and Sherby (1969) have suggested, rightly, we believe, that this might explain the lower activation energy for creep at lower temperatures. This contribution of core diffusion can be included by defining an effective diffusion coefficient:

$$D_{eff} = D_v f_v + D_c f_c$$

where D_c is the core diffusion coefficient, and f_v and f_c are the fractions of atom sites associated with each type of diffusion. The value of f_v is essentially unity. The value of f_c is determined by the dislocation density, ρ, as $f_c = a_c \rho$, where a_c is the cross-sectional area of the dislocation core in which fast diffusion is taking place. The rather sparse data for $a_c D_c$ (the quantity measured by experiments) have recently been reviewed by Balluffi (1970): the diffusion enhancement varies with dislocation orientation (being perhaps 10 times larger for edges than for screws), and with the degree of dissociation and therefore the arrangement of the dislocations; even the activation energy is not constant. In general D_c is about equal to D_b, the grain boundary diffusion coefficient, if a_c is taken as about $5b^2$. By using the common experimental observation that $\rho \approx 10/b^2.(\sigma/\mu)^2$ the effective diffusion coefficient becomes:

$$D_{eff} = D_v \left\{ 1 + \frac{10a_c}{b^2} \left(\frac{\sigma}{\mu}\right)^2 \frac{D_c}{D_v} \right\}$$

which is inserted into eqn. (4.4). At high temperatures, the lattice-diffusional contribution is dominant: for later use we call this the regime of <u>high temperature creep</u>. At lower temperatures, the core-diffusional contribution becomes dominant; this regime we call <u>low temperature creep</u>.

4.2.5 *Diffusional flow*

At sufficiently low stresses, dislocation motion stops, or becomes so slow that it can be ignored. Plastic deformation continues by <u>diffusional flow</u>: the diffusive motion of single atoms from sources on grain or phase boundaries which carry a compressive traction to sinks on boundaries which carry a tensile one. The process involves three steps: the detachment of an atom (or vacancy) from a source-boundary; its diffusive motion across a grain by lattice diffusion or by diffusion in the boundary of the grain itself; and its reattachment at the sink-boundary.

When the boundaries are perfect sinks and sources, the diffusive step is the rate-controlling one. Flow is then Newtonian-viscous: the strain-rate is proportional to the stress. The two alternative flow paths represent independent, additive contributions to the overall strain-rate. The most recent re-analysis of this problem yields the following combined constitutive equation (Raj and Ashby, 1971):

$$\dot{\gamma}_4 = 42 \frac{\sigma\Omega}{kTd^2} \{D_v + \frac{\pi\,\delta}{d} D_b\} \qquad (4.5)$$

Here Ω is the atomic volume, d is the grain size, δ is the effective thickness of a boundary for diffusional transport, and D_v and D_b are the lattice and boundary diffusion coefficients. The volume-diffusion controlled term is known as <u>Nabarro-Herring creep</u>; the boundary diffusion term is known as <u>Coble creep</u>.

But boundaries are not always perfect sinks and sources; then the kinetics of attachment or detachment at the interface or boundary may become rate controlling. Deviations from perfection are caused either by a paucity of sites on a boundary at which attachment or detachment can take place (as at twin boundaries) or by the inhibition, or poisoning, of sites even when they are plentiful. Recent work on this problem (Ashby, 1969, Burton, 1971, Verrall and Ashby, 1975) shows that the creep-rate then becomes a non-linear function of the stress; typically:

$$\dot{\gamma} \propto \sigma^m \qquad (m \geq 1) \qquad (4.6,a)$$

although, under certain circumstances, poisoning can introduce a threshold stress, τ, for additional flow, such that:

$$\dot{\gamma} \propto (\sigma - \tau) \qquad (\dot{\gamma} \geq 0) \qquad (4.6,b)$$

4.2.6 *Coupled Mechanisms*

Certain other mechanisms, or coupled mechanisms, exist which lead to flow. At high stresses the power-law creep eqn. (4.4) describes experiments only if the power (m) is allowed to increase with stress. This power-law breakdown regime is the transition from a diffusion-limited dislocation motion to the obstacle-limited glide regime. Empirically, this can be adequately described by a rate equation of the form:

$$\gamma \propto \sinh \alpha\sigma \text{ , or}$$

$$\gamma \propto (\sinh \alpha\sigma)^m$$

but no satisfactory model has been proposed to describe it.

There exists a second transition regime between diffusional flow and homogeneous power law creep. Diffusional flow is regarded as grain boundary sliding with diffusional accommodation: the incompatibilities generated by the sliding are removed, at a steady rate, by diffusion from sources on some grain boundaries to sinks on others. If the strain-rate is increased sufficiently, diffusion can no longer accommodate the sliding completely - and a non-uniform, power-law creep within the grains (called 'folds' in the creep literature) appears. On further raising the strain-rate, flow becomes increasingly uniform, and the folds disappear: we are now in the regime of homogeneous power-law creep. Macroscopically, this transition region in which sliding is accommodated by glide-plus-climb creep is described by a power law of the form of eqn. (4.4), and is more important in considerations of fracture than of flow (Crossman and Ashby, 1975).

Superplastic flow, as far as can presently be judged, is a form of diffusional flow in which the kinetics of the interface reaction are important. We include it under the heading of "coupled mechanisms" because the disappearance of superplasticity as the strain-rate is increased is caused by the increasing contribution of power-law creep (eqn. 4.4) to the flow behavior.

Finally, we include dynamic recrystallization as a coupled mechanism. For pure metals (Pb, Ni, Cu have been studied) and ceramics (ice, for example) there exists a wide regime of stress and temperature over which a sample recrystallizes as it deforms. The recrystallization per se produces no deformation, but it alters the structure of the sample by lowering the defect content (metals) and by inducing a preferred orientation (ice) such that the material flows more rapidly. Our understanding of this phenomenon is still two rudimentary to be able to describe it by a rate equation - yet it is important in many metal forming operations, in the service behavior of nuclear fuels, and in geophysical problems such as glacial flow.

4.3 DEFORMATION-MECHANISM MAPS

When a polycrystal is loaded at a stress σ and a temperature, T , the net strain-rate is some superposition of the contributions of the several mechanisms described above. Dislocation creep ($\dot{\gamma}_3$) and diffusional flow ($\dot{\gamma}_4$) are independent flow mechanisms involving different defects. At a first approximation, they superimpose linearly, that is, their strain-rates add. Dislocation creep ($\dot{\gamma}_3$) and glide ($\dot{\gamma}_1$ or $\dot{\gamma}_2$) do not. Both processes involve the same defect: they describe the different behavior of dislocations under different conditions. As the stress is raised, the gliding part of the motion of a dislocation becomes more important, and the climbing part less important, until climb is not necessary at all. As a first approximation one may treat dislocation creep and glide as alternative mechanisms, choosing always the faster one. (It could be argued that this choice maximizes the rate at which the free energy of the system decreases, or at which its entropy increases). Finally, there is the problem of the superposition of strengthening agents discussed earlier: as a first approximation, that leading to the slowest strain-rate, is rate controlling. In summary, the net strain-rate of a polycrystal subject to a stress σ at a temperature T is, to a first approximation:

$$\dot{\gamma}_{net} = \dot{\gamma}_4 + \text{Greatest of} \left\{ \begin{array}{l} \dot{\gamma}_3 \\ \text{Least of } \dot{\gamma}_1 \text{ and } \dot{\gamma}_2 \end{array} \right\} \qquad (4.7)$$

This information is most usefully presented as a <u>map</u>, <u>in stress-temperature space</u>, showing the area of dominance of each flow mechanism, and the total strain rate that all of them produce - combined in the manner of eqn. (4.7) (Ashby, 1972). Consider a two-dimensional space with normalized shear stress, σ/μ, as one coordinate and homologeous temperature, T/T_M, as the other (T_m is the absolute melting point). Let σ/μ range from 10^{-8} to 1 and T/T_M range from 0 to 1 as shown in fig. 4.1; these ranges cover all possible values of the variables encountered in practice.

The construction of the maps involves two stages. We first ask: in what field of stress/temperature space is a single mechanism dominant - that is, where does it contribute more to the total strain rate than any other mechanism? The boundaries of the fields are obtained by equating pairs of rate equations (e.g. eqns. 4.1 to 4.6) and solving for stress as a function of temperature. At a field boundary, the two mechanisms which meet there contribute equally to the strain-rate; where three fields meet, three contribute equally. The fields obtained in this way are shown in fig. 4.1.

Given a stress and a temperature, the total strain can be calculated from eqn. (4.7). This allows us to plot onto the diagram contours of constant strain-rate. Figure 4.1, modified in this way, is shown in fig. 4.2; other examples are shown further below in this chapter. The considerable amount of data needed to plot them is listed in Table 4.1.

4.4 PROPERTIES AND LIMITATIONS OF THE MAPS

The two deformation maps shown in figs. 4.2 and 4.3 are for pure nickel with grain sizes of 1000μm and 10μm. There is a large region on the maps at the lower left of the diagrams, below the lowest strain-rate contour, where the behavior is <u>elastic and</u>

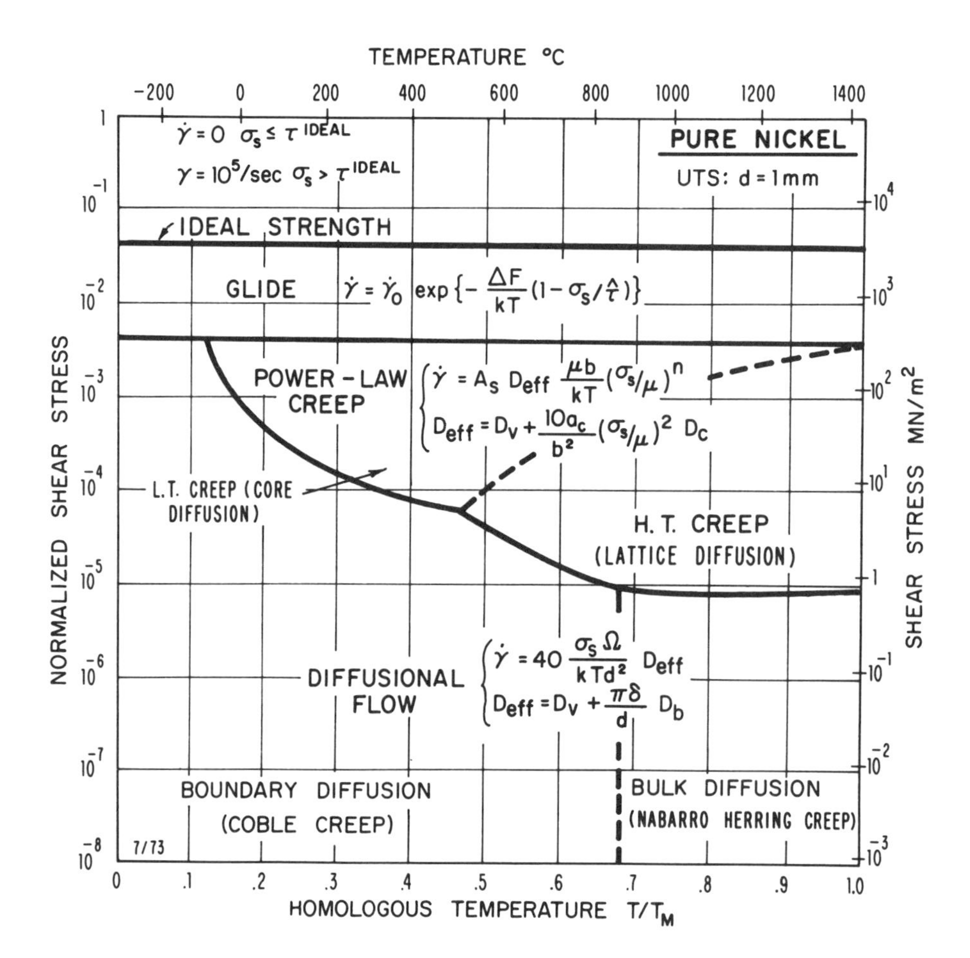

Fig. 4.1 The fields in which a particular mechanism of flow is dominant. The field boundaries are obtained by equating pairs of rate equations.

anelastic. The rate-equations can be used to predict strain-rates in this region, but the values are too low to be observed experimentally. The extent of this elastic region depends on the choice of the minimum observable strain-rate; we have generally chosen 10^{-10}/sec.

The stress-dependence of any mechanism is reflected in the vertical spacing of the strain-rate contours. Mechanisms with greater dependence on stress will dominate at higher stresses, as is shown by the relative positions of dislocation creep and diffusional flow. Similarly, the temperature-dependence of any

TABLE 4.1
DATA FOR DEFORMATION MAPS

		Pure Nickel (ni)	Zirconium Carbide (ZrC)	Ice (H_2O)
Ω	Atomic or Molecular Volume, cm^3	1.09×10^{-23}	2.64×10^{-23}	3.27×10^{-23}
b	Burgers Vector, cm	2.49×10^{-8}	3.34×10^{-8}	4.52×10^{-8}
T_m	Melting Point, ^oK	1726	3530	273
μ_0	Shear Modulus at 300°K, dynes/cm^2	7.89×10^{11} (1)	1.72×10^{12} (5)	3.01×10^{10}
$\frac{1}{\mu}\frac{d\mu}{dT}$	Temperature Dependence of Modulus, dynes/cm^2 ^oK	3.7×10^{-4}	8.0×10^{-5}	7×10^{-4} (10)
D_{ov}	Volume Diffusion, cm^2/sec†	1.9 (2)	1.03×10^{-3} (6)	10 (11)
Q_v	Volume Diffusion, kcal/mole	66.8	172.0	14.2
D_{oB}	Boundary Diffusion, cm^2/sec†	0.07	1×10^{-3}	10
Q_B	Boundary Diffusion, kcal/mole	27.4 (2)	112 (7)	9.4 (7)
δ	Grain Boundary Thickness, cm	5×10^{-8}	6.7×10^{-8}	9×10^{-8}

Symbol	Property	Pure Nickel (ni)	Zirconium Carbide (ZrC)	Ice (H_2O)
D_{oc}	Dislocation Core Diffusion cm^2/sec †	3.1	1×10^{-3}	10
Q_c	Dislocation Core Diffusion, kcal/mole	40.6 (3)	112 (7)	9.4 (7)
A_c	Core Cross-section Area, cm^2	10^{-15}	2×10^{-15}	4×10^{-15}
n	Constants of Dorn Equation	4.6 (4)	5.0 (8)	3.0 (12)
A	Dimensionless	2.56×10^5	4×10^{11}	5.74×10^2
$\hat{\tau}_o$	Obstacle Controlled Flow Stress at 0^oK $dyne/cm^2$	6.5×10^9	5.75×10^9	-
ΔF	Activation Energy (ergs) for Cutting of Obstacles	1.2×10^{-11}	6.4×10^{-11}	-
$\hat{\tau}_p$	Peierls Stress, $dyne/cm^2$	-	6×10^{10} (9)	3.4×10^{10} (13)
ΔF_K	Activation Energy (ergs) for Double Kink Formation	-	3.8×10^{-12}	5.4×10^{-13}

* $\mu(T) = \mu_0 \left[1 - \frac{1}{\mu}\frac{d\mu}{dT}(T-300)\right]$

† $D_v = D_{ov} \exp(-Q_v/kT)$; $D_B = D_{oB} \exp(-Q_B/kT)$; $D_c = D_{oc} \exp(-Q_c/kT)$

Continuation of Table 4.1

References to Table 4.1

(1) G. A. Alers, J. R. Neighbours, and H. Sato, J. Phys. Chem. Solids 13, 40 (1960).

(2) Ahmed Rassem Wazzan, J. Appl. Phys. 36, 3596 (1965).

(3) R. F. Canon and J. P. Stark, J. Appl. Phys. 40, 4361 (1969); 40, 4366 (1969).

(4) A. K. Mukherjee, J. E. Bird and J. E. Dorn, Trans. ASM 62, 155 (1966).

(5) R. Chang and J. J. Graham, J. Appl. Phys. 37, 3778 (1968).

(6) R. A. Andrievskii, V. V. Klimenko, and Yu F. Khormov, Fiz. Metal Metalloved 28, 298 (1969).

(7) Obtained by scaling lattice diffusion data:
$D_{oB} = D_{oc} = D_{ov}$; $Q_B = Q_c = 0.65\ Q_v$.

(8) Derived as best fit to data of D. W. Lee and J. S. Haggerty, J. Am. Ceram. Soc. 52, 641 (1969); and of F. Keihn and R. Kebler, J. Less Common Metals 6, 484 (1969).

(9) Derived hardness data of A. G. Atkins and D. Tabor, Proc. Roy. Soc. A 292, 441 (1966).

(10) G. Dante, Phys. Kondens. Materie 7, 3909 (1968).

(11) R. O. Ramaseier, J. Appl. Phys. 38, 2553 (1967).

(12) Derived from the data of P. Barnes, D. Tabor, and J. C. F. Walker, Proc. Roy. Soc., London A 324, 127 (1971); and of R. H. Thomas, Nature 232, 85 (1971).

(13) Derived from the energy of formation of Bjerrum defects. See J. W. Glen, Phys. Kondens. Materie 7, 43 (1968).

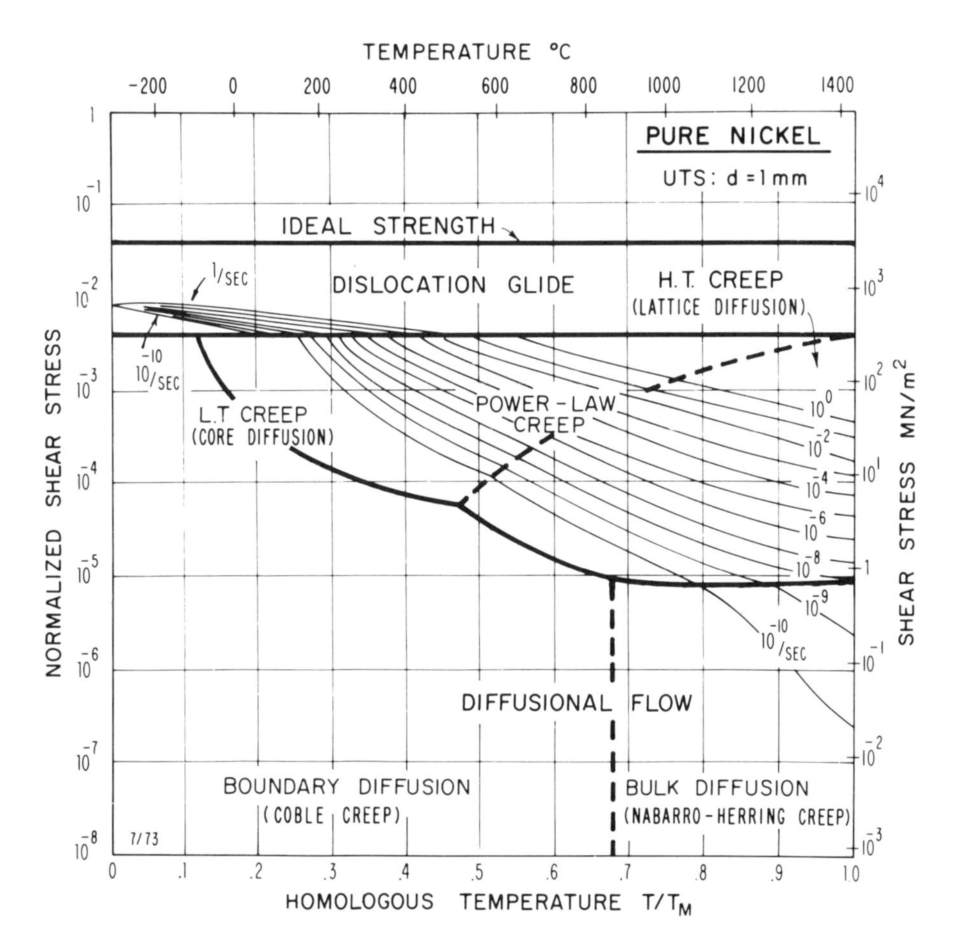

Fig. 4.2 The same as fig. 4.1, with contours of constant strain-rate superimposed. This is a map for nickel with a grain size of 1 mm.

mechanism is shown by the horizontal spacing of the strain-rate contours. The temperature dependence of both diffusional flow and dislocation creep results primarily from that of a diffusion coefficient. High-temperature dislocation creep and Nabarro-Herring creep have the same dependence on D_v and the boundary between them is therefore a line of constant stress. The same is approximately true of the boundary between Coble creep and low-temperature dislocation creep. There is a definite temperature dependence in the glide region, but it is made less prominent by the logarithmic stress scale.

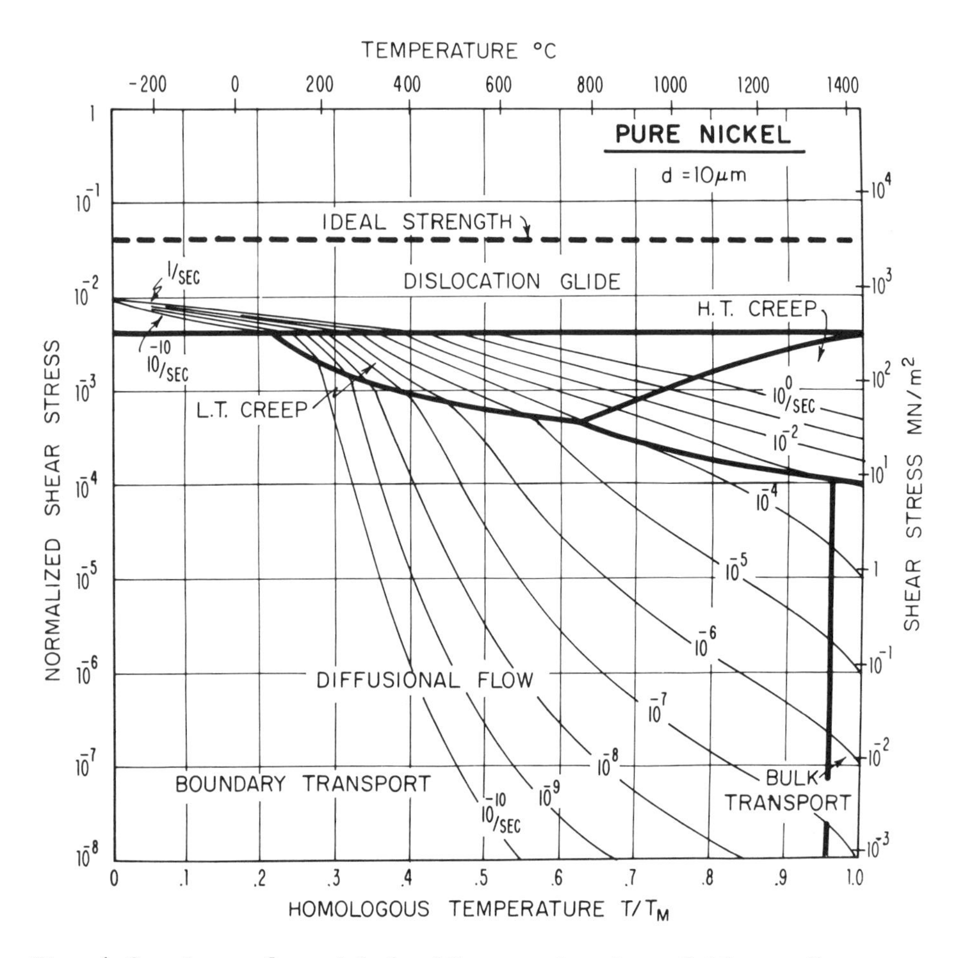

Fig. 4.3 A map for nickel with a grain size of 10μm. Change of grain size not only shifts the contours of constant strain rate, but also moves the field boundaries.

The effect of grain size in these maps is confined to the diffusional creep fields. The Nabarro-Herring creep-rate is proportional to d^{-2}, and the Coble creep-rate is proportional to d^{-3}. This meant that these fields will expand into the dislocation creep fields as the grain size is decreased, as figs. 4.2 and 4.3 illustrate. We have not included any grain-size dependence in the dislocation-creep region. There may be grain-size effects in this region, especially when the grain-size is small - comparable to the dislocation cell size - but this is poorly characterized, and in most cases is not very important. Experimentally there is often a grain-size dependence in the glide region, typically following the Hall-Petch relation, the flow stress (at a given $\dot{\gamma}$) varying as $d^{-1/2}$. This we have omitted because we assumed that the glide behavior is characterized by either a Peierls stress or by a sufficient density of obstacles to overshadow the grain-size effect.

The most obvious limitation of the maps in their present form is their limitation to steady-state flow. Time- and strain-dependent effects are not included. One can construct maps using time or strain-dependent constitutive equations, though displaying the results in a useful way is more difficult, and the results will not be discussed here.

The assumption of steady-state flow is quite adequate for diffusional creep. At least in pure metals, the defect structure (vacancy concentration, dislocation density) does not change with time. This is the sense in which we use the term "steady-state". The overall state of the polycrystal may not be stationary: the grain shape and grain boundary area, and the external surface area, will change as the sample deforms. These changes do, in fact, alter the response of the polycrystal to stress, but the effect is almost always trivially small. Operationally, the material behaves as if at steady state.

The assumption of steady-state flow is acceptable for dislocation creep. Most materials exhibit a well defined steady-state creep-rate in the appropriate range of temperature and stress: the microstructure adjusts itself (during a primary or transient period of creep) to a steady structure characteristic of the stress and temperature applied to the sample. The constitutive equation we use for dislocation creep assumes that the polycrystal is in this steady structural state. This means that a different microstructure exists for every point in the dislocation creep field.

The assumption of steady-state flow is least accurate for dislocation glide. Low temperature experiments produce work-hardening and yield point phenomena. The closest approximation to steady-state glide is at saturation work-hardening, at which the work-hardening rate has dropped to zero (see Chapter 3). In tensile tests this state is not usually achieved (because necking intervenes first), but it can be reached in compression or torsion tests. To approximate it we have used a high obstacle density in the obstacle controlled glide equation, this density being calculated from the experimentally observed saturation flow stress.

Fracture has been ignores in constructing the maps. At low temperatures brittle materials will fracture in tension at a lower stress than that needed for flow, so that the low temperature glide region may be accessible only to tests carried out under a confining pressure. Creep rupture may also cause confusion: if internal cavities form in a material before it enters secondary creep, then it may exhibit a minimum creep-rate that does not represent a true steady-state.

A final word of caution. The diagrams shown in this paper are the best we can construct at present. They are only as good as the data (Table 4.1) and equations used to construct them. Both are still poor. The diagrams can be used for guidance but should not at present be treated as exact. But the diagrams do present in a compact way the known plastic behavior of a material of a particular grain size. A group of such diagrams, spanning the normal range of grain size, summarizes the steady-state flow behavior of the material.

4.5 A RULE OF CORRESPONDING MECHANICAL STATES

Diagrams of an approximate nature are now available for quite a wide range of materials (Frost and Ashby, 1973). Figure 4.4, for example, shows Zirconium carbide - it is typical of the transition-metal carbides. This particular diagram refers to a grain size of 10μm, a common grain size in a compacted and sintered sample of ZrC. The high-temperature end of the diagram resembles those for metals: at low stresses diffusional flow is the dominant mechanism, and, at rather higher stresses, power-law creep becomes dominant. At low temperatures the diagram does differ from those for metals. This is because a large <u>Peierls stress</u> or lattice resistance opposes the motion of dislocations through ZrC at temperatures below about 800°C. The strain-rate contours all converge at absolute zero to a point which, in this case, is close to the 'ideal' shear strength of the material. This appears to be a common observation with oxides, carbides, and borides, and in materials like diamond, silicon and germanium; in short, in materials in which there is a considerable covalent contribution to the bonding. Although dislocations can exist in these materials, their motion is only possible with the aid of thermal fluctuations; at sufficiently low temperatures they simply become frozen in.

Figure 4.5 shows another ceramic material - ice. Its structure differs from that of ZrC; but, like ZrC, dislocations move at low temperatures only when the stress approaches the ideal strength. At high temperatures it exhibits power law creep, much as ZrC does.

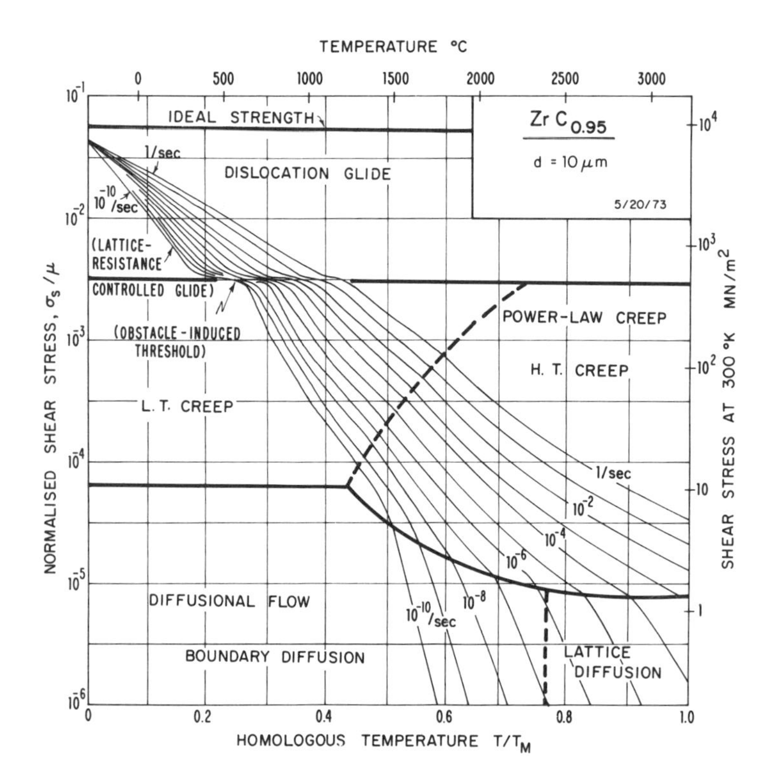

Fig. 4.4 A map for ZrC with a grain size of 10μm. The rapid rise of flow stress below 700^oC is due to the large lattice resistance, or Peierls stress, of ZrC at low temperatures.

How can we compare the mechanical behavior of pure elements and compounds? the obvious thing to do is to compare them under standard conditions. We choose a grain size of 50μm, and a standard strain-rate of 10^{-8}/sec. This corresponds to one contour on the diagram for a material, namely that for the strain rate, 10^{-8}/sec. This one contour, for a variety of materials, is shown in fig. 4.6. The axes on this plot are not normalized stress and temperature, but simply the absolute magnitude of the stress and the temperature. This allows a direct comparison between materials and illustrates how diverse their

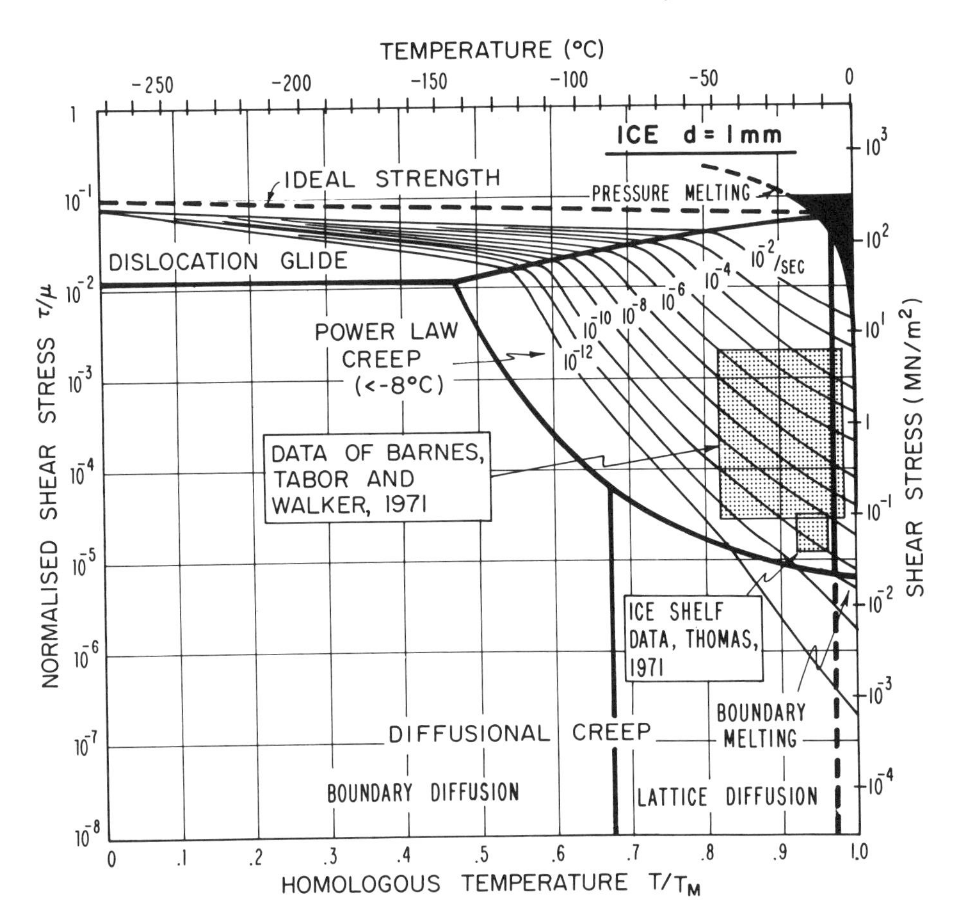

Fig. 4.5 A map for ice. The shaded areas are those which have been investigated experimentally. (The pressure melting line is appropriate for a simple compression test). Above 8°C a new, poorly understood, mechanism of flow is found. It appears to involve a liquid phase at the grain boundaries of the ice crystals.

mechanical strengths really are. At about 200°C, for example, the strengths of Cd and TiC differ by a factor of 10^6. Even within a sub-group of materials with the same crystal structure and the same type of bonding (face-centered cubic metals, for example) the differences are large: Pb and Ni, for example, differ in strength by a factor of almost 10^4 at 260°C.

These differences are directly related to the nature of the bonds which hold the atoms together. The melting point (T_m)

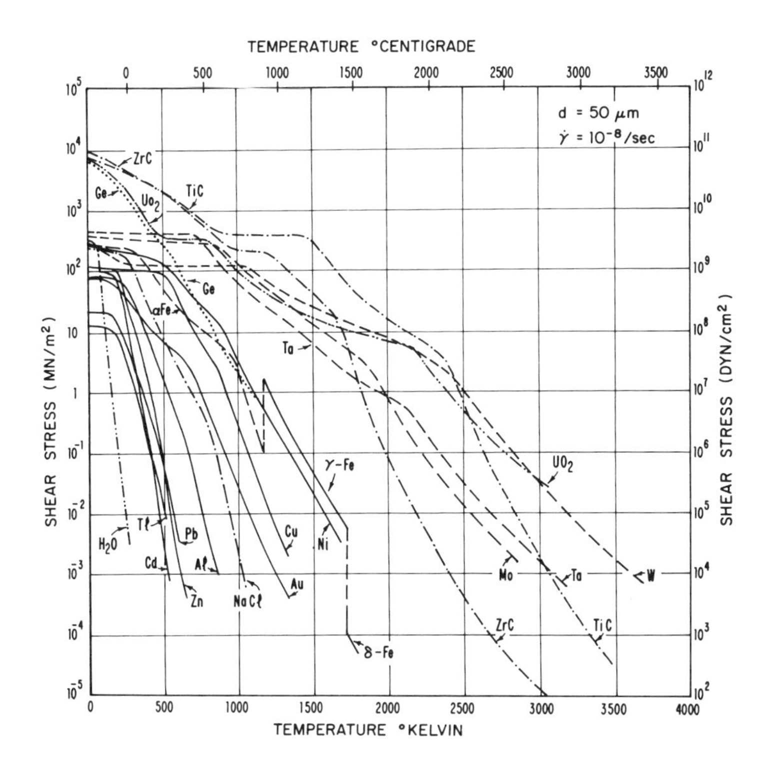

Fig. 4.6 The flow stress at $\dot{\gamma} = 10^{-8}$/sec for a number of materials with a standard grain size of 50μm.

and the moduli (μ) are a measure, respectively, of the cohesive energy, and of the curvature of the force-distance curve for the atoms or ions that make up the structure. By using normalized variables (σ/μ) and homologous temperature (T/T_m), the data are brought together somewhat, as shown in fig. 4.7 - though there is still a spread in σ/μ of about 10^2 at absolute zero, and about 10^5 at the melting point. Perhaps more important is that a grouping of materials by structure and bond type starts to appear: covalently bonded elements and compounds emerge as particularly strong at all temperatures; the FCC metals group together - though Pb remains exceptional; the BCC metals tend to form a separate group; and so on.

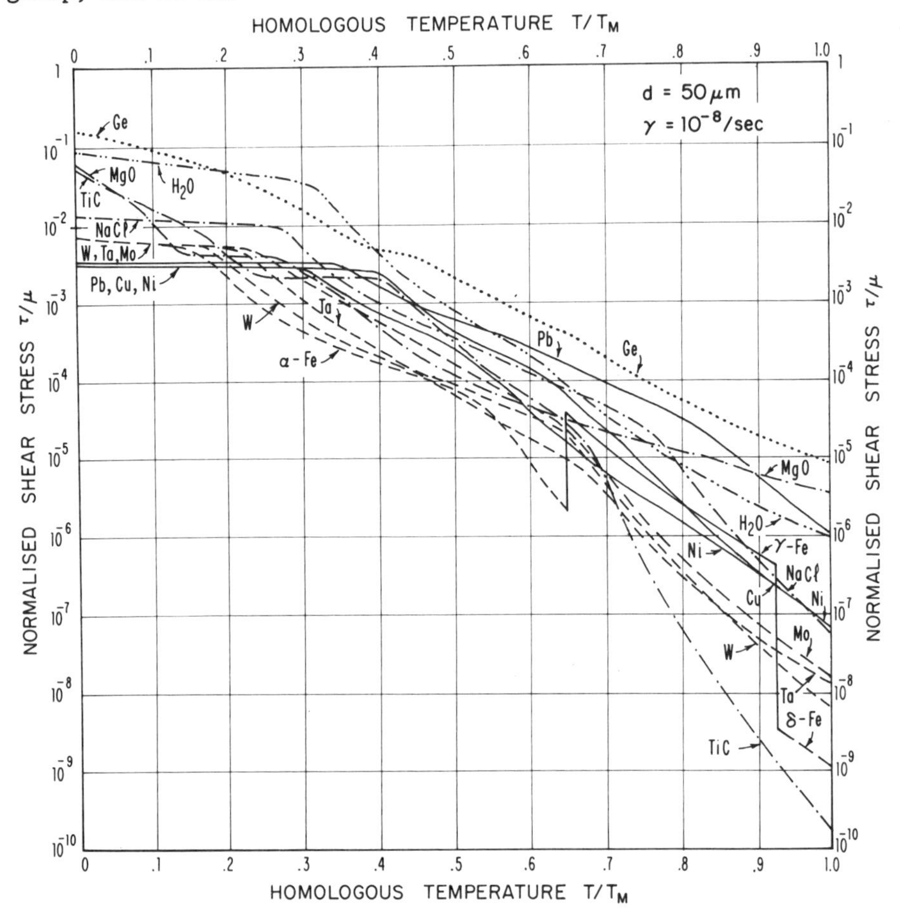

Fig. 4.7 The data of fig. 4.6, plotted using the normalized variables σ/μ and T/T_m.

At high temperatures, near the melting point, the residual differences between materials - and they are large - can be traced principally to the difference in the coefficient for mass transport within them (Table 4.2). Ge, for example, has a melting-point diffusivity of about 10^{-12} cm^2/sec. Titanium carbide, on the other hand, has a melting point diffusivity which is about 10^5 times greater than this. This is almost exactly the factor by which their strengths differ at the

TABLE 4.2

APPROXIMATE MELTING-POINT DIFFUSIVITIES, D_{TM}

MATERIAL	D_{TM} (cm^2/sec)
Ge, Si	3×10^{-12}
Ice	5×10^{-11}
Pb	4×10^{-10}
Ni, Cu, Ag, etc.	4×10^{-9}
Mo, W, Ta, etc.	5×10^{-8}
TiC, ZrC	$\sim 10^{-7}$

melting point. There is a systematic difference between the mechanical strength of BCC metals and FCC metals, which, at high temperatures, have the same origin: the average melting point diffusivity for FCC metals is about a factor of 10 lower than that for BCC metals. Even the anomalous creep behavior of lead has this origin: diffusion in lead is slower than in other FCC metals at the same homologous temperature.

The diffusivity is not the only physical constant that determines the mechanical strength at high temperatures. But it does appear to be the most important one. This suggests a further normalization: if, instead of plotting contours of constant strain-rate, we plot contours of constant strain-rate divided by melting point diffusivity, then the curves will be brought more nearly into superposition*. This is shown in fig. 4.8. This results in superposition of the high temperature ends of the curves, almost to within experimental error. The grouping of materials by structure and bond type is improved by this normalization: lead, for example, is brought

*A more logical comparison is between materials at constant normalized structure, (d/b = const; ρb^2 = const. etc), using reduced variables σ/μ, T/T_M and $\dot{\gamma}b^2/D_{TM}$ where b is the Burger's vector or atom size. In our detailed studies we have done this - it does not alter any of the conslusions stated here.

into registry with the other FCC metals, and the kinks in the curve for iron, due to phase changes, are largely removed. At low temperatures, members of a single structural and bonding group cluster together, but there is a significant and systematic difference between groups. This is mostly due to the lattice resistance or Peierls stress. In FCC metals there is none. In BCC metals the Peierls stress is relatively small. But in oxides and in covalently bonded materials, it is large.

All this suggests a rule of corresponding states: properly normalized the members of one isostructural group, in some standard state (same grain size, etc.) can be described by a single diagram. At the same T/T_m and σ/μ, the members of the group are in corresponding (steady) mechanical states, meaning that they deform by the same mechanism (or combination of mechanisms) and at the same normalized strain rate, $\dot{\gamma}/D_{TM}$.

The physical origin of this rule can be demonstrated - but that is beyond the scope of this chapter. The definition of 'isostructural group' does, however, require elaboration. For the rule to operate, members of a group must not only have the same crystal structure, they must also be electronically similar (have the same bonding type). Both copper and solid CO_2 have the FCC structure, but they differ mechanically because one is a metal while the other is a van-der-Waals solid. NaCl and TiC have the same structure, but the first is predominantly an ionic solid while the second is not, making them mechanically dissimilar. Indeed, the bonding is more important than the crystal structure. Germanium and silicon carbide, for example, have different structures, but very similar bonding - and they appear to belong to the same group from a mechanical point of view.

Table 4.3 shows an attempt to divide elements and some simple inorganic compounds into mechanically equivalent groups. Within a group with the same crystal structure (the NcCl structure, for example) there exist sub-groups with similar bond type. It is to these sub-groups that the rule applies. This means that if the mechanical properties of ZrC, for example, are well understood, then those for NbC can be obtained by the appropriate scaling - provided only that T_m, μ, and D_{TM} are known for it. If further investigation confirms this rule, it provides a powerful tool for predicting, at least approximately, the mechanical behavior of materials for which only the bare minimum of data is available.

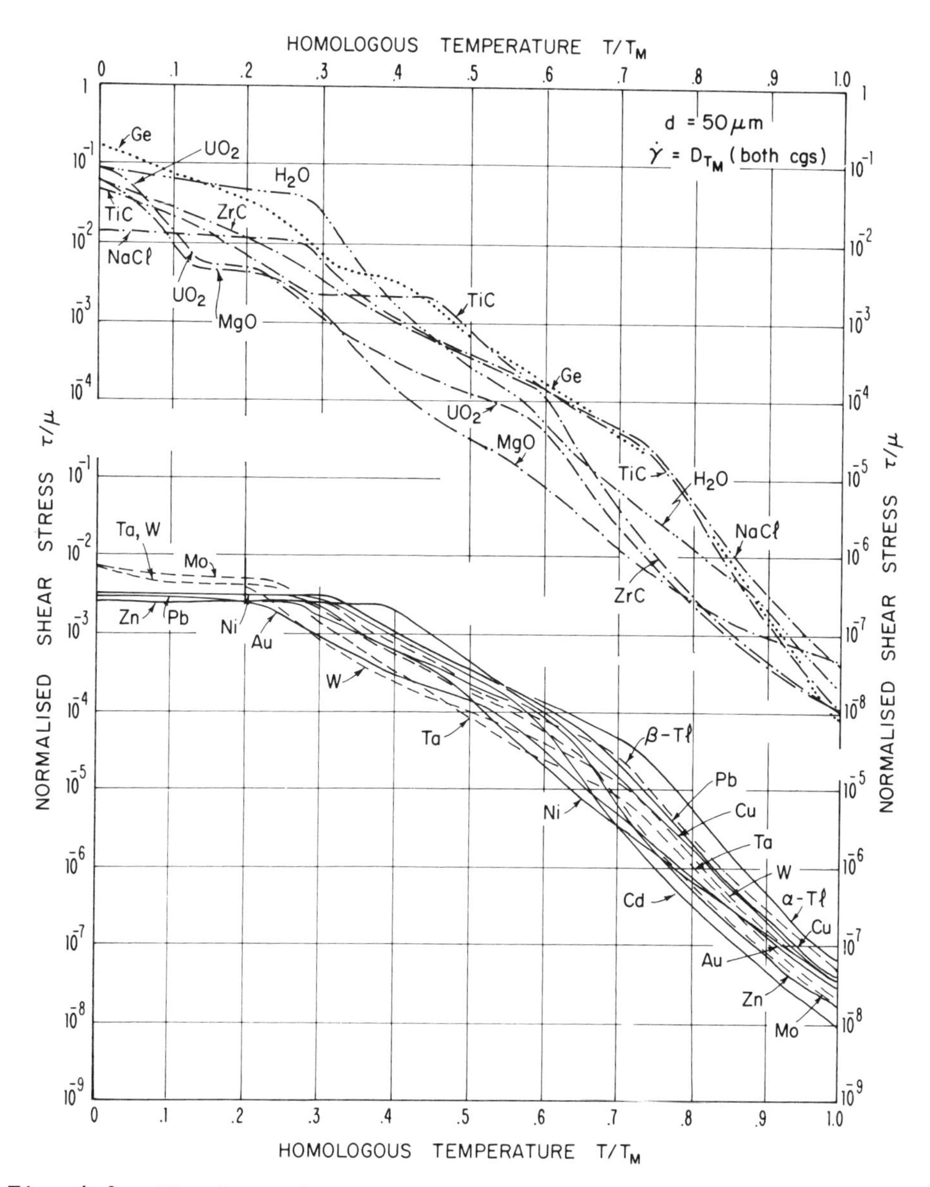

Fig. 4.8 The data after a further normalization: the lines are now contours of constant $\dot{\gamma}/D_{TM}$ instead of $\dot{\gamma}$. The vertical axis has been split for clarity.

TABLE 4.3
ELEMENTS AND SIMPLE COMPOUNDS GROUPED BY MECHANICAL TYPE

Crystal System	Structure Type	Point Groups	Examples
Cubic	FCC	m3m F	Al, Cu, Ag, Au, Pt, Pb, Ni γ-Fe, FCC solid solutions
	Diamond Cubic	m3m F	C, Si, Ge, α-Sn
	Rock Salt	m3m F	(K, Na, Li, Rb) (F, Cl, Br, I) AgCl, AgBr
			MgO, MnO, CaO, CdO, FeO
			PbS, PbTe
			TiC, ZrC, UC, TaC, VC, NbC
	Flourite	m3m F	UO_2, ThO_2, CaF_2, BaF_2, CeO_2
	Spinel	m3m F	$MgAl_2O_4$
	BCC	m3m I	W, Ta, Mo, Nb, α-Fe, [illegible] Fe, γ-U, BCC solid solutions. β-CuZn etc

Crystal System	Structure Type	Point Groups	Examples
	Caesium Chloride	m3m P	CsCl, CsBr, LiTl, MgTl, TlI, AuZn, AuCd, NH_4Cl, NH_4Br
	Zinc Blende	$\bar{4}$ 3m F	α-ZnS, InsB, β-SiC, AlAs, AlSb, GdSb, GaSb, BeS, HgS, AlP, BeTe
Tetragonal	β-Sn	4/mmm I	β-Sn
	Rutile	4/mmm P	TiO_2, SnO_2, PbO_2, WO_2, MnO_2, VO_2, NbO_2, TaO_2, MgF_2, MnF, ZnF, COF, NiF, FeF
Hexagonal	Zinz	6/mmm P	Cd, Zn, Co Mg, Re Te, Be, Zr, Ti, Hf, Y, Gd, Dy, Ho, Er; solid solutions
	Wurtzite	6/mmm	β-ZnS, β-CdS, α-SiC, AlN, In N
	Graphite	6/mmm	C
	Ice		H_2O
Trigonal	α Alumina	$\bar{3}$m R	α-Al_2O_3, Cr_2O_3
	Bismuth	$\bar{3}$ m R	Bi, Sb, As, Hg

4.6 SUMMARY

1. In the temperature range between $0^{o}K$ and its melting point, a polycrystalline material can deform by many mechanisms. The dominant one in any one range (that giving the largest strain-rate) depends on the current values of stress and temperature in that range, and on the structure of the material.

2. Approximate constitutive laws, based on microscopic kinetic models, exist or are currently under development to describe the rate of flow, (usually the steady-state rate) as a function of stress and temperature for each mechanism.

3. The results are conveniently displayed by constructing deformation mechanism maps which show the regime of stress and temperature in which each mechanism is dominant. Strain-rate contours on the maps show the net strain-rate which an appropriate superposition of mechanisms supplies.

4. Though approximate, these maps are useful. They summarize the flow behavior of a material, help in the design and interpretation of experiments, and - if constructed for engineering materials - should help the engineer to choose the correct constitutive law for deformation-governed design. They also highlight the mechanical similarities and differences between materials, and point to a method of normalizing mechanical properties such that all materials of a given crystal structure and bond type are described by a single map. This allows the prediction of the mechanical behavior of materials for which no mechanical data other than the shear modulus is available.

ACKNOWLEDGEMENT

This work was supported by the Scientific Research Council of the U.K., under Contract No. B/RG/4020.

REFERENCES

Ashby, M. F., (1969) Scripta Met., 3, 837; (1972) Acta Met., 20, 887.

Balluffi, R. W., (1970) Phys. Stat. Sol., 42, 11.

Burton, B., (1972) Mat. Sci. and Eng., 10, 9.

Crossman, F. and Ashby, M. F., (1975) to appear in Acta Met.

Frost, H. J. and Ashby, M. F., (1973) Harvard Univ. Final Report, August.

Guyot, P. and Dorn, J. E., (1967) Canad. J. Phys., 45, 983.

Kelly, A., (1966) "Strong Solids", (Clarendon Press: Oxford).

Kocks, U. F., Argon, A. S. and Ashby, M. F., (1975) Prog. Mat. Sci., edited by B. Chalmers, J. W. Christian and T. B. Massalski (Pergamon: New York), Vol. 18, in the press.

Mukherjee, A. K., Bird, J. E. and Dorn, J. E., (1969) Trans. A.S.M., 62, 155.

Raj, R. and Ashby, M. F., (1971) Met. Trans., 2, 113.

Robinson, S. L. and Sherby, O. D., (1969) Acta Met., 17, 109.

Verrall, R. A. and Ashby, M. F., (1975) to appear in Acta Met.

ABSTRACT. *A phenomenological theory of plastic deformation and the experimental approach that it suggests are described. The experimental data that have been developed so far are analyzed in terms of the theory. It is shown that the constitutive law implied by the observations is a plastic equation of state. Each deformation state of the material is a unique state of plastic hardness that can be characterized by a well defined state variable, the hardness. All states of hardness are shown to be related through an analytical scaling law. The stress-strain rate curves that are characteristic of each state of hardness are expressible in simple analytical terms, and the rate of hardness with strain increments (absolute strain hardening) is also given simple form. It is shown that some phenomena that are generally considered to be due to special high temperature mechanisms are already implied by the low temperature data. Implications for micro-structural theory are discussed.*

5. PHENOMENOLOGICAL THEORY: A GUIDE TO CONSTITUTIVE RELATIONS AND FUNDAMENTAL DEFORMATION PROPERTIES

E.W. Hart

C.Y. Li, H. Yamada, and G.L. Wire

5.1 INTRODUCTION

Constitutive laws for the non-elastic deformation behavior of metals that are adequate to describe and predict the deformation response to fairly complex histories of time dependent load application and temperature exposure are valuable for several purposes. The most direct and obvious use is as an essential ingredient in mechanical design calculations. An equally important use is as a concise way to relate deformation behavior to varieties of metallurgical micro-structure. Constitutive laws also provide a model for the various deformation phenomena that must be explained by fundamental theory.

In fact, constitutive laws of such adequacy are not yet known for any metal system. The need is customarily met at the practical level by empirical tabulations of "yield points", strain hardening exponents, "steady state" creep rates, and many other indices of deformation behavior that can be extracted from conventional tests. The empirical data is seldom treated in a unified manner or critically analyzed. At the scientific level it is common to place undue reliance on microstructurally oriented theory and to restrict the domain of experiment to those properties that are addressed by the theories.

The theoretical and experimental approaches described in this article lie somewhere intermediate to these two extremes. Theoretically we seek a framework within which to organize and critically examine the experimental phenomena without forcing it into a predetermined mold. And we pose some new experimental questions that can lead to new information.

The theoretical approach, that we term *phenomenological theory*, is concerned with an analysis of the observed phenomena with special attention to the *type* of relationships involved and to the experimental questions that can be posed. Some of those questions have been inadequately pursued in earlier treatments, and so some extension of experimental technique has been made in order to provide the necessary information. Microstructural understanding plays an important role in this investigation of suggesting the form for some of the theory, but the analysis is wholly in terms of the observable phenomena.

Application of the method yields, naturally, constitutive laws for the materials investigated that cover broad ranges of time dependent deformation histories. These laws are directly applicable for use in predicting deformation behavior. In addition, because of the character of those laws, rather strong requirements can be placed on the theoretical explanations of their origin from micro-structural mechanism.

5.2 THE PHENOMENOLOGICAL THEORY

The phenomenological theory that forms the basis for the analysis of the data is that which was earlier described by Hart (1970). We shall restate the principal features of that theory here.

Our analysis is restricted for the present to the description of simple polycrystalline metals and alloys in rather moderate temperature ranges and for which the deformation is homogeneous and stable. Some relaxation of these restrictions are possible, but we shall not consider them extensively in the present article. We are not concerned, therefore, with the three stage behavior of single crystals but rather with the more highly statistical and stable *régime* of polycrystalline behavior. The restriction in temperature is only to avoid the effects associated with large scale recovery and recrystallization.

Since we are concerned only with the non-elastic deformation we shall ignore the elastic strain component in our constitutive relations. Such strain is simply additive to any non-elastic strain.

It is essential to make a preliminary distinction between two components of non-elastic strain-rate that result from the application of a stress. One of these components is recognizably transient, and we designate it by the symbol $\dot{a}$ since it has some of the features of an anelastic strain-rate. The other component is relatively steady and accounts for the major amount of flow. We shall call that the plastic strain-rate and employ for it the symbol $\dot{\varepsilon}$. The transient strain-rate is appreciable only for relatively short times following abrupt changes of stress or total strain-rate. Its relationship to anelastic creep strain recovery was clearly stated by Bayce, Ludemann, Shepard, and Dorn (1960) in connection with stress changes during creep. It was further shown by Hart (1970) that the load change measurements of Mitra and McLean (1967) were readily rationalized on the same basis. The transient strain component is not negligible and is technologically important. Our main concern in the present work, however, is with the completely unrecoverable plastic strain. We, therefore, restrict our principal discussion and the experiments to the plastic strain.

We formulate the problem initially in terms of uniaxial tensile deformation. In a later section we indicate how the conclusions can be expressed in multi-axial terms.

5.2.1 *The General Phenomenological Relationships*

In the course of any deformation history at constant temperature, and in the absence of static recovery, the variations of the applied stress σ, the specimen plastic strain-rate $\dot{\varepsilon}$, and the strain ε in some current increment of time must be connected through a differential relationship of the form

$$d\ln\sigma = \gamma d\varepsilon + \nu\, d\ln\dot{\varepsilon} \qquad (5.1)$$

The coefficients γ and ν are measurable phenomenological parameters whose values at any time depend in general on the prior deformation history up to that point. Our choice of $\ln\sigma$ and $\ln\dot{\varepsilon}$ as basic variables, rather than σ and $\dot{\varepsilon}$, is justified only by the simplicity of the results that this leads to experimentally and by the non-dimensionality of γ and ν.

The relationship given by eqn. (5.1) is in fact a constitutive relationship if sufficient information is available about how γ and ν depend on the deformation history. If that information is developed only from constant $\dot{\varepsilon}$ stress-strain curves and constant σ creep curves, eqn. (5.1) can lead only to highly path dependent integration. Hart (1970) discussed this problem of integrability at length and proposed that the material could be tested for the existence of a state property that might considerably simplify the form of the constitutive parameters.

He pointed out that the total accumulated strain ε over some deformation path could not be considered to be an identifiable property of the deformed specimen and so could not be expected to be a good state variable of the system. On the other hand, the variables σ and $\dot{\varepsilon}$ at any stage of the deformation history necessarily characterized the current strength or hardness of the material. In fact, if at some instant a specimen is tested to determine the flow stress σ over a broad range of values of $\dot{\varepsilon}$ without strain hardening the specimen appreciably during the test, the resultant relationship $\sigma(\dot{\varepsilon})$ would fully characterize the current state of hardness.

Now a crucial question arises. If such a characteristic is determined for a specimen after a given deformation history, e.g. the curve labeled y_1 in fig. 5.1., and, if a $\sigma(\dot{\varepsilon})$ characteristic is determined for another specimen of the same material after a different deformation history but such that the point $(\sigma_1, \dot{\varepsilon}_1)$ is common to both characteristics, <u>will the two $\sigma(\dot{\varepsilon})$ characteristics be the same or will the second one be some other curve</u> y_1' ? If the two curves coincide we shall have the important conclusion that each state of plastic hardness of the material as a result of deformation is unique and is independent of the deformation path by which it was reached. This relationship can be called a <u>plastic equation of state</u>.

In this paper we shall generally refer to the $\sigma(\dot{\varepsilon})$ characteristic at any level of strain hardening as the current <u>hardness curve</u> for the specimen. The hardness curve is operationally well defined whether the plastic behavior satisfies an equation of state or not. When all hardness curves are unique in the sense described above, and a plastic equation of state holds, we shall call each state, as characterized by its unique hardness curve, a state of <u>plastic hardness</u> or more simply a <u>hardness</u> state.

When an equation of state exists for a material, the coefficient ν is a function only of σ and $\dot{\varepsilon}$. Although it does not follow directly from the equation of state, it might be expected that γ also might be a state function and depend

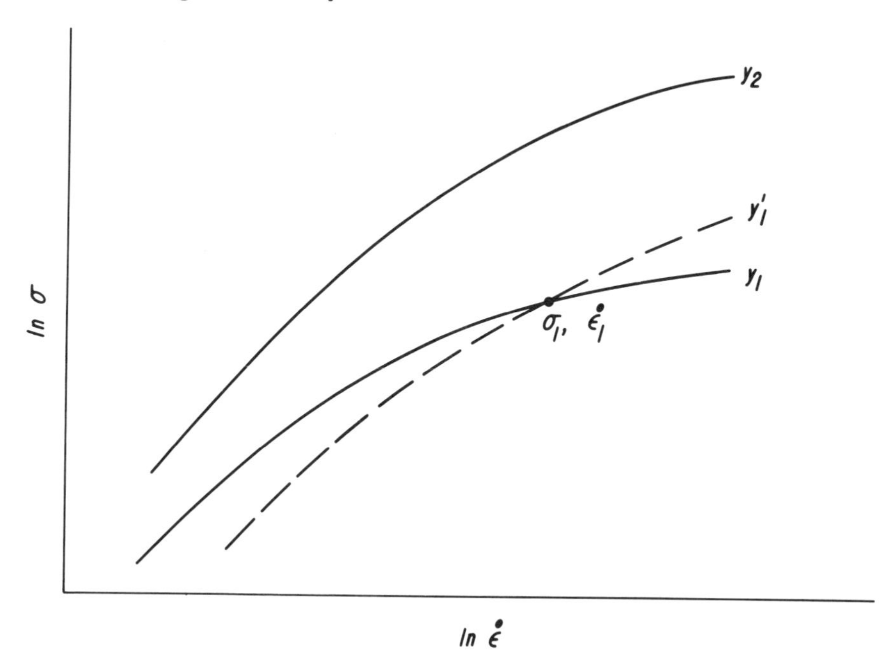

Fig. 5.1 A schematic plot of two stress-strain rate characteristics (solid lines). The sashed characteristic y_1' through the point $\sigma_1, \dot{\varepsilon}_1$ illustrates the question concerning the uniqueness of the characteristic y_1 .

only on σ and $\dot{\varepsilon}$. This conclusion, however, requires separate experimental verification. If ν and γ are both state functions of this type, the use of an incremental constitutive relation like eqn. (5.1) becomes considerably simpler in practical application. Examples of computations with an assumed constitutive equation of that type were given by Hart (1970).

The experimental results described below demonstrate the existence of a plastic equation of state for several materials. The state property does in fact extend to γ as well as ν . Furthermore, we shall show that most of the results can be very well represented in analytical form. This is especially valuable for testing theoretical predictions.

The influence of temperature T , in temperature ranges where thermal recovery is unimportant, is easily included by the addition of an appropriate term to eqn. (5.1), resulting in

$$d\ln\sigma = \gamma d\varepsilon + \nu\, d\ln\dot{\varepsilon} + q d(1/RT) , \quad (5.2)$$

where q is a new phenomenological coefficient and R is the gas constant. It is sometimes convenient to rearrange the equation as

$$d\ln\dot{\varepsilon} = n\, d\ln\sigma - g d\varepsilon - Q d(1/RT) , \tag{5.3}$$

where we define

$$n \equiv 1/\nu , \tag{5.4}$$

$$g \equiv \gamma/\nu , \quad \text{and} \tag{5.5}$$

$$Q \equiv q/\nu . \tag{5.6}$$

In this form, which is appropriate for the consideration of creep behavior, the coefficient Q is the familiar incremental activation energy for the plastic strain rate.

5.2.2 *The Hardness as a State Property*

A consequence of the existence of an equation of state is that all states of constant hardness of the material form a one parameter family of characteristic σ-$\dot{\varepsilon}$ curves (isothermally) or σ-$\dot{\varepsilon}$-T surfaces more generally. We have indicated by the letter y in fig. 5.1 such a parameter. There is no unique way of assigning numerical values to such a parameter. Any reasonable choice results in an explicit form for the equation of state of the type

$$y = y\,(\sigma,\dot{\varepsilon},T) . \tag{5.7}$$

Each value assigned to y determines a different surface of the family. We shall see below that the experimental results suggest a rather natural choice for the hardness parameter y, that can probably be put into one-to-one correspondence with a hardness indenter scale under suitably standard conditions.

For completeness we now summarize the formal consequence of the existence of an equation of state in terms of the integrability of eqn. (5.2).

We may rewrite eqn. (5.2) in the form

$$\gamma d\varepsilon = d\ln\gamma - \nu d\ln\dot{\varepsilon} - g d(1/RT) , \tag{5.8}$$

where at least ν and q are functions only of σ, $\dot{\varepsilon}$, and T. If an equation of state can be shown to exist, then there exists an integrating factor $F(\sigma,\dot{\varepsilon},T)$ for the right hand member of eqn. (5.8). Thus

$$d\ln y = F\gamma\, d\varepsilon \,, \tag{5.9}$$

where y is the hardness parameter discussed above. If one integrating factor exists then there is an infinite number of other ones possible. The integral surfaces, y = const. , as in eqn. (5.7), are always the same family of surfaces that satisfy eqn. (5.8) with $\gamma d\varepsilon = 0$, but the assignment of values of y to each surface depends on the choice of F .

Experimentally, the most convenient procedure is to test for the uniqueness of the $\gamma d\varepsilon = 0$ surfaces and then to choose a convenient measure for the value of y for each hardness state. The form of F is then given immediately from eqn. (5.9). Thus

$$d\ln y/d\varepsilon = F\gamma \,. \tag{5.10}$$

Then, if γ is a function only of $\sigma, \dot{\varepsilon}$, and T , F will be determined as a function of the same variables. The quantity $F\gamma$ then appears as what might be called an absolute strain hardening coefficient, and we shall designate it by the symbol Γ , and write

$$\Gamma(\sigma,\dot{\varepsilon},T) \equiv d\ln y/d\varepsilon \,. \tag{5.11}$$

From a fundamental standpoint Γ will be a measure of the rate of change with straining of the significant micro-structure that determines the hardness. It is, of course, experimentally measurable as a mechanical variable just as the other phenomenological coefficients are.

5.3 THE EXPERIMENTAL PROBLEM

If a crucial test of this phenomenological theory is to be made, one must be able to measure the constant hardness characteristics, shown schematically in fig. 5.1, over significantly large ratios of strain-rate and must be able to test the stability of those characteristics against recovery and variation of deformation histories. Furthermore, in the testing, it must be possible to separate the transient behavior from the more steady plastic strain-rate in a reliable way.

Concisely stated, it is desired to measure for a specimen, after some arbitrary deformation and thermal history, the plastic strain-rate for each of a broad range of stresses such that each measurement introduced an insignificant strain increment. In addition it will be desirable to verify the reliability of such results by repetition and to determine whether they are altered by holding times at various higher temperatures.

Measurements of this sort have been made earlier by various methods. We can unfortunately exploit few of those measurements since they generally covered a fairly narrow ratio of strain rates, and since little attention was generally given to distinguishing the levels of strain hardening of the materials. The techniques that were employed are (1) abrupt load changes during creep, (2) strain-rate changes during constant extension rate tests, and (3) load relaxation tests although the results of such tests have rarely been analyzed for the pertinent σ-$\dot{\varepsilon}$ data that they generate.

The main difficulty with the first two of the methods is that transient behavior is prominent at each finite change of σ or $\dot{\varepsilon}$. In fact much of the prior work employing those methods has been concerned principally with the transient.

The load relaxation method subjects the specimen to a smooth reduction of load, generally at load reduction rates that are slow enough to continuously accommodate the time dependent strain change that is responsible for the transient. Since this anelastic strain change associated with each decrement of stress is proportional to the stress change, the effect of such strain is merely to modify the elastic modulus of the material. Because of this feature, and because of its potential sensitivity, the load relaxation test was chosen for the present work as a method for measuring the σ-$\dot{\varepsilon}$ characteristics.

5.3.1 *The Nature of the Load Relaxation Test*

In a load relaxation test, as generally performed on a screw-driven tensile machine, the specimen is pulled at a predetermined extension rate to some desired extension or load level at which point the machine cross-head motion is stopped. The specimen continues to strain plastically under the action of the residual load exerted by the load train. As the specimen extends plastically, the applied load relaxes elastically. The time rate of change of the applied load $\dot{P}$ is a direct measure of the plastic extension rate $\dot{L}$ of the specimen, and, of course, P at each instant determines the operative applied stress. Therefore, if, during such a test, P is measured as a function of time t with sufficient precision that P can be differentiated with respect to t, one can generate for the relaxation history $\dot{\varepsilon}$ as a function of σ. If the elastic compliance of the load train including specimen is small enough, the entire relaxation will be accomplished with very little plastic strain of the specimen.

The relationships for analyzing the data and a description of an improved technique were given by Lee and Hart (1971). With suitable refinement the load relaxation test is capable of generating reliable data over as much as seven decades of strain-rate in a single relaxation run. The method has been employed successfully now for several materials and by several investigators (Hart and Solomon, 1973; Yamada and Li, 1973, 1974b; Cook, 1973).

The modifications of technique that are necessary in order to obtain this level of sensitivity for the test have been described in considerable detail in the references cited in the previous paragraph. The most important of these modifications are are follows:

(a) The use of digital instrumentation to measure and record the load cell readings. This provides good time resolution of the load readings at rates as high as ten readings per second, and strain rates can be measured as much as a factor of 10^2 higher than is possible with the usual chart recorders. Of course, the digital data facilitates the numerical data processing as well.

(b) Careful temperature stabilization of the entire load train. This minimizes the signal fluctuation caused by the thermal expansion and contraction of elements of the load train at the low strain-rate end of the run when the load reduction rate is very small, and rate measurements can be extended to rates as much as 10^3 slower than is otherwise possible.

There is, of course, nothing magical about the load relaxation test relative to the desired data. It is simply a convenient self-programmed way of subjecting the specimen to a range of applied loads in such a way that the rate of load change is smoothly controlled, and the total strain during the test is safely limited. Furthermore, the strain rate at each load level can be deduced from the load-time record. Some of the experimental problems associated with the test could be further minimized by separately measuring the specimen extension rate by auxiliary extensometers, however, this was not done in any of the work described here.

5.3.2 *The Control of the Transient Strain*

Transient strain phenomena are prominent only at relatively high homologous temperatures for simple materials. Roughly, this means at temperatures greater that about $0.2T_m$, where T_m

is the melting temperature.

In a later section we shall present something of a rationalization for the transient strain phenomena that we believe are associated with time dependent changes of the mobile dislocation distribution in response to changes of applied stress. But for the purposes of the present section we shall describe it only in phenomenological terms.

We consider the total inelastic strain-rate $\dot{\varepsilon}_t$ at any instant to be the sum of the unrecoverable plastic strain-rate $\dot{\varepsilon}$, as discussed above, and of an anelastic strain-rate $\dot{a}$ that is the time rate of change of a recoverable strain a . Thus we write

$$\dot{\varepsilon}_t = \dot{\varepsilon} + \dot{a} . \tag{5.12}$$

The component $\dot{a}$ is transient in the sense that it depends not only on the applied stress σ but also on the current value of a . As a result of the load change measurements of Mitra and McLean (1967) we conclude tentatively that there is a linear relationship between σ and the value of a that is in equilibrium with σ , i.e., the saturation value a_s for which $\dot{a} = 0$. Such a linear relationship defines a modulus M_a such that

$$\dot{a} = 0 , \quad \text{when} , \tag{5.13}$$

$$a = \sigma/M_a \tag{5.14}$$

$$\equiv a_s \tag{5.15}$$

We expect, in line with most traditional approaches to anelastic strain, that $\dot{a}$ depends principally on the difference between σ and $M_a a$ and so we write

$$\dot{a} = \dot{a}(\sigma - M_a a) , \tag{5.16}$$

where the function $\dot{a}$ is not necessarily linear in its argument but does have the property that

$$\dot{a}(o) = 0 . \tag{5.17}$$

Now, since we always observe $\dot{\varepsilon}_t$, we can make conclusions about $\dot{\varepsilon}$ only if we know the value of $\dot{a}$, or if we know it to be negligibly small.

In practice we are able to determine enough of this information because, generally, the time required to relax from one saturation value of a at one value of σ to that at a changed value of σ is small compared to time in which $\dot{\varepsilon}$ would change due to strain hardening. Over most of a relaxation run, once a has reached a saturation value it continuously remains close to the value σ/M_a. Then,

$$\dot{a} = \dot{\sigma}/M_a \tag{5.18}$$

over most of the run and the entire effect is only a contribution to the apparent elastic constant of the apparatus.

The only point of the test where $\dot{a}$ becomes specially important is in a rapid loading. We find then that, if initial loading is always carried past the reloading transient strains typical of interrupted tensile testing (c.f. fig. 5.2), the component $\dot{a}$ is entirely absorbed into the elastic background. If the loading is terminated in the transient region, the component $\dot{a}$ is clearly evident until the point that the falling stress σ becomes equal to $M_a a$ for the current value of a, and the data from then follows eqn. (5.18).

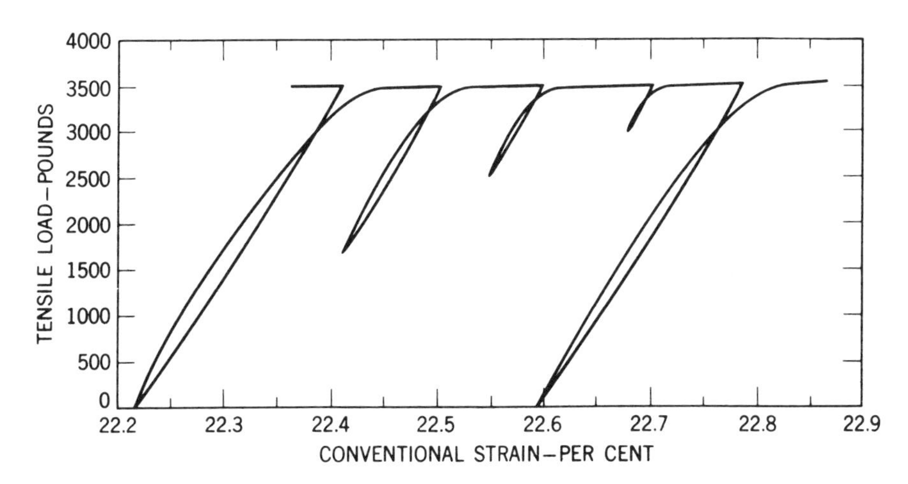

Fig. 5.2 Repeated unloading and reloading of OFHC Copper at constant extension rate at room temperature (Lubahn, 1955). The curves illustrate the persistent loading transient and its distinguishability from the more steady plastic strain.

Examples of deliberate transient response will be shown below as well as some additional evidence for the absence of transient in those tests so designated.

And so, we do not at present investigate the transient itself but rather devise ways of avoiding its interference with the measurement of $\dot{\varepsilon}$.

5.4 SURVEY OF EXPERIMENTAL RESULTS

Since most of the experimental results that have been obtained so far are either published or are to be published, we present here only enough of the accumulated data to show the current status of the investigations. Furthermore, because this approach is rather recent in origin, there are many questions that have not yet been adequately treated by experiment.

The $\log\sigma - \log\dot{\varepsilon}$ curves exhibited in this section were all obtained by load relaxation techniques, and, except where specially noted, all curves are from runs that were preceded by sufficient loading time to saturate the transient strain. Detailed analysis of these and other data will be presented in Section 5.5.

Where curves of several levels of hardness are shown, the intervening strain hardening was accomplished by various amounts of tensile strain between the adjacent hardness tests. The strain hardening was generally accomplished at the highest strain-rate at the same temperature, but in some tests the interim straining was done at higher temperatures or at lower strain-rates.

5.4.1 *"High" and "Low" Temperature Behavior*

The principal distinction among different metals with respect to the form of the measured hardness curves seems to be associated with the homologous temperature range in which the tests are carried out. For curves obtained over strain rate ranges from 10^{-3} to 10^{-10} sec^{-1} the homologous temperature that divides the "high" from the "low" regimes is about 0.25. The only material that has so far been tested both above and below this temperature is commercial purity Zr . Those results (Lee and Hart, 1971) are shown in fig. 5.3. The three curves shown are at substantially the same hardness and correspond to values of T/T_m of .22, .27, and .32. The distinctions in form are as follows: The lowest temperature curve is slightly concave upward. The middle curve has lower slope than the other two and is substantially straight. The highest temperature curve is concave downward. In fig. 5.4 (from Hart, 1974a)

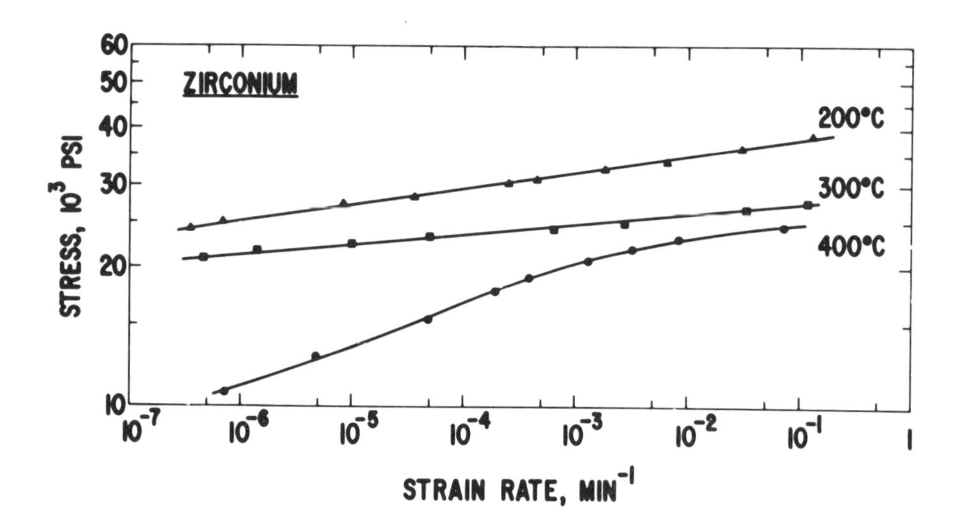

Fig. 5.3 Stress-strain rate characteristics for commercial purity Zirconium from load relaxation tests (Lee and Hart, 1971). The curves, obtained at the test temperatures shown, are at substantially the same hardness levels.

three such curves are plotted schematically on a single temperature compensated strain-rate scale. As we discuss below, we believe that the significant distinction between the "high" and "low" temperature régimes is actually concerned with the fact that the stress levels lie respectively below and above the stress indicated roughly by the horizontal dashed line. There is evidence of the same phenomenon at a single temperature of measurement from some measurements at quite high strain rates. Holt, Babcock, Green, and Maiden (1967) show, for example, a concave-upward behavior for curves of stress vs. log strain-rate for high purity Al at room temperature and at strain rates above 10^{-1} sec^{-1}. Although those measurements are not entirely comparable to the constant hardness curves, the effect is clear.

5.4.2 *The "Low" Temperature Results*

As examples of behavior in the low homologous temperature region we show results at room temperature for a b.c.c. metal, Niobium, and for a f.c.c. alloy, 304 Stainless Steel. The data for Nb (Yamada and Li, 1974,) is shown in fig. 5.5, and the 304 Stainless Steel (Yamada and Li, 1973) appears in fig. 5.6. Data for Nickel and TD Nickel (Yamada and Li, 1974a) have also

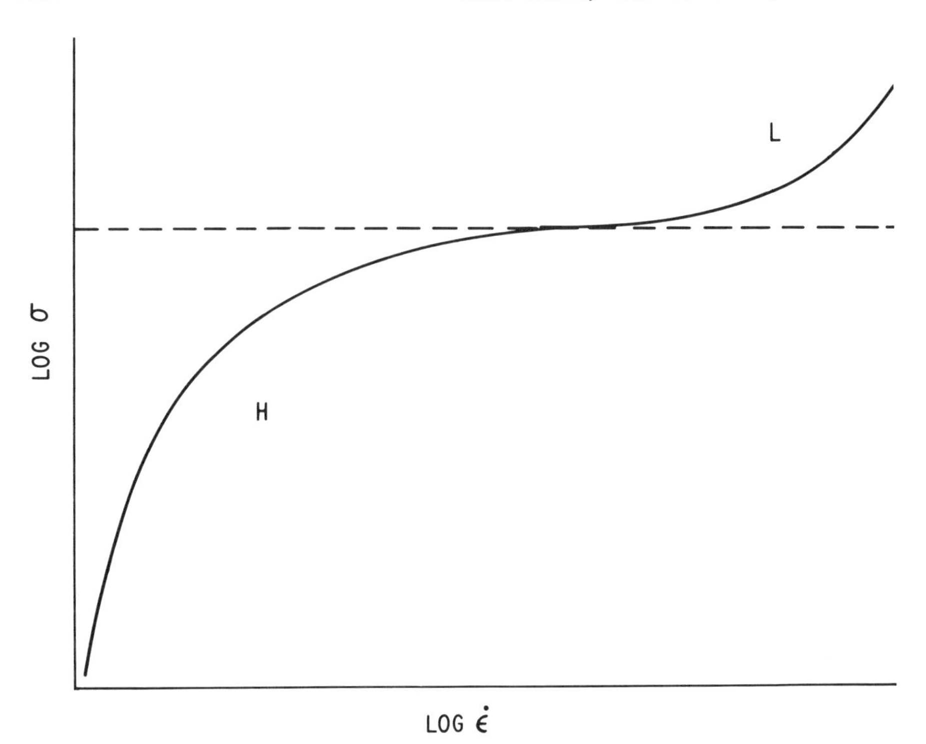

Fig. 5.4 A schematic plot of both "high" and "low" temperature behavior on a continuous temperature compensated strain-rate scale (from Hart, 1972).

been reported.

The principal feature of these curves is that they are concave upward and that the curves tend to become less steep as the hardness level increases. Another feature that distinguishes the "low" temperature data from the "high" temperature is that there is little or no detectable transient associated with the initial loading.

It is remarkable that despite the differences of crystal structure and purity there are no real qualitative differences among these different metallic species.

5.4.3 *The "High" Temperature Results*

The materials tested at high homologous temperatures were 1100 Aluminum and high purity (zone refined) Aluminum. For these materials $T/T_m = .32$ at room temperature (25^oC) and tests have been carried out at several additional temperatures up to $T/T_m = .58$.

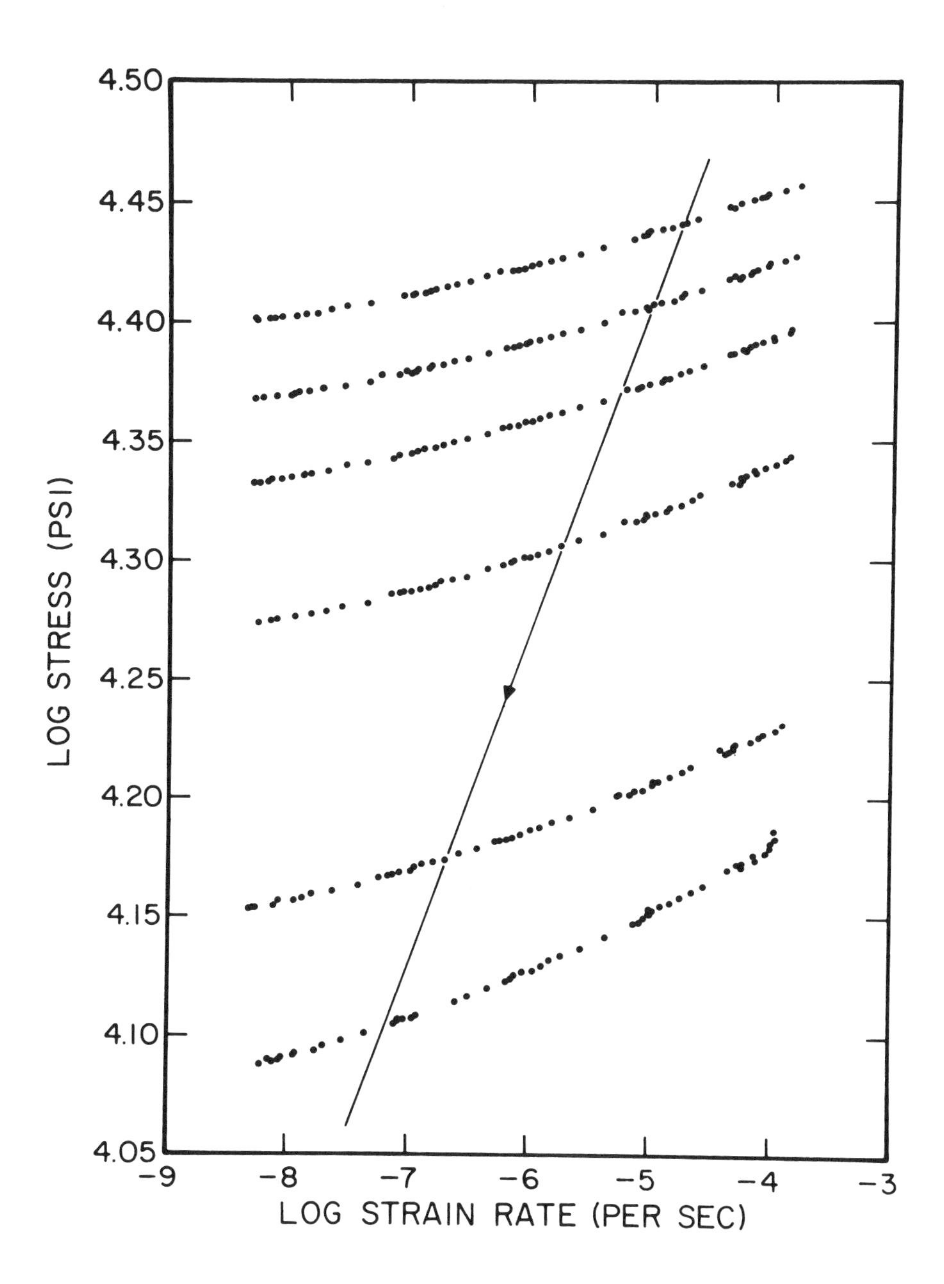

Fig. 5.5 A series of hardness curves for high purity Niobium at room temperature (Yamada and Li, 1974). The diagonal straight line is the direction of hardness scaling as described in Section 5.5.

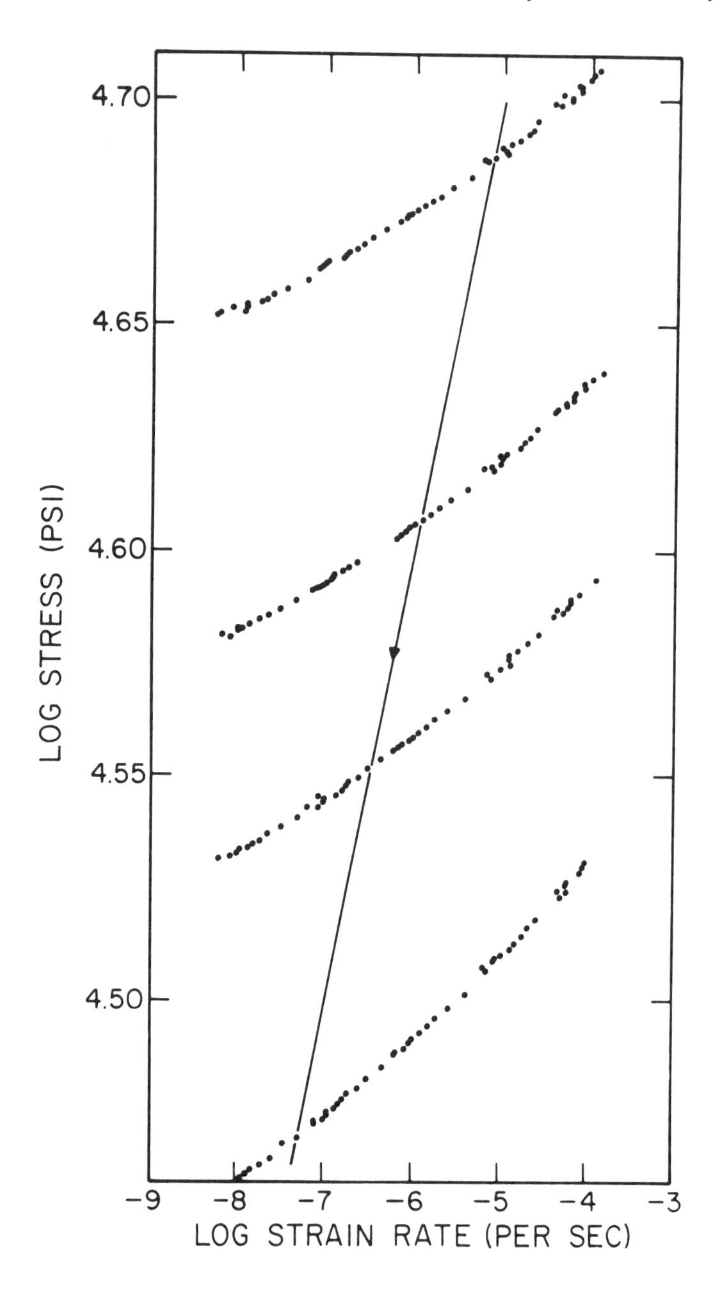

Fig. 5.6 A series of room temperature hardness curves for 304 Stainless Steel (solution treated at 1100°C) after Yamada and Li (1973). The diagonal straight line is the direction of hardness scaling as described in Section 5.5.

A selection of constant hardness curves for high purity Al at room temperature is shown in fig. 5.7. The two curves in fig. 5.8 represent a hardness curve at 270°C for high purity Al (upper curve) and a run made immediately afterward by reloading to a lower initial load level. This pair of curves illustrates how the anelastic component disappears when the anelastic strain reaches equilibrium with the current applied stress and that the hardness curve envelope is then followed. A similar behavior can be exhibited for the same material at room temperature (Hart and Solomon, 1973).

The data for 1100 Al is strikingly similar to that for the high purity Al. A few curves at room temperature (Ellis, Wire, and Li, 1973) are shown in fig. 5.9.

5.4.4 *The Strain Hardening Behavior*

The principal result under this heading is that γ does in fact depend at any single temperature only on the current values of σ and $\dot{\varepsilon}$ and has no further dependence on deformation history. This has been demonstrated by Wire, Ellis, and Li (1973) for 316 Stainless Steel, Niobium, and 1100 Aluminum. The measurements for 1100 Al have been extended to 250°C and the further conclusions from that work will be described in terms of a tentative analysis in Section 5.5.

5.4.5 *Uniqueness of Hardness Curves*

For all tests, the hardness curves obtained at any hardness level were in fact independent of variations in prior history. The only effect of prior history was the level of hardness itself. The hardness curves were always members of the same family. It will be shown in Section 5.5 that it was possible to test very sensitively for the family membership property because of a simple family relationship. Hart and Solomon (1973) showed that even very extensive three-dimensional cold work produced a resulting hardness curve that was the same that would be produced from a monotonic uni-axial extension history.

5.5 ANALYSIS OF THE MEASUREMENTS

In this section we shall show that the data demonstrates the existence of a plastic equation of state for at least some of the materials tested and probably for all of them. Furthermore, we shall show that there are some unexpected simplicities about the explicit form of the equation of state, and that there is a characteristic set of natural units for the deformation variables.

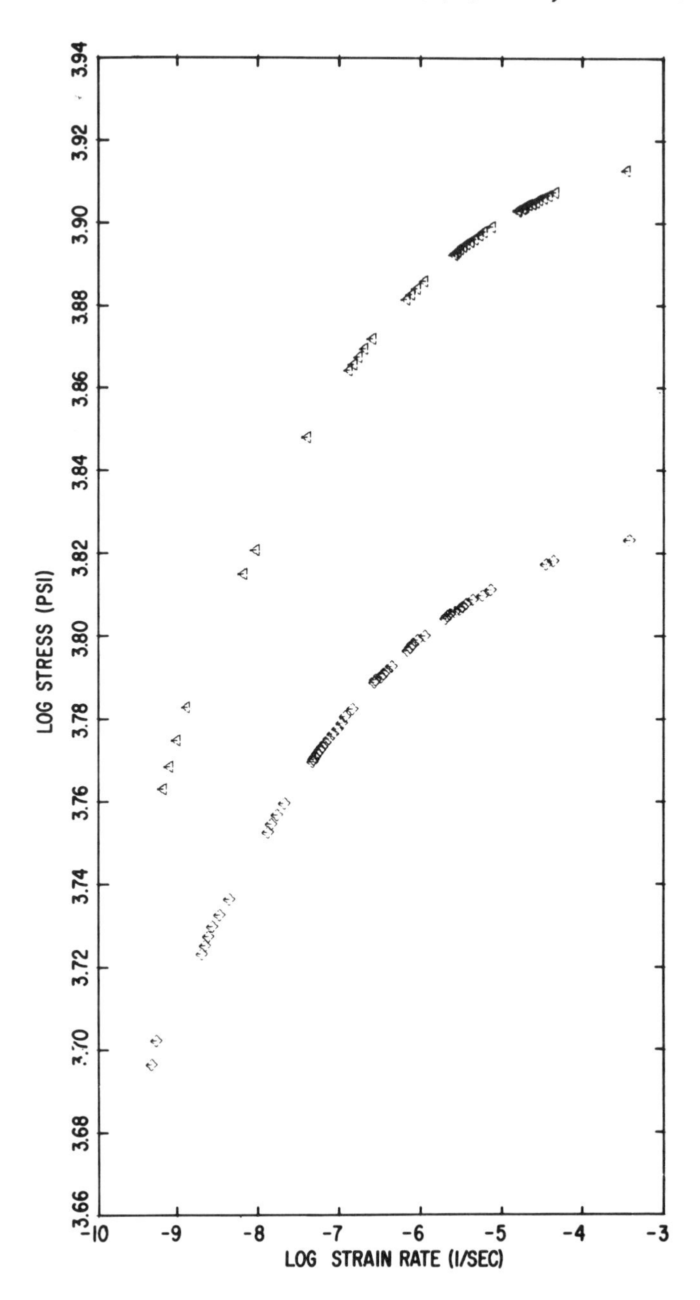

Fig. 5.7 A set of hardness curves at room temperature for high purity Aluminum after Hart and Solomon (1973).

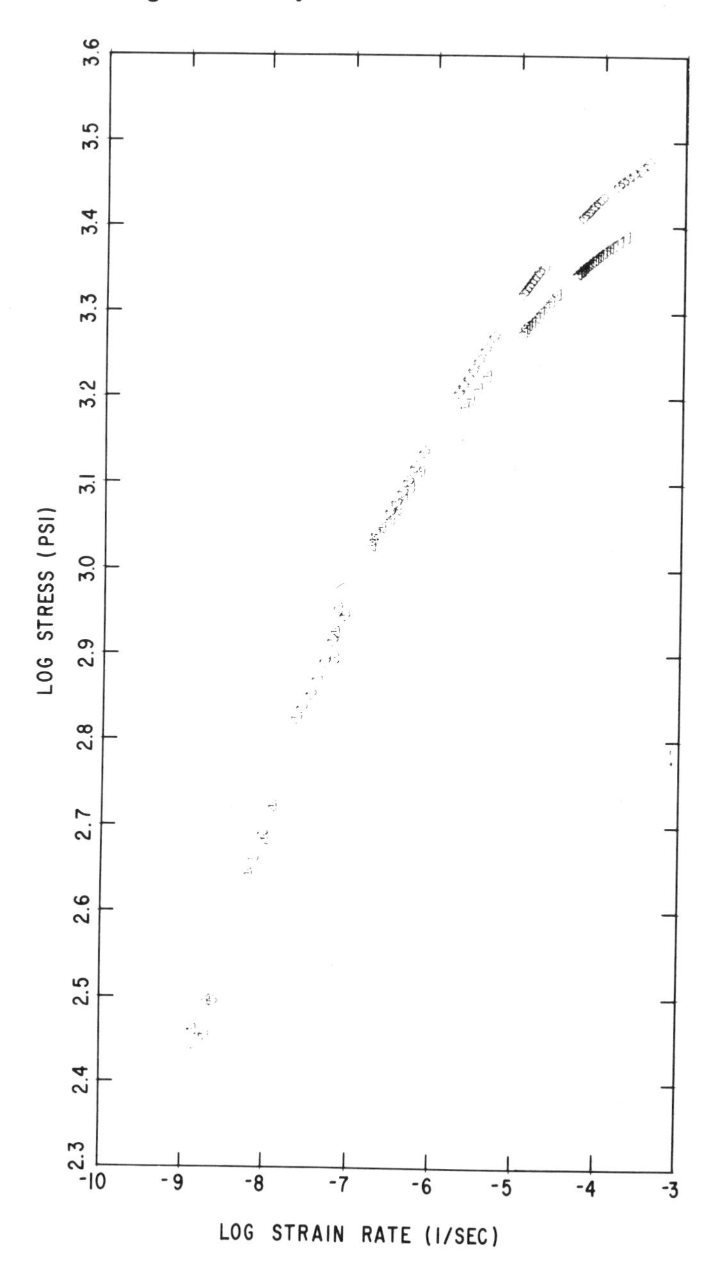

Fig. 5.8 A hardness curve (diamonds) for high purity Aluminum at 270°C and a stress-strain rate curve (triangles)from reloading the same specimen below the previous maximum stress. The reload curve shows the initially high transient strain-rate and the eventual decay of the transient.

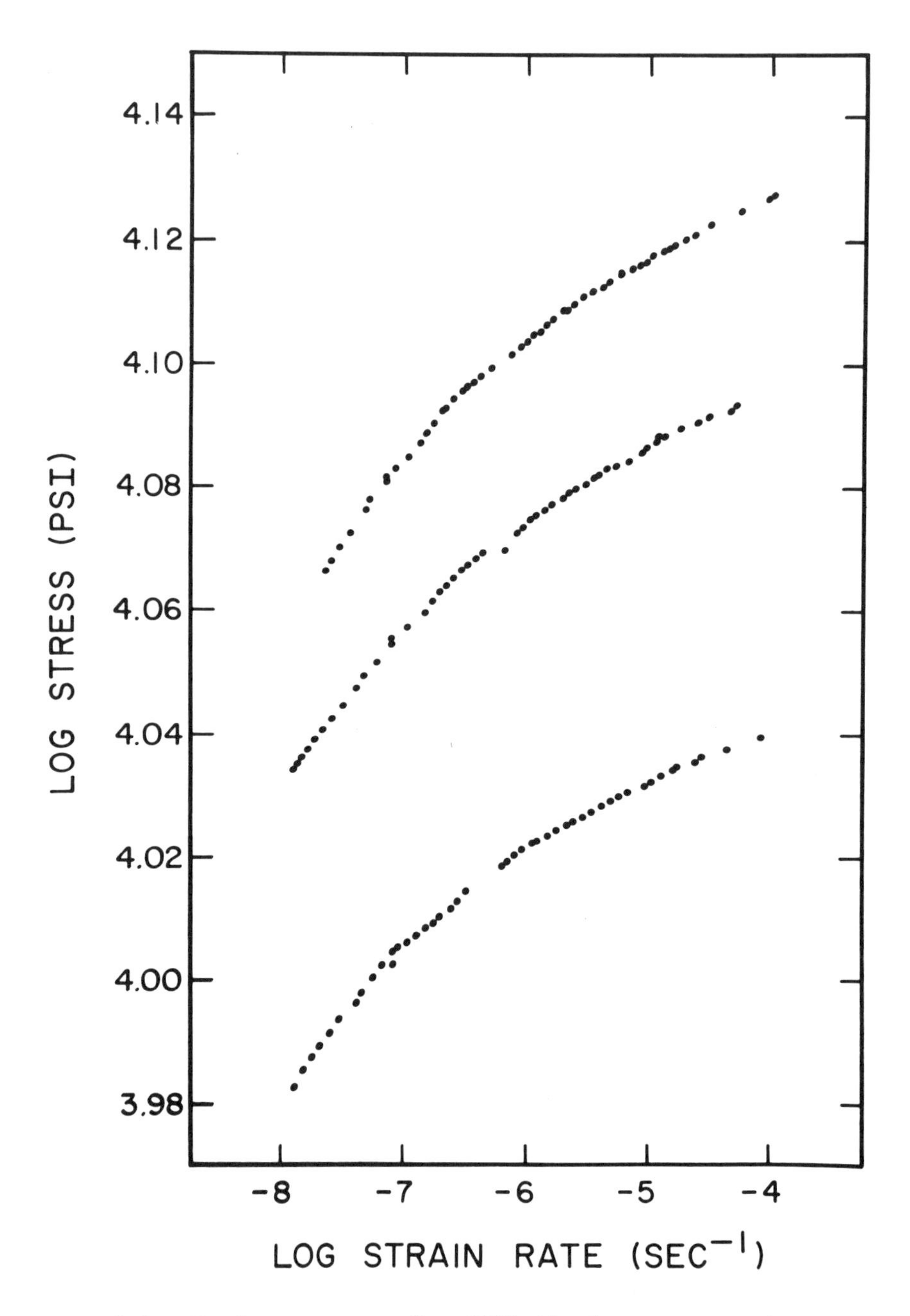

Fig. 5.9 Hardness curves for 1100 Aluminum at room temperature (Ellis, Wire, and Li, 1973).

5.5.1 *The Stress-Strain Rate Scaling of Hardness Curves*

For each material tested at any single temperature, the stress-strain rate curves generated at all levels of hardness did in fact form a single unique family of curves. Verification of this behavior was greatly facilitated by the discovery of a stress-strain rate scaling relationship among curves of various hardness levels. This scaling was first described in detail for the high purity Al by Hart and Solomon (1973) and was found to hold also for Ni, Nb, α-Fe, and two Stainless Steels by Yamada and Li (1973, 1974a, 1974b) and for 1100Al by Ellis, Wire, and Li (1973).

Because of the large range of strain rate investigated for the high purity Al, that data provides the most critical test for the scaling. It was found for those tests that any hardness curve, plotted as $\log\sigma$ vs. $\log\dot{\varepsilon}$ could be translated (without rotation) in a particular direction (oblique to both axes) so that it would coincide with very good precision with the overlapping segment of any other hardness curve to which it was translated. Stated somewhat differently, it appeared that there was a single <u>master hardness curve</u> which could generate all the observed curves from segments of the master curve, and that the observed curves were generated simply by rigid translation of the master curve in a single fixed direction. This is shown schematically in fig. 5.10 for high temperature data like that for Al. In principle, such a master curve represents what any one hardness curve would be like if it could be measured over a very extended range of strain rate. We show, in figs. 5.11 and 5.12, master curves generated by the Nb and 316 Stainless Steel measurements respectively. The master curve for Al is shown in fig. 5.13. In that figure we have also included a curve that corresponds to an analytic curve fit for that data and which we discuss below.

The direction of translation, as shown in fig. 5.10, is specially interesting. It is a line connecting corresponding points of the same value of ν on different hardness curves. Its slope may be written then as

$$\mu \equiv (\delta\log\sigma/\delta\log\dot{\varepsilon})_{\nu} , \qquad (5.19)$$

and we shall further define its reciprocal as

$$m \equiv 1/\mu . \qquad (5.20)$$

The value of m as determined by Hart and Solomon (1973) is $4.6 \pm .25$. This value is remarkably close to the generally

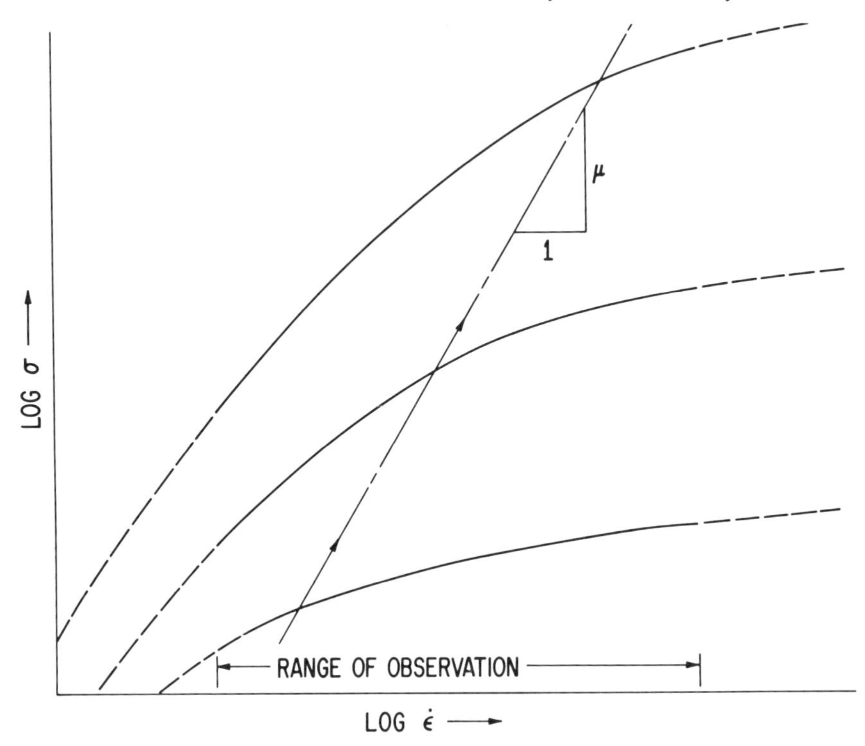

Fig. 5.10 The scaling of stress and strain-rate with change of hardness. A single master curve is shown displaced to several positions in the direction shown by the diagonal line. The observed hardness curves at each level of hardness appear in the strain-rate segment labelled "range of observation" (after Hart and Solomon, 1973)

accepted value of the high temperature creep exponent for Al (Bird, Mukherjee, and Dorn, 1970). We show below that this circumstance may not be accidental. Similar correspondence is observed for the slope m for the translation lines of the other materials discussed here.

5.5.2 *The Influence of Grain Boundary Sliding*

Before we can proceed to a discussion of the analytic fit of the hardness curves or of the effects of temperature change, we must make some comments on the rôle of grain boundary sliding (gbs). This phenomenon is of importance only in the "high" temperature régime.

Since this problem will be treated at much greater length in another publication, we give only a brief account here.

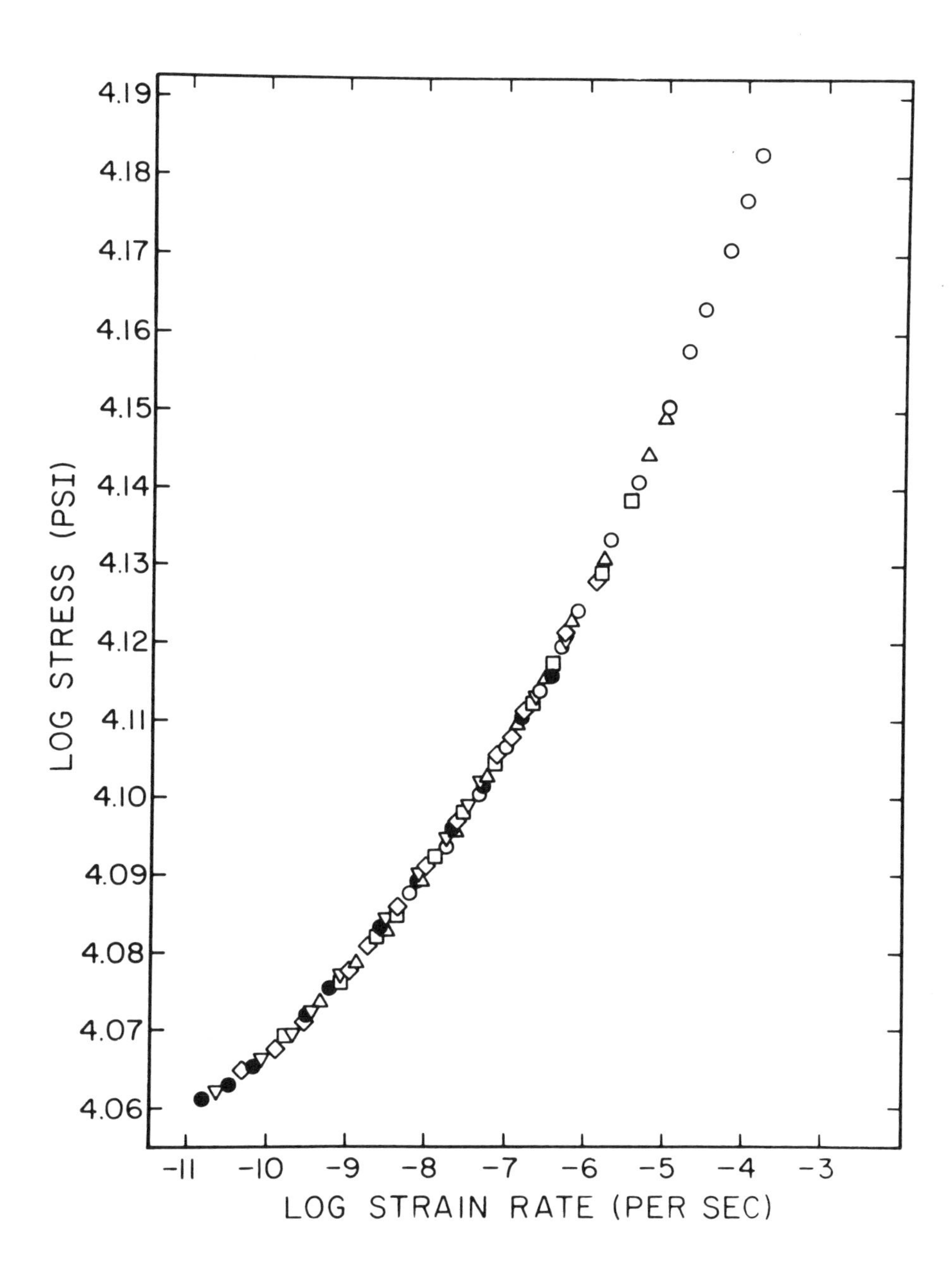

Fig. 5.11 Master hardness curve for high purity Niobium generated from six room temperature curves (Yamada and Li, 1974).

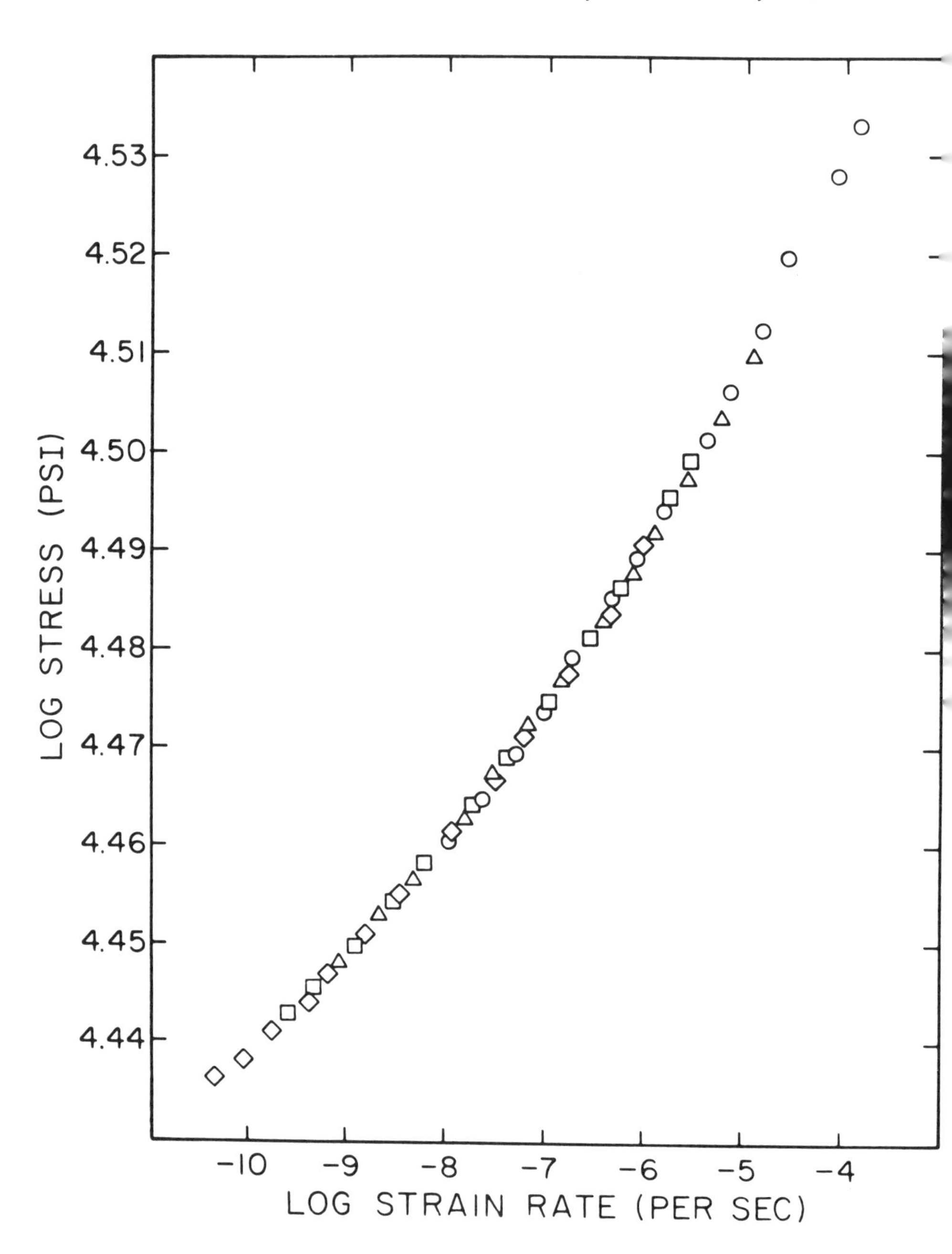

Fig. 5.12 Master hardness curve for 316 Stainless Steel generated from four room temperature curves (Yamada and Li, 1973).

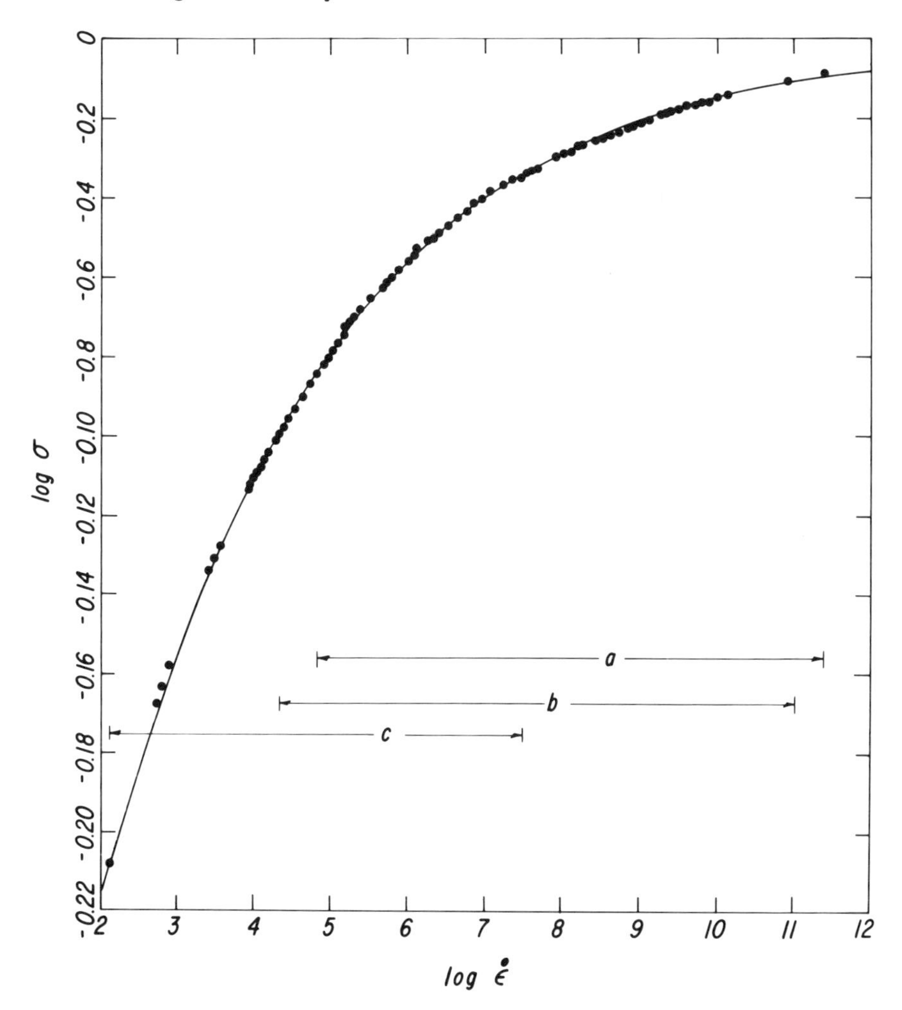

Fig. 5.13 Master hardness curve for high purity Aluminum generated from three room temperature curves. The continuous curve is a plot of the analytic hardness curve of eqn. (5.22) with $\lambda = 0.150$.

Strong evidence for the apparent reduction of creep strength in a variety of polycrystalline Aluminum alloys due to gbs was presented by Servi and Grant (1951), and subsequent studies have shown the generality of gbs in metals above critical temperatures and below certain strain rate levels according to the grain size of the material. Now, our concern with this problem in the present paper is to identify the rôle played by gbs in the mechanical measurements that we make. We wish to be able to distinguish the grain matrix deformation behavior even in the presence of gbs. Ideally, we should be able to do this by supplementing the polycrystal measurements with single crystal experiments. Such work is, in fact, currently in progress, but results are not yet adequate for our purpose. We must rely then on somewhat incomplete theoretical considerations. Fortunately, this approach has been adequate for most of the analysis to date.

By somewhat crude continuum mechanics calculations Hart (1967; 1970b) developed a model for the mechanical behavior of a material (grain matrix) with a non-Newtonian (power law) constitutive law that contained a network of plate-like shear flaws (grain boundaries) that could sustain finite velocity discontinuities across them with Newtonian friction (gbs).

The principal results of the calculation are as follows: Above some sufficiently high strain-rate (dependent on the grain size) the boundaries slide very little compared to the matrix deformation rate. Below some sufficiently low strain-rate (about three decades lower than the upper critical strain-rate) the boundaries are no longer limited by their Newtonian friction and are fully relaxed. The boundary sliding rate is then controlled entirely by the matrix accommodation at the grain corners and therefore reflects the matrix flow law; but the composite appears weaker because it contains a network of shear cracks. This appears, relative to the matrix flow law, as though any typical $\log\sigma$-$\log\dot{\varepsilon}$ point of the matrix flow curve is displaced downward by an amount $\Delta\log\sigma$ (strength loss) and to higher strain-rate by an amount $\Delta\log\dot{\varepsilon}$ (strain-rate enhancement). The theory provides that, if the matrix is Newtonian $(n = 1)$, then $\Delta\log\sigma = \Delta\log\dot{\varepsilon}$, but, as n becomes greater than 1, $\Delta\log\sigma$ rapidly decreases and $\Delta\log\dot{\varepsilon}$ increases. Between the upper and lower critical stresses the flow curve goes through a transition region that reflects both matrix and grain boundary constitutive behavior. It is principally in the transition range that the behavior is sensitive to the grain boundary friction law.

Our measurements of Al and 1100 Al showed mechanical evidence of transition behavior. For the high purity Al this appeared at about 125°C. There was substantial evidence that by at most 200°C the fully relaxed boundary _régime_ was reached, and so that data could again be taken to be characteristic of the matrix flow law with due attention to the $\Delta\log\sigma$-$\Delta\log\dot{\varepsilon}$ shifts noted above. We avoid at present any attempt to analyze the transition _régime_ since there the theory is weakest. Nevertheless, we note that that _régime_ was remarkably consistent with the model.

5.5.3 *The Analytic Form of the Hardness Curve*

The development of a reliable analytic form of the hardness curve was made from the data for high purity Al. It provides a representation only for the "high" temperature data. No attempt has been made to fit the "low" temperature curves since that data is so far known over only moderate strain-rate ranges.

The purpose in such a curve fit was not simply to fit the already extensive range generated at room temperature but to produce a function that might be reliably employed outside of that range. To this end it was desirable to use the curves obtained at higher temperature since these would extend the data to lower strain-rates as temperature compensated strain-rates. Since the gbs transition region had to be avoided, this left a gap in the extended strain-rate scale of unknown width. However, because of the particular mathematical form that the curve satisfies, it was possible to determine the fitting function with very high sensitivity.

The clue was provided by noting that a hardness curve at 270°C (see fig. 5.8), when plotted with a $\log\sigma$ scale that was very compressed compared to the room temperature plots, strongly resembled the room temperature curves as plotted on their scale. In fact it was possible to fit such compressed high temperature curves precisely onto the room temperature master curve. This behavior uniquely identified the larger extended master curve as an exponential of $\log\sigma$ vs. $\log\dot{\varepsilon}$ relative to some point of origin. The exponential e.g. $y = a^{ax}$, is the only curve for which a segment, when compressed (or expanded) in its ordinate, becomes another segment of the same curve at a different abscissa location. Note that this is a purely mathematical test and has nothing to do with the scaling behavior discussed in Section 5.5.1 above. The form of any hardness curve is then found to be given by (we use natural logarithm at this point) (Hart, 1974b)

$$\ln(\sigma^*/\sigma) = \exp[\lambda \ln(\dot{\varepsilon}^*/\dot{\varepsilon})] \quad (5.21)$$

$$= (\dot{\varepsilon}^*/\dot{\varepsilon})^{\lambda} \quad (5.22)$$

where λ is a constant. The peculiar upside down form simply means that the exponential is upside down and right to left. The parameter σ^* can now be taken to be the hardness parameter (the y of eqn. 5.7). Its mate, $\dot{\varepsilon}^*$, is not independent of σ^* but must be related to it because of the scaling law discussed in Section 5.5.1. Thus, we may write, at a single temperature

$$\dot{\varepsilon}^* = A\sigma^{*m} , \tag{5.23}$$

and the constant A is determined if any one pair $(\sigma^*, \dot{\varepsilon}^*)$ is known. Now at any single temperature for which A is known (λ is independent of temperature) all measurable hardness curves can be generated from eqns. (5.23) and (5.22) by varying σ^* .

Because of the special form of the function, the value of σ^* and of $\dot{\varepsilon}^*$ for any measured curve of sufficient extent can be determined rapidly by graphical methods. We do not describe this in detail here because of lack of space.

In fig. 5.13 the data points for three hardness curves are shown translated into congruence, and a plot of eqn. (5.21) is drawn through those points for the parameter value $\lambda = 0.150$. The ordinate and abscissa scales are for the analytic curve with $\log\sigma^* = 0$ and $\log\dot{\varepsilon}^* = 0$. The component hardness curves cover the indicated strain-rate spans labelled a, b, and c.

Note, incidentally, that the fact that the high rate end of curve c fits smoothly along the lower rate ends of the other curves is further evidence for the absence of a transient contribution in curve c . The same comment applies for the curves in the master curves of figs. 5.11 and 5.12 for the low homologous temperature data.

It will be noted that eqn. (5.22) predicts that $\dot{\varepsilon} \to \infty$ as σ approaches σ^* . In fact as σ approaches σ^* the equation breaks down, for the hardness curve will flex upward as in fig. 5.4. This fact is not evident in the high temperature data, however, until σ gets very close to σ^* . The stress σ^* looks then, from the high temperature (low strain rate) viewpoint, as a prospective limiting stress. It is likely that it can be put into one-to-one correspondence with conventional low temperature yield point concepts and possibly even with a suitably controlled hardness indenter index. It is very appropriate and natural then to select σ^* as a hardness parameter. It can now be seen why we used quotation marks with "low" and "high" temperature identifications. The real distinction is that σ , for values of $\dot{\varepsilon}$ that are usual, is either greater than or less than σ^* . The horizontal dashed line in fig. 5.4

is, of course, supposed to be σ^* .

Finally, we note that the data for 1100 Al fits the same type of function as the high purity Al.

5.5.4 *The Effect of Temperature*

The influence of the temperature of testing appears in two ways in the plastic equation of state. It appears in the temperature dependence of the rigidity modulus, which acts as the natural unit for the stress, and it appears in an Arrhenius factor that accounts for thermal activation. These two effects of temperature have been recognized for some time in the literature. They are introduced in the current treatment in a simple way.

The measurements of Sherby, Lytton, and Dorn (1957) found no influence of either stress level or state of strain hardening on the activation energy for flow in Al. The activation energy Q , however, did depend on temperature. In the treatment presented here we can incorporate all temperature effects conveniently in a final form for the auxiliary state parameter $\dot{\varepsilon}^*$. Thus

$$\dot{\varepsilon}^* = (\sigma^*/G)^m f \exp(-Z/RT) , \tag{5.24}$$

where Z is a general activation free energy that incorporates all pre-exponential factors other than those that are explicit, and Z depends only on temperature and the nature of the material. The prefactor f is any arbitrary fixed unit of frequency.

Our experiments verify that, if load relaxation tests are run at two different temperatures for a specimen at constant hardness, σ^*/G is in fact the same at both temperatures.

The differential form of eqn. (5.24) leads to an expression for the familiar incremental activation coefficient Q .

$$d\ln\dot{\varepsilon}^* = m\, d\ln(\sigma^*/G) - d(Z/RT) , \quad \text{and,} \tag{5.25}$$

$$d(Z/RT) \equiv Q\, d(1/RT) , \tag{5.26}$$

$$= [Z - dZ/d\ln RT]d(1/RT) . \tag{5.27}$$

By means of the equation of state it is possible, in principle, to measure Z as a function of T by one relaxation test at each value of T . For each hardness curve generated the value of σ^* and $\dot{\varepsilon}^*$ can be determined by fitting to eqn. (5.22). Then, once the value of m has been determined, Z can be obtained from eqn. (5.24) for each pair of $\sigma^*, \dot{\varepsilon}^*$. If more than one hardness curve is generated at each temperature the accuracy of determination of Z is sharpened.

This program has not been systematically carried out yet for the materials described here because of the grain boundary sliding problem. However, methods of dealing with that problem will be found in due course.

We comment only briefly on the subject of thermal recovery of strain hardening. A natural way to do this is to determine how σ^* changes with time due to exposure at various temperatures. In these tests σ^* can be measured by relaxation tests at some convenient reference temperature after annealing at any elevated temperature.

In the tests with 1100 Al no recovery was observed for any level of hardness at temperatures as high as 250°C. Recovery was observed for the high purity Al at 270°C only after considerable prestrain at room temperature. At lower hardness levels σ^* was stable even at 270°C. No recovery was observed from any hardness level for the high purity material at 125°C. Systematic study of the high temperature recovery rate was not pursued because of the unavoidably long times required to change temperature with apparatus designed primarily for temperature stability.

5.5.5 *The Strain Hardening Rate as a State Property*

Strain hardening measurements have been carried out over a range of temperatures and strain-rates for 1100 Al and in a more restricted range for Nb and 316 Stainless Steel. These measurements and their analysis have not yet been published, and only a brief account of them will be given here.

The principal result is that for all the materials the strain hardening coefficient γ is in fact a function only of σ and $\dot{\varepsilon}$ at each temperature. In other words, γ is a state property. The data for 1100 Al from room temperature to 250°C were analyzed in the following way:

The measured values of γ were converted to the absolute hardening rate Γ by formulas shown in the following section, and these were determined as functions of σ^* and σ. Next, plots of constant Γ were made for σ^* vs. σ/σ^*. Such a set of plots for 1100 Al is shown as fig. 5.14. The experimental scatter for these points is fairly large, but a tentative fit is given by

$$\Gamma \equiv d\ln\sigma^*/d\varepsilon \ , \tag{5.11}$$

$$= C\ (\sigma/G)^{7.8}/(\sigma^*/G)^{12.5} \ , \tag{5.28}$$

where C is a constant.

The striking feature of this result is that the difference of the exponents is equal to the value of m to within the experimental error. The implication is that, if the strain increment $d\varepsilon$ at any value of σ^* is measured in units of a characteristic strain proportional to $(\sigma^*/G)^m$ the resultant <u>normalized absolute hardening rate</u> Γ^* is certainly a function only of σ/σ^*. Thus

$$\Gamma^* \equiv (\sigma^*/G)^m\ \Gamma \ , \tag{5.29}$$

$$= \Gamma^*(\sigma/\sigma^*) \ , \tag{5.30}$$

and, if the power law fit of eqn. (5.28) proves to be at all reliable,

$$\Gamma^* = C\ (\sigma/\sigma^*)^k \ , \tag{5.31}$$

where k is a constant.

The importance of this circumstance is that it confirms, what already seemed apparent, that the hardness scaling of $\dot{\varepsilon}^*$ (eqn. 5.24) was concerned with the elementary unit of strain that was associated with each elementary slip process rather than with the time rate of that process.

5.5.6 *Summary of the Equation of State*

Before proceeding to further discussion we shall summarize the form of the plastic equation of state for the high homologous temperature <u>régime</u> as it applies to Al and 100 Al.

In all the following the state of hardness of the material is completely specified by the single state parameter σ^* (or, better, σ^*/G). The relationships that have been developed express the dependence of $\dot{\varepsilon}$ on σ, T, and σ^* and

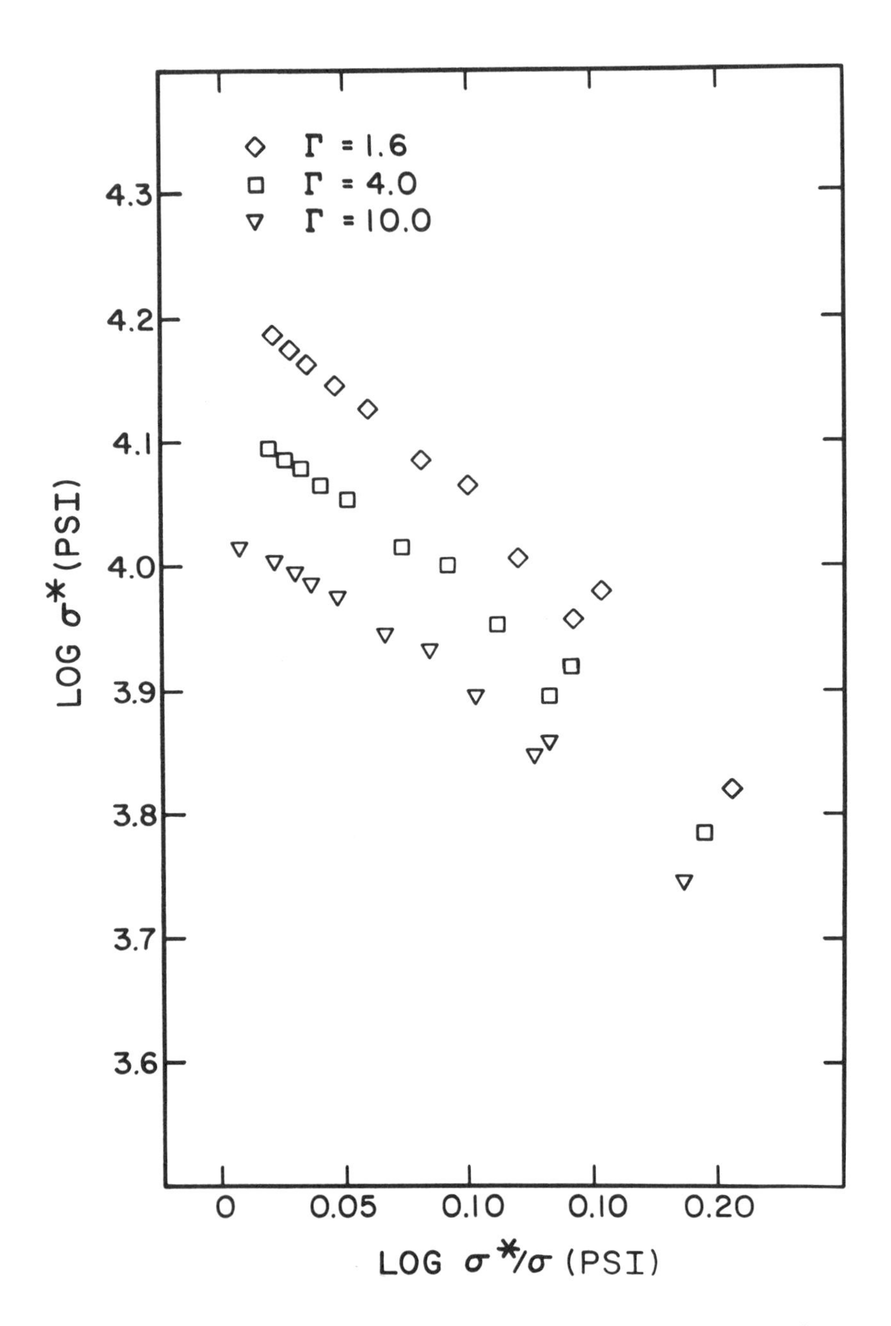

Fig. 5.14 Plots for 1100 Aluminum of points corresponding to three values of Γ . The points are for 25°C, 150°C, 200°C, and 250°C and for values of $\log\dot{\varepsilon}$ from -2.6 to -5.2.

describe how σ^* changes with deformation history. The incremental observable phenomenological parameters ν and γ are uniquely determined by σ and σ^*, and Q (or its integral form Z), is a function only of T.

The equations are:

(a) equation of state-

$$\ln(\sigma^*/\sigma) = (\dot{\varepsilon}^*/\dot{\varepsilon})^\lambda . \tag{5.22}$$

(b) stress and temperature scaling-

$$\dot{\varepsilon}^* = (\sigma^*/G)^m f \exp(-Z/RT) . \tag{5.24}$$

(c) strain hardening-

$$\Gamma \equiv d\ln\sigma^*/d\varepsilon , \tag{5.11}$$

$$= C (\sigma/\sigma^*)^k (\sigma^*/G)^{-m} \tag{5.32}$$

(d) normalized strain hardening-

$$\Gamma^* \equiv (\sigma^*/G)^m \Gamma , \tag{5.29}$$

$$= C (\sigma/\sigma^*)^k \tag{5.31}$$

(e) temperature activation-

$$Z = Z(T) \tag{5.33}$$

From these the values of the phenomenological coefficients can be expressed in terms of σ, σ^*, and T. Thus

$$Q = Z - dZ/d\ln RT , \tag{5.34}$$

$$\nu = \lambda \ln(\sigma^*/\sigma) \tag{5.35}$$

$$\gamma = (1 - m\nu)\Gamma . \tag{5.36}$$

In all these equations λ, m, k, and C are constants, and f is an arbitrary coefficient with dimensions of frequency. Note incidentally, that the final equation relating γ

and Γ gives an explicit expression for the integrating factor F discussed in connection with eqn. (5.9).

Of course, ν and γ can be expressed explicitly in terms of $\dot{\varepsilon}$, σ , and T by eliminating σ^* among eqns. (35, 36, 22 and 24). This cannot be done in simple analytic terms, and so it is better practically to carry out any calculations in terms of the state variable σ^* .

5.5.7 *Three-Dimensional Plastic Constitutive Relations*

The relationships that we have described for uniaxial behavior can be expected to hold for triaxial configurations in terms somewhat similar to those at present employed for theories of incremental plasticity. However, this has not yet been verified by explicit tests, and so we can present only tentative formulation at present.

Let us first consider the problem of isotropy. We are not concerned here with anisotropy that results from crystallographic texture. That is a problem that requires a separate solution. Our concern is with anisotropy that might result from prior deformation alone (still exclusive of real texture development). We believe that such anisotropy of flow relationships are primarily associated with the transient, recoverable strain a . We shall therefore ignore anisotropy in presenting a three-dimensional form for the plastic component $\dot{\varepsilon}$. Of course, a complete constitutive law for the inelastic strain will have to include the anelastic strain rate as well.

We shall use the symbol σ_{ij} for the deviator of the stress tensor τ_{ij} , and designate the strain rate tensor by $\dot{\varepsilon}_{ij}$. Thus

$$\sigma_{ij} \equiv \tau_{ij} - (1/3)\, \tau_{kk}\, \delta_{ij} , \tag{5.37}$$

where a repeated index implies summation over its three values and δ_{ij} is the familiar Kronecker delta. Then

$$\sigma_{kk} = 0 , \tag{5.38}$$

and, of course,

$$\dot{\varepsilon}_{kk} = 0 \tag{5.39}$$

We prescribe the general isotropic relationship between σ_{ij}

and $\dot{\varepsilon}_{ij}$ of the form

$$\dot{\varepsilon}_{ij} = (\dot{\varepsilon}/\sigma)\,\sigma_{ij} \,, \tag{5.40}$$

where σ and $\dot{\varepsilon}$ are the tensor invariants given by

$$\sigma \equiv +\sqrt{\sigma_{ij}\,\sigma_{ij}} \,, \quad \text{and} \,, \tag{5.41}$$

$$\dot{\varepsilon} \equiv +\sqrt{\dot{\varepsilon}_{ij}\,\dot{\varepsilon}_{ij}} \,. \tag{5.42}$$

We further define

$$d\varepsilon \equiv \dot{\varepsilon}\,dt \,. \tag{5.43}$$

It is easy to see now that all the equations considered prior to this section may be considered to be equations relating the stress, strain-rate, and strain increment invariants of eqns. (5.41-43).

The quantities ν and q as measured in uniaxial behavior will retain their same values when the invariant formulation is employed. The quantity γ in the equations for $\sigma, \dot{\varepsilon}$, and $d\varepsilon$ invariants will have a value equal to $\sqrt{2/3}$ times the γ measured in uniaxial terms. These connections follow from our choice of normalization for σ and $\dot{\varepsilon}$.

5.6 IMPLICATIONS FOR CREEP BEHAVIOR

The form of the plastic equation of state as demonstrated for Aluminum has some striking implications for high temperature creep.

We should emphasize once more that upon initially loading a creep specimen there will generally be a contribution to the early creep strain-rate from the anelastic component a. This will account for a considerable amount of what is generally termed the primary creep. Much of that strain is recoverable as <u>creep strain recovery</u> upon unloading the specimen. This is discussed in detail by Lubahn and Felgar (1961) who show that much of the anelastic creep can be identified and subtracted off from the total creep strain leaving substantially the component of strain resulting from what we have called here the plastic (unrecoverable) strain. It is only that component that is described by our equation of state, and that is of principal concern in most fundamental high temperature creep study.

Now, at any point in a constant stress creep history after the anelastic strain-rate $\dot{a}$ has fallen to zero, the rate of change of $\dot{\varepsilon}$ with respect to increments of strain will be described by the phenomenological coefficient g. We can express the current value of g in terms of the explicit form for γ given by eqn. (5.35). Thus

$$g \equiv \gamma/\nu , \quad (5.5)$$

$$= n\gamma , \quad (5.44)$$

$$= (n-m)\,\Gamma . \quad (5.45)$$

From eqn. (5.32) it is evident that Γ is always positive. Then the sign of g (as well as of γ) is determined by the sign of the coefficient $n-m$. <u>At sufficiently low values of σ/σ^* that coefficient in fact becomes negative</u>. This can be seen as follows:

From eqn. (5.35) in reciprocal form (where $l \equiv 1/\lambda$)

$$n = l/\ln(\sigma^*/\sigma) . \quad (5.46)$$

For the high purity Al, $l = 6.74$ and $m = 4.5$ within experimental error. And, as a bit of speculative numerology, we shall write $l = (3/2)m$. Then

$$n = (3/2)m/\ln(\sigma^*/\sigma) , \quad \text{and} \quad (5.47)$$

$$n-m = m[(3/2)/\ln(\sigma^*/\sigma) - 1] . \quad (5.48)$$

Thus, $n-m \leq 0$ when $\ln(\sigma^*/\sigma) \geq 3/2$ or when $\sigma^*/\sigma \geq 4.48$.

Stated slightly differently, $n = m$ when $\ln(\sigma^*/\sigma) = 1.5$ or when $\log(\sigma^*/\sigma) = 0.65$. This effect is directly exhibited by the 270°C curve for Al shown in fig. 5.8. For that constant hardness curve $\log\sigma^*$ is about 3.70 and at $\log\sigma$ of about 3.05 the slope of the hardness curve ν is the same as μ (or $n = m$), and for lower σ the slope is steeper, i.e. $\nu > \mu$ or $n < m$.

The phenomenon we are discussing can be displayed graphically as in fig. 5.15. In that figure we have displayed curves for three successive values of hardness. We show the portion of these hardness curves at values of σ that are in the range of

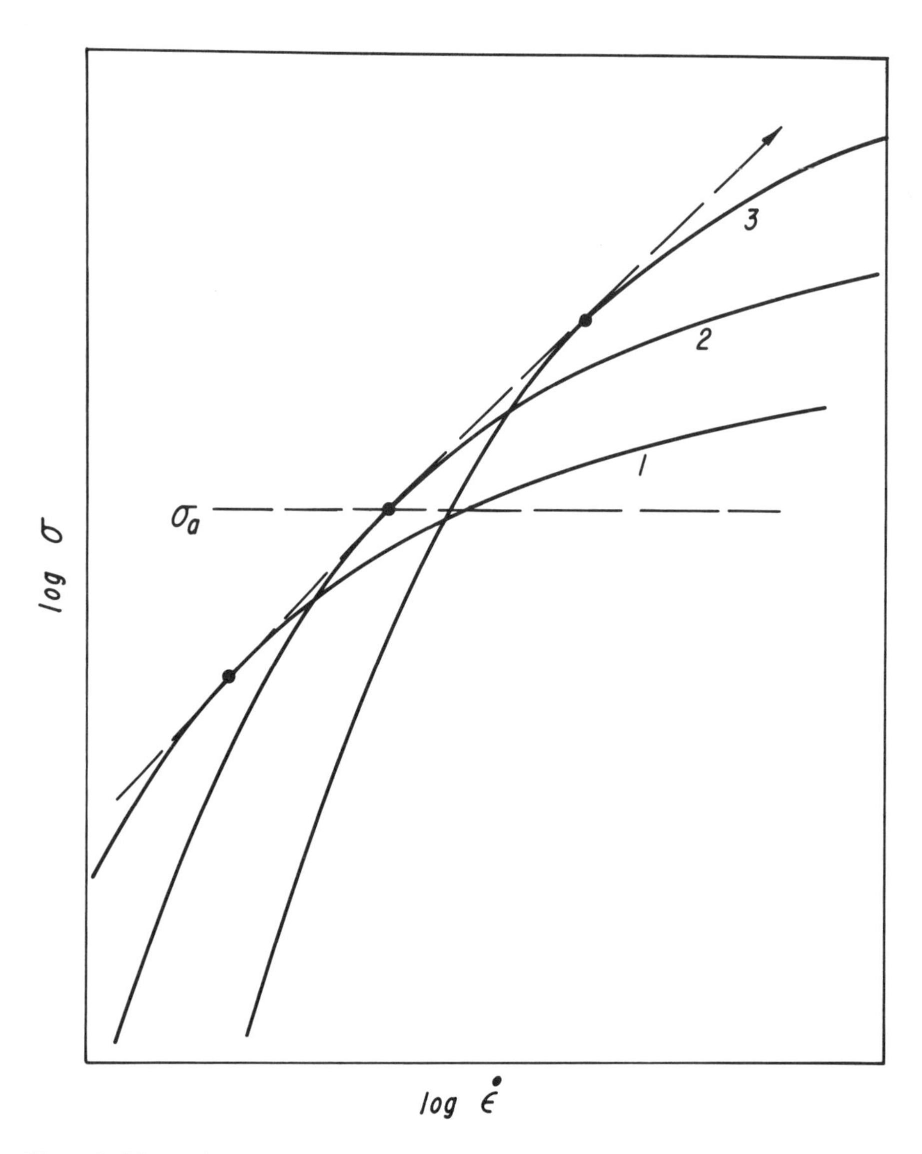

Fig. 5.15 Three hardness curves shown at low stress values. The diagonal line represents a scaling translation line that is, in this stress range, tangent to each hardness curve. The horizontal dashed line, labelled σ_a, represents the stress for a discussion of creep, in Section 5.6.

one-fifth to one-quarter of σ^*. In this range the slope of each hardness curve is comparable to the slope of the scaling translation line. In fact, the dashed line shown tangent to each curve has the slope of the scaling direction. Because of the σ-$\dot{\varepsilon}$ scaling relation upon hardening, any two curves must intersect in one point. For each state of hardness, γ is positive for stresses above the point of tangency and negative below it. This condition holds even though Γ is always positive and would hold even if the absolute hardening rate Γ were constant and independent of stress or hardness. The effect can be directly visualized on the plot. If the three successive hardness states shown are designated 1,2, and 3 in order of increasing hardness, and if, in creep, the applied stress is σ_a as shown, then, as the material hardens continuously from state 1 to 3, the value of $\dot{\varepsilon}$ will <u>decrease</u> until hardness state 2 is reached and will thereafter <u>increase</u>. A similar behavior (with "increase" and "decrease" exchanged) will be observed for σ at constant $\dot{\varepsilon}$.

This brings us to a remarkable result. <u>A creep test at a value of σ that exceeds about 0.22 times the initial value of σ^* for the virgin specimen will display a strain rate minimum even in the complete absence of any plastic inhomogeneity (necking) or recovery of hardness</u>. Furthermore, the locus of all such minimum strain-rates at a variety of stresses will be a line with the slope m.

Of course, just such behavior has been observed for many years of testing, but it is generally assumed that the strain-rate increase after the minimum is due to necking, and the minimum strain-rate is commonly termed a steady state strain-rate. Our results show that a quite different characterization is equally consistent with the creep data.

The form of the equation of state has a further consequence that seems to be supported by the creep data. If the creep test is carried out at stress levels below about 0.22 times the initial value of σ^*, the predicted plastic strain-rate increases continuously from the beginning. However, the initially large anelastic strain-rate will decrease rapidly and reach a value of zero in less than 1% of strain. The total observed strain-rate will then go through a minimum at a quite small value of creep strain. The plastic strain-rate at that point will still be practically that appropriate to the starting value of σ^* for the specimen and so a locus of such pseudo-minimum strain-rates will actually conform fairly well with the low stress end of the initial hardness curve. At low enough stresses (about $10^{-2}\sigma^{*}$) the slope of that curve reaches a value of n about unity and will remain close to that value for about another decade of stress change. Thus the equation of state predicts both the form of creep curve typical of very low stress levels and also a region of behavior for the strain-rate minimum like that which is termed Harper-Dorn creep (Harper and Dorn, 1957).

To summarize, the results of our investigations show that a strain-rate minimum is predicted for creep above a critical stress level, and, at low enough stress levels, Harper-Dorn creep is required. The creep predictions are foreshadowed by the room temperature behavior, and no special change of mechanism is required at higher temperature.

5.7 THEORETICAL IMPLICATIONS

The explicit form of the plastic equation of state that has been demonstrated experimentally in the framework of the phenomenological theory has strong implications for the development of a micro-structural theory of plastic deformation. No elementary theory can be considered adequate unless it can account for at least the following principal features of the observed deformation phenomenology:

(a) The existence of the unique hardness states.

(b) The one-parameter dependence of the hardness.

(c) Stress-strain rate scaling of the states.

These features are fundamental and concern the most general characteristics of the equation of state at both "high" and "low" temperatures. Further features that might be unique to the "high" temperature deformation or that are more detailed are as follows:

(d) The explicit form of the "high" temperature master hardness curve.

(e) The magnitude and character of the transient strain.

(f) The strain hardening behavior with its characteristic strain unit.

(g) Dependence of all parameters on crystal structure and metallurgical microstructure.

It is important to recognize that the high temperature creep behavior is completely consistent with the moderate temperature behavior and does not indicate any special high temperature mechanisms. Any static recovery of hardness by thermal annealing still preserves the state property of the deformation, and, while recovery is certainly present at sufficiently high temperatures and hardness levels, most of the phenomena are explicable even without consideration of thermal recovery. Probably a more complete form of the phenomenological theory should include a purely time-temperature dependent recovery term. This could take the form of a modification of eqn. (5.9) of the type

$$d\ln\sigma^* = \Gamma\, d\varepsilon - \mathcal{R}\, dt , \tag{5.49}$$

where Γ is extrapolated from temperatures where $\mathcal{R}$ is negligible. It would be expected that $\mathcal{R}$ would depend principally on T and σ^*.

5.7.1 *Some General Indications*

We believe that the results to date indicate that a distinction must be made between the mobile dislocations that produce the observed strain-rate and the dislocations that form the obstacle network (forest dislocations) and that limit the motion of the mobile dislocations. This view is consistent with many earlier theoretical ideas.

The most likely origin for the single parameter σ^* that measures the hardness is some significant moment of the spatial distribution of the obstacle dislocations. In simplest form this would be some measure of the mean planar spacing of the obstacles ℓ. (We use the script ℓ to distinguish this symbol from the parameter *1* in the equation of state.) We need not decide exactly what measure of mean spacing we use since all such measures can differ from any other only by numerical constant factors. The stress σ^* is then the value of applied stress σ for which most mobile dislocations can mechanically pass the obstacle structure. This relates σ^* to the mechanical threshold stress $\hat{\tau}$ of slip planes used by Kocks in Chapter 3. Note that σ^* which is defined entirely by a "high" temperature phenomenological observation implies $\hat{\tau}$ which is a property at 0^oK.

The uniqueness and scale covariance of the hardness master curve implies that in some sense all states of obstacle structure are similar and differ from each other only by their value of ℓ.

The very existence of the unique hardness states implies that for any structure given by ℓ the mobile dislocation distribution is uniquely determined at each value of σ by some self-consistent statistical relationship. In other words, at least for $\sigma < \sigma^*$, the mobile dislocation distribution is in a kind of equilibrium with the obstacle structure at each value of σ. This corresponds to the existence at each stress σ of an effective internal strain throughout the material. It is probably the change of that mobile distribution and of the resulting internal strain that is the source of the transient (anelastic) strain that results from a change of σ. An explicit model for this type of behavior was discussed by Hart and Solomon (1973). When $\sigma < \sigma^*$, the probable rate determining process is the activation of blocked dislocations through the principal obstacles. There is then adequate time to form equilibrium dislocation pile-ups behind such blocked dislocations. Those pile-ups would be essentially in mechanical equilibrium because at steady state (at constant ℓ) the intrinsic dislocation friction is not limiting. The effects of such friction and of other interactions would appear as rate controlling mechanism during relaxation from one equilibrium mobile array at one value of σ to another at a changed value of σ.

When $\sigma > \sigma^*$ one expects the obstacles to no longer be limiting (although they might contribute to the effective friction force seen by each mobile dislocation) and now the mobile distribution and its rate of flux is primarily determined by individual dislocation friction and by some mutual stress interaction. This is the régime in which the ideas of "dislocation dynamics" are probably pertinent (see discussion by Hart, 1974a).

It is a general picture of this type that has been behind some of our phenomenological considerations. It is important, however, that the phenomenological theory and the experimental approach stand independent of such heuristics.

5.7.2 *A Model with* σ-$\dot{\varepsilon}$ *Scaling*

We now consider a model of the type discussed by Hart and Solomon (1973) that exhibits stress-strain rate scaling with hardness. Our only purpose in this is to show that such scaling can arise in a natural way. This is not a theory for our equation of state but only a feature of a proper theory. The model does not, in fact, agree quantitatively with our results because it is too simple. We formulate the model in terms of a single slip system and in terms of a planar projection containing the slip direction.

Let the plane of projection contain a distribution of free slip zones terminated at each end by obstacles (Hart and Solomon, 1973). In the absence of mutual interaction among the zones, the stress τ on the lead dislocations at each zone end under the action of the applied shear stress (we shall still call it σ) and of a symmetrical dislocation pile up within each zone (a fully relaxed slip zone) is given in continuum approximation by

$$\tau = 2(1 - \nu)(\ell/b)(\sigma^2/G) , \tag{5.50}$$

where ν is Poisson's ratio and b is the dislocation Burgers vector. We shall employ the same measure ℓ for every linear dimension of the obstacle and slip zone distribution although they might differ from each other by constant geometrical factors. The important thing is that we must exploit the ubiquity of ℓ that is required by feature (b) above.

The force on the lead dislocation is

$$\tau b = 2(1 - \nu) \ell\sigma (\sigma/G) , \tag{5.51}$$

which is independent of b . We assume now that the time rate r of barrier penetration by the lead dislocation depends on the ratio of τb to a critical force $\tau^* b$ that is determined by ℓ as

$$\tau^* b = Gb^2/\ell . \tag{5.52}$$

Note that we again employ ℓ as a measure of obstacle strength. Then

$$\tau/\tau^* = 2(1 - \nu)(\sigma/G)^2(\ell/b)^2 . \tag{5.53}$$

We shall define σ^* to be the value of σ when $\tau/\tau^* = 1$, and so

$$\sigma^* = [2(1 - \nu)]^{-1/2} Gb/\ell . \tag{5.54}$$

Now each time a dislocation leaks through a barrier and annihilates with a nearby one of opposite sign a new dislocation must sweep through the zone to restore the equilibrium pile-up. This is similar to the strain producing step in Weertman's climb control theory (Weertman, 1957). The strain produced in each such event is the slip moment ℓb times the density of slip zones per unit area which we take to be ℓ^{-2} . Thus the mean strain-rate $\dot{\varepsilon}$ is

$$\dot{\varepsilon} = r \cdot b\ell/\ell^2 , \tag{5.55}$$

$$= r \cdot (b/\ell) . \tag{5.56}$$

Now, since r has been assumed to depend only on τ/τ^* and T , it can be taken to depend on σ/σ^* and T . The factor b/ℓ in eqn. (5.56) can be replaced by σ^*/G (eqn. 5.54), and we shall ignore the constant factor. Then

$$\dot{\varepsilon} = (\sigma^*/G) \cdot r(\sigma/\sigma^*,T) , \tag{5.57}$$

and we have produced a strain rate-stress relationship for which both $\dot{\varepsilon}$ and σ scale according to the hardness parameter σ^* . For this very simple model $m = 1$. The form of the master hardness curve depends on the functional form of r .

5.8 CONCLUSIONS AND PROSPECTS

We have tried to show in this Chapter that a consistent phenomenological theory can lead to novel experimental results and to significant new characterizations of those results. The strength of the method is underlined by the circumstance that it facilitated the discovery of the hardness scaling law although the scaling behavior was not an intrinsic part of the theory itself.

Since the principal results of the investigations to date have been summarized in Section 5.5, the present section will be devoted principally to a discussion of the problems that remain.

5.8.1 *The Transient Strain Phenomena*

The strain that we have termed anelastic loading strain requires investigation as a behavior that is in some sense distinct from the unrecoverable plastic strain. Most prior considerations of transient behavior have treated the phenomena as evidence for thermal recovery of strain hardening. Our results show that such an interpretation is unlikely. The problem would be considerably simpler if the transient strain were directly recoverable upon unloading a specimen. However, it has been reported by Hart and Solomon (1973) and by Matlock, Harrington and Nix (1972) that no more than 10% of the transient strain is recoverable by simply releasing the applied stress. Hart and Solomon (1973) discussed possible reasons for this assymetry in connection with their model which is similar to the one we have described in Section 5.7 above. They showed that some loss of the stored strain a could be attributed to continued leakage of lead dislocations through the zone end barriers, especially if some appreciable applied stress remains. Such leakage would convert that part of a to unrecoverable plastic strain. Thus, under small finite stress drops, a can reduce to the equilibrium value at each lower stress by the same process responsible for $\dot{\varepsilon}$ itself. But such reduction of a is clearly much too slow at zero applied stress. It is likely, then, that it is the remaining stored strain that is responsible for the Bauschinger effect. This conclusion is supported by the observations of Buckley and Entwistle (1956) of Bauschinger strain in Al. They observed a linear relationship between the Bauschinger strain and the prior forward stress. The magnitude of such strain is of the order of ten times the elastic strain which agrees with the magnitude of the transient. They discussed a model for the effect that is not too different from ours.

It is clear that this problem must be pursued further both experimentally and theoretically.

5.8.2 *The Grain Boundary Sliding*

This is more a problem of the solid mechanics of a matrix with planar inhomogeneities. It is probable that theoretical investigation will play a major rôle in its solution. Such theoretical results must lead, however, to reliable criteria for analyzing the experimental behavior.

5.8.3 *The "Low" Temperature Regime*

There is at present too little data available to determine the explicit character of the equation of state for values of σ greater than σ^*. Although it seems so far that the slope of the scaling direction μ is the same, for any one material, in both temperature régimes, this must be verified in greater detail. Experimental work is needed over a broad range of temperature that includes the low temperature régime. This is probably done most easily for a material for which room temperature corresponds to a very low homologous temperature.

5.8.4 *More Complex Alloys*

It is not clear that the behavior we have found to be so regular for relatively simple systems will be characteristic of more complex multi-phase systems as well. The answer to this question will rest with the results of further experimental work, but at present we are inclined to be optimistic about the more extended applicability of the general methods. It would certainly be desirable to characterize the influence of metallurgical microstructure on deformation in terms of well-defined parameters of the sort that have been found so far for the simpler systems.

5.8.5 *Explicit Time Dependent Phenomena*

Little work has been done thus far in the measurement of the kinetics of recovery in terms of the state variables. We discussed this briefly in Section 5.7 where it was pointed out that it seemed that the equation of state would provide a good reference framework for the quantitative description of static recovery. A further kinetic phenomenon of considerable importance is solute strain aging.

5.8.6 *The Constitutive Relationship*

Finally we see that the implication of this work is that the non-elastic strain-rate $\dot{\varepsilon}_t$ for a relatively simple polycrystalline metal or alloy is expected to be the sum of at least three components: a quasi-anelastic strain-rate $\dot{a}$, a plastic strain-rate $\dot{\varepsilon}$, and a contribution from grain boundary sliding $\dot{\varepsilon}_g$. These are expected to be linearly additive as tensors, which we designate by bold face symbols. Thus

$$\underline{\dot{\varepsilon}}_t = \underline{\dot{a}} + \underline{\dot{\varepsilon}} + \underline{\dot{\varepsilon}}_g \tag{5.58}$$

The component $\underline{\dot{\varepsilon}}$ is the plastic strain-rate that satisfies the equation of state that we have described at length. The component $\underline{\dot{a}}$ is not yet well described, but it probably has at least the properties we have attributed to it. It of course already satisfies a state equation for which a is a good state variable and we shall write formally that

$$\underline{\dot{a}} = \underline{\dot{a}}(\underline{a}, \underline{\sigma}, T) . \tag{5.59}$$

The grain boundary sliding contribution $\dot{\varepsilon}_g$, from the limited information at present available, has the form

$$\dot{\underline{\varepsilon}}_g = p \cdot \dot{\underline{\varepsilon}}(q\underline{\sigma}) , \qquad (5.60)$$

where the scalar functions p and q depend on σ and T and describe the stress decrement and strain rate enhancement that we discussed above. In particular, $p = 0$ at stresses over a critical stress and $p = \text{const.}$ at sufficiently low stress. A rough determination from the high purity Al data showed that for that material $q = 1$ over the entire range measured and that p reached a constant value of about 2.

Note well that it was only when this three part structure of $\dot{\varepsilon}_t$ was recognized that it was possible to identify a simple behavior for $\dot{\varepsilon}$. These components have different origins, and each obeys a different law. There can be no reason to expect to find a single simple behavior for the composite, $\dot{\varepsilon}_t$. We advocate, therefore, a careful analytical approach in the collection and organization of test data with the goal of discovering simple relationships for the strain-rate components that might be associated with various microstructural aspects of the material tested. The development of constitutive laws should then proceed by synthesis of these elements. Such an approach should lead to fairly reliable constitutive laws even when only limited test information is available

With a synthesis of this nature and with the deeper understanding derived from it we may finally expect to produce: (a) constitutive relations of real use for mechanical calculations, (b) a framework and a set of fundamental parameters in terms of which we might express the mechanical effect of metallurgical microstructure, and (c) a description of the fundamental relations in deformation that require theoretical explanation.

ACKNOWLEDGEMENT

C.Y. Li, H. Yamada, and G.L. Wire wish to acknowledge support from the National Science Foundation through the Materials Science Center at Cornell University and from the U. S. Atomic Energy Commission.

REFERENCES

Bayce, A. E., Ludemann, W. D., Shepard, L. A., and Dorn, J. E., (1960) Trans. A.S.M., 52, 451.

Bird, J. E., Mukherjee, A. K., and Dorn, J. E., (1970) in "Proc. Int. Conf. on Quantitative Relations between Properties and Microstructure at Haifa" (Haifa, Israel: Israel Univ. Press), p. 255.

Buckley, S. N. and Entwistle, K. M., (1956) Acta Met., 4, 352.

Cook, J. H., (1973) Doctoral Dissertation, University of Utah.

Ellis, F. V., Wire, G. L., and Li, C.Y., (1973) to be published.

Hart, E. W., (1967) Acta Met., 15, 1545; (1970a) Acta Met., 18, 599; (1970b) "Ultrafine-Grain Metals", edited by J. J. Burke and V. Weiss (Syracuse: Syracuse Univ. Press), Chap. 11; (1974a) in "Rate Processes in Plastic Deformation", edited by J. C. M. Li, (Plenum Press; New York); (1974b) to be published.

Hart, E. W. and Solomon, H. D., (1973) Acta Met., 21, 295.

Harper, J. G. and Dorn, J. E., (1957) Acta Met., 5, 654.

Holt, D. L., Babcock, S. G., Green, S. J., and Maiden, C. J., (1967) Trans. A.S.M., 60, 152.

Kocks, U. F., (1975) this volume, Chap. 3.

Lee, D. and Hart, E. W., (1971) Met. Trans., 2, 1245.

Lubahn, J. D. and Felgar, R. P., (1961) "Plasticity and Creep of Metals" (Wiley: New York).

Lubahn, J. D., (1955) J. Metals, (September 1955), p. 1031.

Matlock, D. K., Harrington, W. C., Jr., and Nix, W. D., (1972) Acta Met., 20, 661.

Mitra, S. K. and McLean, D., (1967) Met. Sci., 1, 192.

Servi, I. S. and Grant, N. J., (1951) Trans. AIME, 191, 909.

Sherby, O. D., Lytton, J. L., and Dorn, J. E., (1957) Acta Met., 5, 219.

Weertman, J., (1957) J. Appl. Phys., 28, 362.

Yamada, H. and Li, C.Y., (1973) Met. Trans., 4, 2136; (1974a) in "Rate Processes in Plastic Deformation", edited by J. C. M. Li, (Plenum Press; New York); (1974b) Acta Met., in press.

Wire, G. L., Ellis, F. V., and Li, C. Y., (1974) to be published.

LIST OF SYMBOLS

a, $\dot{a}$	transient non-elastic strain and strain rate
ε, $\dot{\varepsilon}$	plastic strain and strain rate
σ	tensile stress [except in Section 5.5.7]
τ_{ij}	full stress tensor
σ_{ij}	deviator of stress tensor
σ, $\dot{\varepsilon}$	stress and strain rate invarients in Section 5.5.7
G	shear modulus
T	absolute temperature
R	gas constant
γ, ν, q	phenomenological coefficients (eqns. 5.1, 5.2)
g, n, Q	phenomenological coefficients (eqns. 5.4-6)
μ, m	scaling coefficients (eqns. 5.19, 5.20)
y	generic hardness variable (eqn. 5.7)
Γ	absolute strain hardening coefficient (eqn. 5.11)
σ^*	hardness variable (eqn. 5.21)
$\dot{\varepsilon}^*$	characteristic strain rate (eqn. 5.21, 5.23)
Z	activation free energy (eqn. 5.24)
Γ^*	normalized strain hardening coefficient (eqn. 5.29)
λ	parameter of equation of state (eqn. 5.21)
k	parameter in strain hardening relation (eqn. 5.31)
f	arbitrary characteristic frequency (eqn. 5.24)

ABSTRACT. *Dislocation structures in different work-hardened metals (and alloys) are surveyed with special emphasis on the information obtained from transmission electron microscopy of specimens in which the dislocations have been pinned in the unloaded or the stress-applied state. A detailed account of the dislocation structure in deformed FCC single crystals is given and correlated with other observations. Its properties and characteristic parameters are evaluated quantitatively. In stage II work-hardening, analysis of the free primary dislocations indicates agreement with the behavior of a spectrum of free dislocation segments in a long-range internal stress field, as expected from a simple model. Some similarities and differences between the mechanical properties of FCC and BCC metals are summarized. Examples are given of the drastic effect of different deformation temperatures on the dislocation structures in BCC metals. The final part deals with the work-hardening behavior and the microstructure of single crystals of high symmetry orientations and of polycrystals. Multiple slip, the formation of cell structures and the effect of grain boundaries are briefly discussed.*

6. DESCRIPTION OF THE DISLOCATION STRUCTURE AFTER UNIDIRECTIONAL DEFORMATION AT LOW TEMPERATURES

H. Mughrabi

6.1 INTRODUCTION

6.1.1 *General Background*

A satisfactory description of plastic deformation and work-hardening must be based on a detailed knowledge of the mechanisms of dislocation glide and interaction. The mechanisms of formation and the properties of the dislocation structure built up during deformation are essential aspects. Many basic dislocation mechanisms have been predicted by dislocation theory before the advent of techniques permitting the direct observation of dislocations. During the past 15 years, the most prominent of these techniques, transmission electron microscopy (T.E.M.), has been applied extensively to the study of dislocation microstructures in deformed materials, ably supported by other methods such as surface observations, X-ray and etch pit studies and magnetic measurements.

Today, it is evident that the observed dislocation structures are generally too complicated to justify a description of plastic deformation and work-hardening by any one simple mechanism developed from first principles. It must be the ultimate goal to characterize complicated microstructures quantitatively in terms of simple physical quantities that can form the basis of comprehensive theoretical descriptions.

The interpretation of T.E.M. observations is usually not straightforward. As a matter of principle, investigations on unloaded specimens do not permit conclusive comparison with theoretical models which necessarily refer to the stress-applied state. This basic difficulty is aggravated by the serious problem of additional dislocation rearrangement and loss during thinning (cf. Hirsch 1963, Seeger 1963). In pure FCC metals, pinning of the dislocations by irradiation with fast neutrons (Essmann 1963) has proved to be a useful (and possibly the only) technique to retain the unmodified bulk dislocation structure in the unloaded (Essmann 1963, 1965a,b, Ramsteiner 1966/67, Essmann and Strunk 1970) and in the stress-applied states (Crump and Young 1968, Mughrabi 1968, 1970a, 1971a,b).

Existing evidence indicates that unpinned and pinned dislocation structures exhibit only minor differences with respect to the gross features. However, a general characteristic of unpinned structures is the almost complete absence of free (and smoothly curved) dislocations. The preferential loss of free dislocations which are not stabilized in dislocation clusters and networks is not surprising. These "mobile" dislocations represent, at the most, 10% of the total dislocation density. Nevertheless, they convey substantially more information on the deformation process than the dense "inactive" dislocation clusters which are difficult to analyse in any case.

6.1.2 *Objectives and Outline*

A number of authors have presented excellent reviews of the microstructural aspects of work-hardening of FCC metals (see, for example, Seeger 1957, Mader 1963, Seeger 1963, Nabarro, Basinski and Holt 1964, Mitchell 1964, Kronmüller 1967, Seeger and Wilkens 1967) and BCC metals (Hirsch 1968, Christian 1970).

In this chapter we describe and discuss selected examples of dislocation structures that have been investigated in reasonable detail. Emphasis will be placed on the evaluation of T.E.M. observations of pinned structures.* For this reason and for the sake of simplicity, it is necessary to restrict

*Pinned dislocation structures observed in cyclically hardened specimens are discussed in Chapter 7.

the scope considerably. We shall deal mainly with single phase FCC metals, in particular copper, deformed at low temperatures where recovery processes involving self-diffusion are insignificant.

The fundamental results of extensive studies of "simple" microstructures in single crystals form the basis for the understanding of more complicated ones. Hence, a major section is devoted to the work-hardened structures in FCC single crystals deformed in single slip. The discussion will be confined almost entirely to stage II work-hardening which has received most attention over the years.

In another section, we present some examples of dislocation structures in (BCC) α-iron single crystals to demonstrate the influence of crystal structure on microstructure and to indicate similarities and differences between work-hardening of FCC and BCC metals.

Finally, some aspects of the work-hardened microstructures in single crystals oriented for multiple glide and in polycrystals are discussed.

6.2 DEFORMATION OF FCC SINGLE CRYSTALS ORIENTED FOR SINGLE SLIP

6.2.1 *General Remarks*

The main features of the dislocation arrangement, as observed by T.E.M., are reviewed qualitatively (Section 6.2.2), correlated with other observations (Section 6.2.3) and evaluated quantitatively (Section 6.2.4). A discussion of the processes governing the flow stress (Section 6.2.5) and work-hardening (Section 6.2.6) is followed by some remarks on work-hardening theories (Section 6.2.6).

In the following, the primary slip system is defined by the primary glide plane (111) and the primary Burgers vector $\underline{b}_p = 1/2\ [\bar{1}01]$. Consequently, $[1\bar{2}1]$ is the line direction of primary edge dislocations. Resolved shear stresses are denoted by τ, the temperature of deformation by T_d. The terms unloaded state and stress-applied state refer to the condition under which the neutron irradiation was performed.*
On the micrographs the radiation damage is visible as a coarse

---

*Pinning in the stress-applied state was performed at temperatures significantly below T_d in order to prevent radiation-enhanced creep (cf. Mughrabi 1969), in all cases below 20 K.

dotty background. Unless otherwise stated, the micrographs refer to copper.

6.2.2 *Qualitative Survey and Discussion of Dislocation Arrangement*

In virgin crystals, dislocation motion and multiplication have been shown to occur at values of τ of only a few g/mm^2, i.e. well below the yield stress beyond which macroscopic yielding is observed. This has been demonstrated particularly clearly by Borrmann X-ray topography of copper crystals under load or pinned in the stress-applied state (cf. Young and Sherrill 1967, Young 1968).

Dislocation behaviors in work-hardening stages I and II have been investigated in detail in a number of FCC metals and alloys (Cu: Fourie and Murphy 1962, Basinski 1964, Essmann 1963, 1964, 1965a,b, Steeds 1966, Cu and Cu-Al: Steeds and Hazzledine, Cu-Al: Pande and Hazzledine 1971a,b, Ni and Ni-Co: Mader, Seeger and Thieringer 1963, Au: Ramsteiner 1966, 1966/67, Ag: Moon and Robinson 1967).

Stage I deformation is characterized by the elastic interaction of (individual) primary dislocations, leading to the formation of (edge) dipoles. In regions of low dislocation density, individual primary dipoles and free curved primary dislocations are observed in the stress-applied state (fig. 6.1) whose curvatures correspond to the applied stress. This indicates the absence of significant long-range internal stresses (Crump and Young 1968, Mughrabi 1968, 1971a). Dislocation dipoles are predominantly of edge character since screw dislocations of opposite sign can obviously annihilate mutually by stress-induced cross slip (Essmann 1964). Figure 6.1 shows a rare example of a screw dislocation dipole which has been retained by pinning at low temperatures. In regions of higher dislocation density extended planar edge multipole bundles (and walls) are usually observed (cf. fig. 6.2).

In stage II more complicated structures are formed, as dislocations on secondary slip systems are activated and interact with those of the primary system. Some planar multipoles, very similar to those in stage I, are observed in early stage II (fig. 6.2). More commonly, primary multipole bundles observed in stage II are shorter, denser and less regular and contain also secondary dislocations (fig. 6.3). Sometimes the bundles link up to form extended walls perpendicular to $\underline{b}_p$. Another feature visible in fig. 6.3 are dislocation segments that are smoothly curved in the unloaded state, indicating the

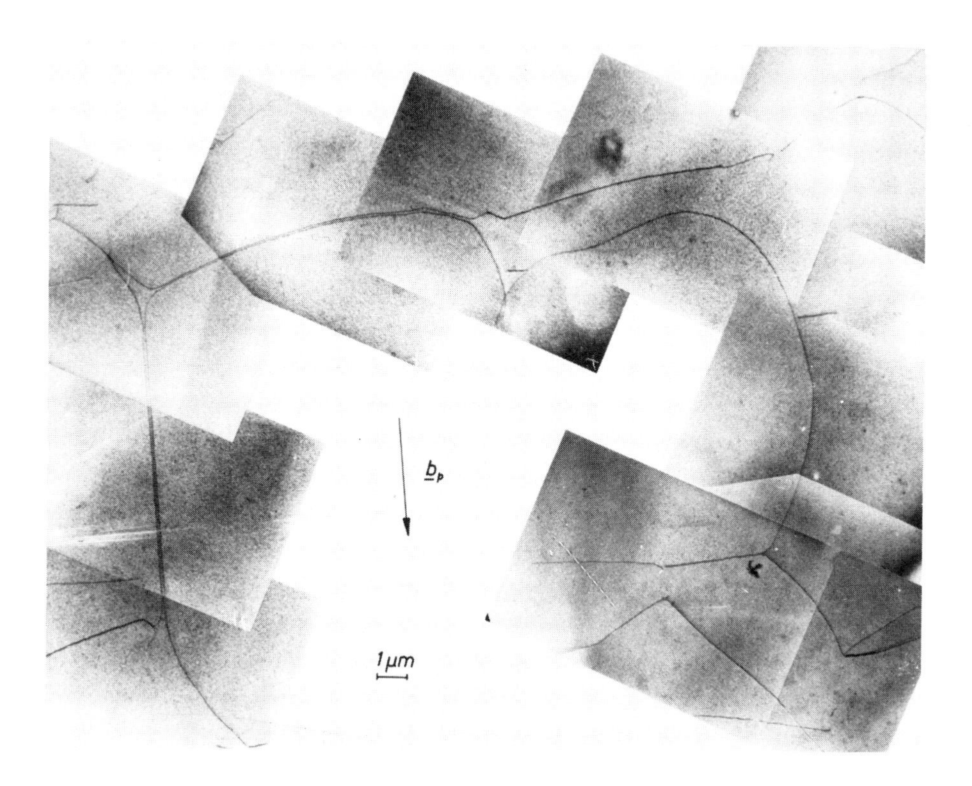

Fig. 6.1 State I, $\tau = 0.11\ kg/mm^2$, $T_d = 78\ K$, (111)-section, stress-applied state. (from Mughrabi 1971a, courtesy of Taylor & Francis)

presence of substantial long-range internal stresses (Essmann 1963).* It is evident from Essmann's work that the free dislocations act as local stress probes that convey valuable information on the distribution of internal stresses on a microscopic scale.

In the stress-applied state, groups of primary dislocations of the same sign are a common feature. The spatial variation of the curvatures of the dislocations of the group shown in fig. 6.4 confirms the considerable changes in the locally acting effective stresses (i.e. external plus internal stresses) over distances of several microns, a qualitative indication of

*The term long-range internal stresses refers to internal stresses whose wavelength is significantly larger than the mean dislocation spacings.

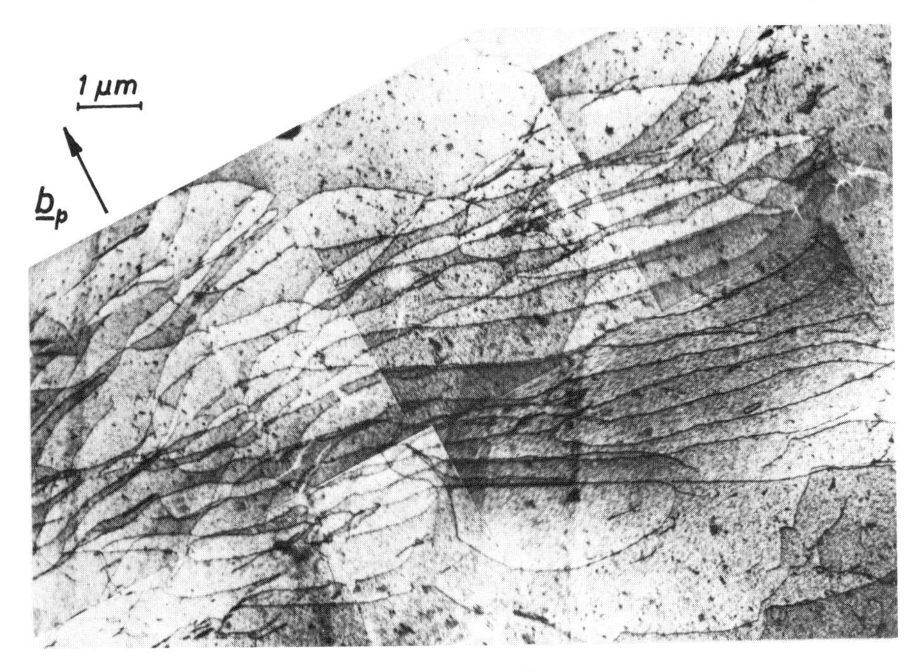

Fig. 6.2 Stage II, τ = 1.2 kg/mm^2, T_d = 78 K, (111)-section, stress-applied state. (from Mughrabi 1971b, courtesy of Taylor & Francis)

the long-range internal stress field. Piled-up dislocations of the same sign, the classical source of long-range internal stresses, are also observed (fig. 6.5).*

Secondary slip occurs preferentially in regions where it is favored by the applied and the internal stresses. Accordingly, secondary dislocations accumulate in the vicinity of the primary dislocations. The observations show that the multipole bundles contain secondary dislocations, but perhaps more important, evidence of extensive secondary slip is observed in regions of internal stress concentration, i.e. at the head of

*During T.E.M. of (111)-sections of pinned stage II specimens, it is generally difficult to obtain an even illumination (Essmann 1965b). This is particularly true in the vicinity of grids, cf. figs. 6.6, 6.7 and 6.8, and, especially in the stress-applied state, in regions containing primary dislocation groups. This experience reflects the elastic deformation of the lattice due to the local excess of dislocations of the same sign (Mughrabi 1971b).

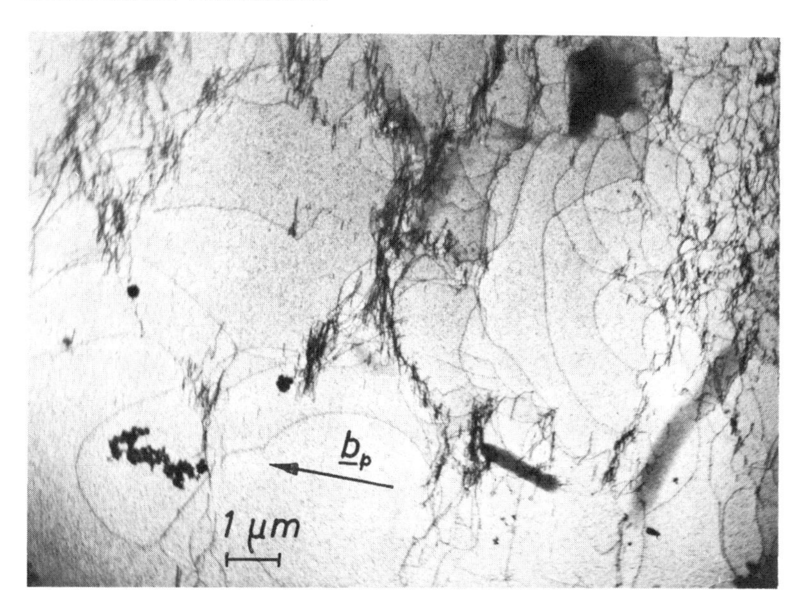

Fig. 6.3 Stage II, $\tau = 1.66\ kg/mm^2$, $T_d = 293$ K, (111)-section, unloaded state. (from Essmann 1965b, courtesy of Akademie-Verlag)

primary dislocation groups.

As a result, complicated networks are formed roughly parallel to the primary glide plane, consisting of primary and secondary dislocations and their reaction products (Basinski 1964, Essmann 1965b, Steeds 1966). These layer-like structures have been given different names: grids (Essmann 1965b), carpets (Steeds 1966), converted pile-ups, sheets (Hirsch and Mitchell 1967). They will be called grids here.

Electron micrographs of the stress-applied state reveal characteristic features of their formation. Figure 6.6 shows an example. In this case dislocations of the conjugate slip system $(\bar{1}\bar{1}1)$ [011] react piecewise with dislocations of a primary group, forming short sessile Lomer-Cottrell (L.C.) segments along the trace of the conjugate slip plane.* There is a marked change of background contrast at the borderline of

---

*In early stage II reactions with dislocations of the critical slip system $(\bar{1}11)$ [110] predominate. It is clear from the work of Steeds (1966) that the critical slip system is favored more by the internal stresses due to primary pile-ups than the conjugate slip system $(\bar{1}\bar{1}1)$ [011].

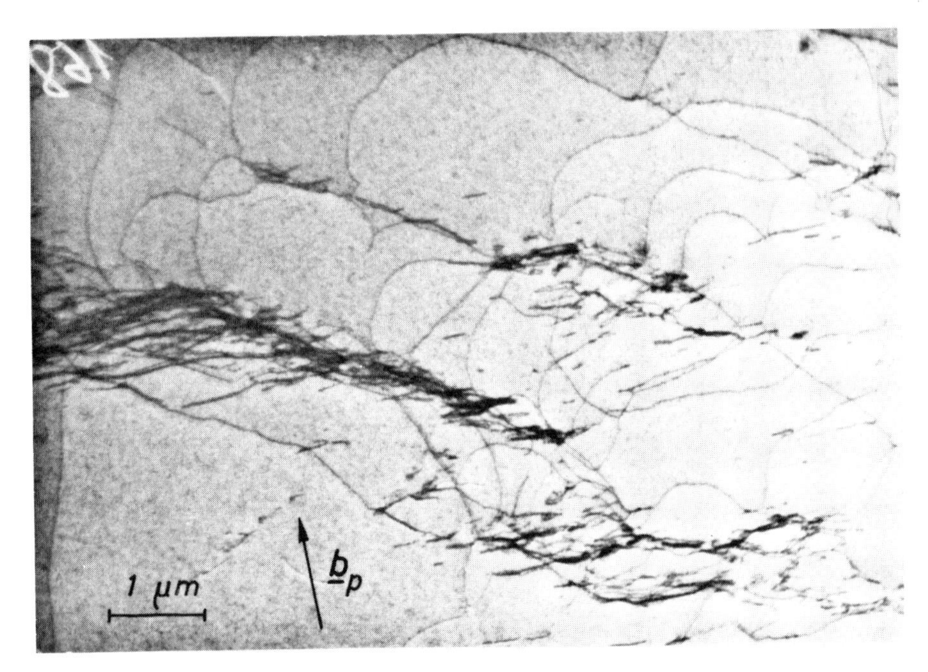

Fig. 6.4 Stage II, $\tau = 1.2\ kg/mm^2$, $T_d = 78$ K, (111)-section, stress-applied state. (from Mughrabi 1970, unpublished)

the grid, indicating a misorientation. The idealized arrangement of dislocations, as deduced from contrast experiments (Essmann 1965b, Essmann and Strunk 1970), is also indicated in fig. 6.6. The fact that these grids act as obstacles to glide dislocations is reflected by their growth to dimensions of several ten microns by the successive incorporation of other dislocation groups. An example of an extended grid is shown in fig. 6.7.

The formation and growth of grids may be looked upon as the consequences of incomplete intersections of primary and secondary dislocations. These form attractice junctions (L.C. segments) which, however, remain stable during subsequent deformation so that intersections do not actually occur. One obvious reason lies in the short meshlengths of the grids which do not allow sufficient dislocation bowing to break down the junctions at low temperatures.

On the other hand, the meshlength in the dislocation multipole bundles is obviously large enough to permit dislocations

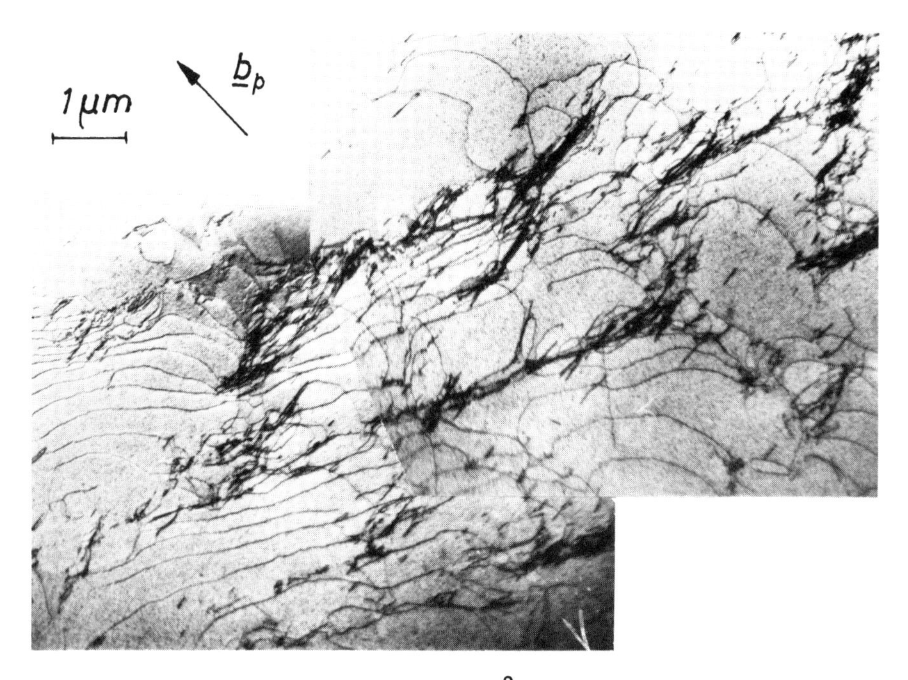

Fig. 6.5 Stage II, $\tau = 1.2$ kg/mm^2, $T_d = 78$ K, (111)-section, stress-applied state. (from Mughrabi 1968, courtesy of Taylor & Francis)

to bow out of the bundles between stable anchoring points.* Figure 6.8 shows an example of dislocation segments of the same sign bowing out of a bundle, while others rest in front of a grid (top right corner). In fact, this seems to be the major dislocation source mechanism in stage II, as observed by T.E.M. (Mughrabi 1970, 1971b). This process is most evident in regions of high local effective stresses where even very short segments can bow out.

Figure 6.8 suggests that dislocations leaving the bundles are "pushed out" by others of the same sign that replace them. This would imply that the bundles are not strong obstacles to dislocation glide (Essmann 1965b, Mughrabi 1970, 1971b). The observation that the structure of the bundles is altered during deformation provides additional evidence that, once formed,

*The anchoring points that subdivide the dislocations forming the bundles into shorter free segments cannot be resolved. In addition to secondary dislocations and reaction products, they could be superjogs, formed on the cross slip plane during the annihilation of the screw segments (Essmann 1964).

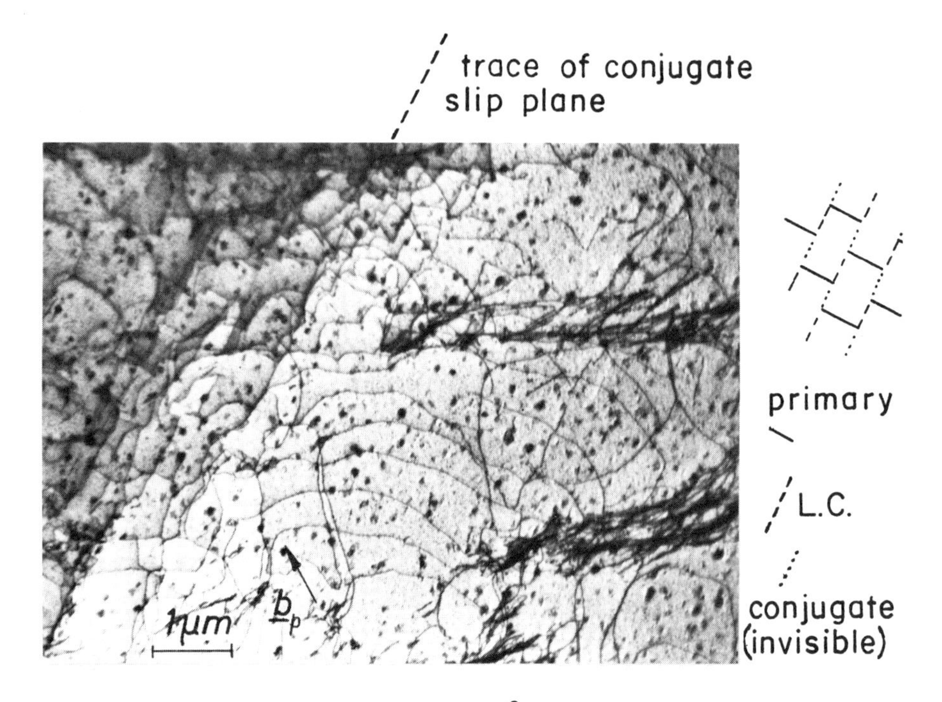

Fig. 6.6 Stage II, $\tau = 1.7\ kg/mm^2$, $T_d = 78\ K$, (111)-section, stress-applied state. (from Mughrabi 1971b, courtesy of Taylor & Francis)

the bundles are not stable enough to resist subsequent structural changes to the extent of dissociation (cf. Hazzledine 1964, 1971, Neumann 1971).

The "trapping" of rather long free primary dislocations of predominantly the same sign in the dislocation-poor regions between bundles and grids (figs. 6.4 - 6.8) is difficult to explain, unless "physically invisible" obstacles are postulated. The most obvious candidates for these obstacles are the mentioned long-range internal stresses.

Sections oblique to the primary glide plane give less direct information on the behavior of primary glide dislocations but are very suitable to reveal the dominating rôle of the primary slip system and the spatial arrangement of wall (cf. also fig. 6.12b) and grid structures. In fig. 6.9, imaged with the diffraction vector $\underline{g} = (111)$, the trace of the primary glide plane (which lies perpendicular to the plane of the figure) is easily recognizable. Primary (edge) dislocations are only visible through residual contrast, since $\underline{g} \cdot \underline{b}_p = 0$. Two kinds

Fig. 6.7 Stage II, $\tau = 1.65 \text{ kg/mm}^2$, $T_d = 4.2$ K, (111)-section, unloaded state. (from Essmann 1965, unpublished micrograph)

of regions can clearly be distinguished: regions of high secondary (and primary) dislocation density corresponding to the grids (G) and others containing predominantly primary dipole (D) (and multipole) bundles and walls (diffuse contrast). This corresponds to a division of the crystal into regions of high (G) and low (D) obstacle density. Figures 6.4 - 6.8 suggest strongly that free dislocations glide predominantly in the "glide regions" containing the multipole structure (cf. Seeger and Wilkens 1967) and become (permanently) immobilized by incorporation into the grid structure (figs. 6.6 and 6.7).

The influence of deformation temperature on the dislocation arrangement is demonstrated in fig. 6.10, showing sections perpendicular to $\underline{b}_p$. The layer structure of the dislocation grids is much more pronounced at $T_d = 78$ K than at $T_d = 473$ K. Changes between light and dark contrast indicate alternating misorientations. Detailed investigations (Essmann and Strunk 1970) suggest that the more frequent occurrence of complete intersections at the higher temperature inhibits the formation of extended grids. Instead, dislocation walls perpendicular to $\underline{b}_p$ are formed by non-conservative glide polygonization

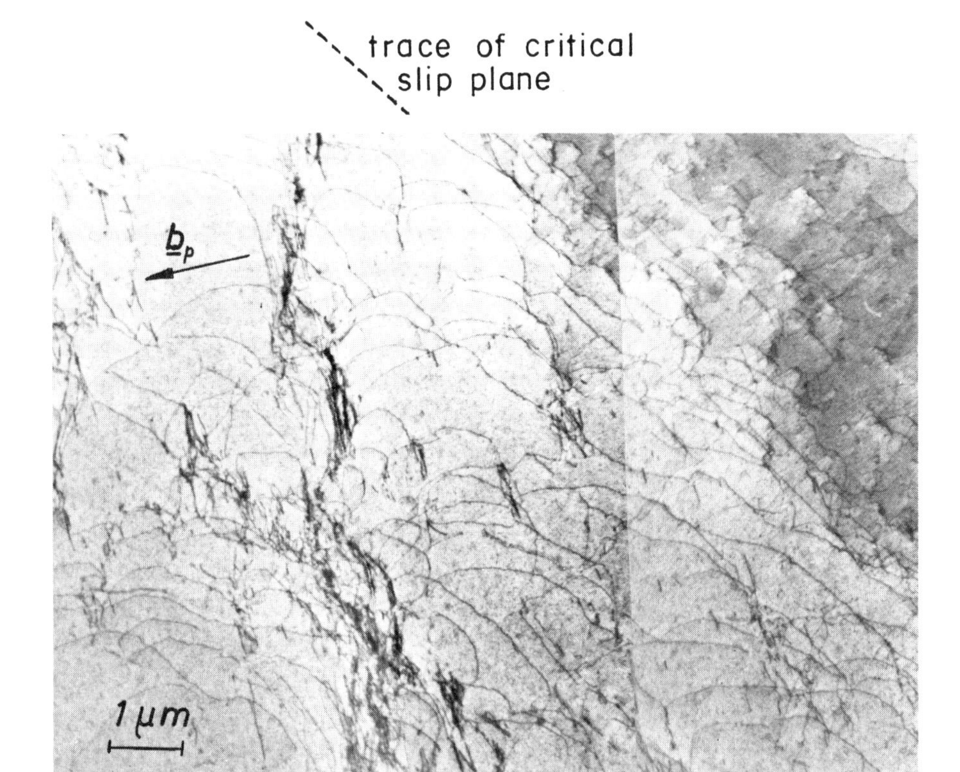

Fig. 6.8 Stage II, τ = 1.2 kg/mm^2, T_d = 78 K, (111)-section, stress-applied state. (from Mughrabi 1971b, courtesy of Taylor & Francis)

(Strunk and Essmann 1969). The structure is somewhat more "amorphous", and secondary dislocations are more evenly distributed than at lower temperatures.

On the other hand, Edington's (1969) T.E.M. observations indicate little influence of strain rate on the dislocation structure over the strain-rate range 10^{-4} = 10^{4} s^{-1}, inspite of well-documented effects on the mechanical properties. At this point it is appropriate to discuss some limitations of T.E.M. in its ability to reveal certain microstructural details.

Direct observation of the interaction of primary and secondary forest dislocations (i.e. those threading the primary glide plane) is difficult in FCC metals, because the forest

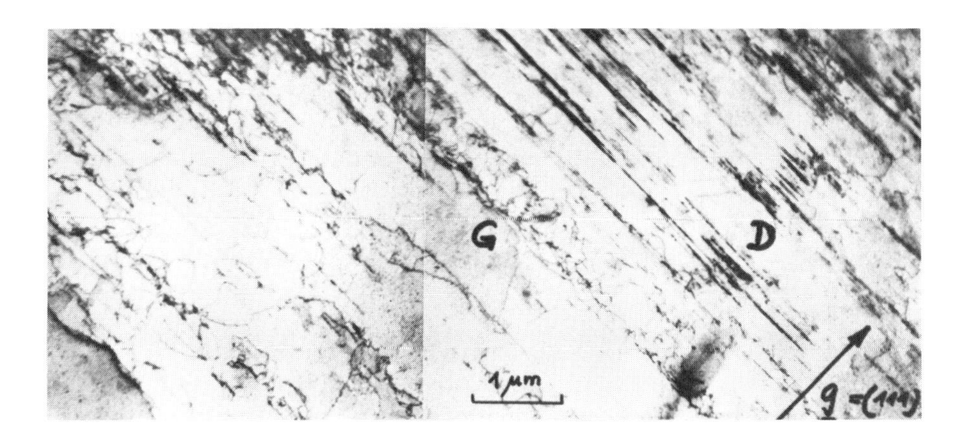

Fig. 6.9 Stage II, $\tau = 1.65\ kg/mm^2$, $T_d = 4.2$ K, (111)-section, unloaded state. (from Essmann 1965b, courtesy of Akademie-Verlag)

Fig. 6.10 $\tau = 2.8\ kg/mm^2$, $(\bar{1}01)$-section, unloaded state. (from Essmann and Strunk 1970, courtesy of Z. Metallk.)

dislocations are not distributed randomly but are hidden in the bundles. By comparison, fig. 6.11 shows the interaction

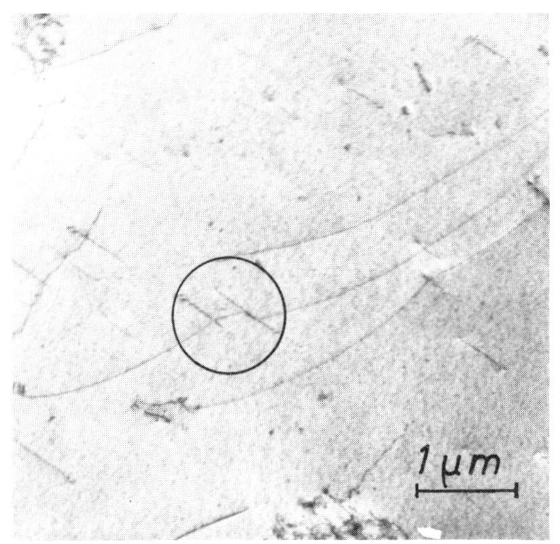

Fig. 6.11 Fatigued α-iron single crystal, ($\bar{1}01$)-section (primary glide system ($\bar{1}01$) [111]), T_d = 293 K, 42 cycles, plastic shear strain amplitude γ_p = ± 0.005, τ = 3.6 kg/mm^2. (Stark and Mughrabi 1973)

of glide dislocations with randomly distributed forest dislocations in a cyclically hardened α-iron single crystal. The encircled region reveals the formation of an attractive junction with one of the two forest dislocations but not with the other one which is of opposite sign (repulsive interaction).

T.E.M. has not resolved basic differences between work-hardening in stages II and III (Essmann 1963, 1965b, Steeds and Hazzledine 1964, Steeds 1966). The overall structure continues to shrink in scale with increasing deformation throughout stages II and III. This applies to the spacings between bundles, walls and grids, the mean free segment lengths in the bundles, the meshlengths in the grids as well as to the wavelengths of the internal stress field. Increasing misorientations, in some cases up to $\sim 30^o$, about the [$1\bar{2}1$] axis have been observed in stage III and related to recrystallization (cf. Buck and Essmann 1964, Steeds and Hazzledine 1964, Steeds 1966). However, microstructural processes typical of dynamical recovery, in particular marked cross slip, as evidenced by slip line studies (Seeger, Diehl, Mader and Rebstock 1957), have not been identified so far by T.E.M.

The influence of stacking fault energy on dislocation structure, as evidenced by T.E.M., is not too significant (Mader, Seeger and Thieringer 1963, Steeds and Hazzledine 1964, Moon and Robinson 1967). Low stacking fault energy favors the formation of more and longer Lomer-Cottrell dislocations. In high stacking fault energy materials edge multipole bundles are more prominent and easy cross slip favors pinching-off of dipoles, as indicated by a high density of short elongated loops. However, in single phase alloys of low stacking fault energy and high friction stress, a markedly different dislocation arrangement is observed, characterized by the interleaving interaction of primary and secondary glide systems deforming in planar slip (Whelan 1958, Pande and Hazzledine 1971b).

6.2.3 *Correlation with Other Observations*

The predominance of primary slip is evident from the early work, summarized by Schmid and Boas (1935), and has been confirmed more recently by surface slip line studies (Mader 1957), measurements of orientation changes (Ahlers and Haasen 1962, Mitchell and Thornton 1964) and shape changes (Kocks 1964, Basinski and Basinski 1970). Dominant primary slip can only be reconciled with the high secondary dislocation density observed in stage II, if the slip path of secondary dislocations is significantly smaller than that of primary dislocations, presumably only of the order of the distance between adjacent grids (Essmann 1965b, Steeds 1966).

While T.E.M. resolves many microscopic details of the dislocation behavior, other techniques with a larger field of view are more suitable to study long-range correlations of the dislocation arrangement. Etch pitting (Livingston 1962, Basinski and Basinski 1964) and X-ray Berg-Barrett topography (cf. Wilkens 1967) have been applied successfully. A distinct advantage of the latter technique is its ability to determine the rotation axes associated with the misorientations of wall and grid structures.

X-ray topography studies indicate that the observed misorientations are due to local excess densities of dislocations of the same sign ($\sim 10^{8}$ cm^{-2}, Wilkens 1967). The dislocation structures perpendicular to $\underline{b}_p$ have been identified as polarized bundles and (kink) walls, misoriented about [1$\bar{2}$1] (Honeycombe 1951, Wilkens 1967, Obst, Auer and Wilkens 1968/69). Observations in the stress-applied state reveal a fine structure associated with a higher degree of polarization (Mughrabi 1970b) which is presumably due to the primary dislocation groups observed between the bundles (Mughrabi 1971b). The grid structure is characterized by alternating misorientations (double layers, Wilkens 1967)*, mainly also about [1$\bar{2}$1],

*At higher deformations, the Berg-Barrett technique is not able to resolve individual grids but rather reveals the contrast due to packets of grids in regions of high grid density separated by regions of low grid density (cf. also fig. 6.9).

sometimes (at lower deformations) with a twist component about [111] (Wilkens and Eckert 1964, Wilkens 1967, Obst et. al. 1968/69, cf. also Essmann 1963, 1965b, Steeds and Hazzledine 1964, Steeds 1966).

Among the methods that do not rely on direct observation, magnetic measurements, in connection with theory, belong to the few techniques that are able to distinguish between different dislocation arrangements (cf. Kronmüller 1967). The magnetic techniques are non-destructive and average over large volumes. They provide clear evidence that in stage II, in contrast to stage I, long-range internal stresses are built up with an amplitude of the order of the applied stress.

Related to the behavior of primary glide dislocations in a long-range internal stress field is the anelastic backstrain upon unloading, typically about 10^{-4} in stage II (Salama and Roberts 1970). By a consideration of the densities of free primary dislocations in the unloaded and in the stress-applied state, it has been shown that the relaxation of dislocations between bundles (which are adequate obstacles during unloading) can account for the backstrain (Mughrabi 1971c, cf. also Kronmüller and Marik 1972).

6.2.4 *Quantitative Characterization and Discussion of the Dislocation Structure*

a) Introductory Remarks. The preceding sections have shown that direct observation of the dislocation microstructure yields little information on the nature of thermally activated processes. While these govern the temperature and strain-rate dependence of deformation (Seeger 1954, also cf. Chapter 3), it is generally accepted that, at least in pure FCC metals, they have an only minor influence on the flow stress and work-hardening in stage II (cf. Diehl and Berner 1960). We shall here deal only with largely athermal properties of the dislocation structure.

b) Length of Slip Lines and Surface Effects. The shrinking scale of the dislocation pattern with increasing deformation is reflected by a decrease in slip line lengths (Blewitt, Coltman and Redman 1955, Mader 1957). Mader's detailed replica studies of the "active length" L of slip lines formed (by dislocation groups)* at a given shear strain γ (referred to

---

*The heights of the slip lines indicate that, in copper, typically about 20 dislocations have emerged per slip line (Mader 1957).

stage II) yielded:

$$L_{\perp,\odot} = \frac{\Lambda_{\perp,\odot}}{\gamma} \tag{6.1}$$

The constants $\Lambda_{\perp}$ and $\Lambda_{\odot}$ refer to lengths $L_{\perp}$ and $L_{\odot}$ of slip lines formed by edge and screw dislocations respectively and were determined as $\Lambda_{\odot} = 4 \times 10^{-4}$ cm and $\Lambda_{\perp} \sim 2\Lambda_{\odot}$ for copper single crystals deformed at room temperature. Since in stage II, $\tau = \theta_{II}\,\gamma$, with $\theta_{II} \sim 13$ kg/mm^2, eqn. (6.1) could also be written as $L \propto 1/\tau$ with appropriate proportionality constants.

The question whether surface slip lines are representative of processes in the bulk material has been discussed repeatedly (cf. Seeger 1963, Nabarro et. al. 1964). Detailed mechanical measurements by Fourie (1967, 1968) and T.E.M. of near-surface regions (Fourie 1970, Mughrabi 1970) have established the existence of a "soft surface" effect. These investigations showed that the region bounded by the exit surface of $\underline{b}_p$ deforms longer in stage I than the center region of the crystal. Figure 6.12 shows an example of the dislocation arrangement near the surface and in the interior, as viewed in an oblique (101) section containing $\underline{b}_p$. It is evident that near the surface the typical stage I structure, i.e. primary edge dipoles and multipoles, still prevails, while the dislocation arrangement in the interior is characteristic of stage II (dense edge bundles and walls as well as grids).

If the crystal deforms homogeneously over the cross section, then the markedly lower dislocation density near the surface requires the slip paths to be (about 2.5 times) larger in these regions than in the center (Mughrabi 1971d). Recent observations of slip lines on surfaces exposed at different distances below the original surface have confirmed this for Cu-5 at. % Al (Fourie 1971) and for copper single crystals (Himstedt and Neuhäuser 1972). Figure 6.13 shows Fourie's observations on Cu-5 at. % Al.

With eqn. (6.1) and considering surface effects, one finds that, in the interior, for $\tau \sim 1.5$ kg/mm^2 ($\gamma \sim 0.1$), $L \sim 30$ µm. Comparing this with typical distances between multipole bundles (~ 5 µm), it becomes evident that glide must be a result of <u>cooperative dislocation motion</u> and that the bundles are not effective obstacles, as concluded previously.

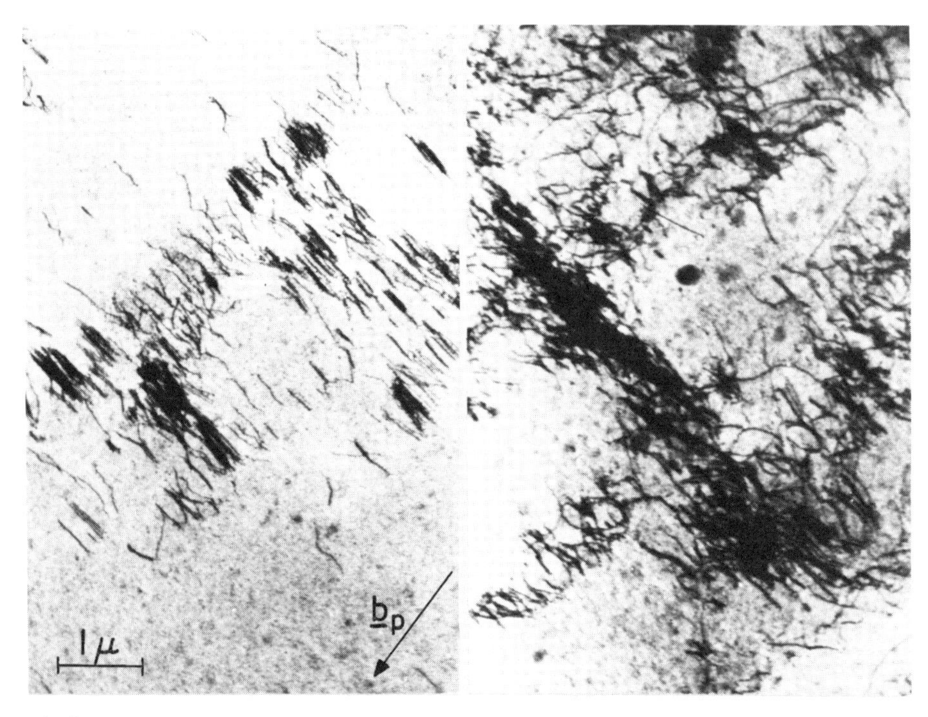

Fig. 6.12 Stage II, $\tau = 1.2\ kg/mm^2$, $T_d = 78\ K$, (101)-section. (from Mughrabi 1970b, courtesy of Akademie-Verlag)

c) Dislocation Density. The dislocation density is one of the important parameters that enters into all theoretical models of flow stress (cf. Nabarro et. al. 1964). It is most reliably determined by T.E.M. observations. Micrographs of the same area imaged with different reflections permit, in principle, the distinction between densities of dislocations of different Burgers vectors. Since this is very tedious and almost impossible at high dislocation densities, statistical counting procedures are usually employed (Smith and Guttmann 1953). The evaluation requires a number of assumptions based on more detailed contrast analysis (cf. Essmann 1966, Steeds 1966).*

*The analysis does not distinguish between dislocations of the same Burgers vector but lying on different glide planes.

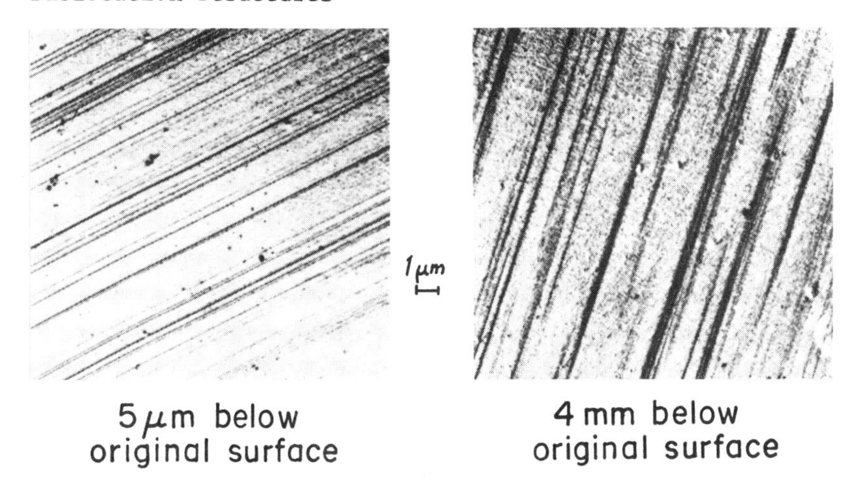

Fig. 6.13 Cu-5.8 at. % Al, end of "stage II", T_d = 293 K, carbon replicas of new slip lines formed in a plastic strain interval of 5%. (from Fourie 1971, courtesy of S. African Mic. Soc.)

Figure 6.14 shows a plot of the dislocation densities ρ_p

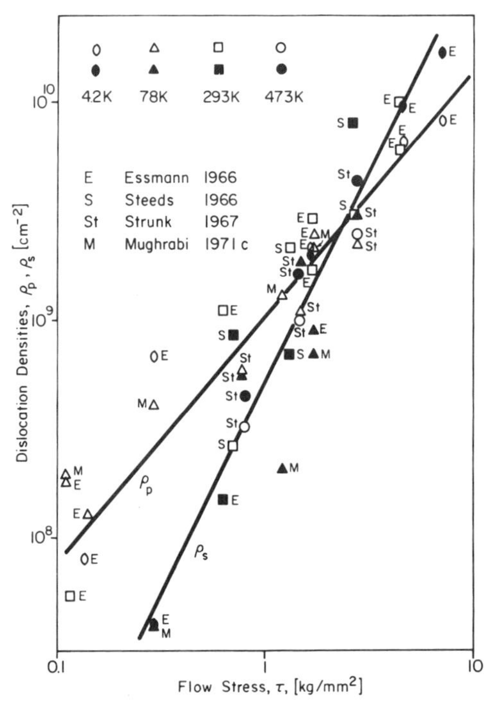

Fig. 6.14 Dislocation densities ρ_p , ρ_s in copper single crystals as a function of flow stress τ.

and ρ_s of primary (Burgers vector $\underline{b}_p$) and secondary (all other Burgers vectors) dislocations as a function of flow stress τ for copper deformed at various temperatures, as evaluated by different workers. The (etch-pit) data of Livingston (1962) and Basinski and Basinski (1964) have not been included. Their values are systematically lower in stage II, presumably because of the limited resolution of the etch pit technique.(Essmann 1966). The densities of free primary dislocations are of the order of $\sim 10^8$ cm^{-2} in stage II, being about twice as high in the stress-applied state as in the unloaded state (Mughrabi 1971c). Total dislocation densities ρ_t, plotted against τ, are shown in fig. 6.15.

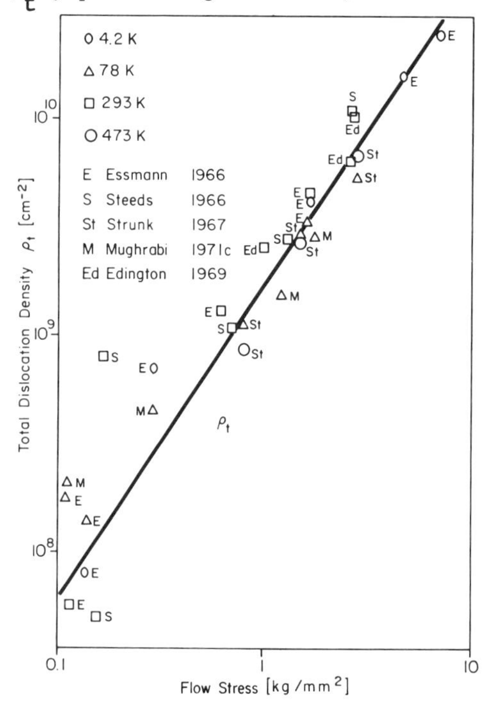

Fig. 6.15 Total dislocation density ρ_t in copper single crystals as a function of flow stress τ.

The scatter in the data reflects the generally low accuracy of these measurements. No influence of temperature can be detected. While the total secondary dislocation densities of different authors are compatible, a detailed comparison of their data with respect to the contributions of individual typed of secondary dislocations yielded contradictory results. There is agreement that all Burgers vectors occur in stage II. Hirth locks ($\underline{b}$ = <100>) have also been found (Strunk 1970). Figure 6.14 shows one feature which is rather surprising. As first observed by Essmann (1966), ρ_p increases more slowly with increasing τ than ρ_s which grows roughly proportional to τ^2. As a consequence, ρ_t also increases somewhat more slowly than proportional to τ^2. Tentatively, we write $\rho_p \propto \tau$. In the evaluation no allowance is made for primary dislocations that have changed their Burgers vector by reactions with secondary dislocations (Essmann 1966, Steeds 1966, Seeger 1968). Hence the density of dislocations that were originally primary is underestimated. Also, high dislocation densities ($\gtrsim 5 \times 10^9$ cm^{-2}) are generally underestimated because of overlapping images. These arguments cannot account fully for the different dependences of ρ_p and ρ_s on τ (Mughrabi 1971c).

Recently, Essmann and Rapp (1973) proposed a model of dislocation accumulation taking into account not only the annihilation of screw but also of edge dislocations. The model is based on the observation of large numbers of broken-up narrow dipoles in weakly neutron-irradiated copper single crystals deformed into stage I and is supported by similar observations on Cu-2.5 at. % Ga crystals (cf. fig. 6.16). Both materials exhibit anomalously long stages I (shear strains $\sim 0.5 - 0.8$), accompanied by high primary dislocation densities ($\sim 3 \times 10^9$ cm^{-2}) which favor annihilation.* By assuming that edge dipoles with a width $y \lesssim 16$ Å annihilate spontaneously (or decompose into debris), Essmann and Rapp could account for the observed dependence of ρ_p (and ρ_s) on strain in stages I _and_ II.

If this interpretation is correct, then the value $y \gtrsim 16$ Å for stable dipoles may have a fundamental meaning. Approximating the _local dislocation density_ ρ_{loc} in dense edge multipole bundles (which cannot be resolved by T.E.M.) by $\rho_{loc} \sim 1/y^2$, an upper limit follows, namely $\rho_{loc} \lesssim 2 \times 10^{13}$ cm^{-2}. This hypothetical value does not depend on τ. On the other hand, if one assumes that the bundles are stable at the maximum effective stress $\tau_{eff,max} = \tau + \tau_{i,max}$, (cf. Section 6.2.4d) and that the stability is governed by the dipole passing stress $\tau_{dip} \gtrsim \tau_{eff,max}$, then

*Preferred sites for annihilation would be expected to be the bundles.

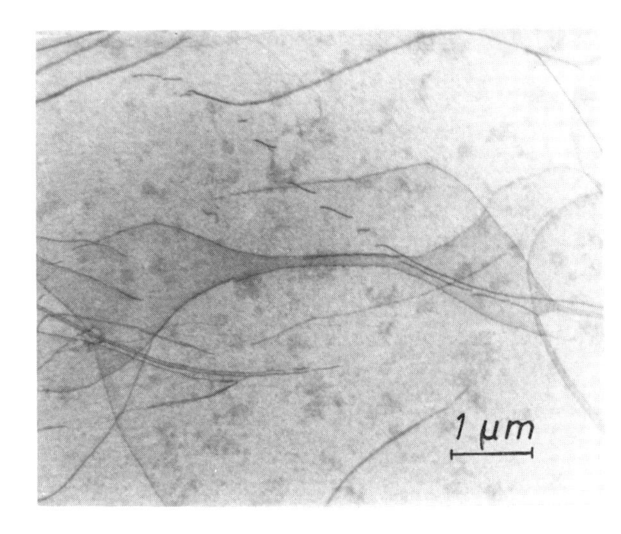

Fig. 6.16 Cu-2.5 at. % Ga single crystal, stage I, T_d = 78 K, (111)-section decomposition of narrow dipoles. (Mughrabi 1968, unpublished).

$$y \lesssim \frac{\mu b}{8\pi(1 - \nu)\, \tau_{eff,max}} \qquad (6.2)$$

where μ is the shear modulus* and ν the Poisson number. With $\tau_{i,max} \sim \tau$, i.e. $\tau_{eff,max} \sim 2\tau$, we obtain $\rho_{loc} \gtrsim 1.1 \times 10^{11} \tau^2$ [cm^{-2}] , with τ in kg/mm^2 . Lower values are possible in regions, where the stability is governed by the Frank-Read bowing stress $\tau_{F.R.}$ (cf. Section 6.2.4e).

d) Long-Range Internal Stress Field. The experimental observations leave no doubt as to the existence of long-range internal stresses τ_i in stage II. Their origin requires further examination. Classical pile-ups and dislocation groups are observed, but not in sufficient number to explain the observations quantitatively beyond early stage II (Mughrabi 1971b). There is a general difficulty in ascribing τ_i only to the primary dislocations, since their density does not increase fast enough during deformation.

In all probability, cooperative glide processes of primary and secondary dislocations are largely responsible for the observed value of τ_i in the primary glide plane. Secondary dislocations are activated locally with the aid of τ_i (re-

*We use μ = 4200 kg/mm^2 in the following.

sulting from the dominant primary system) but fail to relax τ_i appreciably. The reason probably is that they do not travel very far before they react with the primary dislocations and form sessile configurations in which they themselves become trapped. They thus "pin down" the stress-active structures.

The formed grids and polarized multipole structures probably contribute significantly to τ_i (Essmann 1965b), since their (local) resultant Burgers vector is far from zero and retains a large component in the direction of $\underline{b}_p$, as evidenced by the misorientations. Also, they do not act as (stress-free) low-angle grain boundaries, (Wilkens 1967, Seeger and Wilkens 1967, Seeger 1968), since they terminate in the crystal (a closed cell structure does not form in stage II)*. In summary, in addition to the primary dislocation groups and pile-ups, the observed primary - secondary structures (Seeger and Wilkens 1967) must be considered as possible sources of τ_i which are effectively stabilized by the secondary dislocations they contain.

From the evaluation of the smallest radii of curvatures of free primary dislocations follows the amplitude $\tau_{i,max}$ of τ_i. It is about 0.7 - 0.8 τ in the unloaded state (Essmann 1965b) and corresponds to the full applied stress in the stress-applied state (Mughrabi 1971b). This indicates that the anelastic back strain by reverse motion of free primary dislocations during unloading relaxes only a small fraction of τ_i (cf. Kronmüller and Marik 1972). The wavelength λ, i.e. the average distance between regions of maximum (or minimum) τ_i (measured in the primary glide plane), has been determined (Mughrabi 1971b). The result, including also more recent evaluations, can be summarized by the relation $\lambda \sim 30/\tau$ [μm], with τ in kg/mm^2. The exact determination of λ is difficult, since τ_i exhibits considerable fluctuations on a smaller scale ($\sim$1 μm), i.e. submaxima and -minima (< $\tau_{i,max}$, subwavelengths $\sim 5/\tau$), possibly due to the bundles.

e) <u>Lengths of Free Dislocation Segments and Local Dislocation Densities in Multipole Bundles</u>. An analysis of the multipole

---

*The tendency to form a closed cell structure increases in stage III.

bundles is essential for an understanding of the activation of dislocation sources and the interaction between primary and secondary dislocations in the "glide regions". Figure 6.17a shows the frequency distribution $H(\ell)$ of the length ℓ of

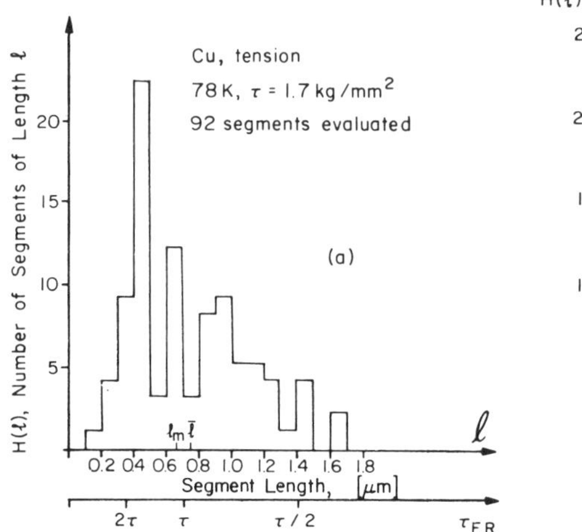

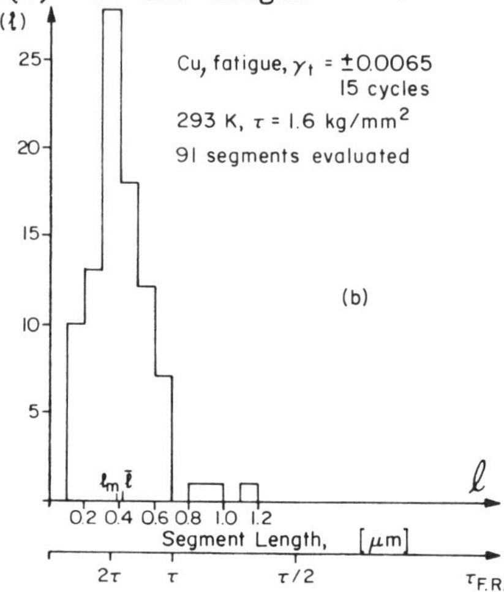

Fig. 6.17 Copper single crystals, frequency distribution $H(\ell)$ of free segment lengths ℓ in bundles. (γ_t = total plastic shear strain amplitude).

free edge segments, determined from observations of segments bowing out of bundles between stable anchoring points (cf. Fig. 6.8). The length ℓ refers to the inter-obstacle spacing, i.e. to the presumable segment length before bowing occurred. Characteristic values of the distribution are its median ℓ_m and the weighted average $\bar{\ell}$.

Plotted also on the abscissa are corresponding values of the Frank-Read stress $\tau_{F.R.}$.* It is evident that many bowed-out dislocations can act as sources at $\tau_{F.R.} < \tau$ and that, with the aid of $\tau_{i,max} \sim \tau$, most of them are potential dislocation sources. For comparison, fig. 6.17b shows the result of

---

* $\tau_{F.R.}$ was calculated for edge segments, assuming that, in the absence of τ_i , dislocation segments are elliptical in shape and that τ is related to the radii of curvature $r_\perp$ and $r_\odot$ of edge and screw segments as $\tau = 0.42/r_\perp = 1.1/r_\odot$, with τ in kg/mm^2 and $r_\perp$ and $r_\odot$ in microns (Mughrabi 1973). In good approximation, $\tau_{F.R.,\perp} \sim 1.1\ \mu b/\ell$ and $\tau_{F.R.,\odot} \sim 1.5\ \mu b/\ell$.

a similar study on cyclically deformed copper single crystals (Mughrabi 1973). Here, Frank-Read bowing without the aid of τ_i is almost impossible.*

Longer bowed-out segments (~ 1.5 μm, cf. fig. 6.17a) that have almost reached the critical Frank-Read configuration, indicate that they have overcome the short range elastic dipole interaction and are merely held back because τ_{eff} is locally not quite as large as $\tau_{F.R.}$, (cf. fig. 6.18). In these cases we can conclude that, locally, $\tau_{F.R.} \gtrsim \tau_{dip}$, i.e.

$$\frac{1.1\ \mu b}{\ell} \gtrsim \frac{\mu b}{8\pi(1-\nu)y}\ ,\ \text{from which follows}\ y \gtrsim \frac{\ell}{8.8\pi(1-\nu)} \sim 0.05\ell$$

Writing again $\rho_{loc} \sim 1/y^2$, $\rho_{loc} \lesssim 400/\ell^2$. At $\tau = 1.7$ kg/mm^2 and with $\ell \sim 1.5$ μm, $\tau_{F.R.}$, $\sim 1.1\ \mu b/\ell \sim 0.5\tau$, i.e. $\ell \sim 2.4\ \mu b/\tau$. Thus, in these loose bundles, $\rho_{loc} \lesssim 6 \times 10^9\ \tau^2$ [cm^{-2}] (with τ in kg/mm^2), as compared to the earlier estimate for dense bundles $\rho_{loc} \gtrsim 1.1 \times 10^{11}\ \tau^2$ [cm^{-2}].

This crude analysis indicates that the dislocation density in the bundles is not homogeneous. There can be little doubt that certain bundle regions, in which both ℓ and y are large, are transparent to glide dislocations, in particular, if τ_{eff} is large too.

f) Summary of Properties of Dislocation Structure. For practical purposes it is convenient to summarize the results of this section by classifying some of the discussed properties P of the dislocation structure into three categories:

i) $P = A/\tau$, II) $P = B\tau$, III) $P = C\tau^2$. The properties are listed in Table 6.1 with the appropriate (approximate) proportionality constants. These values apply to copper at low temperatures ($\lesssim 293$ K) and have been taken or evaluated from the work of Essmann (1965b,c), Strunk (1967) and Mughrabi (1971b,c).

The reader is cautioned against overinterpretation on two counts. Firstly, these are average values which can easily distort the physical relevance, if the distribution is not considered. This has been demonstrated by including $\tau_{F.R.}$ for

---

*The fact that figs. 6.17a and b refer to different temperatures has probably no effect on our arguments, since, if the pinning points are thermally surmountable, then ℓ in fig. 6.17a should be even larger at higher temperatures (fig. 6.17b).

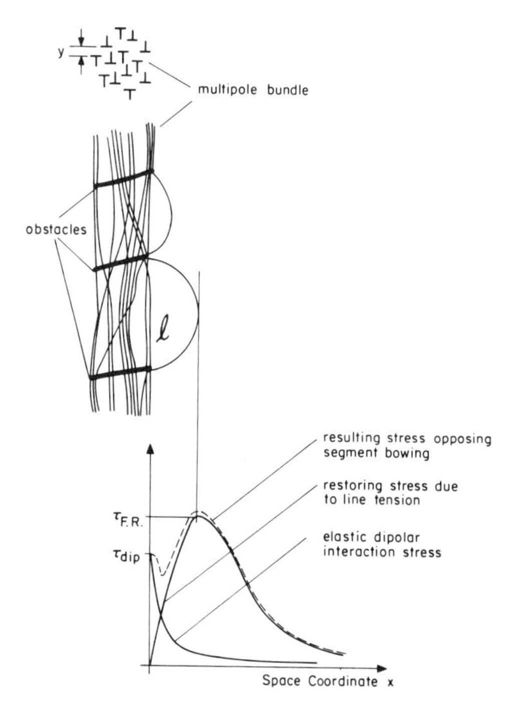

Fig. 6.18 Dislocation bowing out of multipole bundle, stress considerations neglect situation at anchoring points.

Table 6.1

Property	P = P(τ)	A/τ	$B\tau$	$C\tau^2$
	Proportionality constants	$A \left[\frac{\mu m\ kg}{mm^2}\right]$	$B \left[\frac{kg/mm^2}{kg/mm^2}\right]$	$C \left[\frac{10^8\ cm^{-2}}{kg^2/mm^4}\right]$
Bundles	a) Spacing	6		
	b) Free Segment length $\bar{\ell}$	1.2		
Grids	a) Spacing	5		
	b) Mean meshlength	0.4		
Slip	a) $L_\perp$ (surface	110		
lines	b) $L_\perp$ (bulk	40		
Wavelength	λ of τ_i	30		
	$\tau_{i,max}$		1	
	$\tau_{eff,max} = \tau + \tau_{i,max}$		2	
	$\tau_{F.R.}(\bar{\ell})$		1	
	ρ_p		10*	
	ρ_s			6.5
	ρ_{loc} in dense bundles			$> 10^3$
	ρ_{loc} in loose bundles			< 50

*Here the units are $\left[\frac{10^8\ cm^{-2}}{kg/mm^2}\right]$

the weighted average $\bar{\ell}$. The close agreement with the flow stress τ is misleading, as is evident from inspection of fig. 6.17a. Secondly, a "principle of similitude" (Kuhlmann-Wilsdorf 1968, cf. also Nabarro et. al. 1964) is not assumed, although most dimensions of the dislocation structure vary as

$1/\tau$. There are important exceptions. For example, dislocation grids (and walls at $T_d \gtrsim 300$ K) do not shrink but grow with deformation and ρ_p does not vary proportionally to τ^2. The principle of similitude does not account for such changes in the character of the dislocation structure (cf. Seeger and Wilkens 1967, Seeger 1968).

6.2.5 *The Flow Stress*

a) Introduction. Almost all models of the athermal component of the flow stress in stage II require that $\rho_p \propto \tau^2$. This applies to the theories involving long-range internal stresses (Seeger et. al. 1957, Hirsch and Mitchell 1967), the meshlength theory (Kuhlmann-Wilsdorf 1962, 1968) and implicitly to forest theories (Basinski 1959, Saada 1962). The experimental result that $\rho_s \propto \tau^2$, whereas ρ_p is not, could indicate that ρ_s correlates better with τ than ρ_p. Models that demand explicitly that $\rho_s \propto \tau^2$ are the forest theory, and, through the principle of similitude, the meshlength as well as the statistical theory (Kocks 1966). In any case, we conclude from T.E.M. that the processes that govern τ must occur in the "glide regions" (cf. Section 6.2.2).

b) Interactions with the Secondary Dislocation Forest (Hirsch - Saada Process). The athermal component of τ in Saada's calculations (1963) arises mainly from the stress $\tau_{i,f}$ necessary to break the junctions formed by the (short range) elastic attractive interaction (Hirsch 1959) of primary and forest dislocations. Recent, more refined calculations (Schoeck and Frydman 1972) of $\tau_{i,f}$ account for about 30% of the flow stress (in copper), assuming that observed values of ρ_s correspond entirely to the density of randomly distributed trees.

As discussed earlier, the most likely sites for primary-secondary forest interactions (in the glide regions) are the bundles. Since most of the secondary dislocations are arranged in the primary glide plane (in the bundles and predominantly in the grids), most do not contribute to the forest density. Still, the local forest density in the bundles could be larger than that of the random distribution assumed by Schoeck and Frydman.

However, a number of experimental facts indicate that the bundles are not stress-controlling obstacles (cf. Sections 6.2.2, 6.2.4b and 6.2.4e). In summary, it does not seem likely that $\tau_{i,f}$ is a major part of the flow stress.* It becomes evident here that other theories such as the meshlength or the statistical theory encounter similar difficulties in relating their critical inter-obstacle spacings to the experimentally observed values (cf. fig. 6.17a).

c) Dislocation Bowing in a Long-Range Internal Stress Field. We shall now discuss the behavior of free (edge) glide dislocations in a fluctuating long-range internal stress field. In an idealized one-dimensional model we can write:

$$\tau_i(x) = \tau_{i,max} \cos\left(\frac{2\pi x}{\lambda}\right), \text{ with } \tau_{i,max} \sim \tau , \qquad (6.3)$$

so that
$$\tau_{eff}(x) = \tau\left(1 + \cos\left(\frac{2\pi x}{\lambda}\right)\right), \qquad (6.4)$$

x being the space coordinate.

First, we assume that free segments of length ℓ are distributed homogeneously, averaged over the whole crystal.** The distribution is allowed to be locally inhomogeneous, as observed in practice, but we are only interested in the spatially averaged behavior. Application of the stress τ_{eff} will bow out the segments, some of them overcritically. In the absence of obstacles, the bowed-out segments will move to new positions of equilibrium, where τ_{eff} and the restoring stress due to line tension compensate (fig. 6.19). Irreversible dislocation bowing is possible only for segments lying between x_1 and x_2 ($\tau_{eff} \gtrsim \tau_{F.R.}$) which bow out to positions x_1' and x_2'.

*Essmann and Strunk (1970) conclude that at higher temperatures (473 K), because of the more frequent occurrence of complete intersections, $\tau_{i,f}$ can make a larger contribution.

**This assumption is justified, since the spacings of the bundles (containing the free segments) are smaller than λ. We need only consider dislocations of one sign, since the same arguments apply to dislocations of opposite sign.

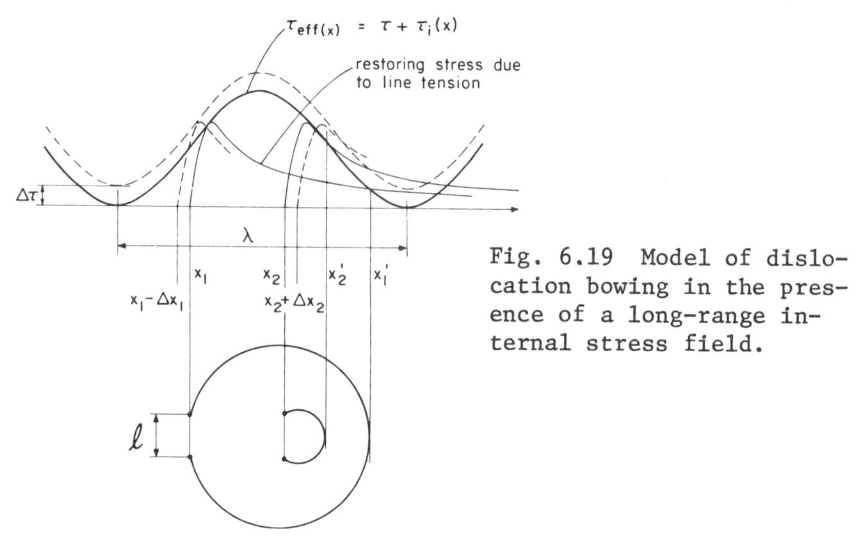

Fig. 6.19 Model of dislocation bowing in the presence of a long-range internal stress field.

Next, we consider a segment spectrum, characterized by equal densities of segments of four different lengths, corresponding to $\tau_{F.R.} = \tau/2$, τ , 2τ and 4τ , as shown in fig. 6.20 for the stress-applied and the unloaded states. The radii of the bowed-out segments correspond to τ_{eff} everywhere along the x-axis, i.e. $r(x) \propto 1/\tau_{eff}(x)$.* We note that in the stress-applied state, and to a less degree in the unloaded state, the dislocations are no longer distributed evenly along the x-axis, being clearly under-represented in regions of high τ_{eff} . An important consequence is that, in the presence of τ_i , the spatial average of reciprocal radii of curvature, i.e. $\langle 1/r \rangle$, corresponds no longer to $\langle \tau_{eff}(x) \rangle$ but to a smaller (experimental) value $\langle \tau_{eff}(x) \rangle^*$, especially if short segments ($\tau_{F.R.} > \tau$) are weakly represented. This result can be generalized qualitatively. For a distribution of segments of length ℓ about a median ℓ_m , for which $\tau_{F.R.}(\ell_m) \lesssim \tau$,

*It is emphasized again that this picture represents the spatially averaged distribution of the locations of free dislocations with respect to $\tau_i(x)$ and not any dislocation configuration one would expect to observe locally in a crystal.

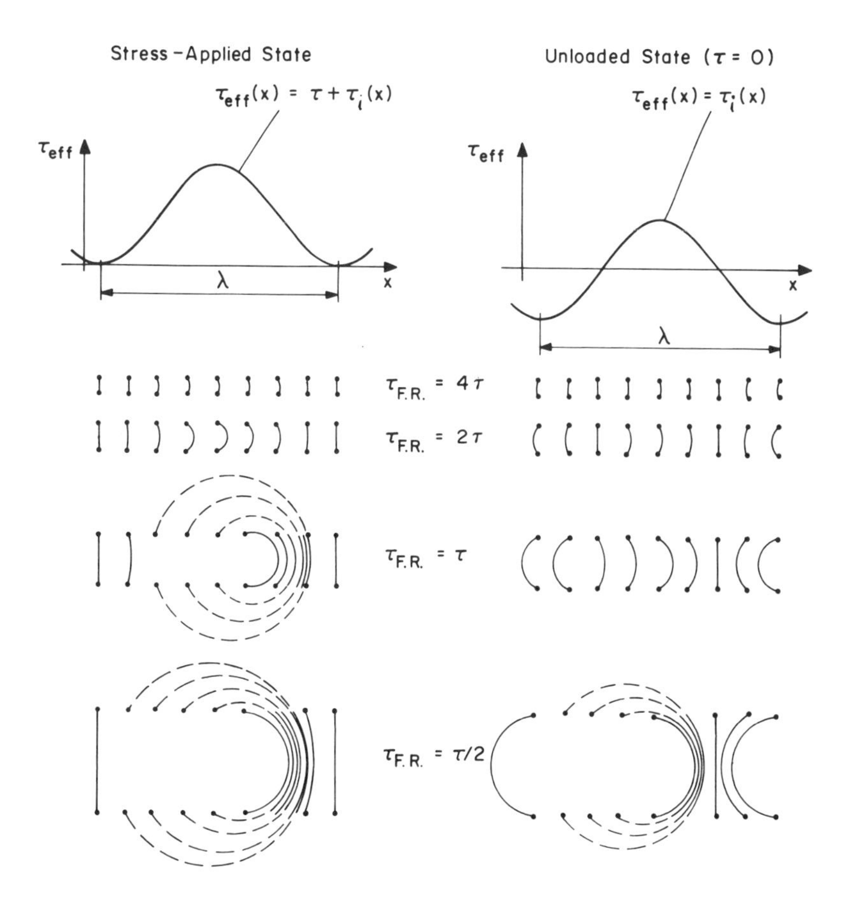

Fig. 6.20 Behavior of a spectrum of dislocation segments of different lengths in the presence of a long-range internal stress field.

$$\langle\tau_{eff}(x)\rangle^{*}/\langle\tau_{eff}(x)\rangle < 1 \tag{6.5}$$

From eqn. (6.4), we can easily calculate $\langle\tau_{eff}(x)\rangle$:

$$\langle\tau_{eff}(x)\rangle = \frac{4}{\lambda}\int_{0}^{\lambda/4} \tau_{eff}(x)\,dx \;^{*}; \quad \text{and obtain} \tag{6.6}$$

---

* Since the experimental analysis does not distinguish between dislocations of different sign, only $\langle|\tau_{eff}|\rangle^{*}$ can be determined in the unloaded state. This is considered by the integration limits $\lambda/4$ and 0 in eqn. (6.6).

$$<\tau_{eff}(x)> \;=\; \tau \qquad \text{(stress-applied state)} \tag{6.7}$$

and $$<|\tau_{eff}(x)|> \;=\; \frac{2}{\pi}\,\tau_{i,max} \qquad \text{(unloaded state, } \tau = 0) \tag{6.8}$$

From T.E.M. observations $<|\tau_{eff}(x,y)|>^*$ can be evaluated. We assume that the qualitative arguments are not affected in the two-dimensional case. To come as close as possible to the discussed one-dimensional model, only radii of curvature $r_\perp$ of edge dislocations are evaluated. Table 6.2 summarizes the results of the analysis for different specimens. In the eval-

Table 6.2

Deformation	Flow stress τ $[kg/mm^2]$	Stress-applied state $<\tau_{eff}>^*$ $[kg/mm^2]$	Stress-applied state $\frac{<\tau_{eff}>^*}{<\tau_{eff}>}$	Unloaded state $<\|\tau_{eff}\|>^*$ $[kg/mm^2]$	Unloaded state $\frac{<\|\tau_{eff}\|>^*}{<\|\tau_{eff}\|>}$
Tension (78 K)	1.2	0.64	0.53	0.28	0.46
	1.7	1.03	0.6	0.36	0.42
Fatigue (293 K)	1.6	1.6	1.0	0.76	0.83

uation of $<|\tau_{eff}|>^*$ for the unloaded state (eqn. (6.8), the experimental values $\tau_{i,max} = 0.8\tau$ (tension) and $\tau_{i,max} = 0.9\tau$ (fatigue) obtained in the unloaded state were used.

From Table 6.2 we infer that in tension $\tau_{F.R.}(\ell_m) \lesssim \tau$, whereas in fatigue $\tau_{F.R.}(\ell_m) > \tau$. This is confirmed convincingly by direct comparison with $H(\ell)$, shown earlier for two of the specimens to which Table 6.2 refers (cf. fig. 6.17). The conclusions drawn from fig. 6.17 (cf. Sections 6.2.4e, 6.2.5b) are thus verified qualitatively by this independent method.

The observed free dislocations have been shown to behave as expected in a long-range internal stress field and to much less degree as would follow from other models. This agrees with the observations of groups of long, weakly curved dislocations of the same sign, held back in obstacle-free regions. It is concluded that the interaction of primary glide dislocations with long-range internal stresses, resulting mainly from primary and secondary dislocation arrangements (parallel to the primary glide plane), accounts for a major part of the flow stress.*

6.2.6 *Work Hardening and Comments on Work-Hardening Theories*

Irreversible dislocation bowing (cf. figs. 6.19 and 6.20) and glide of dislocations that entangle in obstacles (here $\tau_{eff} > \tau_{F.R.}$ is not imperative) contribute to plastic deformation. So far, the given description refers to static equilibrium dislocation configurations in the stress-applied state and does not describe the dynamic deformation process.

Deformation can only proceed, if τ is increased, say by an increment $\Delta\tau$ (cf. fig. 6.19). τ_{eff} may be envisaged to increase momentarily by the same amount. Then shorter segments in the intervals $(x_1 - \Delta x_1)$ and $x_2 + \Delta x_2)$ can bow out irreversibly. At the same time, segments previously bowed out in regions between x_1' and x_2' can propagate further. Some will overcome the maximum opposing internal stress. Some other segments in regions of low $\tau_{eff} + \Delta\tau$ may achieve the same, with the aid of stress fluctuations due to glide activity in their vicinity. The accompanying structural changes (increased dislocation and obstacle density), resulting also from secondary slip, cause the dislocation structure to adapt itself to the increased stress level, i.e. $\tau_{i,max}$ increases, while λ, L and $\bar{\ell}$ decrease incrementally, and the situation becomes static again. The continuous, repeated occurrence of these processes in a dynamic, spatially fluctuating fashion is held responsible for work-hardening.

In this model, the crystal deforms preferentially in soft regions (Kocks 1966, Seeger and Wilkens 1967, Hirsch and Mitchell 1967, Mughrabi 1971b, Alden 1972). As these work-harden, formerly hard regions become new soft ones (cf. also fig. 6.9). Long and short segments contribute to the plastic strain, depending on the local value of τ_{eff}. Obviously, longer segments ($\tau_{F.R.} < \tau$) make a larger contribution, since a larger fraction of them can bow out irreversibly and these can travel further than shorter segments ($\tau_{F.R.} > \tau$).

---

*Measurements (by the decremental unloading technique, cf. Gibbs 1966) of the internal stress effectively opposing dislocation glide yield about 0.87 for copper single crystals in stage II (MacEwen, Kupcis and Ramaswami 1969).

Therefore, the longer segments are thought to be largely responsible for the long-range cooperative glide processes, extending over the bundle spacings. Such processes are expected to operate in local bursts, related to the partial decomposition of the bundles (cf. fig. 6.8) in regions of strong glide activity (Mughrabi 1971b, cf. Neumann 1971, Hazzledine 1971 and also Chapter 12), in agreement with surface slip line (Mader 1957) and T.E.M. observations (Mughrabi 1971b), showing that dislocations glide in groups.

This qualitative model bears remarkable similarities to the long-range stress theory in its original form (Seeger et. al. 1957), devised before T.E.M. became available and also to the theory of Hirsch and Mitchell (1967). It is recognized that structural details, such as "disguised pile-ups" (cf. Nabarro et. al. 1964), will have to be considered in a more refined treatment of the former theory (Seeger and Wilkens 1967, Seeger 1968). The theory of Hirsch and Mitchell (1967), in some respects a long-range stress theory, is able to predict eqn. (6.1), advocates a cooperative source mechanism and uses all the contributions to τ discussed above, i.e. $\tau_{F.R.}$, $\tau_{i,f}$ and τ_i, without specifying their relative contributions. It is, however, implicit to the theory that τ_i is relaxed substantially by secondary slip, contrary to observation.

The complexity and non-randomness of the dislocation structure make it evident that the ideal description requires a statistical theory.* While this is recognized (Seeger, Kronmüller, Mader and Träuble 1961, Kocks 1966, Argon 1969, Alden 1972, 1973, Schoeck and Frydman 1972), available theories apply essentially to the simpler case of hardening by point obstacles** (Kocks 1966, Alden 1972, 1973) or of solid solutions (Labusch 1970). Still, a general feature of these theories appears to be their ability to explain (qualitatively) the smooth elastic-plastic transition, a finite asymptotic yield stress and aspects of recovery. Zarka's statistical theory (1973) gives perhaps the most complete description of dislocation mechanisms, although it neglects secondary reaction products and cannot account for the formation of clustered slip

---

*It should also be pointed out that thermodynamical considerations can provide many fundamental aspects of deformation (Kronmüller 1967, Lambermont 1973).

**Argon's stage I theory (1969) considers the elastic interaction of primary dislocations.

lines. Nevertheless, stress-strain curves of the correct form are obtained (for details, we refer to Chapter 9).

As essential result of a recent stochastic model of work-hardening (Feltham 1973) is that the rate of linear stage II work-hardening is expected to depend only weakly on temperature, as is observed experimentally. In its present form, the model does not lend itself easily to a comparison with T.E.M. observations.

6.3 AN EXCURSION TO BCC METALS (α-IRON SINGLE CRYSTALS)

Essential differences in the plastic behavior of BCC metals with respect to that of FCC metals are the strong temperature and strain-rate dependence and the ease of cross-slip. In simple terms, this can be ascribed to an effective friction stress that is partly intrinsic (of the Peierls-Nabarro type) and, to the other part, very sensitive to small amounts of impurity atoms (cf. Christian 1970, Šesták 1970). The extended core structure of screw dislocations plays a decisive role. In their low-energy configuration, screw dislocations are <u>sessile</u> (Mitchell, Foxall and Hirsch 1963). The transformation to the <u>glissile</u> configuration (which immediately permits cross slip) and thus dislocation glide itself are thermally activated processes.

At intermediate temperatures ($\sim$0.1-0.2 T_m, T_m being the melting temperature), single crystals of BCC metals behave similarly to FCC crystals, exhibiting "three-stage" work-hardening and roughly comparable microstructures (cf. Hirsch 1968, Šesták and Seeger 1971)*. Figure 6.21, from the work of Keh, Spitzig and Nakada (1971) shows an example of the dislocation arrangement in an α-iron single crystal strained into "stage II". Features similar to those in copper can be recognized, i.e. edge bundles and primary-secondary grid structures.

The deformation characteristics change drastically at low temperatures. Hardening is parabolic and the observed dislocation pattern is strikingly different (cf. fig. 6.22, Spitzig and Thomas 1972). As a result of the strongly reduced mobility of screw dislocations, long screw segments are drawn out by the more mobile edge dislocations (Solomon and McMahon 1968). The relative absence of extended dense tangles is related to the fact that mainly screw dislocations are formed and that annihi-

---

*Šesták and Seeger (1970) have pointed out that BCC metals do not exhibit typical stage II behavior, since the macroscopic yield stress <u>coincides</u> with the onset of cross slip.

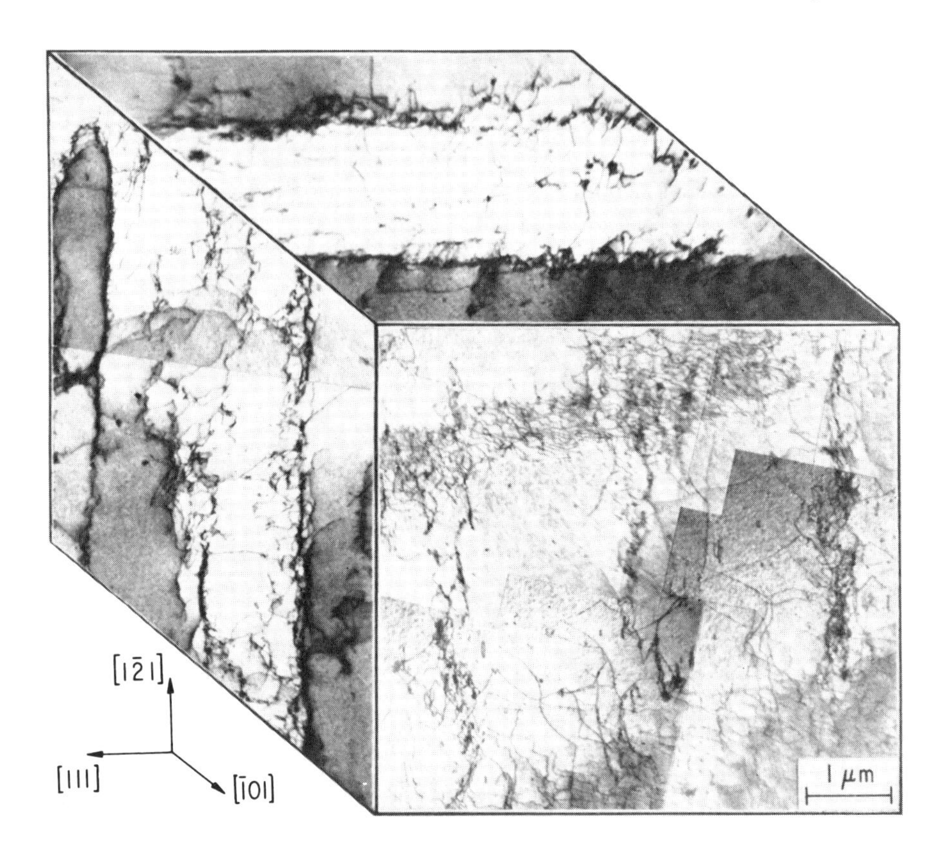

Fig. 6.21 Three-dimensional dislocation structure in α-iron single crystal deformed into stage II at 295 K ($\tau \sim 4.5$ kg/mm^2). Pinned by strain ageing at half the flow stress at 373 K, primary glide system ($\bar{1}$01) [111]. (from Keh et al. 1971, courtesy of Taylor & Francis).

lation by stress-induced cross-slip still plays a rôle (Ströhle 1973).

The flow stress is characterized by a thermal component that increases strongly with decreasing temperature and varies only little with strain. Work-hardening is essentially due to an increasing internal stress (Spitzig and Keh 1970).

6.4 WORK-HARDENED STRUCTURES IN SINGLE CRYSTALS ORIENTED FOR MULTIPLE SLIP AND IN POLYCRYSTALS

The work-hardening behavior of single crystals generally depends on the orientation of the tensile axis (cf. Diehl 1956).

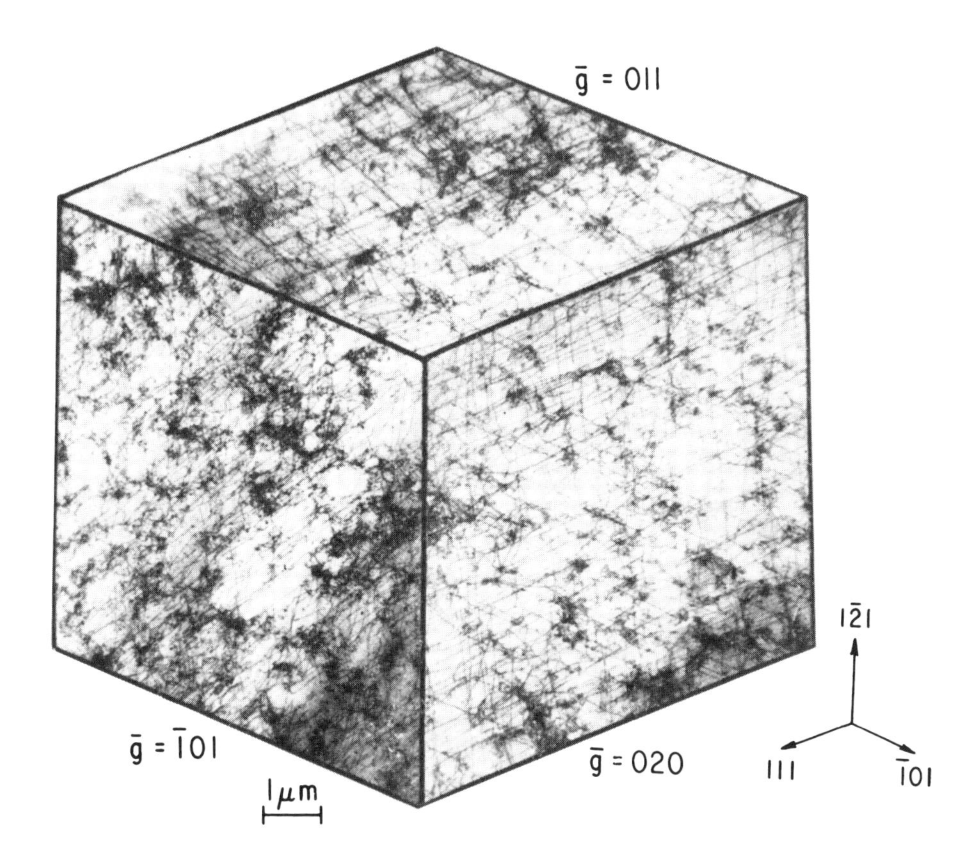

Fig. 6.22 Three-dimensional dislocation structure in α-iron single crystal deformed at 173 K ($\tau \sim 12$ kg/mm^2). Pinned by strain ageing at half flow stress at 373 K, primary glide system ($\bar{1}$01) [111]. (from Spitzig and Thomas 1972, courtesy of Taylor & Francis.

Crystals of higher symmetry (corner) orientations exhibit a general tendency to avoid intersecting multiple slip as far as possible. Thus FCC crystals of [110] orientation, having 4 equivalent glide systems (2 pairs of coplanar systems) deform on one glide plane in coplanar slip, the stress-strain curves and slip line pattern closely resembling those of single-slip orientation crystals (cf. Vorbrugg, Goetting and Schwink 1971). This is in contrast to [110] orientation BCC crystals (4 equivalent non-coplanar glide systems {111} <$\bar{1}$01>) which exhibit interpenetrating multiple slip (Keh 1965). On the other hand, [100] - oriented NaCl crystals (4 equivalent non-coplanar glide systems {110} <1$\bar{1}$0>) deform in single slip, or at least in block glide, (cf. Kear, Silverstone and Pratt 1966) as confirmed recently by T.E.M. (Strunk 1972).

Multiple slip is observed in [100] and [111] FCC (cf. Kocks 1970, Vorbrugg et. al. 1971) and BCC metals (Keh 1965). The stress-strain curves begin steeply and resemble those of polycrystals (cf. Keh 1965, Kocks 1970). Still, single slip may prevail in regions (Keh 1965, Vorbrugg et. al. 1971). In the curved region of the stress-strain curves of FCC metals, the slip line pattern reveals abundant cross slip (Vorbrugg et. al. 1970).

The dislocation structures in [100] -oriented copper single crystals have recently been investigated in detail, also after pinning in the unloaded state (Göttler 1973). A three-dimensional cell structure was observed. Dislocations of all 6 Burgers vectors occurred in comparable densities, also those that were not activated by the external stress. Figures 6.23 and 6.24 give an impression of the structure at two different

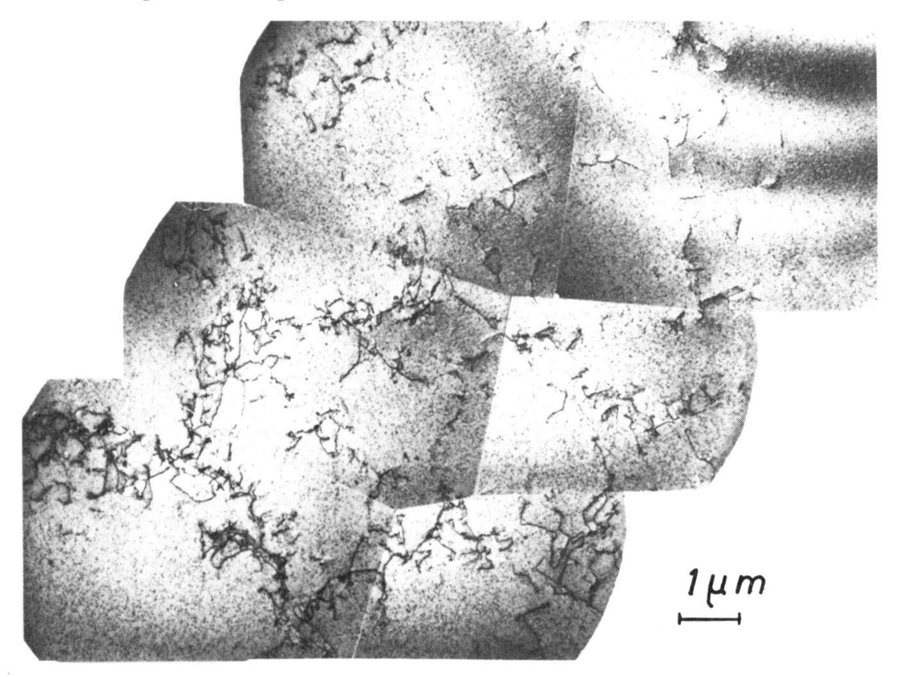

Fig. 6.23 Cu, [100] -orientation, τ = 1.66 kg/mm^2, T_d = 293 K , {111} - section, unloaded state. (from Göttler, 1975, courtesy of Taylor & Francis).

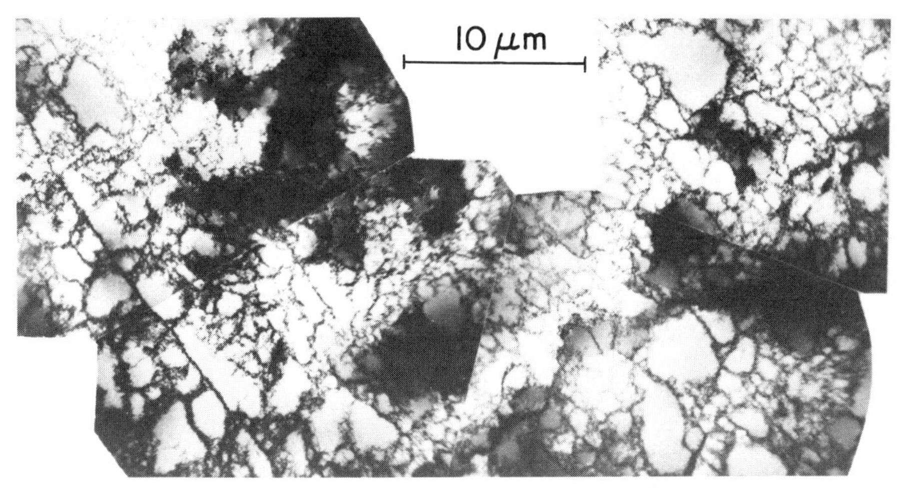

Fig. 6.24 Cu, [100] orientation, τ = 6.9 kg/mm^2, T_d= 293 K, {111} section, not pinned (Göttler 1973).

stages of deformation. In fig. 6.25 the average cell diameter $\bar{D}$ and the slip line length L (from the work of Vorbrugg et. al. 1971) are plotted against τ (resolved on the equivalent {111} <$\bar{1}$01> systems). This figure indicates that at low stresses, the (poorly developed) cell structure is transparent to glide dislocations but determines dislocation glide paths at higher stresses. Surface effects (cf. Section 6.2.4b) have not been investigated and cannot be excluded.

In addition, Göttler obtained some other quantitative relationships:

$$\tau = 0.33\ \mu b \sqrt{\rho_t} \tag{6.9}$$

$$\bar{\ell} = \frac{0.7}{\tau}\ [\mu m]\ ,\quad \tau \text{ in kg/mm}^2 \tag{6.10}$$

$$\tau_{i,max} \sim \tau \tag{6.11}$$

Comparing these results with those listed in Table 6.1, it appears that multiple slip leads to smaller ℓ-values but does not relieve internal stresses significantly. For details of a work-hardening model we refer to the original paper (Göttler 1973).

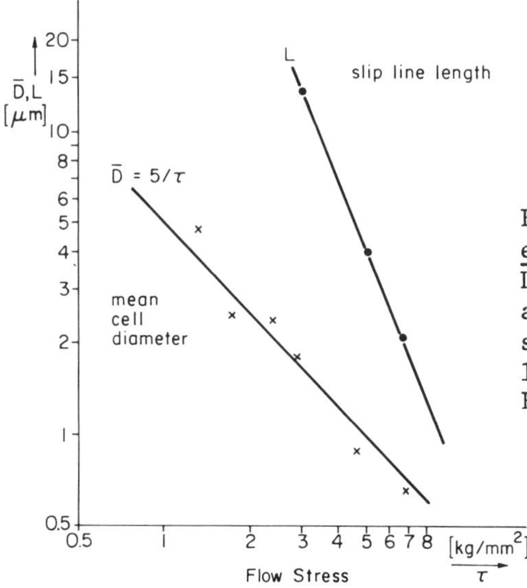

Fig. 6.25 Cu, [100] -orientation, mean cell diameter $\bar{D}$ and slip line length L as a function of flow stress τ . (from Göttler, 1975, courtesy of Taylor & Francis).

Polycrystal deformation has been reviewed many times (cf. Macherauch 1964, Kocks 1970). Since most of the grains are not in high symmetry orientations with respect to the tensile axis, it is not so surprising that Essmann, Wilkens and Rapp (1968) report that, in copper, after an initial deformation to strains $\varepsilon \lesssim 1\%$, characterized by strong work-hardening and multiple slip, most grains deform mainly on one dominant glide plane (cf. fig. 6.26).* Supporting evidence, obtained on previously neutron-irradiated specimens which develop strongly enhanced slip lines that are easy to investigate, is shown in fig. 6.27. The parallel dark and bright slip lines indicate coplanar slip on the dominant glide plane.

Multiple slip, ensuring compatibility, does occur near the grain boundaries, where considerable lattice misorientations are observed. Earlier detailed slip line studies on nickel (cf. Zankl 1963, Schwink and Vorbrugg 1967) support the work of Essmann et. al. (1968). These observations indicate that dislocation slip paths are determined by the grain size at low deformations and decrease during deformation.

Essmann et. al. attribute work-hardening to athermal and thermally activated intersection processes (in the initial region) plus superimposed increasing internal stresses, com-

*Larger grains sometimes split into sub-regions having different dominant glide planes (cf. also Schwink and Vorbrugg 1967).

Fig. 6.26 Grain in Cu showing slip on dominant glide plane, tensile stress $\sigma \sim 10$ kg/mm^2, $\varepsilon \sim 3\%$, $T_d = 78$ K. (from Essmann et. al. 1968, courtesy of Pergamon Press)

patible with their observations, during subsequent deformation. The latter interpretation follows a model by Schwink (1965). The relation obtained by Essmann et. al. between the tensile flow stress σ and ρ_t is

$$\sigma = 0.95\ \mu b \sqrt{\rho_t} \quad (\text{with } \mu = 4600 \text{ kg/mm}^2) \tag{6.12}$$

At larger deformations a 3-dimensional cell structure forms. In a detailed study of cell structures in copper over a wide range of temperatures, Staker and Holt (1972) found that

Fig. 6.27 Shadowed replica of neutron-irradiated (10^{18} n/cm^2) copper strained 4% at 78 K. (from Essmann et. al. 1968, courtesy of Pergamon Press)

$$\bar{D} = 16\ \rho_t^{-1/2} \tag{6.13}$$

Using the temperature-corrected shear moduli and standardizing the high temperature flow stresses to room temperature by dividing by the Cottrell-Stokes ratio, they obtained (with $\tau = \sigma/2$)

$$\tau = 0.57\ \mu b\ \sqrt{\rho_t}\ ^{*} \tag{6.14}$$

using $\mu = 4750\ kg/mm^2$ at room temperature. The relation between τ and D, with μ at temperature, was found to be

$$\tau = 10.5\ \frac{\mu b}{\bar{D}} \tag{6.15}$$

The forms of the relationships (6.13) and (6.15) support an analysis of cell formation by Holt (1970).

---

*Comparison of eqns. (6.12) and (6.14) could indicate a loss of dislocations in the unpinned specimens of Staker and Holt.

The influence of grain boundaries (cf. Hirth 1972) remains a central question. Grain boundaries can act as both dislocation sources and obstacles to dislocation glide. The latter forms the basis for the frequently used Hall-Petch relation. Modifications of this description have been proposed. Kocks (1970) has considered the different glide behavior in the interior of grains and near the boundaries by treating the polycrystal as a composite. An important distinction has been introduced by Ashby (1970) between "statistically stored" dislocations that accumulate during straining of pure crystals and "geometrically necessary" dislocations that ensure compatibility in plastically non-homogeneous materials. Using these ideas with some modifications, Thompson, Baskes and Flanagan (1973) have most recently been able to account for the influence of grain size on stress-strain curves of copper, aluminum (that did not follow the Hall-Petch relation) and brass.

6.5 SUMMARY

The major aim of this review was to describe and characterize quantitatively the dislocation structure in single phase metals and to relate it to the mechanical properties. For this purpose, priority was given to the experimental evidence obtained by transmission electron microscopy of dislocation structures that had been pinned in the unloaded or in the stress-applied state.

A central part was devoted to stage II work-hardening in FCC single crystals. The observed dislocation arrangements were discussed in considerable detail, special emphasis being placed on the significant rôle of free primary dislocations. Extensive use was made of the fact that these dislocations act as local microscopic stress probes. Related observations, obtained with other techniques, were surveyed.

The microstructure was evaluated quantitatively. Surface slip lines and their relation to dislocation glide paths in the interior were discussed. Dislocation densities of primary and secondary dislocations were summarized as a function of flow stress. The long-range internal stress field was characterized by an amplitude corresponding to the applied stress and by a wavelength much larger than the mean dislocation spacing. It was attributed to primary dislocation groups and pile-ups and to extended primary-secondary dislocation structures.

The experimental evidence indicates that the edge dislocation multipole bundles are not effective obstacles, that bowing of free segments out of the bundles is probably the major source mechanism and that the bundles are the most likely sites for intersections of primary glide and secondary dislocations. An estimate was given of the local dislocation densities in bundles. The frequency distribution of free segment lengths in the bundles was evaluated. It followed that, with the aid of the internal stress, most free segments are potential dislocation sources. A number of arguments indicated that neither the interaction with the dislocation forest nor Frank-Read bowing can account for the major part of the flow stress.

The experimentally observed spatial distribution of dislocation curvatures was found to be compatible with the behavior of a spectrum of dislocation segments of different lengths in a long-range internal stress field. A qualitative model of work-hardening, involving long-range cooperative slip processes, was described. It was concluded that, among the available work-hardening theories, only those of the long-range internal stress type were applicable.

A short description was given of the dislocation properties in BCC metals. At intermediate temperatures, BCC crystals have been found to exhibit similar stress-strain curves and dislocation structures as FCC crystals. At low temperatures, however, a markedly different deformation behavior is observed, characterized by a high "lattice friction" stress and the strongly reduced mobility of screw dislocations.

In the final part, the mechanical properties and the work-hardened structures of single crystals of high symmetry orientations and of polycrystals were briefly reviewed. A number of observations indicated that intersecting multiple slip is improbable, unless crystals or grains are in high symmetry orientations with respect to the tensile axis. Coplanar slip on a dominant glide plane has been reported by different authors. Some recent quantitative data on observed cell structures were presented. Finally, the significant differences between the glide behavior in the interior of grains and near grain boundaries, reflecting the condition of compatibility across the grains, was mentioned.

ACKNOWLEDGEMENT

This work has benefitted by the continued interest and support of Professor A. Seeger and Dr. M. Wilkens throughout the years. Dr. U. Essmann, R. Frydman, Professor H. Kronmüller, Dr. H. Strunk and many other colleagues have contributed in numerous discussions. Dr. E. Göttler has kindly made available his results prior to publication. This help is deeply appreciated. The financial support of the Deutsche Forschungsgemeinschaft is gratefully acknowledged.

Note added in proof:

More recently, Himstedt and Neuhäuser (1972) and Garner and Alden (1974) observed significantly larger slip lines on deformed copper single crystals than those reported earlier by Mader (1957). These authors used material of higher purity and, in the work of Garner and Alden, lower deformation temperatures. It is therefore to be expected that the quantitative microstructural data obtained from the T.E.M. observations reported here for high purity copper single crystals deformed at 78 K should compare better with the results of these more recent slip line studies.

Garner, A., and Alden, T.H., (1974) Phil. Mag. 29, 323.

Ahlers, M., and Haasen, P., (1962) Acta Met., 10, 977.

Alden, T. H., (1972) Phil. Mag., 25, 785; (1973) Met. Trans., 4, 1047.

Argon, A. S., (1969) Physics of Strength and Plasticity, edited by A. S. Argon (Cambridge, Mass: MIT Press), p. 217.

Ashby, M. F., (1970) Phil. Mag., 21, 399.

Basinski, Z. S., (1959) Phil. Mag., 4, 393; (1964) Discuss. Faraday Soc., 38, 93.

Basinski, Z. S., and Basinski, S. J., (1964) Phil. Mag., 9, 51.

Basinski, S. J., and Basinski, Z. S., (1970) Proc. 2nd Int. Conf. on Strength of Metals and Alloys (Metals Park, Ohio: ASM), p. 189.

Blewitt, T. H., Coltman, R. R., and Redman, J. K., (1955) Rep. Conf. Defects in Solids, (London: Physical Society), p. 369.

MacEwen, S. R., Kupcis, O. A., and Ramaswami, B., (1969), Scripta Met., 3, 441.

Buck, O., and Essmann, U., (1964) Phys. Stat. Sol., 4, 143.

Christian, J. W., (1970) Proc. 2nd Int. Conf. on Strength of Metals and Alloys (Metals Park, Ohio: ASM), p. 31.

Crump, III, J. C., and Young, Jr., F. W., (1968) Phil. Mag., 17, 381.

Diehl, J., (1956) Z. Metallk., 47, 331.

Diehl, J., and Berner, R., (1960) Z. Metallk., 51, 522.

Edington, J. W., (1969) Phil. Mag., 19, 1189.

Essmann, U., (1963) Phys. Stat. Sol., 3, 932; (1964) Acta Met., 12, 1468; (1965a and b) Phys. Stat. Sol., 12, 707, 723; (1966) Phys. Stat. Sol., 17, 725.

Essmann, U., and Strunk, H., (1970) Z. Metallk., 61, 667.

Essmann, U., Wilkens, M., and Rapp, M., (1968) Acta Met., 16, 1275.

Essmann, U., and Rapp, M., (1973) Acta Met., 21, 1305.

Feltham, P., (1973) J. Phys. D: Appl. Phys., 6, 2048.

Fourie, J. T., (1967) Canad. J. Phys., 45, 777; (1968) Phil. Mag., 17, 148; (1970) Phil. Mag., 21, 977; (1971) Proc. South African Electron Microscopy Soc., (Pretoria: SAEMS) p. 57.

Fourie, J. T., and Murphy, R. J., (1962) Phil. Mag., 7, 1617.

Gibbs, G. B., (1966) Phil. Mag., 13, 317.

Göttler, E., (1973) Doctorate Thesis, Technical University Braunschweig.

Hazzledine, P. M., (1964) Discuss. Faraday Soc., 38, 184; (1971) Scripta Met., 5, 847.

Himstedt, N., and Neuhäuser, H., (1972) Scripta Met., 6, 1151.

Hirsch, P. B., (1959) Internal Stresses and Fatigue in Metals, edited by G. M. Rassweiler and W. L. Grube (Amsterdam: Elsevier), p. 139; (1963) The Relation between the Structure and the Mechanical Properties of Metals, Vol. I, (London: S.H.M.O.) p. 39; (1968) Proc. Int. Conf. Strength of Metals and Alloys, Trans. Japan Inst. Metals., Suppl. 9, XXX.

Hirsch, P. B., and Mitchell, T. E., (1967) Canad. J. Phys., 45, 663.

Hirth, J. P., (1972) Met. Trans., 3, 3047.

Holt, D. B., (1970) J. Appl. Phys., 41, 3197.

Honeycombe, R. W. K., (1951) J. Inst. Met., 80, 49.

Kear, B. H., Silverstone, C. E., and Pratt, P. L., (1966) Proc. Brit. Ceram. Soc., 6, 269.

Keh, A. S., (1965) Phil. Mag., 12, 9.

Keh, A. S., Spitzig, W. A., and Nakada, Y., (1971) Phil. Mag., 23, 829

Kocks, U. F., (1964) Trans. AIME, 230, 1160; (1966) Phil. Mag., 13, 541; (1970) Met. Trans., 1, 1121.

Kronmüller, H., (1967) Canad. J. Phys., 45, 631.

Kronmüller, H., and Marik, H. J., (1972) Phil. Mag., 26, 523.

Kuhlmann-Wilsdorf, D., (1962) Trans. AIME, 224, 1047; (1968) Work-Hardening, edited by J. P. Hirth and J. Weertman (New York: Gordon and Breach), p. 97.

Labusch, R., (1970) Phys. Stat. Sol., 41, 659.

Lambermont, J. H., (1973) unpublished work.

Livingston, J. D., (1962) Acta Met., 10, 229.

Macherauch, E., (1964) Z. Metallk., 55, 60.

Mader, S., (1957) Z. Phys., 149, 73; (1963) Electron Microscopy and Strength of Crystals, edited by G. Thomas and J. Washburn (New York: Interscience Publishers), p. 183.

Mader, S., Seeger, A., and Thieringer, H. M., (1963) J. Appl. Phys., 34, 3376.

Mitchell, T. E., (1964) Progr. Appl. Mat. Res., 6, 117.

Mitchell, T. E., Foxall, R. A., and Hirsch, P. B., (1963) Phil. Mag., 8, 1895.

Mitchell, T. E., and Thornton, P. R., (1964) Phil. Mag., 10, 315.

Moon, D. M., and Robinson, W. H., (1967) Canad. J. Phys., 45, 1017.

Mughrabi, H., (1968) Phil. Mag., 18, 1211; (1969) J. Sci. Instr., 2, 351; (1970a) Proc. VIIth Int. Conf. on Electron Microscopy, Grenoble, Vol. II, p. 267; (1970b) Phys. Stat. Sol., 39, 317; (1971a, b and c) Phil. Mag., 23, 869, 897, 931; (1971d) Phys. Stat. Sol. (b), 44, 391; (1973) Proc. 3rd Int. Conf. on Strength of Metals and Alloys, (Cambridge: Inst. Metals) p. 407.

Nabarro, F. R. N., Basinski, Z. S., and Holt, D. L., (1964) Adv. Phys., 13, 193.

Neumann, P., (1971) Acta Met., 19, 1233.

Obst, B., Auer, H., and Wilkens, M., (1968/69) Mater. Sci. Eng., 3, 41.

Pande, C. S., and Hazzledine, P. M., (1971a, b) Phil. Mag., 24, 1039, 1393.

Ramsteiner, F., (1966) Mater. Sci. Eng., 1, 206; (1966/67) Mater. Sci. Eng., 1, 281.

Saada, G., (1963) Electron Microscopy and Strength of Crystals, edited by G. Thomas and J. Washburn (New York: Interscience), p. 651.

Salama, K., and Roberts, J. M., (1970) Scripta Met., 4, 749.

Schmid, E., and Boas, W., (1935) Kristallplastizität (Berlin: Springer).

Schoeck, G., and Frydmann, R., (1972) Phys. Stat. Sol. (b), 53, 661.

Schwink, Ch., (1965) Phys. Stat. Sol., 8, 457.

Schwink, Ch., and Vorbrugg, W., (1967) Z. Naturf., 22a, 626.

Seeger, A., (1954) Phil. Mag., 45, 771; (1957) Dislocations and Mechanical Properties of Crystals, edited by J. C. Fisher, W. G. Johnston, R. Thomson and T. Vreeland, Jr. (New York: John Wiley), p. 243; (1963) The Relation between the Structure and the Mechanical Properties of Metals, Vol. I (London: S.H.M.O.), p. 4; (1968) Work Hardening, edited by J. P. Hirth and J. Weertman (New York: Gordon and Breach), p. 27.

Seeger, A., Diehl, J., Mader, S., and Rebstock, H., (1957) Phil. Mag., 2, 323.

Seeger, A., Kronmüller, H., Mader, S., and Träuble, H., (1961) Phil. Mag., 6, 639.

Seeger, A., and Wilkens, M., (1967) Realstruktur und Eigenschaften von Reinststoffen, edited by E. Rexer (Berlin: Akademie-Verlag), Vol. III, p. 29.

Šesták, B., (1972) 3rd Int. Conf. Reinststoffe in Wissenschaft und Technik, Dresdcn 1970, Akademie-Verlag, Berlin, p. 221.

Šesták, B., and Seeger, A., (1971) Phys. Stat. Sol. (b), 43, 433.

Smith, C. S., and Guttmann, L., (1953) Trans. AIME, 197, 81.

Solomon, H. D., and McMahon, Jr., C. J., (1968) Work Hardening, edited by J. P. Hirth and J. Weertman (New York: Gordon and Breach), p. 311.

Spitzig, W. A., and Keh, A. S., (1970) Met. Trans., 1, 3325.

Spitzig, W. A., and Thomas, L. E., (1972) Phil. Mag., 25, 1041.

Staker, M. R., and Holt, D. L., (1972) Acta Met., 20, 569.

Stark, X., and Mughrabi, H., (1973) unpublished work.

Steeds, J. W., (1966) Proc. Roy. Soc. A, 292, 343.

Steeds, J. W., and Hazzledine, P. M., (1964) Discuss. Faraday Soc., 38, 103.

Ströhle, D., (1973) Doctorate Thesis, University of Stuttgart.

Strunk, H., (1967) Diplom Thesis, University of Stuttgart; (1970) Phil. Mag., 21, 857; (1972) Phys. Stat. Sol. (a), 11 K, 105.

Strunk, H., and Essmann, U., (1969) Z. Metallk., 60, 367.

Thompson, A. W., Baskes, M.I., and Flanagan, W. F., (1973) Acta Met., 21, 1017.

Vorbrugg, W., Goetting, H.Ch., and Schwink, Ch., (1971) Phys. Stat. Sol. (b), 46, 257.

Whelan, M. J., (1958) Proc. Roy. Soc. A, 249, 114.

Wilkens, M., (1967) Canad. J. Phys., 45, 567.

Wilkens, M., and Eckert, K., (1964) Z. Naturf., 19a, 1459.

Young, Jr., F. W., (1968) Dislocation Dynamics, edited by A. R. Rosenfield, G. T. Hahn, A. L. Bement and R. I. Jaffee (New York: McGraw-Hill), p. 313.

Young, Jr., F. W., and Sherrill, F. A., (1967) Canad, J. Phys., 45, 757.

Zankl, G., (1963) Z. Naturf., 18a, 795.

Zarka, J., (1973) J. de Mécanique, 12, 275.

LIST OF SYMBOLS

x,y	space variables
ε	strain
γ	resolved shear strain
γ_t	resolved total shear strain amplitude in cyclic deformation
γ_p	resolved plastic shear strain amplitude in cyclic deformation
σ	tensile stress
τ	(external) resolved shear stress, flow stress
τ_{dip}	edge dipole passing stress
$\tau_i = \tau_i(x,\ldots)$	space-dependent (long-range) internal stress-field
$\tau_{eff} = \tau_{eff}(x,\ldots)$	space-dependent "effective" shear stress, sum of τ and τ_i
$\tau_{i,max}$	maximum value (amplitude) of τ_i
$\tau_{eff,max}$	maximum value of τ_{eff}, sum of τ and $\tau_{i,max}$
$\tau_{i,f}$	shear stress necessary to overcome (short-range) elastic interaction during intersection of forest dislocations
$\tau_{F.R.}$	Frank-Read bowing stress
L	slip paths of dislocations
r	radius of curvature of dislocations
λ	wavelength of τ_i (and τ_{eff})
y	edge dipole width
D	cell diameter

ℓ	length of "free" dislocation segments between anchoring points in dislocation bundles and networks
$H(\ell)$	frequency distribution of lengths ℓ
ℓ_m	median of $H(\ell)$
ν	Poisson's ratio
θ_{II}	work-hardening coefficient in stage II
$\underline{b}$	Burgers vector
$\underline{b}_p$	primary Burgers vector
u	shear modulus
T_d	temperature of deformation
T_m	melting temperature
$\underline{g}$	diffraction vector
ρ_p	primary dislocation density
ρ_s	secondary dislocation density
ρ_t	total dislocation density
ρ_{loc}	local dislocation density
$\perp$	subscript symbol for an edge dislocation
$\odot$	subscript symbol for a screw dislocation

ABSTRACT. *A detailed description of the cyclic work-hardened state in Cu, Al, and α-Fe is given as a function of strain amplitude. The saturation structure in Cu cycled at low strain amplitudes consists of walls of dislocation associated with persistent slip bands (PSB's) and veins of dislocation between the PSB's. At high strain amplitudes, 2 and 3-dimensional cell structures predominate in Cu with the walls composed of edge dislocation dipoles, multipoles and point defect clusters. Cyclic work-hardened structures in Al are similar, but the transition between regions of wall, vein, and cell structure is more gradual. Cyclic work-hardened structures in α-iron are strain rate sensitive. At low strain rates and amplitudes, the structures are comparable to those found in the FCC metals, while at high strain rates dislocations are usually jogged and of a screw character. At high strain amplitudes 3-dimensional cell structures are common in α-Fe.*

The structural parameters for the three metals are tabulated as a function of strain amplitude, and a model is given for the cyclic flow stress based on the dislocation structures. Finally the stability of the cyclically work-hardened state is discussed.

7. DESCRIPTION OF THE WORK-HARDENED STRUCTURE AT LOW TEMPERATURE IN CYCLIC DEFORMATION

J.C. Grosskreutz and M. Mughrabi

7.1 INTRODUCTION

The work-hardened structure induced by cyclic deformation, although containing much that resembles the structure induced by monotonic deformation, must be treated on its own terms, which is the approach we take in this chapter. At the conclusion, we shall devote some space to a discussion of the similarities and differences between cyclic and unidirectional work-hardened structures.

In this chapter we restrict the scope of the discussion in three important ways. First, we describe only those structures which are produced at low temperatures; i.e., temperatures for which there are no time-dependent changes in the work-hardened structure, only cycle-dependent changes. In general, this temperature regime falls below about one-half the melting temperature, T_m. Second, we concentrate the discussion to structures which are representative of the equilibrium or saturation stage of cyclic deformation (see Section 7.2.2). Deformation structures produced during the transient stages are discussed only in the light of showing how saturation eventually occurs. Finally, we confine our discussion to those metals for which sufficient information is available for a fairly complete description. The metals are copper, aluminum, and iron.

The general objective of this chapter is, therefore, to give a qualitative and, where possible, a quantitative description of the equilibrium work-hardened structures produced by cyclic deformation of the three metals just named. In support of this objective, we shall describe the effects of stress or strain amplitude on these structures and, where known, the effects of temperature and strain rate. Having described the defect structure in this way, we shall then offer a semi-quantitative model for the formation of these structures and for the cyclic flow stress in copper. The chapter concludes with a discussion of the stability of the cyclic work-hardened structure in copper and a comparison of cyclic and unidirectional structures.

We have drawn extensively on the work of others as well as our own and references to the original literature are listed at the end of the chapter. The reader is also referred to the more recent reviews of the subject by Segall (1968), Plumbridge and Ryder (1969), and Grosskreutz (1971).

7.2 BASIC DEFINITIONS AND OBSERVATIONAL METHODS

In this section, we give some of the more commonly accepted ways of describing cyclic deformation behavior of metals. The experimental methods by which the work-hardened structure is usually observed are also discussed.

7.2.1 *The Cyclic Hysteresis Loop*

The cyclic stress-strain response of a well annealed metal is shown in fig. 7.1. Under constant amplitude plastic strain control, fig. 7.1a, the stress amplitude required to maintain the strain limit increases rapidly at first; i.e., the metal hardens. Under constant amplitude stress control, fig. 7.1b, the corresponding strain amplitude decreases rapidly during the first few cycles as the metal hardens. Note also that the metal <u>creeps</u> during this period of hardening under constant stress control.

In this chapter, we shall limit discussion to the results obtained under completely reversed cycling.

7.2.2 *Cyclic Hardening and Softening*

In the study of cyclic hardening under constant amplitude loading, the strain controlled test, fig. 7.1a, has been used

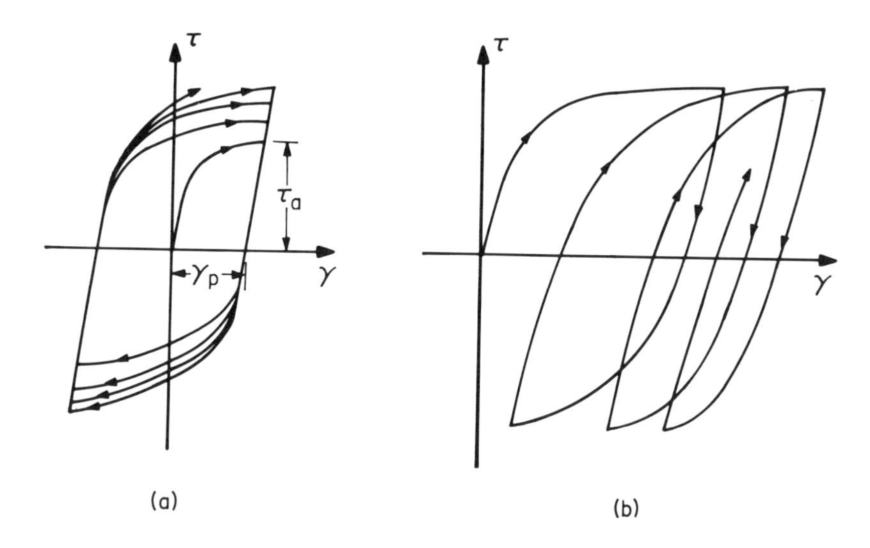

Fig. 7.1 Cyclic stress-strain loops: a) strain controlled; b) stress controlled.

extensively. The behavior of the material is followed by plotting the peak stress τ_a for each cycle against the number of cycles N. For well-annealed metals like Cu, Al, and Fe, the cyclic hardening curve is shown schematically in fig. 7.2a. For initially cold-worked metals, the stress needed to enforce the strain limit will decrease with increasing cycles and cyclic softening occurs, fig. 7.2b. In either case,

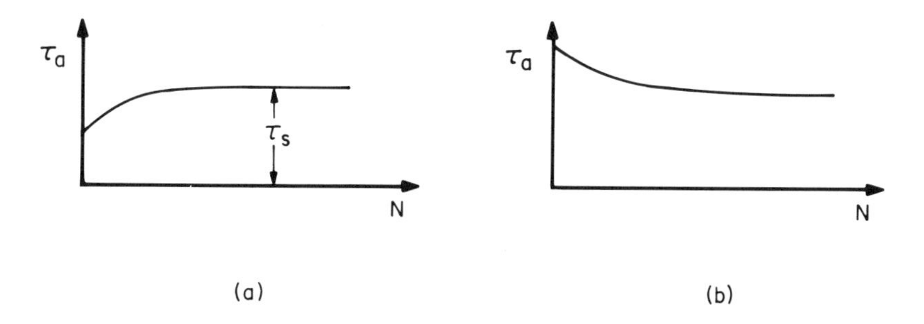

Fig. 7.2 Strain controlled cyclic hardening/softening curves: a) hardening; b) softening.

the rapid hardening or softening stage (a few per cent of total fatigue life for Cu, Al and Fe) is followed by a steady state, or saturation stage in which the peak stress remains fairly constant.* During the saturation stage, the hysteresis loop essentially retraces itself. It is interesting to note that a compressive mean stress is required to completely reverse the strain amplitude on a metal like copper in the saturation stage. That is, the steady state hysteresis loop is asymmetrical. The compressive stress required is approximately 5% larger than the tensile stress. The significance of this asymmetry is discussed further in Section 7.6.**

7.2.3 *Cyclic Stress-Strain Curve*

If a series of cyclic deformation tests are performed at different, constant strain amplitudes, and the peak, saturation stress is recorded in each case, then a plot of this saturation stress versus the applied strain amplitude is called the <u>cyclic stress-strain curve</u>. Stated another way, the cyclic stress-strain curve is the locus of the tips of the steady-state hysteresis loops, fig. 7.3a.

For most materials, the cyclic stress-strain curve is dependent on the initial state, fig. 7.3b. However, wavy slip mode materials such as Cu, Al, and Fe, have <u>unique</u> cyclic stress-strain curves which are independent of the initial work-hardened structure. The cyclic stress-strain curves for copper and iron crystals are shown in fig. 7.12.

The uniqueness of the cyclic stress-strain curve for these materials suggests that the internal microstructure at any point on the curve should be independent of prior deformation history. Experimental observation of copper cycled at fairly high strain amplitudes supports this suggestion. However, it does not generally hold true for the case of high amplitude cycling followed by low amplitude cycling. We return to this point in Section 7.3.

---

*For most materials, the peak stress increases (or decreases) at a much slower rate in the saturation stage than in the rapid hardening (or softening) stage. Kettunen and Kocks (1972) and Woods (1973) report slight softening in saturation for Cu single crystals.

**The cyclic creep illustrated in fig. 7.1b persists into the saturation stage and is a consequence of the fact that the stress cycle <u>is</u> symmetric in the case of a constant stress controlled test.

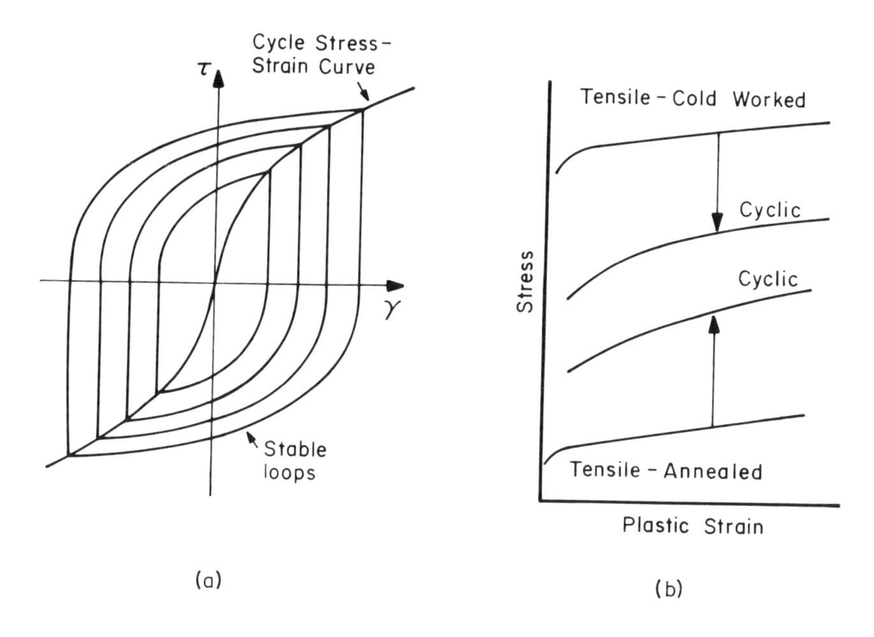

Fig. 7.3 Schematic illustration of cyclic stress-strain curves: a) construction from steady-state cyclic loops; b) comparison of tensile and cyclic stress-strain curves (after Feltner and Laird, 1967a).

7.2.4 *Homogeneity of Deformation*

Careful microscopy of the surface of Cu single crystals during rapid hardening has revealed a uniform distribution of fine slip (Basinski, et. al., 1969). After prolonged cycling, however, it is well known that slip on the surface of copper is concentrated into bands which persist well into the material. At low strain amplitudes, these bands are spaced a few microns apart, and the region between them contains only the fine slip which is characteristic of the rapid hardening stage. Therefore, when Cu is cycled at low amplitudes (and other wavy slip materials such as Al and Fe as well), it appears that the deformation is homogeneous during rapid hardening, but becomes inhomogeneous in the saturation stage. As the stress or strain amplitude is increased, the spacing of slip bands decreases until at some higher amplitude the entire surface is covered with persistent slip bands. Above this amplitude, the deformation appears from surface observations to be homogeneous in the saturation stage as well.

Whether or not deformation in the interior of cyclically deformed copper (or Al and Fe) is homogeneous in the saturation stage is extremely important to the problem of writing constitutive equations. The experimental evidence necessary to settle this question is limited at present, but it appears that for copper at least, cyclic work-hardened structures consist of two phases, even at high amplitudes. We treat this problem in more detail in Section 7.3.2c but wish to make the point here that cyclic deformation of Cu, Al, and Fe in the saturation stage is not likely to be homogeneous, especially at low strain amplitudes. Deformation structures observed in the interior of these materials should be viewed in this light and future experiments should be designed to reveal the extent of the inhomogeneity.

7.2.5 *Surface and Interior Observations*

Cyclic deformation structures have been observed on the surface in the form of slip line and etching patterns. The difficulty with these techniques is their limited ability to provide detailed information about internal microstructure. This limitation may be partially removed by sectioning and polishing techniques. Taper sectioning reveals the surface topography produced by slip lines and bands to a remarkable degree. Successive polishing off of thin surface layers reveals the penetration depth of features like persistent slip bands. Etching of cross-sections can be used to construct a three-dimensional picture of the microstructure revealed by the etchant.

X-rays have been used to deduce the internal microstructure mainly through use of spotty diffraction ring techniques and small angle (double-Bragg) scattering. The existence of small, misoriented subgrains or cells can be revealed in this manner and their average size and degree of misorientation measured.

Direct observation of internal microstructures has been achieved by transmission electron microscopy (T.E.M.) of thin foils cut from bulk specimens. The early observations were flawed by the fact that dislocations are lost or rearranged in the act of thinning specimens for observation. Pinning of the structures by neutron bombardment prior to specimen thinning has greatly improved the accuracy of the observations. Recently, structures representative of the <u>stress-applied</u> state have been observed by irradiating specimens under load with neutrons prior to thinning for observation in the electron microscope. This last technique, while tedious and time-consuming, produces results most representative of the actual internal structures produced during cyclic deformation.

To obtain the best description of the work-hardened structure, correlation of the results obtained by all the above techniques is necessary. However, the greatest emphasis here will be placed on those results obtained by T.E.M., especially under stress-applied conditions.

A complete description of the microstructure is possible only by constructing a three-dimensional model from T.E.M. of carefully oriented foils cut from the original specimen. Very few such models are available to us at this time, and these only for copper. The lack of three-dimensional models has led to some ambiguity in nomenclature. Perhaps the most overworked word has been "cell", which has been used in many cases to describe 2-dimensional networks of dislocation walls without regard to (or with only weakly supported arguments regarding) their 3-dimensional nature. In this chapter, the word "cell" will be used with the proper modifier to indicate whether the structure referred to is 2- or 3-dimensional in nature.

7.3 QUALITATIVE DESCRIPTION OF THE CYCLIC WORK-HARDENED STATE AT SATURATION

In this section, we give first an overview of cyclic hardening processes in FCC and BCC metals and their contribution to the attainment of saturation and the shapes of the cyclic stress-strain curves. Following this discussion, a detailed description of the cyclic deformation structures in Cu, Al, and α-Fe is given.

7.3.1 *Rapid Hardening and Attainment of Saturation*

a) General Remarks on Dislocation Properties and Behavior in FCC and BCC Metals

In the past, research work on the dislocation microstructures formed during cyclic deformation was primarily concerned with studies on different FCC metals and alloys. In the middle 1960's, the results and ideas that had evolved from a decade of intensive study of the microstructures formed in FCC metals during unidirectional deformation (see, for example, the reviews by Seeger 1957, Nabarro, Basinski and Holt 1964, Seeger and Wilkens 1967) found application and led to more systematic investigations of the cyclically-hardened state (cf. Lukáš and Klesnil 1967, 1970, 1971; Feltner and Laird 1967 a,b, 1968) and to a distinction between planar and wavy slip mode materials (Feltner and Laird 1967 a,b; McEvily and Johnston 1967). The former are characterized by a low stacking fault energy and difficult cross slip and the latter by a high stacking fault energy and easy cross slip (see Schoeck and Seeger 1955). This description is the key to understanding the unique cyclic stress-strain curves of wavy slip materials (Feltner and Laird 1967 a,b).

By contrast, the investigations of the dislocation structures of cyclically-hardened BCC metals are far less complete in spite of intensified research in recent years on unidirectionally work-hardened structures and dislocation properties of the BCC structure in general (see reviews by Hirsch 1968, Christian 1970, Šesták 1970). The similarities and differences between the dislocation mechanisms responsible for the work-hardening of unidirectionally deformed BCC and FCC metals have been worked out and reviewed (Šesták and Seeger 1971). It is our aim to try to incorporate these fundamental aspects in a general description of the dislocation structures formed in cyclically deformed FCC and BCC metals.

In the BCC lattice, screw dislocations have an extended core structure because of their three-fold symmetry (Burgers vector 1/2 <111>) and are sessile in their low energy configuration. The transformation to the glissile form is aided by both thermal activation and stress which help to overcome the effective lattice friction stress (of the Peierls-Nabarro type) associated with this core structure. Since this friction stress is the dominant resistance, screw dislocations, *once mobile, can readily cross slip* on other low-index planes of the <111>-zone.

This behavior explains the strong temperature-dependence of the flow stress of BCC metals. Below a certain temperature which is around room temperature for α-iron, the flow stress increases sharply. Above that temperature the flow stress depends only weakly on temperature at moderate strain rates (10^{-5} - 10^{-4} s^{-1}). In this higher (intermediate) temperature range (well below 0.5 T_m), the strong difference in the mobilities of edge and screw dislocations is much less pronounced and the dislocation behavior, aside from the mentioned ease of cross slip, becomes more comparable to that of FCC metals (Šesták and Seeger 1971). This is also reflected by the fact that, in this intermediate temperature range, BCC metal single crystals oriented for single slip exhibit similar unidirectional "three-stage" work-hardened curves similar to FCC metals (Mitchell, Foxall and Hirsch 1963, Hirsch 1968).

For our purpose, BCC metals can be characterized by a very high stacking fault energy (easy cross slip) and a strongly temperature *and* strain rate dependent lattice friction stress. At intermediate temperatures the dislocation behavior is similar to that in FCC metals and markedly different at low temperatures, where screw dislocations are almost immobile. In addition, the extreme sensitivity of the mechanical properties of

BCC metals to the smallest amounts of interstitial impurities (C, N, O) should be mentioned. We shall not be able to discuss this complication in any detail.

The dislocation structures formed in cyclically-hardened BCC metals in that temperature and strain rate range in which analogies exist between dislocation behavior in BCC and FCC metals will now be discussed together with those in cyclically deformed FCC metals. In the low temperature (and/or high strain rate) regime, BCC metals will be considered separately afterwards.

Assuming that edge and screw dislocations are equally mobile, plastic deformation can be described by the irreversible bowing out of dislocation segments under the action of the applied stress. Encounters of dislocations of opposite sign on closely neighboring glide planes lead to the formation of dipoles because of the attractive elastic interaction. In most cases, screw dislocation dipoles can annihilate by stress-induced cross slip (Essmann, 1964), so that mainly edge dipoles and multipole structures remain. The passing distance within which annihilation can occur is larger for materials of higher stacking fault energy. In such materials screw dislocations will not be able to travel very far before being annihilated. Repeated cross slip favors the formation of large numbers of rather short edge dipoles and loops (dislocation debris). Tetelman (1962) and Seeger (1963) have suggested different mechanisms of dipole formation. Feltner (1963, 1965) has emphasized the importance of debris formation by moving screw dislocations dragging a jog (Gilman and Johnston 1962) and subsequent pinching-off by cross slip. Another possible mechanism is based on the fact that, through repeated stress-induced double cross slip, dislocations acquire a high density of superjogs on the cross slip plane (Essmann 1964, Argon and East 1968). Encounters of such edge dislocations of opposite sign and subsequent annihilation at the intersections with jogs also lead to the formation of (jogged) edge dipole loops elongated in <121> (see fig. 7.4). Mechanisms of the formation of sessile loops, involving jogs formed by intersections, have been proposed by Hancock and Grosskreutz (1969). Regardless of the relative contributions of these different mechanisms, the important fact remains that a high (local) dislocation density can be accumulated with increasing deformation, since edge dipoles cannot annihilate by conservative glide.

In the range of low temperatures and/or high strain rates BCC metals behave quite differently. Considering screw dislocations to be much less mobile than edge dislocations, it follows that moving edge dislocations will draw out screw dislocations behind them. This has been confirmed in unidirectional

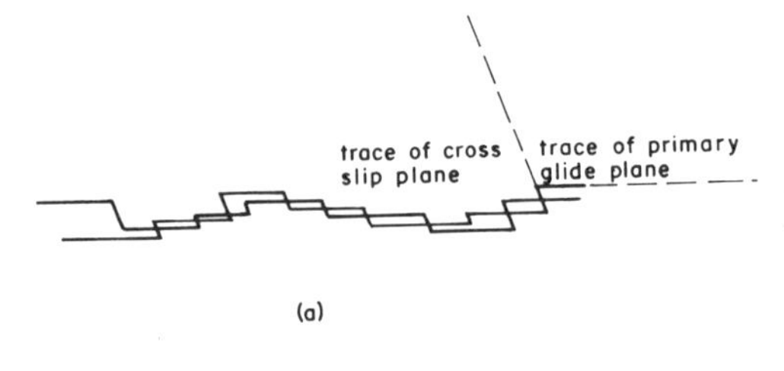

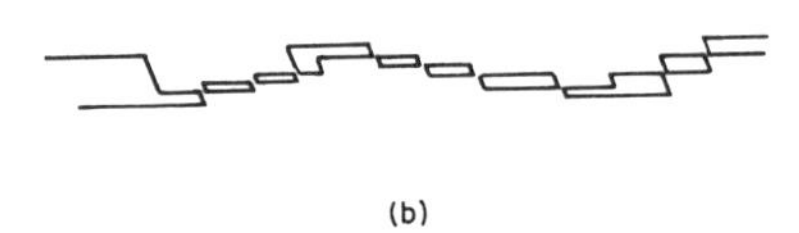

Fig. 7.4 Dipole formation by encounters of segmented edge dislocations of opposite sign. Section perpendicular to Burgers vector.

deformation experiments at low temperatures (Solomon and McMahon 1968). The edge dislocations can form dipoles in the same way as described before. The increase in dislocation density is thus mainly due to the growing density of the screw dislocations. However, annihilation by stress-induced cross slip sets a limit to the possible increase in (local) dislocation density.

b) Rapid Hardening

During cyclic deformation the dislocation behavior discussed above leads in all cases to a loss of mobile dislocations and to the formation of an obstacle structure which impedes dislocation glide. Higher stresses become necessary to continue deformation. The following mechanisms can be important:

i) Creation of new mobile dislocations by breaking up of dipole structures and/or by bowing out of dislocation segments beyond the critical Frank-Read radius.

ii) Glide of dislocations past obstacles. These can be dislocations of the same and of other glide systems activated at higher stresses and by internal stresses due to the built-up dislocation structure.

Figures 7.5 and 7.6 show examples of cyclic hardening curves of copper (Herz 1973) and high purity α-iron (Kütterer 1973) single crystals oriented for single glide. These curves were obtained in controlled plastic strain amplitude tests, the plastic strain amplitude γ_p being defined as the hysteresis loop half width at zero stress. The resolved shear stresses τ

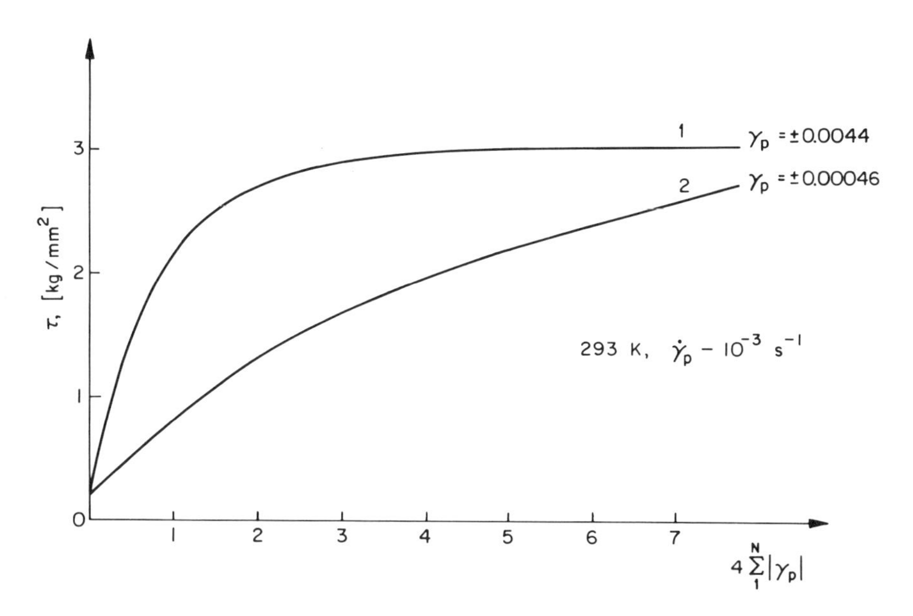

Fig. 7.5 Cyclic hardening curves of ccpper single crystals.

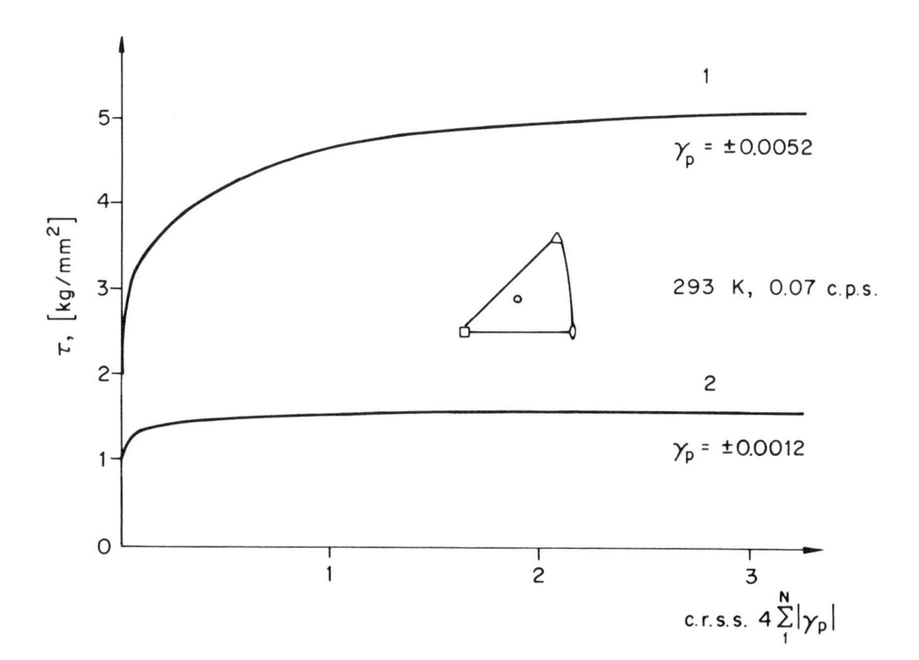

Fig. 7.6 Cyclic hardening curves of α-iron single crystals.

and the cumulative resolved shear strain $4\sum_{1}^{N}|\gamma_p|$ (c.r.s.s.) refer to the primary slip systems (111) [$\bar{1}$01] (copper) and ($\bar{1}$01) [111] (iron).

It follows from fig. 7.5 that, although the hardening rate at low strain amplitudes is much smaller than at high strain amplitudes for metals like copper, the stresses become comparable after a large number of cycles.* This remarkable ability to harden at very small strain amplitudes reflects the building up of a high dislocation density. Because of the reduced glide distance at low strain amplitudes the structural changes (including internal stresses and secondary dislocation activity) per unit strain and hence the hardening rates are much smaller than at higher amplitudes (cf. Kettunen and Kocks 1972).

Figure 7.7 shows an example of the dislocation structure in the primary glide plane (111) of a copper single crystal cycled at room temperature into the rapid hardening stage. The dislocation structure of the stress-applied state has been retained by pinning with a dose of $2 \times 10^{18}/cm^2$ fast neutrons under stress at $4.2^{\circ}K$. The curvatures of the bowed-out free primary dislocations between the dislocation bundles give an indication of the spatial variation of the local stresses. A detailed analysis revealed the existence of a spatially oscillating long-range internal stress field with an amplitude of the order of the peak stress. Curved dislocations observed in unloaded specimens confirm this (fig. 7.8). The evaluation shows furthermore that the average Frank-Read stresses necessary to bow free edge dislocations segments irreversibly out of the multiple bundles are larger than the peak stress. Source activation is hence only possible in regions where the internal stresses aid the external stress (Mughrabi 1973, cf. also Chapter 6, Section 6.2.2d).

In the case of α-iron (fig. 7.6), the ability to harden is reduced drastically at small strain amplitudes and comparable (high) shear strain rates $\dot{\gamma}_p$ ($10^{-3}s^{-1}$)**. It is important

*The observations of Kettunen and Kocks (1972) show that the true tensile flow stress is always somewhat higher than the peak cyclic stress, especially at low amplitudes.

**Initial softening has occasionally been observed in α-iron at low amplitudes. This is probably caused by the unlocking of dislocations (yield point phenomenon), since softening has not been observed in high-purity metals (low interstitial content).

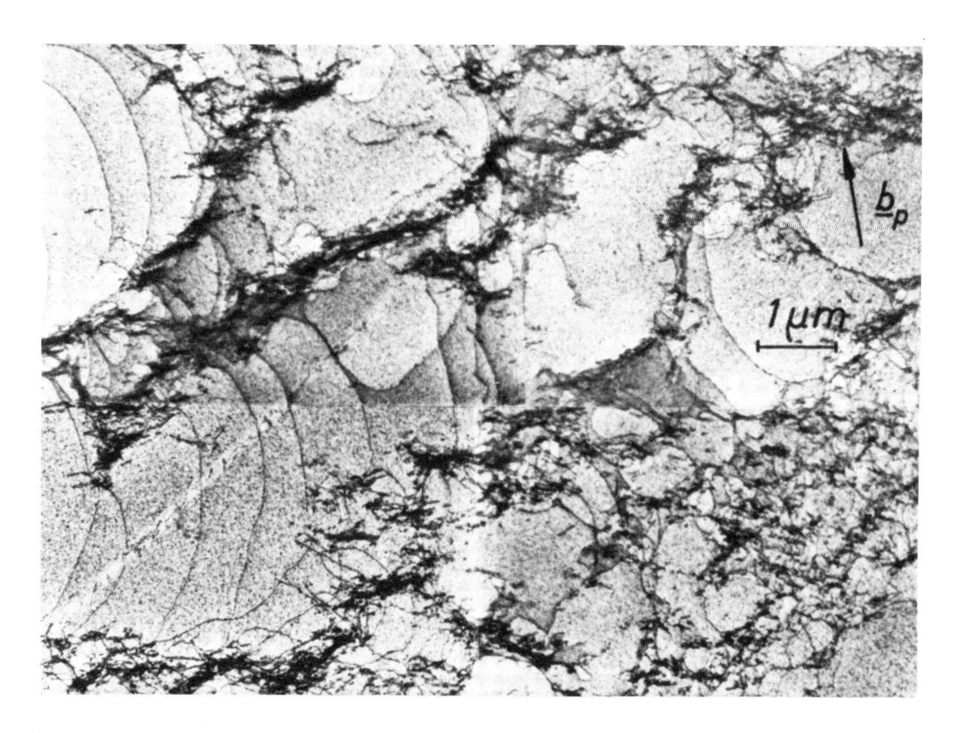

Fig. 7.7 Dislocation structure in copper single crystals in stress-applied state (293K). 15 cycles, $\gamma_t = \pm 0.0065$ (γ_t = total strain amplitude), $\tau = 1.6$ kg/mm^2. $\underline{b}_p$ is primary Burgers vector.

to recognize that, depending on the strain rate, dislocations in α-iron at room temperature can exhibit "high" or "low" temperature behavior. This is further demonstrated in fig. 7.9, showing cyclic hardening curves at two strain rates differing by a factor of only 40 (Stark and Mughrabi 1973). At the lower strain rate the peak stress increment $\Delta\tau$ per unit c.r.s.s.

$4 \sum_{N=1}^{1/4\,\gamma_p} |\gamma_p|$, normalized with respect to the shear modulus μ ($\mu_{Fe} = 7.8 \times 10^3$ kg/mm^2, $\mu_{Cu} = 4.2 \times 10^3$ kg/mm^2); i.e.,

$\Delta\tau/(\mu)\ 4 \sum_{N=1}^{1/4\gamma_p} |\gamma_p|$ (0.48 for iron, cf. fig. 7.9), approaches that of copper (0.55, cf. fig. 7.5) at comparable strain amplitudes; it is much lower (~0.2) at the higher strain rate. This simply confirms that, at room temperature, the building up of

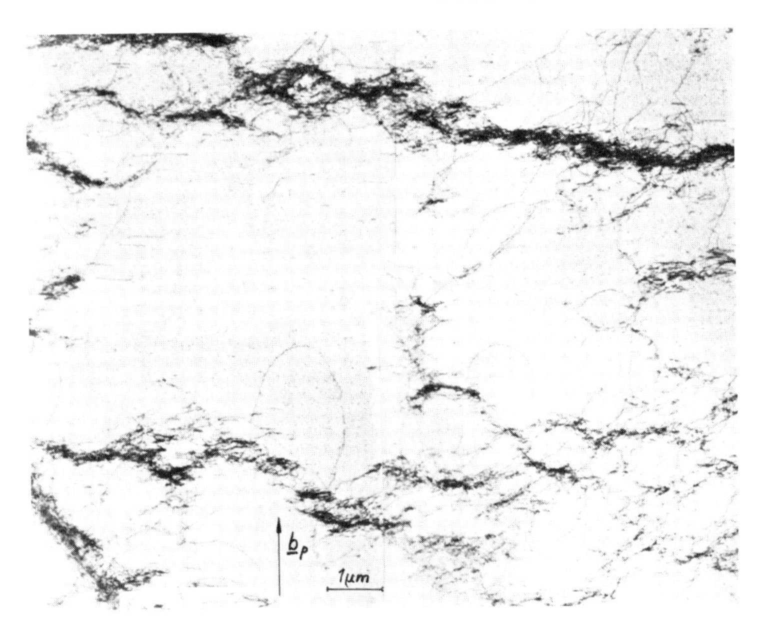

Fig. 7.8 Same as fig. 7.7, unloaded state.

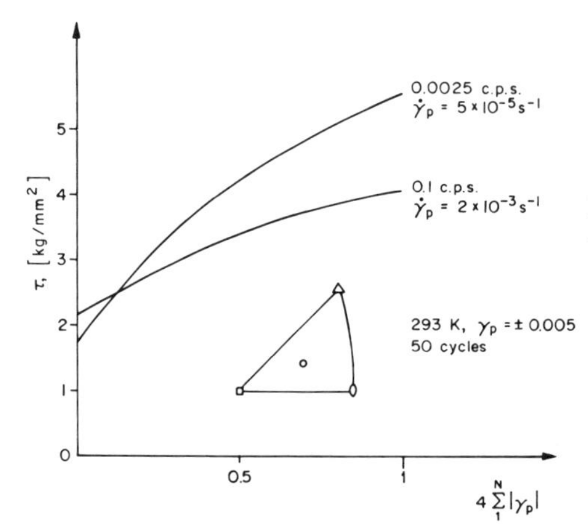

Fig. 7.9 Effect of strain rate on cyclic hardening curves of α-iron single crystals.

high dislocation densities (edge multipole structures) is favoured by low strain rates (cf. Section 7.3.1a). X-ray Berg-Barrett topographs of the primary glide plane ($\bar{1}01$) reveal a well-developed microstructure after cycling at the lower strain rate (fig. 7.10a) but not after high strain rate cycling (fig. 7.10b). The bands of extinction contrast are perpendicular to the primary Burgers vector $\underline{b}_p$ = 1/2 [111] in agreement

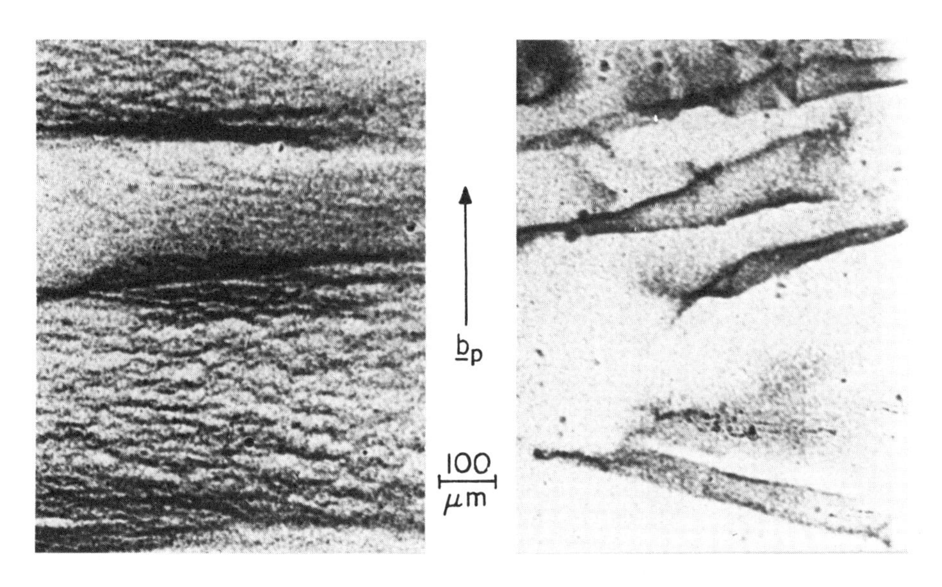

Fig. 7.10 X-ray Berg-Barrett topographs of α-iron single crystals after cyclic hardening at different $\dot{\gamma}_p$. 50 cycles, $\gamma_p = \pm 0.005$, 293 K. (002)-plane and reflection, trace of primary glide plane vertical.

with an edge multipole structure. Observations by Yoshikawa and Okamoto (1968) on cyclic hardening of iron single crystals showing that, at a given strain amplitude, the peak stress increment after a certain number of cycles at low temperatures is smaller than at room temperature, can be interpreted along the same lines.*

An example of the dislocation structure in the primary glide plane ($\bar{1}01$) of an α-iron single crystal (containing 15 wt. ppm C in solution) that was cyclically deformed well into the rapid hardening stage according to curve 1 of fig. 7.6 is shown in fig. 7.11 (Stark and Mughrabi 1973). The dislocation structure was pinned in the unloaded state by strain ageing at 80 C.

---

* A discussion of the systematic investigations of Abdel Raouf and Plumtree 1971) on low carbon steel at different cyclic strain rates <u>and</u> temperatures, involving not only cycle-dependent but also time-dependent structural charges (dynamic strain ageing) lies outside the scope of this chapter.

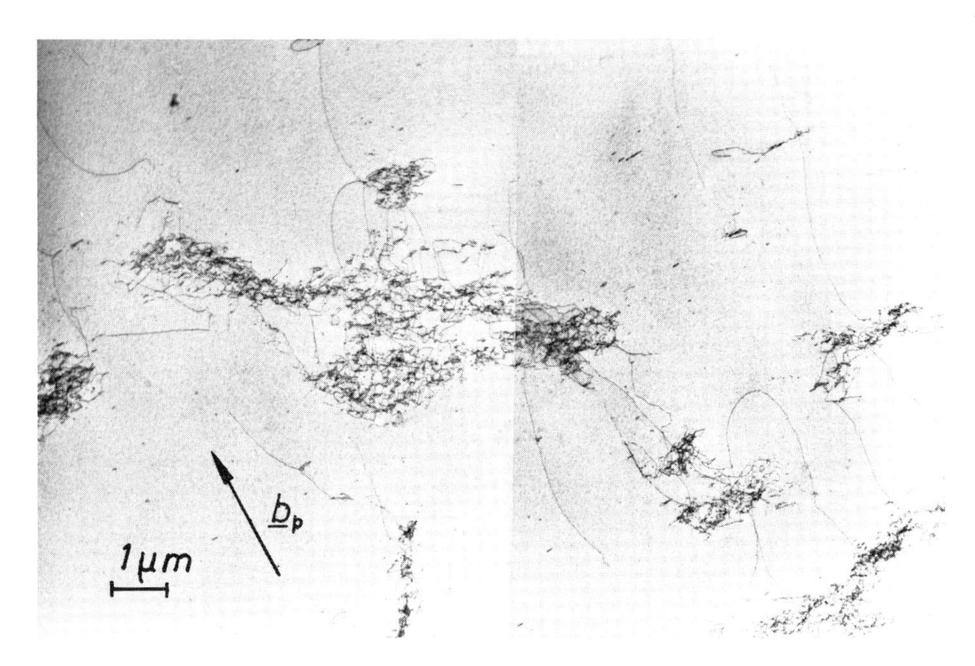

Fig. 7.11 Dislocation structure in the rapid hardening stage of α-iron single crystals (293 K). $\gamma_p = \pm 0.005$, 42 cycles, 0.1 c.p.s., $\tau = 3.6$ kg/mm^2.

In addition to some poorly developed edge multipole bundles, networks composed mainly of primary and secondary screw dislocations are observed. The elliptical shape of the free dislocations results from the equilibrium of internal stress, the lattice friction stress and line tension; dislocations of screw character predominate. The overall dislocation density ($\sim 1.5 \times 10^9$ cm^{-2}) is about one order of magnitude smaller than in the copper single crystal discussed earlier (fig. 7.7), as would be expected for "high" strain rate cycling.

c) Saturation, Cyclic Stress-Strain Curves of FCC and BCC Single Crystals (Copper and α-Iron)

Figures 7.5 and 7.6 indicate that, after a large number of cycles (the number being smaller for higher strain amplitudes) the cyclic hardening rates approach zero. This implies that a dislocation structure has formed that can accommodate the imposed plastic strain without significant hardening (cf. Section 7.5). However, the attainment of saturation need not necessarily mean that the dislocation structure undergoes no further changes. A constant saturation stress τ_s is compatible with a slowly increasing dislocation density and a gradual rearrangement of the dislocation structure into one of lower energy (cf. Section 7.4.4). Typical dislocation structures in the saturation stage range from dipolar and multipolar configurations (low amplitudes) to three-dimensional cell structures (high amplitudes). These structures are described in Sections 7.3.2-4.

Feltner and Laird (1967a, 1969) have shown that copper and α-iron polycrystals have unique cyclic stress-strain curves. Cyclic stress-strain curves for copper and α-iron single crystals oriented for single glide are shown in fig. 7.12. The

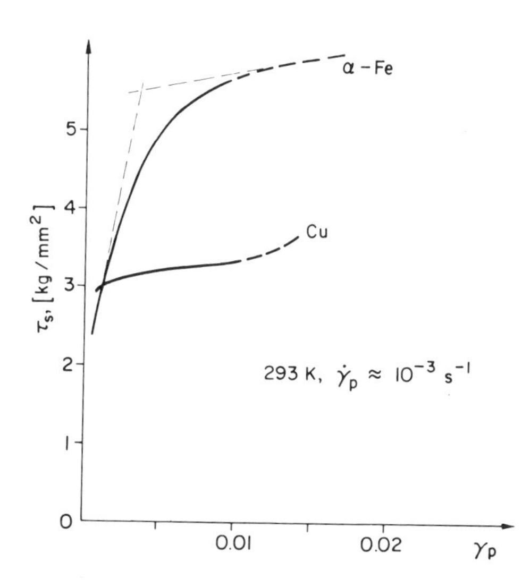

Fig. 7.12 Cyclic stress-strain curves of copper and α-iron single crystals.

curve for copper was constructed on the basis of our own work (Hancock and Grosskreutz 1969, Herz 1973 and Mughrabi 1973) and data of other workers (Atkinson, Brown, Kwadjo, Stobbs, Winter and Woods 1973, quoting also unpublished work of Watt (1967) and Gostelow and Segall (1969)). The corresponding curve for α-iron, representative of cyclic deformation "high" strain rates, is preliminary and is based on the work of Kütterer (1973) and Stark and Mughrabi (1973).

The two curves exhibit significant differences in the range $\gamma_p < 0.01$. At very low amplitudes, ($\gamma_p < 0.001$) there is hardly any available experimental data. Atkinson et al. (1973) express doubt whether copper crystals can be saturated at all in this range (cf. also fig. 7.5). Considering also recent work on polycrystalline copper (Lukáš and Klesnil 1973), it appears safe to conclude from fig. 7.12 that in the range $0.0001 < \gamma_p < 0.01$, τ_s depends only weakly on γ_p for copper single crystals (cf. Section 7.3.2c), whereas it increases strongly with γ_p for α-iron single crystals at the given strain rate.

The strongly reduced cyclic hardenability of α-iron with decreasing strain amplitude can be explained by taking into account that, under the given experimental conditions (cf. Sections 7.3.1a,b), screw dislocations are much less mobile than edge dislocations. Small strains (microstrains) can be accommodated by the motion of edge dislocations over short distances at stresses of the order of the friction stress. A certain increase in (screw) dislocation density results. Since the friction stress effectively prevents large scale (unstable) dislocation glide at low stresses, the overall structural changes (obstacle density) and the associated hardening increments are small. Higher strains require glide of edge <u>and</u> screw dislocations (cross slip!) over larger distances and hence higher stresses. This leads to more severe structural changes, higher obstacle densities, and larger dislocation hardening.

On the basis of these arguments, a division of the cyclic stress-strain curve of α-iron into ranges of weak (low amplitude) and strong (high amplitude) hardenability (as indicated by dashed lines in fig. 7.12) is justified from a fundamental point of view. This relates well to the fact that α-iron exhibits a fatigue limit (low amplitude range). The fatigue limit in pure α-iron could thus be that stress or strain level above which screw dislocations can move over larger distances (and hence cross slip, cf. Section 7.3.1a) <u>at the strain rate of the fatigue test</u>.

A similar distinction does not follow convincingly for copper from the shape of the cyclic stress-strain curve. This agrees with the fact that dislocation behavior in copper does not change fundamentally with stress amplitude. Of course, mechanisms such as cross slip, secondary dislocation activity, etc. do depend on stress or strain amplitude and influence the dislocation structure which is formed. However, on the basis of metallographic and T.E.M. evidence, it is common practice for copper and other FCC metals to distinguish between high and low amplitudes (Avery and Backofen 1963) with respect to lifetimes as well as to hardening mechanisms (e.g. Nine and Bendler

1964, Kettunen 1967, Feltner and Laird 1967 b, Nine and Kuhlmann-Wilsdorf 1967).*

7.3.2 *Copper*

The saturation work-hardened structures produced by cyclic deformation of copper are described most conveniently in terms of the strain amplitudes which produce them.

a) Saturation Structures at Low Strain Amplitudes. "Low" strain amplitudes are those amplitudes which produce fatigue lives of the order of 10^6 cycles or greater. For copper single crystals, this corresponds approximately to $\gamma_p \lesssim .002$. The work-hardened structures produced in single crystals at low amplitude have been observed by Laufer and Roberts (1964, 1966), Lukáš, et. al. (1968), Roberts (1969), and by Woods (1973), using TEM. Polycrystalline copper deformed in this range has been the subject of TEM study by Lukáš et. al. (1966), and Lukáš and Klesnil (1973). The most common structures are walls of dislocation (ladder structure) associated with persistent slip bands (PSB's), and veins (bundles) of dislocation associated with the material between PSB's. It is usually assumed that the majority of plastic deformation in the saturation stage occurs within the PSB's.

A three-dimensional, composite micrograph of the vein structure has been given by Basinski, Basinski and Howie (1969), fig. 7.13. Although their specimens were not cycled into the saturation stage, they were sufficiently hardened to illustrate this type of low amplitude structure. The interlocking veins, consisting mostly of primary edge dislocations, are visible on the $(\underline{1}11)$ slip plane. The veins are shown in cross-section on the $(\bar{1}11)$ cross-glide plane in fig. 7.13. A true cross-section is obtained by viewing the $(1\bar{2}1)$ plane. This view is shown in fig. 7.14 which is taken from a crystal cycled into saturation at $\gamma_p = .001$ (Woods, 1973). At this strain amplitude, about 60-90 percent of the crystal volume is occupied by the vein structure. Within these regions, the veins themselves occupy roughly 50 percent of the available volume. Increasing the

* The stress level that divides high and low amplitude fatigue has been associated with τ_{III} , the stress at the onset of stage III in unidirectional deformation. It is evident from the work of Kettunen and Kocks (1969, 1972) that this distinction is more appropriate to constant stress than constant strain cycling.

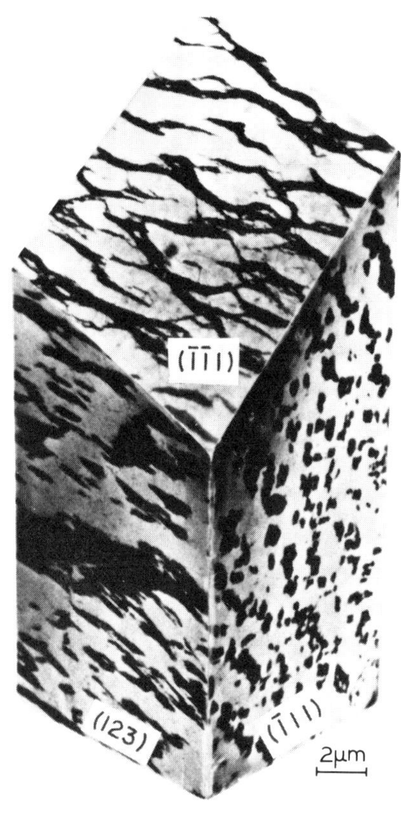

Fig. 7.13 Three-dimensional representation of the dislocation arrangement in a nearly hardened copper single crystal, $\gamma_p = \pm 0.003$. (from Basinski, et. al., 1969, courtesy of Taylor & Francis)

strain amplitude to $\gamma_p = 0.003$ seems to produce better organized or more tidy vein structures. Figure 7.15 shows a (111) section containing these veins, and fig. 7.16 shows a ($1\bar{2}1$) cross-section through the vein structure at this amplitude. (Woods, 1973, refers to these structures as "uncondensed walls".)

The wall or ladder structure (associated with persistent slip bands) is shown in fig. 7.17 as a ($1\bar{2}1$) section where it is enclosed on both sides by the vein structure. These walls consist mainly of primary edge dipoles. The distance to which these walls extend in the ($1\bar{2}1$) direction (into the paper, fig. 7.17) is subject to controversy. Lukáš, et. al. (1968) present evidence that the walls exhibit a 2-dimensional cell structure when viewed on the (111) plane. The average length of the wall (cell size) is about 3 μm in the [$1\bar{2}1$] direction. Woods (1973), on the other hand, maintains that walls may extend for hundreds of microns in the [$1\bar{2}1$] direction and that a two-dimensional cell structure is not representative of PSB's

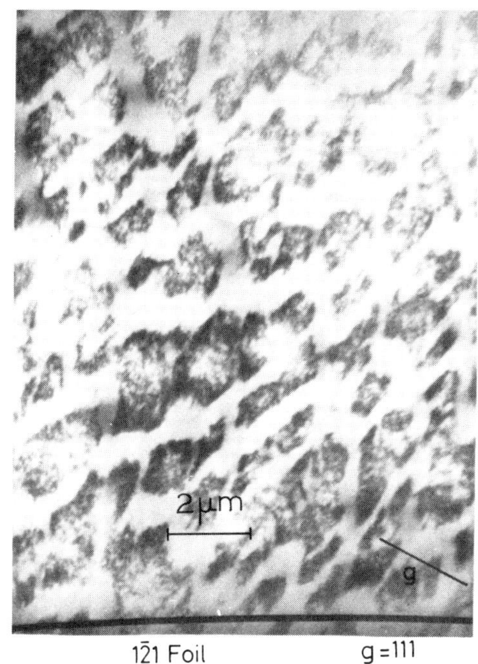

Fig. 7.14 Cross-section through vein structure in copper single crystal. (1$\bar{2}$1) section, $\gamma_p = \pm 0.001$ (from Woods, 1973, courtesy of Taylor & Francis)

in copper.* The normal to these walls is close to [$\bar{1}$01]. The extent of the walls in the [111] direction varies from 1-5 μm before they change direction or are interrupted by a "channel" or ladder boundary (fig. 7.18). Total extent of the wall structure in the [111] direction depends on the strain amplitude. For $\gamma_p = 0.001$, fig. 7.18, they extend 10-50 μm. At higher amplitudes, they extend further (i.e., fill more of the crystal volume) and the walls themselves are more compact, fig. 7.19.

Secondary dislocations are present in the vein and wall structures, but the high density of primary edge dislocations, usually in the form of long dipoles and multipoles, prevents positive identification of the Burgers vectors. Weak beam electron microscopy has shown that appreciable numbers of point defect clusters with mean diameter 44Å exist within the walls (Piqueras, et. al., 1972), fig. 7.20. The regions between the

*It may be significant that the observations of Lukáš, et. al. relate to stress-controlled cycling, whereas Woods employed strain controlled cycling.

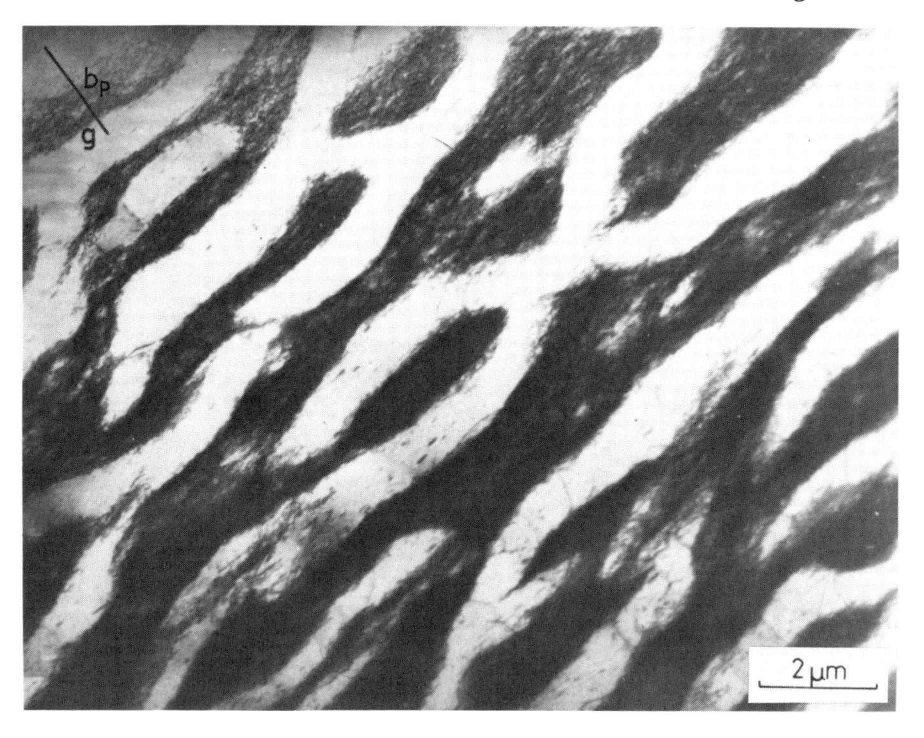

Fig. 7.15 Vein structure (uncondensed walls) as viewed on the primary glide plane of copper, (111) section, $\gamma_p = \pm 0.003$ (from Woods, 1973, courtesy of Taylor & Francis).

veins and walls contain primary screw segments, with some anchored on the boundaries. The density of these screws is greater between the walls than between veins. Point defect clusters are also found in the interwall regions. The misorientation (rotations) across walls and veins is too small to be measured.

In summary, two types of work-hardened structures in Cu single crystals are observed at low plastic strain amplitudes: interlocking veins and well-ordered walls consisting mostly of primary edge dislocations. At plastic strain amplitudes $\lesssim 0.001$, the vein structure is dominant. (Long range stresses are probably absent in these structures.) Raising the strain amplitude appears to convert the less ordered vein structure into wall structure, which is dominant at the higher end of the strain region (~ 0.003). The structure associated with PSB's is the wall structure, and it is contained within slabs of material having [111] normals. Primary screw dislocations are found between the walls and, to a lesser extent, between the

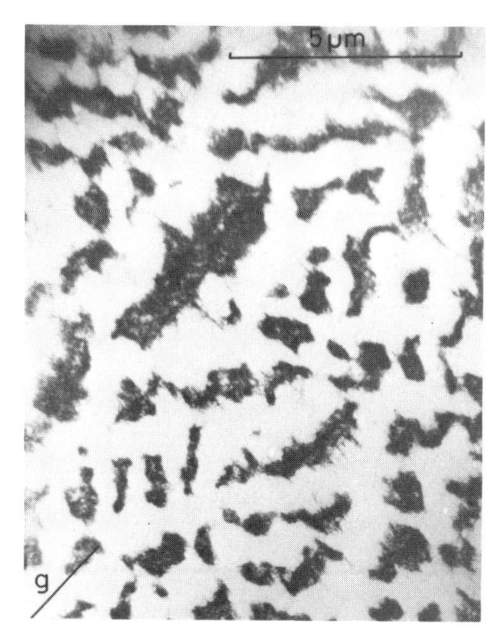

Fig. 7.16 Cross-section through vein structure (uncondensed walls) in copper single crystal. $(1\bar{2}1)$ section, $\gamma_p = \pm 0.003$ (from Woods, 1973, courtesy of Taylor & Francis).

veins. Point defect clusters are found within the walls, and, to a lesser extent, between the walls.

It should be clear that deformation at low amplitudes in copper is inhomogeneous, with the majority of the strain being accommodated in the PSB wall structures. At the upper end of the low amplitude range, the majority of the crystal is filled with wall structures, some more condensed or "tidy" than others.

b) Saturation Structures at High Strain Amplitudes. "High" strain amplitudes are those which produce fatigue lives of the order of 10^5 cycles or less. For copper single crystals this corresponds approximately to $\gamma_p \gtrsim 0.004$. The work-hardened structures produced in single crystals at high amplitudes have been observed by Hancock and Grosskreutz (1969), Gostelow (1971), Lukáš and Klesnil (1971), and Mughrabi (1973) by means of TEM. Polycrystalline copper has been observed by many workers, but the most comprehensive work is that by Feltner and Laird (1967b). The common structures are walls of dislocation and 2-, and 3-dimensional cells.

Three-dimensional, composite micrographs comparable to that shown in fig. 7.13 have not been reported for single crystals at high amplitudes. However, single micrographs have been published by different workers which show the structures on the primary (111) glide plane, on $(1\bar{2}1)$ planes, and on $(\bar{1}01)$ planes.

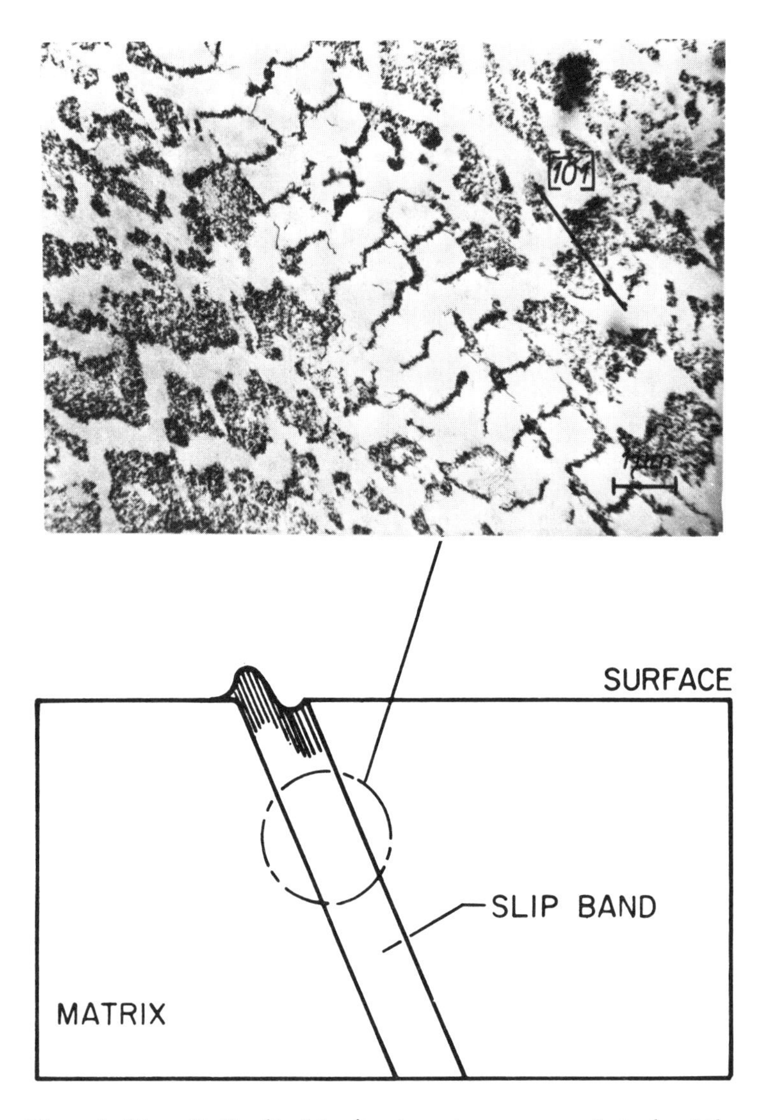

Fig. 7.17 Wall (ladder) structure associated with persistent slip band in copper enclosed on both sides by vein structure. ($1\bar{2}1$) section. (from Lukáš, et. al., 1968, courtesy phys. stat. sol.)

Unfortunately, no one study has reported all three orientations for the same specimen. Figure 7.21 shows a ragged, open wall structure on the primary glide plane for $\gamma_t = 0.0075$. Figure 7.22 shows that some of these walls lie in the ($\bar{1}01$) plane as

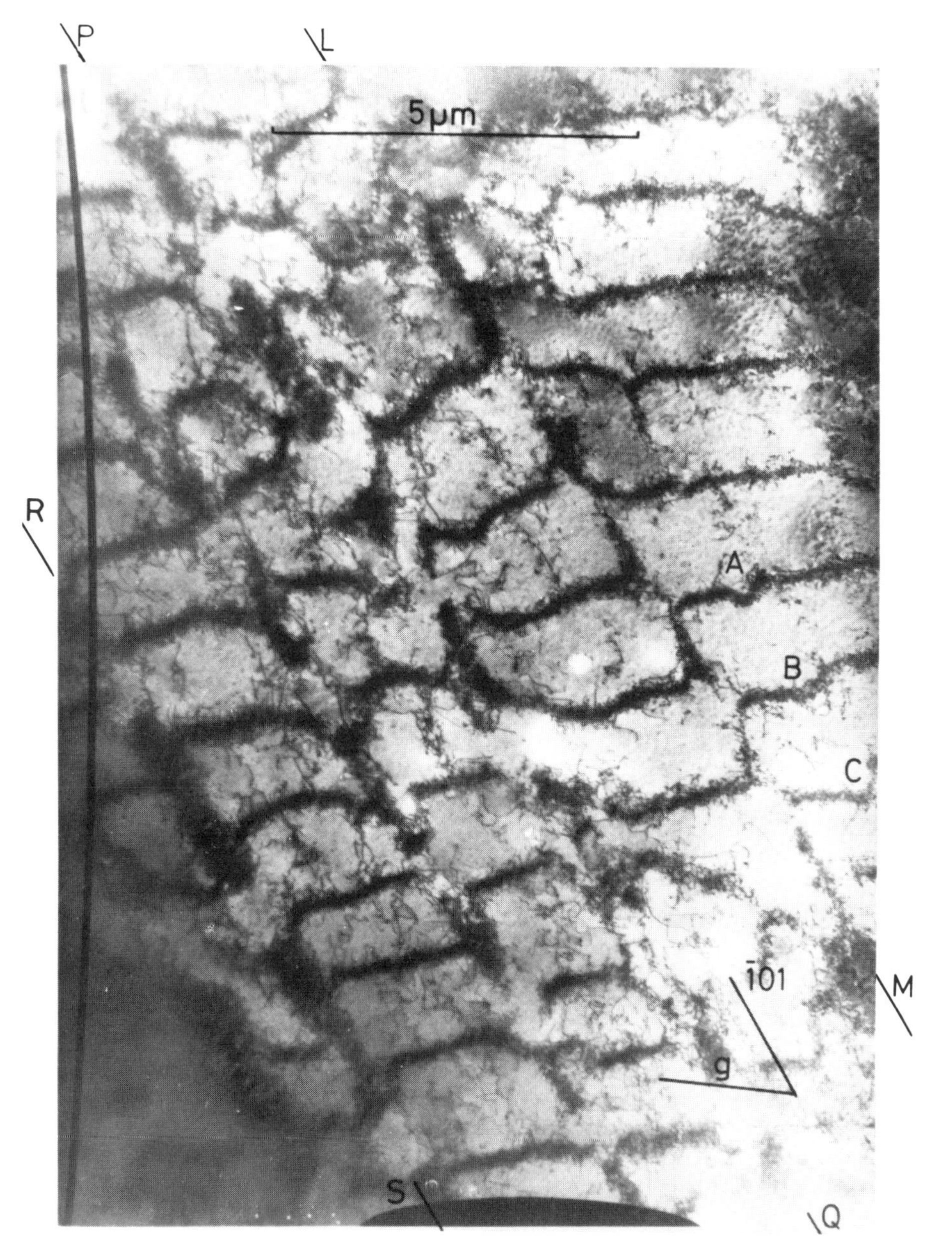

Fig. 7.18 Wall (ladder) structure in copper single crystal. LM, PQ, RS indicate "channel" boundaries. $(1\bar{2}1)$ section, $\gamma_p = \pm\ 0.001$. (from Woods, 1973, courtesy of Taylor & Francis).

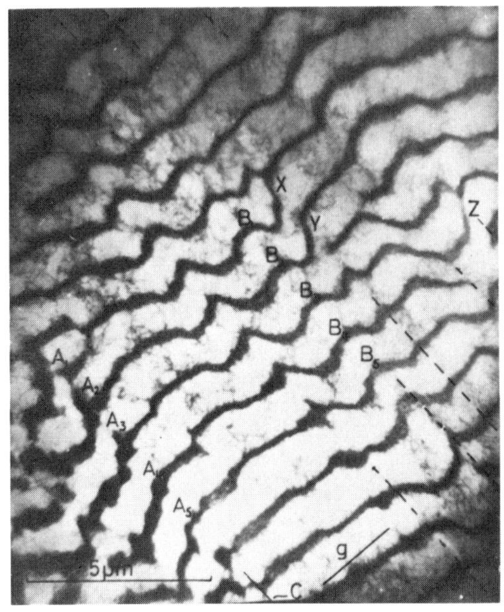

Fig. 7.19 Wall (ladder) structure in copper single crystal. $(1\bar{2}1)$ section, $\gamma_p = \pm 0.003$ (from Woods, 1973, courtesy of Taylor & Francis).

in the case of low amplitudes. However, other $(\bar{1}01)$ micrographs show that some walls have [111] normals, indicating the possibility of a 2-dimensional cell structure whose walls have [111] and $[\bar{1}01]$ normals (Hancock and Grosskreutz, 1969). No observable rotations about these walls exist.

Mughrabi (1973) has observed work-hardened structures on the (111) plane of single crystals cycled into the beginning of saturation at strain amplitude $\gamma_p = 0.005$. In this work the dislocations were pinned by fast neutron irradiation at the peak stress. Two characteristic micrographs of the primary glide plane (111) are shown in fig. 7.23. Areas of comparably low dislocation density are separated by dense primary multipole bundles containing a fairly high secondary dislocation density. Stereo micrographs indicate that the structure is intermediate between veins and walls. These form a two-dimensional "cell" structure elongated in $[1\bar{2}1]$. The volume fraction of veins and walls is estimated to be about 10%.

Inside these cells, groups of free primary dislocations of the same sign and having predominantly screw character (fig. 7.23a) are observed over larger areas ($\sim$20 µm x 20 µm). In thicker foil regions (5,000 - 10,000 Å), the elastic interaction of such dislocation groups of opposite sign can be recognized (fig. 7.23b).

The evaluation of dislocation curvatures indicates that local effective stresses (cf. 7.3.1c) vary spatially between zero and about twice the peak stress τ_s. High internal

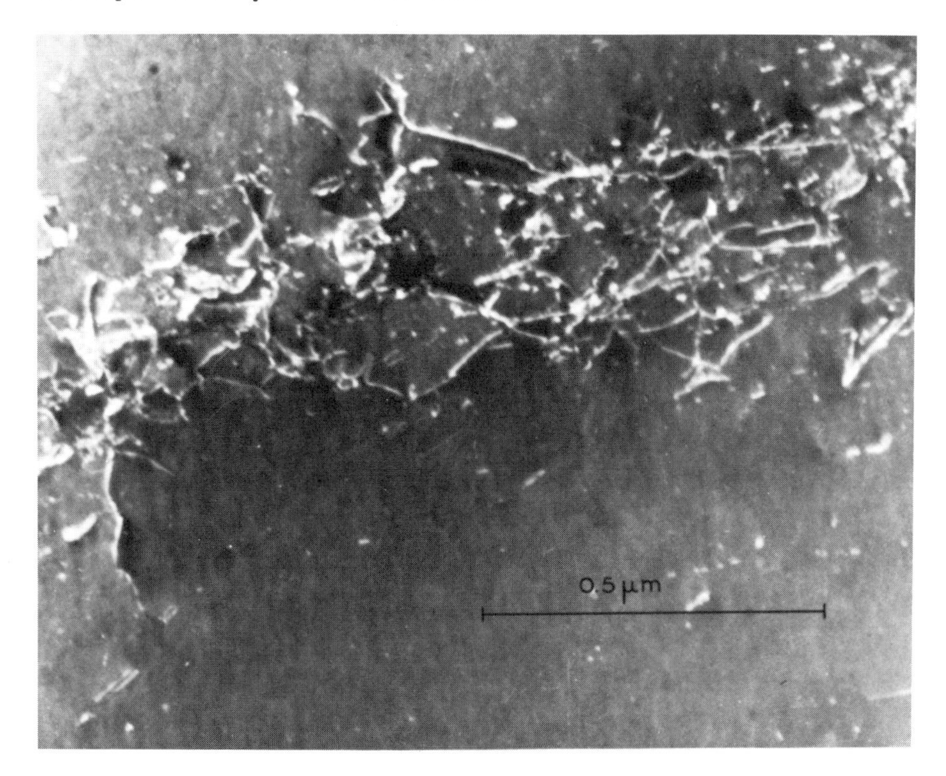

Fig. 7.20 Weak beam electron micrograph of copper single crystal which shows point defect clusters in dislocation walls. (111) section, $\gamma_p = \pm 0.0015$ (after Piqueras, et. al., 1972).

stresses τ_i ($\sim \pm \tau_s$) are apparently confined mainly to the vicinity of regions of high dislocation density. In some of these areas ($\tau_i \sim + \tau_s$), edge dislocation segments can bow out of the bundles and "cell" walls, reaching almost critical Frank-Read configurations (cf. fig. 7.23). Because of the fluctuating nature of the internal stresses (in time and space) we assume that the dislocation structure has been built up in such a way to be stable up to local stresses of roughly twice the peak stress.

Single crystals cycled at constant stress amplitude to produce lives of 10^5 cycles were examined by TEM on the (111) and $(1\bar{2}1)$ planes (Lukáš and Klesnil, 1971). Two-dimensional cells were observed on both planes, thereby leading to the conclusion that a true, three-dimensional cell structure exists. No other type structure was observed.

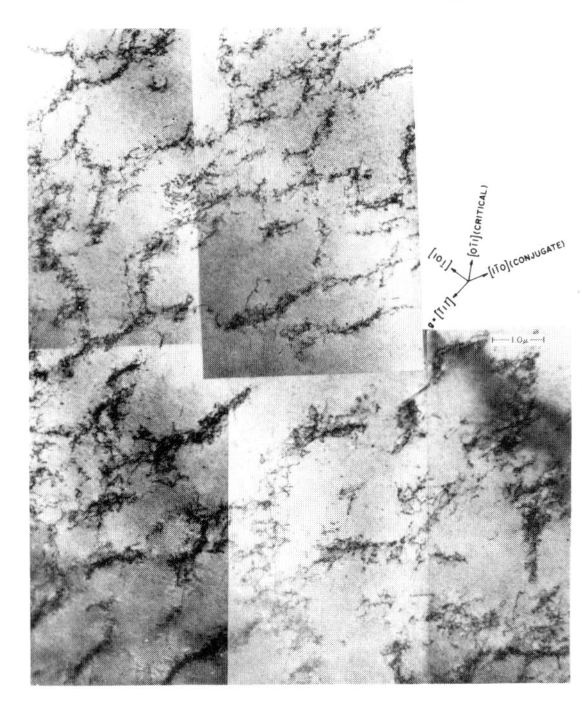

Fig. 7.21 Ragged, 2-dimensional cell wall structure in copper single crystal. (111) section, $\gamma_t = \pm .0075$, (after Hancock and Grosskreutz, 1969).

The highest strain amplitude applied to single crystals, $\gamma_p = 0.01$, was utilized by Gostelow (1971). Two types of structures were observed after reaching saturation: a "tidy" region of wall structures exhibiting two-dimensional cells on $(1\bar{2}1)$, and an "untidy" region of ill-formed three-dimensional cells. These regions exist as slabs whose boundaries are $(\bar{1}01)$ planes. This two-phase structure is shown in fig. 7.24.

Polycrystalline copper, cycled at $\gamma_p = 0.006$ and 0.06, exhibited a three-dimensional cell structure (Feltner and Laird, 1967b). No other structures were observed. The fact that this microstructure does not depend on prior history is shown in fig. 7.25. Identical two-dimensional cell structures were produced by cycling both annealed and cold worked (5 percent reduction in diameter) specimens.

The dislocation walls produced at high amplitude, regardless of whether they form cell walls, are composed of edge dislocation dipoles, multipoles and point defect clusters (Piqueras, et. al., 1972). The higher the amplitude, the more fragmented are the dipoles and the higher the density of point defect clusters. In polycrystals cycled at very high amplitudes, rotations about the cell walls of a few degrees become observable. Above $\gamma_p \simeq 0.007$ the number of secondary dislocations within walls equals or exceeds the number of primary

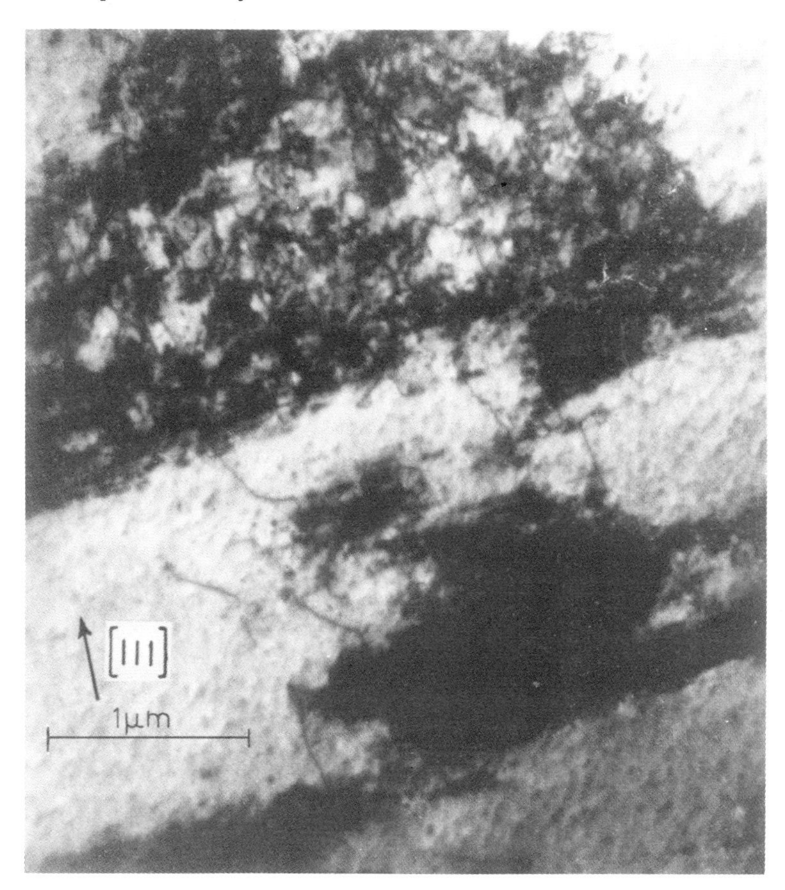

Fig. 7.22 Dislocation walls as viewed on ($\bar{1}01$) plane of copper single crystal. ($\bar{1}01$) section, $\gamma_t = \pm\ 0.0075$ (after Hancock and Grosskreutz, 1969).

dislocations. Cell walls which are simple twist boundaries have also been observed (Laufer and Roberts, 1966; Feltner and Laird, 1967b). Dislocations in the cell interiors are mostly primary screws, although isolated edge dipoles and point defect clusters exist.

In summary, two types of work-hardened structures in Cu single crystals are observed at high amplitudes: dislocation walls and two- or three-dimensional cell structures. The latter structure develops at the higher amplitudes, although a two-phase structure still persists at $\gamma_p = 0.01$. Isolated, curved primary screw dislocations are observed in the regions between walls or in the cell interiors when the stress-applied

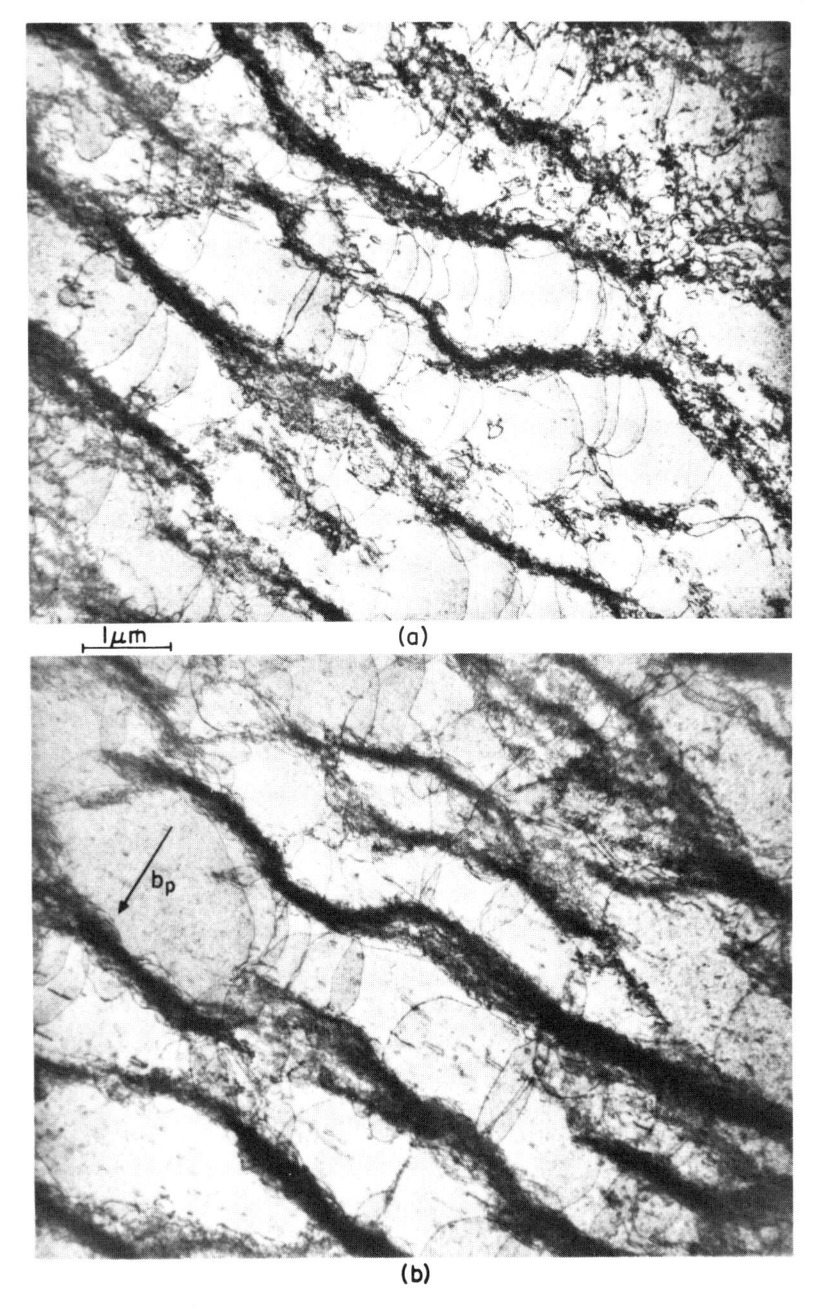

Fig. 7.23 Dislocation structure in copper single crystals cycled into saturation. (111) section stress-applied state. $\gamma_p = \pm 0.005$, 300 cycles, $\tau = 3.2\ kg/mm^2$.

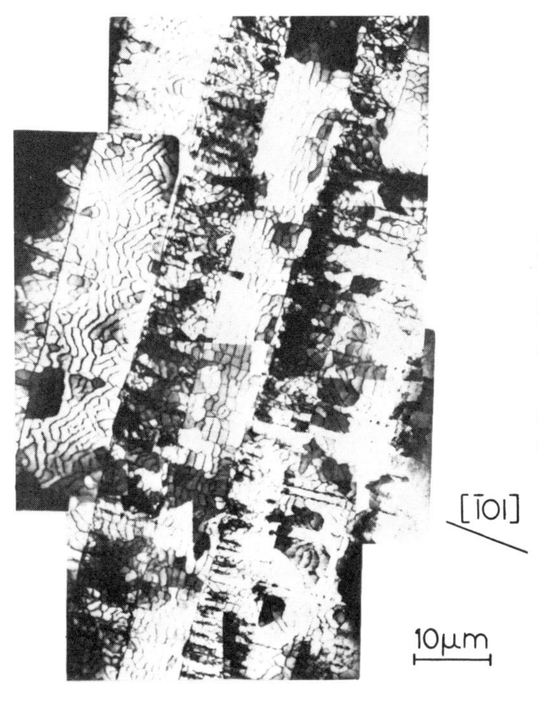

Fig. 7.24 Two-phase dislocation structure in copper single crystal: dislocation walls and 3-dimensional cell structure. (121) section, $\gamma_p = \pm 0.01$, 1500 cycles, (from Gostelow, 1971, courtesy

state is observed.

c) Homogeneity of Deformation in Copper Single Crystals. The question of homogeneity of deformation deserves attention, especially in a book on constitutive equations. The evidence for inhomogeneity at low strain amplitudes is fairly strong, based on the increasing volume fraction occupied by the wall structure (PSB's) as the amplitude increases. The flatness of the cyclic stress-strain curve (fig. 7.12) over a wide range of amplitudes is also consistent with the picture of a crystal gradually filling with regions of high strain activity which requires no increase in flow stress.

The real question is whether deformation ever becomes homogeneous at some threshold strain amplitude. Winter (1973) has estimated that the surface of copper is completely filled with PSB's when $\gamma_p \approx 0.0085$. However, the TEM evidence presented by Gostelow (1971) shows that the internal deformation structure formed at higher amplitudes ($\gamma_p = 0.01$), although consisting of homogeneous wall structure at the beginning of saturation, later breaks up into two phases, a wall structure and a three-dimensional cell structure. (See fig. 7.24). These observations are consistent with those made at lower amplitudes

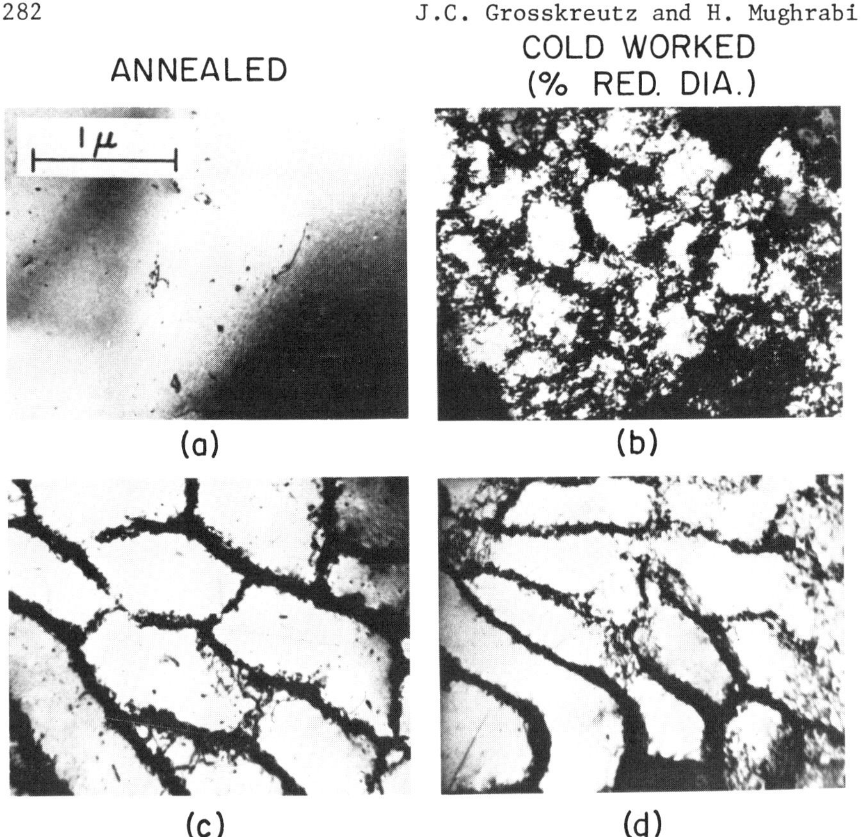

Fig. 7.25 Dislocation structures in polycrystalline copper before cycling, (a) and (b); and after 1000 cycles at $\gamma_p \sim 0.006$, (c) and (d) (from Feltner and Laird, 1967b, courtesy Pergamon Press).

by Laufer (1969) who reported that continued cycling caused the initial wall structure in PSB's to break up into a three-dimensional cell structure which was bounded by fresh wall structures at the edges of the growing PSB. Woods (1973a) has suggested that this transition occurs when the wall structure cannot sustain further cumulative strain. If three-dimensional cells are the final product of continued cyclic straining, then one would assume that they are a harder structure than the wall or ladder structure, and that a larger fraction of the applied strain amplitude is accommodated in the relatively larger dislocation-free regions of the wall structure. In this sense then, deformation would continue to be inhomogeneous in copper at plastic strain amplitudes higher than 0.0085. The degree with which strain is partitioned between the two deformation phases has yet to be determined.

As a final comment on the subject of deformation homogeneity in copper, we call attention to the fact that no observations have been made of deformation structures at very low amplitudes ($\gamma_p \lesssim 0.0005$) or at very high amplitudes ($\gamma_p > 0.01$). In both these regimes, the slope of the cyclic stress-strain curve (fig. 7.12) is larger than in the central region of inhomogenous deformation, a fact which would be consistent with more homogeneous deformation at the extremes.

7.3.3 *Aluminum*

The available information to describe the cyclic work-hardened structures in aluminum is much less detailed than that just presented for copper. Much of the work was performed in the early days of TEM before the techniques for oriented sectioning of single crystals became available. Moreover, dislocation loss and rearrangement during thinning of Al is much more severe than for Cu due to the higher stacking fault energy of Al. Nor have neutron irradiation and stress-applied experiments been carried out to prevent rearrangement and losses of the structure. Nevertheless, a fair description of the structures can be given. In fact, because of the similar FCC structure and wavy slip nature of Cu and Al, we should expect the work-hardened structures to be similar. The higher stacking fault energy of Al, and hence easier cross-slip of screw dislocations, should have the effect of producing the typical "high" amplitude structures seen in Cu at lower strain amplitudes in Al.

The work-hardened structures produced in single crystals of aluminum at low stress amplitudes ($5 \times 10^5 < N_f < 10^7$ cycles) have been observed by Levine and Weissmann (1968) and by Mitchell and Teer (1969, 1970). Two types of structure were observed: bands or patches of tangled dislocations and elongated dislocation loops which were found in the interior, and two-dimensional cell or ladder structures associated with PSB's near the surface. In contrast to Cu, the transition from the ladder structure to the band structure is not sharp, but is gradual.

Similar low amplitude structures have been observed in polycrystalline Al by Grosskreutz and Waldow (1963) and by Krejci and Lukáš (1971). Figure 7.26 illustrates the band or patch structure, and fig. 7.27 depicts well-developed cells; both structures were observed in a specimen cycled for 1.5×10^6 cycles at a tensile strain amplitude of $\pm$ 0.0003.

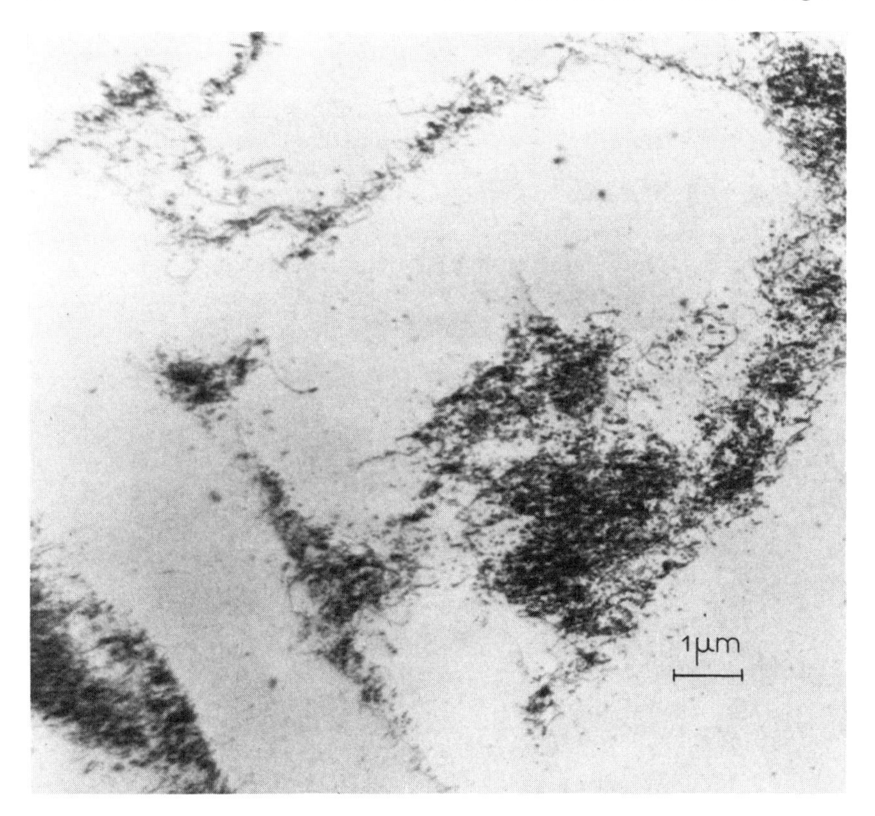

Fig. 7.26 Dislocation bands or patches in polycrystalline aluminum cycled into saturation. $\gamma_p \sim \pm 0.0006$ (from Grosskreutz and Waldow, 1963, courtesy of Pergamon Press).

Work-hardened structures produced at high amplitudes ($N_f < 5 \times 10^5$ cycles) have been observed by Grosskreutz (1962) and Krejci and Lukáš (1971). Only one type of structure was observed: two-dimensional cell structures. An example is shown in fig. 7.28. Misorientation between the cells is evident from the contrast effects.

Dislocations are seldom observed in the regions between bands or cell walls, primarily because they are lost or move to the walls during thinning of the specimens.

7.3.4 *Alpha-Iron*

a) General Remarks. The effect of strain rate (i.e., frequency at a given strain amplitude) was ignored in our description of

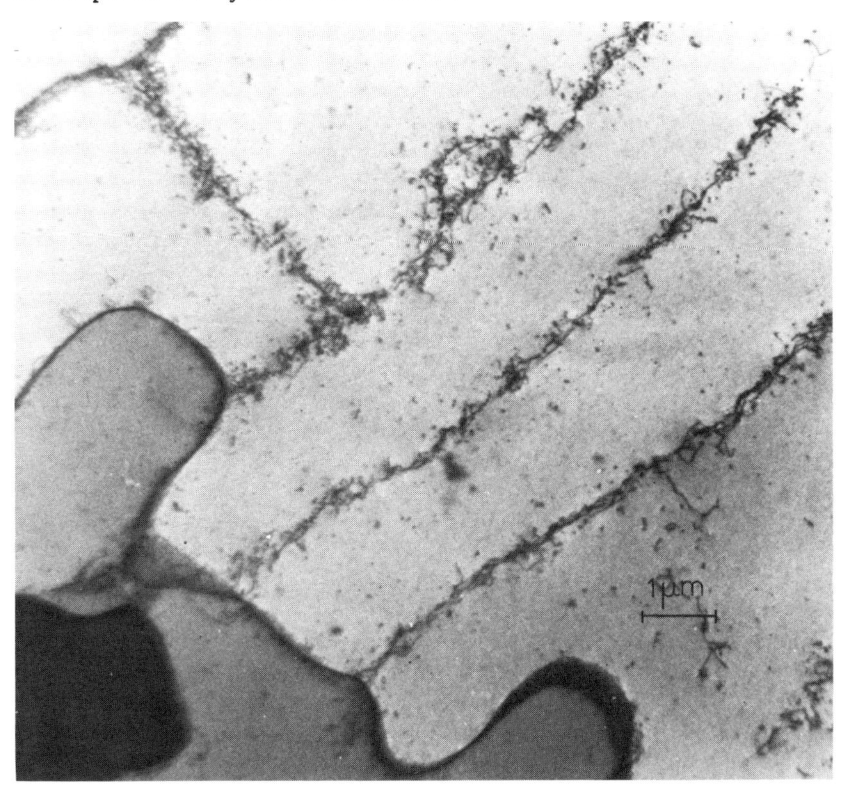

Fig. 7.27 Two-dimensional cell or ladder structure in polycrystalline aluminum. $\gamma_p \sim \pm 0.0006$ (from Grosskreutz and Waldow, 1963, courtesy of Pergamon Press).

copper (Section 7.3.2) and aluminum (Section 7.3.3). This effect is important in cyclic deformation of α-iron (cf. Section 7.3.1). In typical published fatigue work on α-iron (polycrystals) at room temperature (frequency ~50 c.p.s. and $\gamma_p \sim 10^{-4}$-10^{-2}) the mean strain rate $\dot{\gamma}_p$ is 2×10^{-2} -2 s^{-1}. It is obvious from Section 7.3.1 that at this "high" strain rate, "low temperature" dislocation behavior is to be expected. This means that the comparison of cyclic deformation data comprising strain amplitudes is meaningful only at comparable cyclic strain rates. The effect of cyclic strain rate on microstructure has not been investigated, aside from a study of cell sizes (Abdel Raouf and Plumtree 1971, cf. Section 7.4.2).

b) Polycrystals. There are a number of investigations on the dislocation structure of high purity α-iron and low carbon

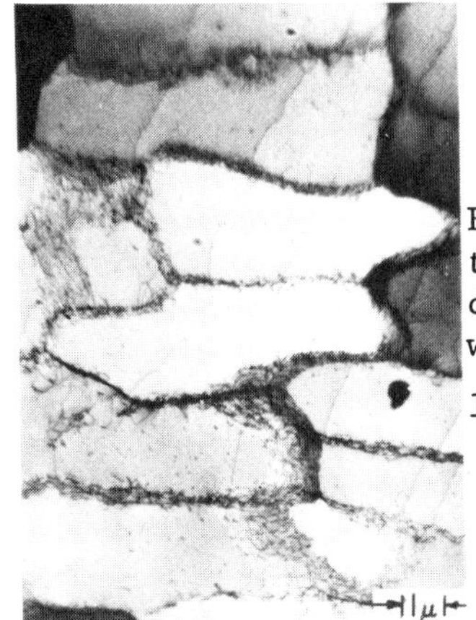

Fig. 7.28 Two-dimensional cell structure in polycrystalline aluminum which displays misorientation across cell walls. $\gamma_p \sim \pm 0.004$ (from Grosskreutz, 1962, courtesy of Am. Inst. Phys.).

steels after cyclic hardening at prescribed stress amplitude (McGrath and Bratina 1965a, b Lukáš and Klesnil 1964, 1965, Lukáš, Klesnil and Rys 1965, Klesnil, Holzmann, Lukáš and Rys 1965, Vingsbo 1970, Bergström, Vingsbo and Lagerberg 1969, 1969/70, Veith 1971, Ivanova, Goritskiy, Orlov and Terent'yev 1972). In a few cases, the microstructure formed during controlled strain cycling was studied (Wei and Baker 1965a, b; Abdel-Raouf and Plumtree 1970, Abdel-Raouf, Benham and Plumtree 1970, 1971). The general result of these observations is that three-dimensional cell structures are formed at high amplitudes, the cell size decreasing with increasing amplitude (Abdel-Raouf, Benham and Plumtree 1970). At lower amplitudes dislocation tangles, dense dislocation clusters, heavily jogged screw dislocations and dislocation loops have been observed (Wei and Baker 1965b, Bergström et. al. 1969/70, Krejci and Lukáš 1971). In the near-surface regions, slip bands (Klesnil and Lukáš 1965a), dense bands of dislocations and parallel rows of dislocation loops (Wei and Baker 1965a, b) were found. Such rows of dislocation loops have recently also been observed in the interior (Bergström et. al.).

At high and low amplitudes, dislocations are usually jogged and predominantly of screw character (Bergström et. al. 1969/70, Krejci and Lukáš 1971). This is probably a consequence of the high strain rates employed, as discussed before. This argument is further supported by the fact that all reported dislocation densities are of the order $\sim 10^{10}$ cm^{-2} (Lukáš and Klesnil 1964, Lukáš, Klesnil and Rys 1965, Vingsbo, Bergström and Lagerberg 1969, Krejci and Lukáš 1971); i.e., distinctly lower than in FCC metals, where dense edge dislocation multipoles predominate.

In one investigation at high stress amplitudes on a low carbon steel a mean cyclic strain rate of $\sim 10^{-3}$ s^{-1} was employed (Veith 1971). The observed dislocation structure is strikingly similar to that reported for copper at high amplitudes (cf. fig. 7.23).

c) Single Crystals. Very little work has been done on single crystals. Lawrence and Jones (1970) reported the formation of cell structures and the increase of cell size with stress amplitude, similar to the mentioned observations on polycrystals. After low stress amplitude cycling (Krejci and Lukáš 1971), diffuse dislocation bands and tangles containing primary and coplanar jogged screw dislocations and dislocation loops were observed as well as long straight screw dislocations between the bands.

Single crystals of α-iron have recently been investigated after cycling at two different plastic strain amplitudes (Stark and Mughrabi 1973).* At the smaller amplitude (and strain rate!) very dense edge multipole bundles (formed early in the rapid hardening stage) were observed, interlinked by loose and less regular networks. A cell structure was not formed (fig. 7.29). On the other hand, a three-dimensional cell structure was observed at the larger amplitude (fig. 7.30). In the interior of the cells a fairly high density of free curved dislocations ($\rho \sim 3 \times 10^{8}$ cm^{-2}) and a large number of (unidentified) irregular loops ($\sim 3 \times 10^{13}$ cm^{-3}), can be recognized. Higher magnification reveals that the free dislocations are meanderlike in shape and "decorated" with loops (fig. 7.31). The latter could be agglomerates of carbon and intrinsic point defects (cf. also Wilson 1973).**

7.3.5 *Planar Slip Alloys*

Although beyond the stated scope of this chapter, the cyclic work-hardening structures in planar slip metals (low stacking

*The dislocation structure was pinned by strain ageing as described previously in Section 7.3.1b.

**These defect clusters could be related to secondary cyclic hardening observed in α-iron containing (interstitial) carbon atoms (cf. Abdel-Raouf, Plumtree and Topper 1973, Stark and Mughrabi 1973).

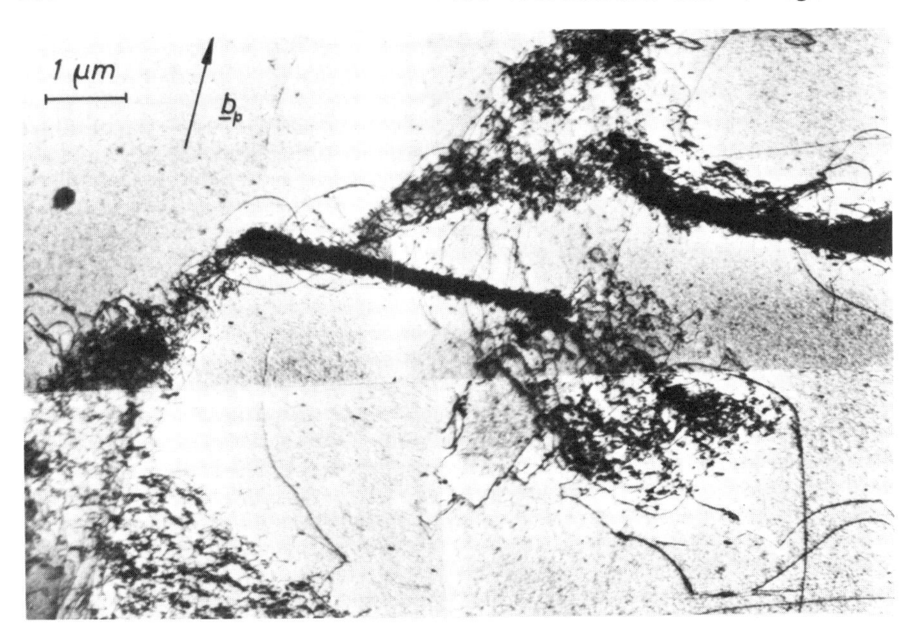

Fig. 7.29 Dislocation structure in α-iron single crystals cycled into saturation. $\gamma_p = \pm\ 0.0018$, 0.1 c.p.s., 1050 cycles, $\tau = 3.3\ kg/mm^2$. T = 293K. (from Stark and Mughrabi, 1973).

fault energies) will be mentioned and referenced briefly to indicate the major difference from the structures discussed above.

Cu-31%Zn has been studied extensively by Lukáš and Klesnil (1970, 1971, 1973) over a range of stress and strain amplitudes giving lives from 4×10^4 to 10^6 cycles. Over this range, the work-hardened structure consisted of only one type: planar arrays of primary edge dislocations lying in regularly spaced bands along primary slip planes. At very high strain amplitudes, Grosskreutz, et. al. (1966) found a cell structure in Cu-30%Zn. Cu-Al alloys have been studied by Woods (1973) and by Feltner and Laird (1967b) with similar results.

Thus, in contrast to the wavy slip materials, planar slip materials exhibit only one type of structure over the usual range of stress or strain amplitude. One may conclude that the deformation is homogeneous in these materials. They do not, however, exhibit unique cyclic stress-strain curves.

7.3.6 *Cyclic Deformation Map*

A convenient summary of the information presented in this

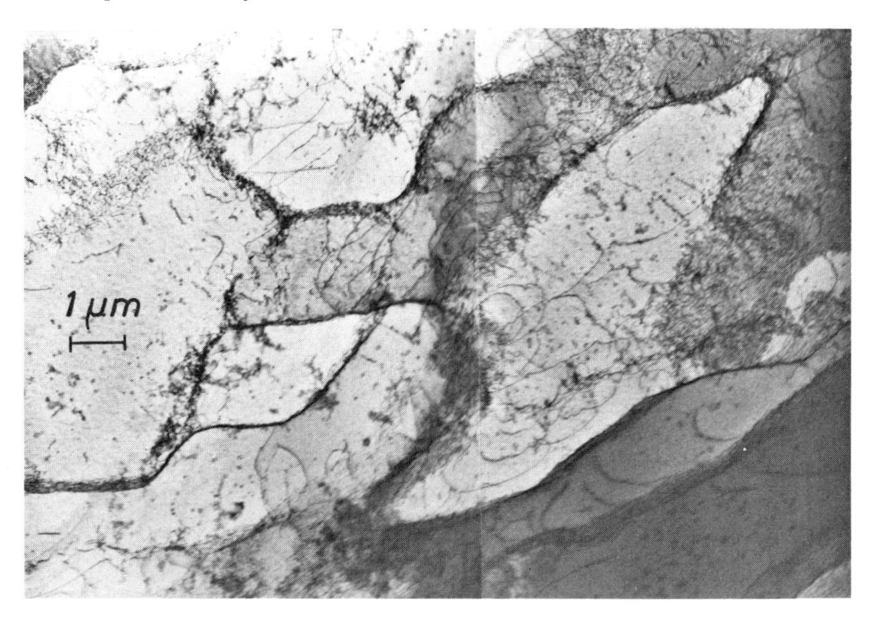

Fig. 7.30 Dislocation structure in α-iron single crystals cycled into saturation. $\gamma_p = \pm\ 0.005$, 0.1 c.p.s., 450 cycles, $\tau = 5.4\ kg/mm^2$, (from Stark and Mughrabi, 1973).

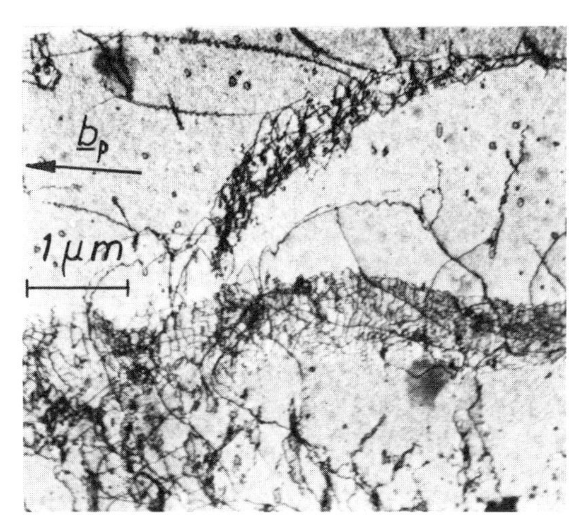

Fig. 7.31 As Figure 7.30, higher magnification.

section is given in fig. 7.32, which is based on a map originally given by Lukáš and Klesnil (1973). For purposes of ref-

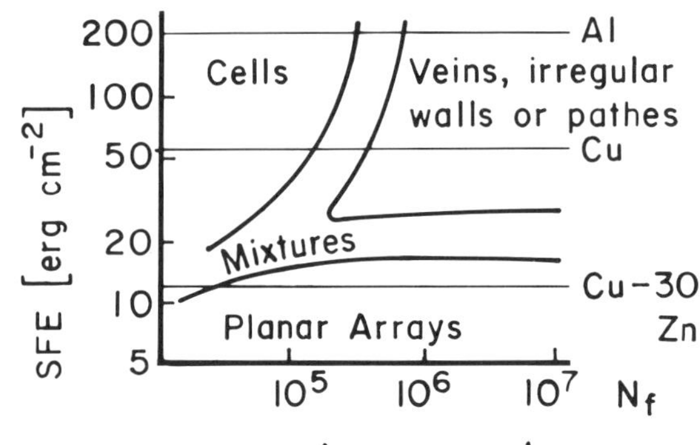

Fig. 7.32 Cyclic deformation map showing regions in which various work-hardened structure predominate, (from Lukáš and Klesnil, 1973, courtesy of Akademic Verlag).

erence, the stacking fault energy for Al is 200 ergs/cm^2, for Cu, 55 ergs/cm^2, and for Cu-30%Zn, 13 ergs/cm^2.

7.4 QUANTITATIVE DESCRIPTION OF THE SATURATION STAGE, WORK-HARDENED STATE

7.4.1 *Vein and Wall Spacings and Thicknesses*

The very nature of the interlocking vein structure in copper makes it difficult to define the spacing. Basinski, Basinski, and Howie (1969) measured the spacing of veins on the (111) glide plane during the rapid hardening stage and found it decreased from about 20 μm near the beginning of cycling to about 2 μm as saturation was approached. The micrographs of Lukáš, et. al. (1968) indicate the vein spacing to be about 1 μm in the saturation stage. Characteristic spacings of the vein and uncondensed wall structure, measured in [$\bar{1}$01]; i.e., the direction of b_p, and the volume fractions of these structures in the regions they occupy are listed in Table 7.1. These values are based on the work of Basinski, Basinski and Howie (1969), Woods (1973) and our evaluation of their published micrographs and our own.

The bands of dislocation observed at low amplitude in Al may be spaced anywhere from 1-5 μm apart, with no apparent regularity.

Dislocation walls associated with PSB's in Cu are spaced with remarkable regularity. Values taken from the work of Woods (1973) are shown in the bottom line of Table 7.1. Strain amplitude seems to have little effect on this spacing in the low amplitude region.

Table 7.1 Characteristic Spacings of Veins and Walls in Copper

Structure	γ_p	Volume fraction occupied %	Mean spacings (μm)	Mean thickness (μm)	Mean width of free region (μm)
Veins (uncondensed walls)	0.001 0.003 0.005	50 30 10	1.4 1.6 1.7	0.8 0.6 0.3	0.6 1.0 1.4
Walls (ladder)	0.001 0.003	10	1.5	0.1-0.2	1.3-1.4

It is obvious that veins and uncondensed walls are more voluminous at low amplitudes. This finds an explanation, if one takes into account that long range internal stresses are much more significant at high than at low amplitudes (Avery and Backofen 1963, Feltner and Laird 1967b, cf. also Section 7.3.1d and 7.6.1). It follows that the dipole structure formed at low amplitudes is expected to be stable at local stresses of the order of the peak stress. This allows for larger mean dipole spacings and a more voluminous structure than at high amplitudes where the structure is expected to be stable at local stresses of twice the peak stress (cf. Section 7.3.2b).

7.4.2 *"Cell" Size*

Two- and three-dimensional cells may be equiaxed or elongated, depending on the plane of observation and the material. It is usual practice to report cell size as some mean spacing of the cell walls. Cell sizes in the range 0.5-5 μm have been reported for Cu, Al and Fe, the larger sizes being associated with lower saturation stress amplitudes.

Two-dimensional cells 1.2 μm in diameter are formed in copper single crystals cycled at $\gamma_t = 0.0075$, $N_f = 5 \times 10^4$ cycles (Hancock and Grosskreutz, 1969). Similar cells 1-1.5 μm in diameter have been reported by Klesnil and Lukáš (1971) in crystals with total fatigue life 10^5 cycles. Cell walls spaced 1-2 μm apart were reported by Mughrabi (1973) for copper crystals observed in the stress applied state (γ_p = 0.005).

An inverse relationship between cell size and the saturation flow stress has been demonstrated for polycrystalline copper, by Pratt (1967) fig. 7.33. Feltner and Laird (1967b) have

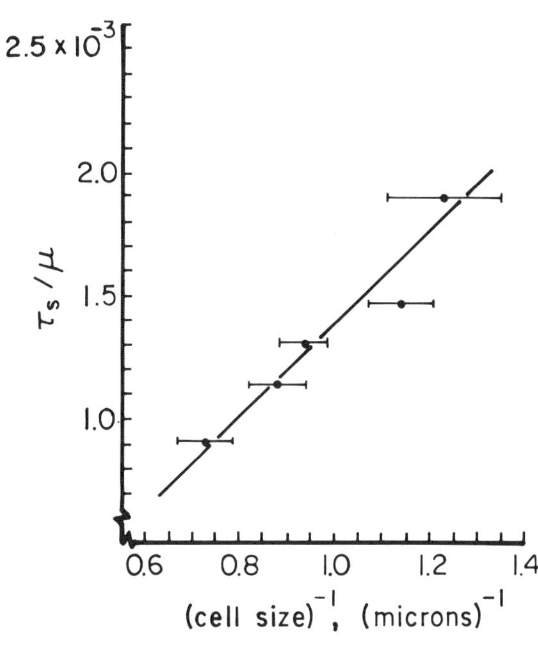

Fig. 7.33 Relationship between cell size and saturation flow stress (normalized to the shear modulus) for cyclically-hardened polycrystalline copper, (from Pratt, 1967, courtesy of Pergamon Press).

found a similar dependence, with cell sizes as small as 0.5 μm for $\gamma_p \approx 0.06$.

The same inverse relationship holds for polycrystalline iron regardless of whether one starts with annealed or cold-worked samples, fig. 7.34. (Abdel-Raouf, et. al., 1971).

The average volume of three-dimensional cells formed in polycrystalline aluminum at various total strain amplitudes has been measured by X-ray techniques and is shown in fig. 7.35. In this case, the average cell size in Al varies from 5 μm to a limiting value of 2 μm.

The effect of temperature on the size of two-dimensional cells formed in polycrystalline copper has been measured by Feltner and Laird (1967b). They report a 40% decrease in cell size in specimens cycled at 78°K compared to specimens deformed at 300°K.

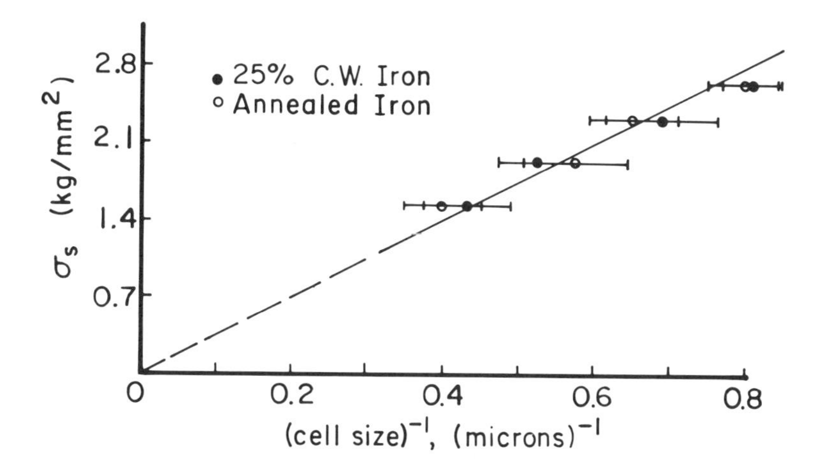

Fig. 7.34 Relationship between cell size and saturation flow stress for cyclically-hardened polycrystalline iron, (from Abdel-Raouf, et. al., 1971, courtesy of AIME-ASM).

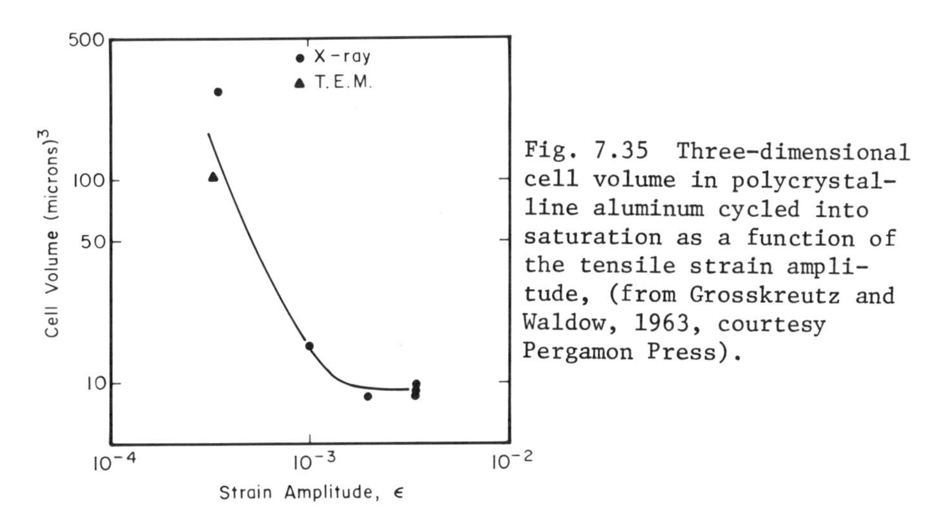

Fig. 7.35 Three-dimensional cell volume in polycrystalline aluminum cycled into saturation as a function of the tensile strain amplitude, (from Grosskreutz and Waldow, 1963, courtesy Pergamon Press).

The effect of strain rate on the saturation flow stress and the cell size in polycrystalline iron has been reported by Abdel-Raouf and Plumtree (1971). As the rate was increased for a given strain amplitude, the saturation flow stress also increased, and the cell size decreased in accordance with the relation shown in fig. 7.34.

7.4.3 *Distribution of Lengths of Free Dislocation Segments in Veins and Walls*

The subdivision of dislocations in the veins and walls into short (edge) segments arises from the segmented shape of dislocations shown in fig. 7.4 (regardless of whether annihilation occurs or not) and from reactions with secondary dislocations. Because of the high dislocation density in these structures, these segment lengths are usually not accessible to evaluation. In the stress-applied state, however, segments bowing out of bundles and walls are observed and the distances ℓ between the pinning points can be measured. The frequency distribution $H(\ell)$ belonging to the structure discussed before (fig. 7.23) is shown in fig. 7.36. This diagram probably underestimates

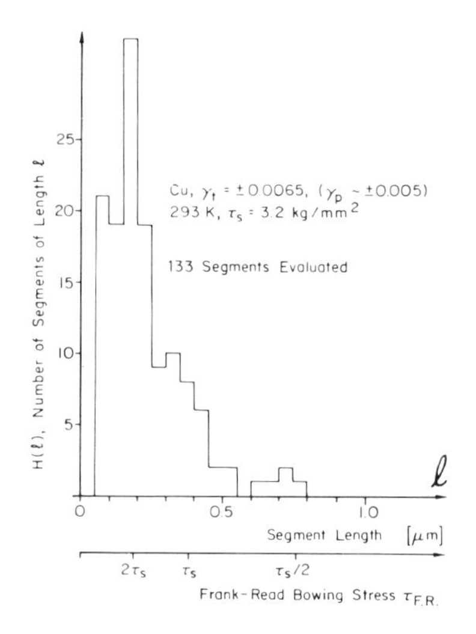

Fig. 7.36 Frequency distribution $H(\ell)$ of free segment lengths ℓ in bundles (and walls) of structure shown in Fig. 7.23.

short segments ($\ell < 0.1$ μm) that could not be resolved. Corresponding values of the Frank-Read bowing stresses $\tau^*_{F.R.}$ (expressed in multiples of τ_s) are also plotted. It follows that in regions of maximum local stresses ($\sim 2\,\tau_s$, cf. Section 7.4.1) short edge dislocation segments down to $\ell \gtrsim 0.2$ μm can be activated as sources.

7.4.4 *Dislocation Densities*

Dislocation densities in cyclically-hardened materials have been determined in a number of ways: 1) directly from TEM micrographs; 2) indirectly from resistivity measurements;

---

* Compare with footnote on page 222.

3) indirectly from magnetic saturation measurements; and
4) indirectly from X-ray measurements.

a) Direct Measurements. A variety of different densities may be measured from a single transmission electron micrograph of cyclically deformed material because of the inhomogeneous distribution of dislocations. A local density may be measured in the veins, walls, or the free regions between these structures. Or, an average density may be measured, averaged over the area viewed in several micrographs. Furthermore, the measurement may be confined to dislocations of a single Burgers vector, or all dislocations may be included; i.e., the total density. In this section, we shall quote only total densities, remembering that the highest contribution comes from the primary dislocations but that the density of all secondary dislocations is of almost the same order of magnitude at high amplitudes.

Local dislocation densities ρ_{loc} in copper cycled to saturation are given in Table 7.2 This table is based on values taken from the work of Woods (1973) and Mughrabi (1973). Values belonging to the dense regions are estimates (cf. Section 7.4.4b). Evaluation of ρ_{loc} in the dense edge multipole bundles observed in α-iron at the lower amplitude (fig. 7.29) indicates values as in copper (i.e., 10^{11}-10^{12}cm^{-2}).

Table 7.2 Local Dislocation Densities in Copper Single Crystals Cycled to Saturation

γ_p	Structure	Densities (cm^{-2}) Dense Regions	Low-Density (free) Regions
0.001	Veins	10^{11}	10^8 - 10^9
0.003	Walls	10^{11} - 10^{12}	2×10^9
0.005	Walls	10^{11} - 10^{12}	2×10^9

It seems clear that the majority of dislocations in the veins and walls are not available to accommodate the strain during cyclic deformation. Most make up the immobile fraction (cf. Section 7.5.1b). Those dislocations which are observed under stress-applied, neutron irradiated conditions (Mughrabi, 1973) to be in the low density regions, fig. 7.23, are the true mobile fraction. It is their density which is important for plasticity considerations.

The average dislocation densities in copper and α-iron crystals for both cyclic and tensile deformation are compared in Table 7.3. The densities are given for several values of the flow stress during hardening. The table is based on average values from the work of Hancock and Grosskreutz (1969), Basinski, Basinski and Howie (1969), Woods (1973) and Mughrabi (1973), whereas the values for α-iron are from the work of Stark and Mughrabi (1973), cf. Sections 7.3.1b and 7.3.4c. Dislocation densities observed after tensile deformation of copper are averages estimated from the work of Essmann (1966), Steeds (1966), Edington (1969), Essmann and Strunk (1970) and Mughrabi (1971) (cf. fig. 6.15). Values for α-iron are based on the data of Keh (1965).

Table 7.3 Average Dislocation Densities in Copper and α-Iron Crystals

Metal	τ (kg/mm^2)	Densities (cm^{-2})			
		Cyclic			Tension
		$\gamma_p = 0.002$	$\gamma_p = 0.003$	$\gamma_p = 0.005$	
Copper	1.5	-	$\sim 10^{10}$	$\sim 5 \times 10^9$	3×10^9
	3.0	-	$\sim 5 \times 10^{10}$	$\sim 10^{11}$	8×10^9
α-Iron	2	2×10^9	-	6×10^8	10^9
	3	$\sim 10^{10}$	-	9×10^8	3×10^9
	5	-	-	6×10^9	10^{10}

In copper, the average dislocation density does not depend strongly on amplitude, probably because dense multipole structures are always dominant. The available values for α-iron show that, at $\gamma_p \sim 0.002$, the dislocation density is higher than at $\gamma_p \sim 0.005$, a consequence of the high contribution of the dense multipole bundles at the lower amplitude (and strain rate).

Direct measurement of dislocation densities in cyclically deformed aluminum is difficult and subject to error. First, the densities are too high to resolve in the bands and walls. Second, rearrangement and loss of dislocations in cell interiors during foil preparation would make any measurement in that region meaningless (Grosskreutz and Shaw, 1964).

b) Indirect Evaluation of T.E.M. Data. The local dislocation densities ρ_{loc} in dense regions cannot be evaluated directly. Estimates can be made as follows. Assuming that multipole structures are stable at the highest local effective stresses $\tau_{eff,max}$, cf. Section 7.4.1, an upper bound of the dipole spacings y follows from the condition that $\tau_{eff,max}$ is smaller than the dipole passing stress; i.e.,

$$\tau_{eff,max} \lesssim \frac{\mu b}{8\pi\,(1 - \nu)y} \qquad (\nu = \text{Poisson number}) \tag{7.1}$$

and

$$\rho_{loc} \gtrsim 1/y^2 \tag{7.2}$$

For copper, at $\tau_s \sim 3.2\ kg/mm^2$, values of $\rho_{loc} > 10^{12}\ cm^{-2}$ (high amplitudes, $\tau_{eff,max} \sim 2\,\tau_s$) and $\rho_{loc} \gtrsim 2.5 \times 10^{11}\ cm^{-2}$ (low amplitudes, $\tau_{eff,max} \sim \tau_s$) are obtained.

In regions, where short segments of dislocations bow out of multipole structures (cf. fig. 7.23), the assumption can be made that there the dipole passing stress is smaller or of the order of the Frank-Read bowing stress (Mughrabi 1973). This procedure yields very similar values as above.

Finally, an upper bound of ρ_{loc} can be estimated, if one assumes that for values of $y \lesssim 16$ Å, edge dipoles annihilate spontaneously (Essmann and Rapp 1973). A reasonable upper limit of $\rho_{loc} \lesssim 2 \times 10^{13}\ cm^{-2}$ is obtained.

c) Resistivity Measurements. Both dislocations and point defects contribute to changes in electrical resistivity during cyclic deformation. However, annealing provides a means of separating these contributions, since point defects anneal out first. Johnson and Johnson (1965), Helgeland (1967), and Polák (1969) have measured the dislocation density in cyclically deformed copper at saturation by means of electrical resistivity. The tests were all performed in the transition region between low amplitude and high amplitude strains ($N_f \sim 5 \times 10^5$ cycles). Values for the total, average dislocation density range from $2 - 5 \times 10^{10}\ cm^{-2}$. These values are in agreement with measurements made by TEM, Table 7.3.

An interesting result of the resistivity measurements is shown in fig. 7.37. Here the number of cycles at which satura-

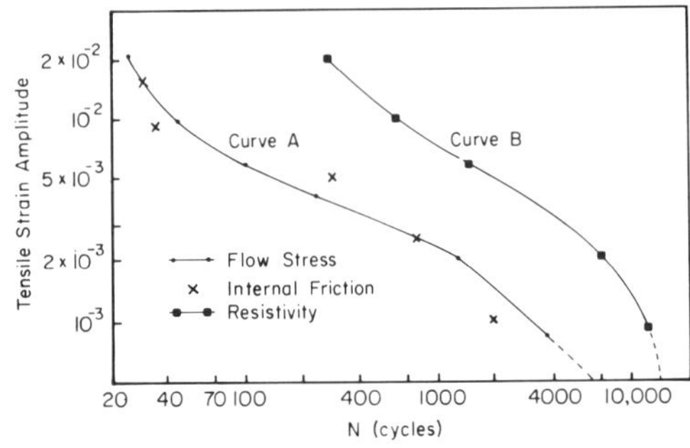

Fig. 7.37 The number of cycles at saturation _versus_ strain amplitude for flow stress (Curve A), internal friction (crosses) and electrical resistivity (Curve B). (from den Buurman and Snoep, 1972, courtesy of Pergamon Press).

tion occurs in polycrystalline copper are plotted _vs._ strain amplitude for the flow stress (Curve A), the internal friction (crosses), and electrical resistivity (Curve B). The internal friction, which is proportional to the free length of dislocation (mainly in the regions between walls or cell interiors), correlates closely with the flow stress, which should be proportional to the mobile dislocation density. However, the resistivity does not saturate until nearly a factor of 10 later in the number of cycles. Den Buurman and Snoep (1972) conclude that the dislocation density in the (cell) walls continues to increase for a period of cycles after saturation of the flow stress is reached.

d) _Magnetic Measurements_. Magnetic measurements have been applied successfully to the study of microstructures formed during unidirectional deformation of ferro-magnetic specimens (cf. Kronmüller 1967). Recently, such measurements have been performed on cyclically-hardened α-iron single crystals (Kütterer 1973). The coercive force H_c (which is roughly proportional to $\sqrt{\rho}$) is shown in fig. 7.38 as a function of τ and N for one specimen in the rapid hardening stage (fig. 7.38a) and for another in the saturation region (fig. 7.38b). In the latter case, H_c continues to increase, although the stress level remains almost unchanged. (cf. similar behavior of resistivity in Cu, Section 7.4.3c). Veith (1971) has made a similar observation on low carbon steels, cycled in the saturation range at prescribed stress amplitude. This suggests that, as the dislocation density continues to increase, the dislocations rearrange into configurations of lower energy (cell formation). Measurements of the susceptibility in the approach to magnetic saturation (Kütterer 1973) which is much

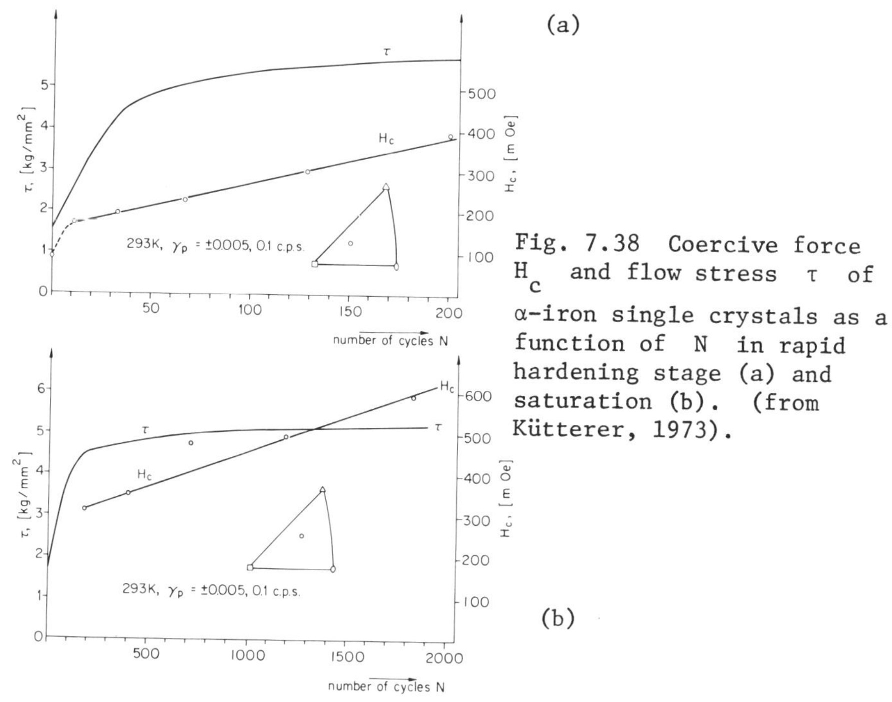

Fig. 7.38 Coercive force H_c and flow stress τ of α-iron single crystals as a function of N in rapid hardening stage (a) and saturation (b). (from Kütterer, 1973).

more sensitive than H_c to the arrangement of the dislocations (cf. Kronmüller 1967) confirm this interpretation.

e) X-ray Measurements. X-ray line broadening and asterism is commonly observed in deformed materials. Both effects are more pronounced after cycling at high amplitudes than after low amplitude fatigue (e.g., Clarebrough, Hargreaves, West and Head 1957, Wood and Segall 1957, Hartmann and Macherauch 1963, Wood, Riemann and Sargent 1964.)

In a recent quantitative study on copper single crystals deformed in tension and fatigue (Herz 1973) the shapes of X-ray line profiles were evaluated according to the theory of Wilkens (1969, 1973). The results, all referring to stresses $\tau \sim 3$ kg/mm^2 can be summarized as follows: 1) X-ray line broadening after cyclic deformation is smaller at low amplitudes and always smaller than after tension. 2) Evaluations of the tails of X-ray line profiles show that a much larger fraction of the dislocation density is arranged in dipole configurations after cyclic deformation than after tension. 3) At low strain amplitudes the contribution of secondary dislocations is distinctly lower than at high amplitudes. 4) After cyclic hardening, dislocation densities of $\sim 5 \times 10^{10}/cm^2$ are obtained as compared to lower values ($\sim 10^{10}/cm^2$) found after tensile deformation. 5) The X-ray line widths indicate clearly that internal stresses are relatively absent at low amplitudes as compared to high amplitudes (Wilkens, unpublished). 6) The X-ray rocking curve half-width continues to increase during cycling in saturation at high amplitudes.

7.4.5 *Some Quantitative Relationships*

An empirical relationship between dislocation density, as derived from resistivity measurements, and "cell" size (collected from a variety of published results) has been proposed by Mandigo (1972). The correlation is shown in fig. 7.39. The

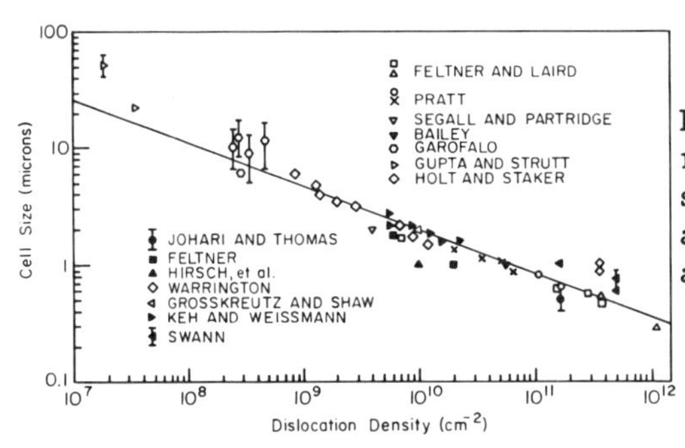

Fig. 7.39 Empirical correlation of dislocation cell size in deformed copper, aluminum and iron with average dislocation density, (from Mandigo, 1972).

straight line drawn through the points has a slope equal to -3/8. If the slope of this line is taken for convenience to be close to -1/2, we can write

$$d = C\rho^{-1/2} \tag{7.3}$$

where C is a constant. But, we also know that cell size and the saturation stress are inversely related (fig. 7.33); i.e.,

$$d = C'\tau_s^{-1} \tag{7.4}$$

Combining eqns. (7.3) and (7.4) gives

$$\tau_s = C''\rho^{1/2} \tag{7.5}$$

where C'' is another constant. A similar set of relations

has been reported for unidirectionally deformed copper polycrystals (Staker and Holt, 1972, cf. Chapter 6).

Equation (7.5) has the form often derived for the relationship between unidirectional flow stress and the dislocation density. In fact, the relationship shown in eqn. (7.5) has been demonstrated empirically by den Buurman and Snoep (1972), using the dislocation density derived from internal friction and modulus changes which occur during cyclic deformation.

There is a well-established empirical relationship between the saturation flow stress and the applied strain amplitude,

$$\tau_s = \tau_o (\gamma_p)^{n'} \tag{7.6}$$

The exponent n', called the cyclic strain hardening exponent, and τ_o are material constants. The value of n' is approximately 0.15 for wavy slip mode polycrystals. For copper single crystals, n' is much smaller, of the order of 0.05. Clearly, it would be desirable to derive the form of eqn. (7.6) in a fundamental way from a relationship such as eqn. (7.5).

7.5 SEMI-QUANTITATIVE MODEL OF THE SATURATION STAGE DEFECT STRUCTURE IN COPPER

7.5.1 *Accommodation of Plastic Strain*

a) Dislocation Mechanisms. A number of dislocation mechanisms that do not produce severe structural changes have been proposed to explain the accommodation of the imposed shear strain:

i - Reversible dislocation bowing below the Frank-Read stress (Avery and Backofen 1963). Feltner (1965) has pointed out correctly that subcritically bowed-out dislocation segments should relax along the same non-linear elastic stress-strain curve along which bowing occurred and should hence not contribute to plastic strain. However, if a friction stress τ^* exists, plastic strain will not recover reversibly until after the stress τ_s has been relaxed by $2\tau^*$ (cf. Cottrell 1953, Segall 1968), and then the recovery will only be partial. Friction stresses are present in BCC metals intrinsically (cf. Section 7.3.1a) and in fatigued FCC metals [dislocation debris, (cf. Section 7.3.1a) and point defect clusters (Section 7.5.2)].

ii - Irreversible Frank-Read bowing of longer segments. The expansion of newly generated loops from the "cell" walls into the "cells" and entangling in the walls (Feltner and Laird 1967b) implies a continuous increase in dislocation density unless considerable annihilation occurs (Basinski, et. al. 1969).

iii - Back and forth motion of the same dislocations in the cells (cell shuttling, Watt and Ham 1966). In the absence of other mechanisms, the required stress is the Frank-Read stress of dislocation segments bowing out between the "cell" walls.

iv - Dipole flipping (Feltner 1965). In this model the plastic strain results from the flip-flop motion of edge dislocation dipoles between the two possible equilibrium configurations. The required stress is the dipole passing stress.

v - Point defect hardening (Broom and Ham 1959, Segall, Partridge and Hirsch 1961, Basinski, et. al. 1969). The strong temperature dependence of the flow stress after cyclic hardening is ascribed to a friction stress due to point defect clusters formed during cycling. The plastic strain would thus be due to the motion of dislocations past these clusters.

All these mechanisms can, in principle, contribute to the plastic shear strain. Bearing in mind that the strains in regions of high and low dislocation density must be compatible, it appears likely that more than just one of these mechanisms is important.

b) Dislocation Motion in Veins (and Walls). At low amplitudes the vein structure is dominant, veins occupying a volume of about 50%. This makes it impossible to ignore the accommodation of strain *in* the veins. Without friction stresses (cf. Section 7.5.2) the only applicable non-hardening mechanism producing *plastic* strain *in* the veins is dipole flipping (Feltner 1965). Although the mechanism certainly operates, it is questionable whether it can quantitatively explain the plastic strain. The mechanism is certainly not applicable in the dislocation-poor regions (Basinski et. al. 1969, Grosskreutz 1971, cf. Section 7.5.1c).

From eqn. (7.1), it follows that, if the stability of veins is governed by the dipole passing stress, then at low amplitudes ($\tau_s \sim 3$ kg/mm^2, $\tau_{eff,max} \sim \tau_s$) $y \lesssim 200$Å. Hence, flipping is possible only if $y \sim 200\text{Å}$. Considering the statistical nature of dipole formation (cf. fig. 7.4), this spacing applies only to a small fraction of the local dislocation density of $\sim 2 \times 10^{11}\ \text{cm}^{-2}$ (cf. Section 7.4.4b), say 5%; i.e., to $\rho_{dip} \sim 5 \times 10^{9}$ dipoles/cm^2. This estimate accounts for $\gamma_p = \rho_{dip}\, b \cdot y \sim 0.00025$, a value that is too low by almost an order of magnitude.

On the other hand, if the stability of veins is controlled by the Frank-Read bowing stress of short segmented dislocations and loops, then larger dipole spacings are possible; i.e., $y > 200\text{Å}$. Thus, with $\rho_{loc} \sim 1/y^2$; i.e., $\rho_{dip} \sim 1/2y^2$, $\gamma_p = \frac{b}{2y}$ (neglecting dislocation bowing)*. For $\gamma_p = 0.001$, an average value of $y \geq 1200\text{Å}$ is necessary, implying an unusually low value of $\rho_{loc} \sim 5 \times 10^{9}\ \text{cm}^{-2}$ (cf. Table 7.2).

If enough plastic strain can be accommodated between the veins (cf. Section 7.5.1c) but not in the veins, then an extra elastic strain is required in the veins to match the total strains at the boundaries. This extra elastic strain could be due to reversible dislocation bowing and the "elastic deformation" of dipoles; i.e., to a reduction in shear modulus μ. However, an extra elastic strain of ~ 0.0005 requires that the reduction in μ in the veins exceed that outside the veins by almost $0.5\ \mu$, whereas typical reductions in bulk shear modulus after fatigue are of the order $0.1\ \mu$ (den Buurman and Snoep 1972)!

Possibly a way can be found out of this dilemma by considering the rather high friction stresses in the veins and walls (Piqueras, Grosskreutz and Frank 1972, cf. Section 7.5.2). Then dislocation bowing and the "deformation" of dipoles would contribute partly to the plastic strain. A more reasonable reduction in shear modulus would follow as well as a thermal component of the saturation stress (cf. e.g., Broom and Ham 1959, Feltner 1965).

c) Dislocation Motion in the "Cells". In dislocation-poor regions; i.e., between veins and (cell) walls of the described structures, cell shuttling may be an important process (Watt and Ham 1966, Hancock and Grosskreutz 1969, Basinski et. al.

---

*Subcritical dislocation bowing, in the absence of a friction stress, does not contribute to plastic shear strain (cf. Section 7.5.1a).

1969, Mughrabi 1973). Figure 7.23 suggests that mobile screw dislocation groups of alternating sign are arranged on neighboring glide planes separated by a distance d (estimated to be about 0.5 μm for copper), as indicated in fig. 7.40.

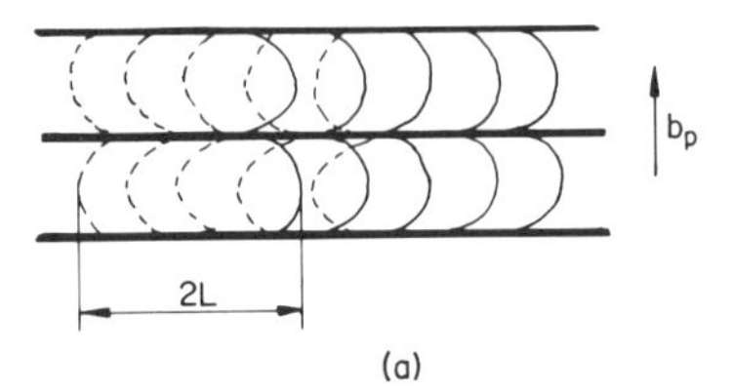

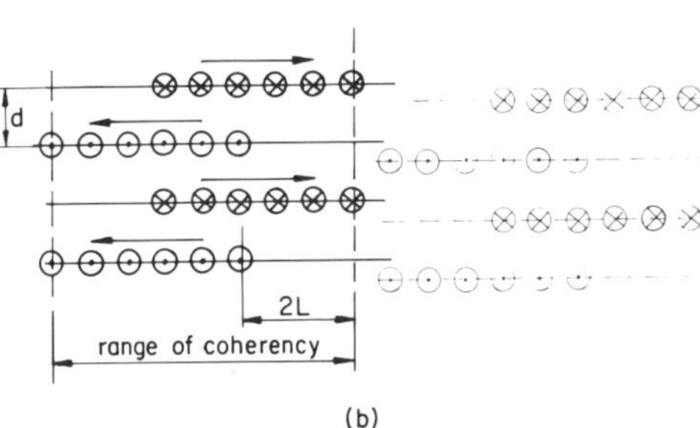

Fig. 7.40 Arrangement of mobile screw dislocations in saturation stage. a) Section parallel to primary glide plane; b) section perpendicular to $\underline{b}_p$.

If d has a physical meaning and is the distance within which screw dislocations of opposite sign annihilate by stress-induced cross slip, then the evolution of such a structure is straight forward. Eventually only those dislocations survive that do not encounter other screw dislocations of opposite sign on neighboring glide planes within a distance d . Figure 7.40 represents the simplest arrangement meeting this requirement. Plastic strain can be accommodated by the non-hardening back and forth motion of the same dislocations.*

Assuming homogeneous slip with glide paths $\pm$ L, we can write

$$\gamma_p = \pm \frac{n}{d} bL , \tag{7.7}$$

where n/d is the dislocation density, and n is the number of mobile dislocations per unit length on the glide plane. With the experimental values $\gamma_p = \pm 0.005$, $n/d \sim 10^9$ cm^{-2} , we obtain $L \sim \pm 2$ μm (Mughrabi 1973). Figure 7.23 suggests

---

* Entanglement of drawn-out edge dislocations in the walls is not expected to lead to a substantial increase in dislocation density, since such dislocations will be annihilated during reversed glide.

that this structure easily permits such motion and also larger slip distances as may be appropriate to strongly localized inhomogeneous slip; e.g., in the PSB structure.

The model explains the almost complete reversal of slip lines (cf. Watt, Embury and Ham 1968, Basinski et. al. 1969). Deviations from this behavior probably reflect deviations from the idealized model. Thus, some annihilation may still occur in saturation. Figures 7.23 and 7.36 suggest that lost dislocations can be replaced readily at an only slightly increased stress by Frank-Read bowing of dislocations out of the "walls". Such processes may also account for "non-hardening" structural changes observed during saturation (cf. Polák 1969, Kütterer 1973, Herz 1973, den Buurman and Snoep 1972).

The discussed mechanism is expected to be less efficient, when the slip distances of screw dislocations are limited by "cell" walls and grain boundaries, especially in materials of higher stacking fault energy (larger d, say $\gtrsim 1\ \mu m$). In this case the "packing" of dislocations on the active slip planes (i.e. n) would have to be very high ($n \gtrsim 10^5/cm$). Cross-slip of piled-up dislocation groups (comparable to unidirectional stage III behavior) would disperse glide*, competing with localization of glide by annihilation due to stress-induced cross slip.

7.5.2 *The Role of Point Defect Clusters*

Point defects and point defect clusters produced by cyclic deformation have been observed by changes in electrical resistivity (Polák, 1969; Helgeland, 1969; Johnson and Johnson, 1965) and by transmission electron microscopy (fig. 7.20). It is generally agreed that the concentration of such defects is larger after cyclic deformation than after unidirectional deformation. Any model of the work-hardened structure and its relation to the saturation flow stress must therefore consider the role played by point defects and their clusters. In general, it would be expected that they would contribute to a temperature-dependent component of the flow stress, given the usual model of thermal activation of dislocations over such barriers. The basic problem then is to determine the relative magnitude of this thermal component compared to the total flow stress. The experiments of Piqueras, et. al. (1972) on copper represent the first direct attempt at such a measurement.

*Wells (1973) has presented a model describing the transition from planar to wavy slip by cross slip at the head of a pile-up.

The temperature dependence of the tensile yield stress of copper crystals which were cycled into saturation at room temperature is given in fig. 7.41. Each curve corresponds to a

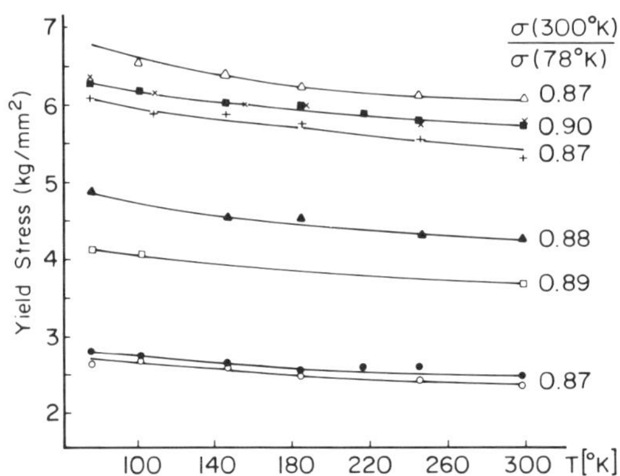

Fig. 7.41 Temperature dependence of the tensile yield stress measured on copper single crystals after cyclic hardening into saturation at 293°K, (from Piqueras, et. al., 1972, courtesy of Akademie Verlag).

different cyclic plastic (tensile) strain amplitude; crystal No. 7 was cycled at ± 0.0003, and crystal No. 5 at ± 0.0046. The ratio of the yield stress at 300°K to that at 78°K is 0.88, independent of the cyclic strain amplitude under which the crystal was deformed. This ratio is to be compared with a value of 0.93 obtained from crystals deformed in pure, unidirectional tension (Adams and Cottrell, 1955). The conclusion is that the thermal component of the flow stress is larger under cyclic deformation.

The magnitude of the thermal component, τ^*, may be calculated as a function of temperature using the Frank-Rühle-Saxlova theory (Frank et. al. 1968) provided the average size of the obstacles (point defect clusters), d and the obstacle spacing ℓ are determined experimentally. These parameters may be found from statistical analysis of weak beam T.E.M. photographs such as fig. 7.20. The final result of a self-consistent evaluation of cell-shuttling requiring mobile dislocation densities ρ_m of the order of $1 - 4 \times 10^9$ cm^{-2}*), is

*) Comparison of these values with experimental observations (cf. fig. 7.23) shows that the latter yield somewhat lower values of $\sim 10^9$ cm^{-2}. The analysis of Section 7.5.3 indicates, however, that weak pile-up effects exist. Thus, as far as the effect on the flow stress is concerned, the lower experimentally determined values of ρ_m is formally equivalent with the higher value required by the self-consistent analysis of cell shuttling in which pile-up effects were not considered.

that the ratio of τ^* to the total saturation flow stress τ_s, taken from fig. 7.41, falls in the range

$$0.15 < \tau^*/\tau_s < 0.25$$

depending on the temperature and applied strain amplitude. The ratio τ^*/τ_s is highest at low temperatures and low strain amplitudes.

7.5.3 *The Flow Stress*

Because of the very high and not well-known defect density in the veins and walls a quantitative distinction between the various dislocation mechanisms (cf. Section 7.5.1) is almost impossible. We therefore confine ourselves to a discussion of the flow stress in the interior of "cells." At high amplitudes this should account for about 90% of the specimen volume. Assuming additivity of different contributions to the saturation flow stress τ_s (~ 3.2 kg/mm^2), we may write:

$$\tau_s = \tau^* (\dot{\gamma}_p, T) + \frac{A}{r_\odot} + 2B \frac{\mu b}{2\pi d} \tag{7.8}$$

The three terms represent the thermal component τ^* due to point defect clusters (cf. Section 7.5.2), the Frank-Read stress $\tau_{F.R.}$ of the screw dislocations bowing out between the walls and the passing stress arising from the elastic interaction of screw dislocation groups of opposite sign (cf. figs. 7.23, 7.40). $\tau_{F.R.}$ is expressed in terms of the radius of curvature $r_\odot$ of screw dislocations (the constant A has been determined empirically to 1.1×10^{-3} kg/mm, cf. Mughrabi 1973). The constant B is a stress-enhancement factor accounting for pile-up effects.

Piqueras et. al. (1972) have established that $\tau^* \sim 0.2$ $\tau_s \sim 0.65$ kg/mm^2. With the experimental value $r_\odot \sim 1$ μm, we find $\tau_{F.R.} = \frac{A}{r_\odot} \sim 1.1$ kg/mm^2. Using $d = 0.5$ μm (cf. Section 7.5.1c), we must assume $B \sim 2$ in order to explain the missing fraction of τ_s.

Neglecting the elastic interaction with the walls (cf. fig. 7.23) is justified to some extent, since the penetration distance of high internal stresses into the cells is small compared to the free spacing between walls (cf. Section 7.3.2b).

At lower amplitudes, glide distances are probably reduced. The smaller free spacings between veins would require a larger $\tau_{F.R.}$ However, $A/r_{\odot}$ can be smaller than $\tau_{F.R.}$, provided dislocation bowing is weak.* Elastic interaction of dislocations is almost certainly smaller, since pile-up effects become less important and since dislocations may not have to glide past one another.* In addition, if dislocations do bow out critically, their strong curvature will resist elastic interaction over a large fraction of their length. Available experimental evidence indicates that the friction stress τ^* makes a larger contribution at lower amplitudes (Piqueras et. al. 1972). This would prevent dislocations from relaxing elastically when the stress is reversed.

Thus, although the individual contributions of the three terms in eqn. (7.8) may vary from case to case, a description of τ_s by eqn. (7.8) should still be possible in most cases.

7.6 STABILITY OF THE CYCLIC WORK-HARDENED STATE

7.6.1 *Strain Bursts*

Neumann (1968) has shown that copper single crystals cycled at slowly increasing stress amplitudes ($\Delta\tau/\Delta N \sim 10^{-4} - 10^{-3}$ kg/mm^2/cycle) exhibit regular sequences of so-called strain bursts. During each strain burst which lasts about 50 cycles, γ_p increases drastically (from about 10^{-4} to 10^{-3}) and decreases again. The next strain burst follows after the stress has increased by a constant fraction ($\sim$10%). Strain bursts have also been found in single crystals of silver, aluminum, zinc and magnesium (Neumann and Neumann 1970). The frequently observed "drifting" during a strain burst; i.e., elongations

---

* The possible importance of these "subthreshold" effects is reflected by the observation that the hardening coefficient at the peak stress is much larger at smaller amplitudes (Kettunen and Kocks 1969). This is related to the fact that, especially at low amplitudes, the unidirectional yield stress is larger than the cyclic peak stress (Kettunen and Kocks 1972).

(in either direction) up to some 10^{-3}, reflects the high degree of structural instability (Kralik and Mughrabi 1973).

Neumann (1968) has given an interpretation in terms of the repeated formation and subsequent avalanche-like dissociation of dipole and multipole structures throughout the crystal (cf. Chapter 12). Recent work (Kralik and Mughrabi 1973) has shown that strain bursting is favored at high frequencies and/or low temperatures, suggesting that the suppression of thermal activation is important.

Dislocation structures in copper single crystals (pinned after unloading) between two strain bursts (fig. 7.42a) and in a strain burst do not exhibit as dramatic differences as one might expect. The reason probably is that "old" and "new" dipoles cannot be distinguished in fig. 7.42b. However, the observed regularly spaced primary edge multipole bundle structure ($\rho \sim 2\text{-}3 \times 10^9$ cm^{-2}) is much more fragmented in fig. 7.42b which also reveals some free primary dislocations. The weak curvatures of the latter indicate the almost total absence of long range internal stresses as compared, for example, to fig. 7.8*.

7.6.2 *Non-Hardening Overstraining and Coarse Slip*

Tensile deformation of cyclically-hardened specimens is characterized by an elongated yield plateau, before hardening sets in (Broom and Ham 1959). This probably reflects the non-hardening strain that can be accomplished by the screw dislocations (cf. fig. 7.23), as they traverse and leave the "safe coherent" range (cf. fig. 7.40). Glide into neighboring regions is accompanied by annihilation and activation of new sources. Hardening sets in.

Copper single crystals, cycled into saturation at $\gamma_p \sim \pm 0.005$, exhibit a yield plateau that extends over a shear strain of ~ 0.013, corresponding to average glide distances of ~ 5 µm (Mughrabi 1973). This distance is interpreted as the mean "range of coherency" of the glide zones to which the crystal has been "trained"**. The latter term has been borrowed from

---

* This is presumably the reason for the very low secondary dislocation activity. It appears justified to infer that long range internal stresses are generally absent when the stress level increases slowly; i.e., also in low strain amplitude tests (cf. Section 7.3.1b).

** It is evident from Section 7.5.1c that the range of coherency should be at least 2L; i.e., $\gtrsim 4$ µm.

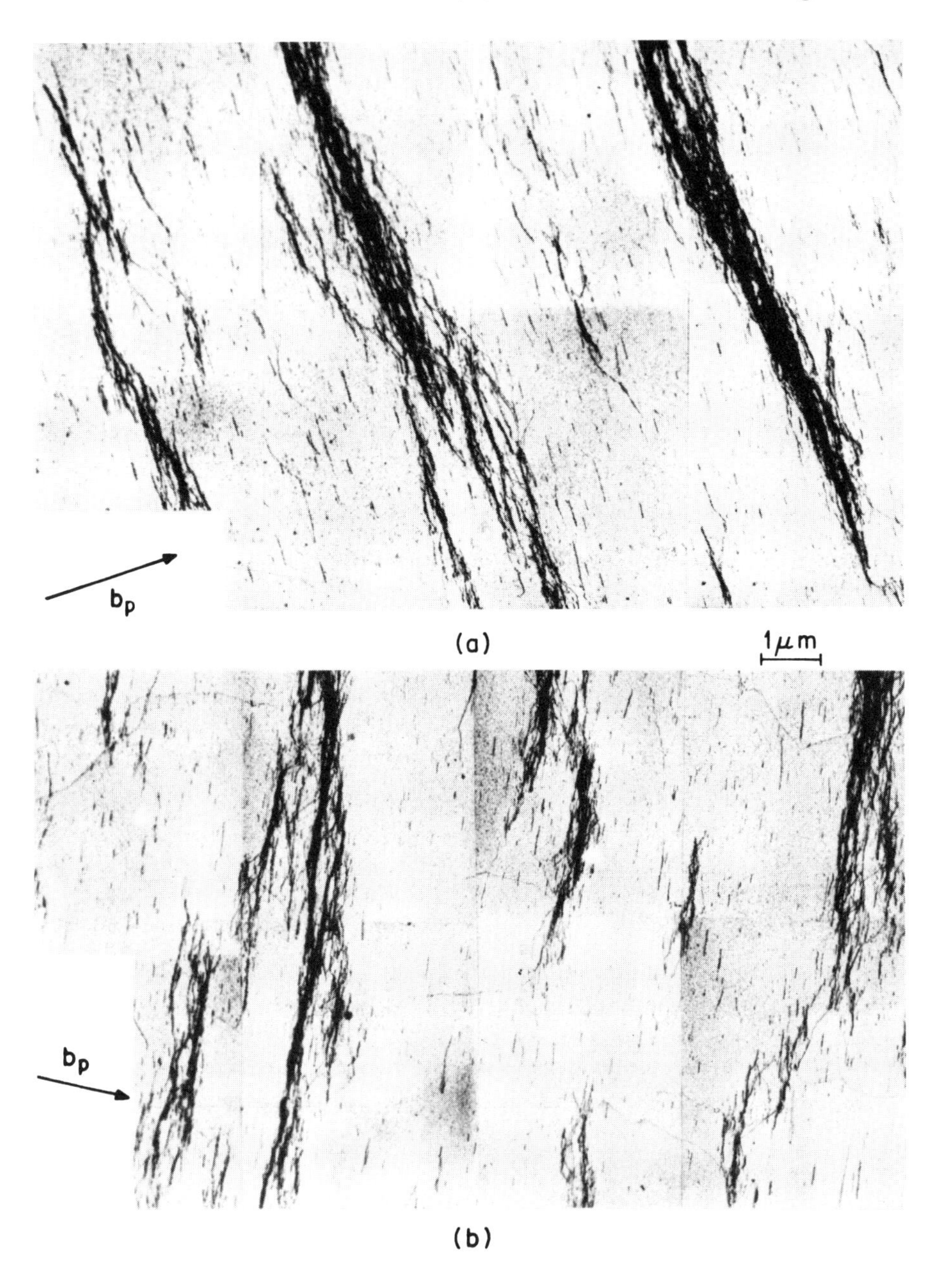

Fig. 7.42 Dislocation structure in primary glide plane (111) of copper between two strain bursts (a), and in a strain burst (b). $\tau \sim 1.2$ kg.mm, 293°K, pinned in unloaded state.

Basinski et. al. (1969).

As pointed out by Neumann (1968), the coarse slip observed during overstraining (Basinski et. al. 1969) and in unidirectional deformation (Broom and Ham 1959) is probably related to the phenomenon of strain bursts (cf. Section 7.6.1). The dissociation of multipole structures is presumably triggered by the locally disturbed equilibrium in regions where new sources are activated as described above. The pointed-out analogy to coarse slip in irradiation- and quench-hardened materials (Broom and Ham 1959, Basinski et. al. 1969) appears uncertain, since both the yield point and coarse slip are largely suppressed in prestrained irradiation-hardened materials.

7.6.3 *Comments on Cyclic Softening*

Softening of previously work-hardened materials by cyclic deformation was mentioned in Section 7.2.2. This phenomenon has been observed for most materials, including complex, multiphase alloys in commercial use. The most general explanation would seem to be that all cyclic or tensile work-hardened structures (and the long and short range stresses which determine the flow stress) are metastable in character, and that repeated cyclic strains "provide" the activation energy to break them down. Detailed mechanisms depend, of course, on the particular microstructures involved.

As a case in point, we consider copper as characteristic of of pure, wavy slip mode materials. At high cyclic strain amplitudes ($\gamma_p \gtrsim 5 \times 10^{-3}$ for polycrystals), the saturation flow stress is a unique function of the applied strain amplitude regardless of whether the metal has been initially strain-hardened (cyclic or tensile), or annealed (Section 7.22). The saturation deformation structures are also independent of prior deformation. Small cells, produced in prestrained copper, soften into larger cells (Feltner and Laird, 1967b). Cells produced by cyclic straining can be grown or shrunk in size simply by varying the applied strain amplitude (Pratt, 1967). In each case, the cell size is characteristic of a given combination of τ_s and γ_p. This same behavior has been demonstrated for iron (Abdel-Raouf, et. al., 1971).

At low strain amplitudes ($\gamma_p \lesssim 10^{-3}$), the situation is more complex in copper (Lukáš and Klesnil, 1973). Although specimens prestrained in tension soften under cyclic straining, the cyclic stress-strain curve is not unique, being displaced upward for greater values of prestrain. (It is as if the cyclic strain were not large enough to activate break-up of the pre-strain structure.) Similarly, the pre-strain cells, while expanding in size during cyclic softening, do not approach the size of cells produced by cyclic straining of annealed specimens at the same amplitude. On the other hand, specimens pre-hardened by cyclic strain were observed to soften to a unique saturation stress value. The saturation flow stress was completely reversible as the cyclic strain amplitude was raised and lowered. But the cyclic deformation structures were not reversible. Cells produced at higher strain amplitudes ($\sim 5 \times 10^{-4}$) did not revert to a vein structure at lower strain amplitudes. We are thus faced with a situation in which the flow stress is reversible, but the deformation structure is not. The only conclusion is that the long range stresses exerted by veins and walls formed at low strains are approximately the same, (Lukáš and Klesnil, 1973), or that long range stresses are nearly absent at low amplitudes (cf. Section 7.3.2a).

A simple model of cyclic softening in copper may be formed based on these observations. Cell walls produced by tensile prestraining are high energy configurations. Cell walls produced by cyclic straining are low energy configurations and contain large numbers of dislocations, larger by a factor of at least 10 than those occurring in crystals strained unidirectionally to the same flow stress (Grosskreutz, 1971). These cyclically produced walls contain edge dislocation multipole arrays in equilibrium with the internal and applied stresses. When crystals hardened either by unidirectional or cyclic pre-straining are cycled at a lower amplitude, softening may occur by action of the repeated cyclic stresses (analogous to thermal fluctuations) which cause rearrangement and annihilation of dislocations in the high energy (tensile) walls, and break-up and annihilation of multipoles in the low energy (cyclic) walls, which contain unbalanced internal stresses. After continued cycling, a new equilibrium is established, consistent with the dislocation density - saturation flow stress relationship in eqn. (7.5). If the cyclic strain amplitude is sufficiently high, cell growth occurs. Otherwise, the annihilation and reduction of dislocation density occurs without major re-arrangement of structure.

Throughout this discussion of cyclic softening, it is assumed that reversal of the plastic strain always occurs - a condition which is necessary for softening to occur (Feltner and Laird, 1967a).

7.6.4 *Asymmetry of the Cyclic Hysteresis Loop Under Completely Reversed Straining*

We return here to a discussion of the asymmetric hysteresis loop first mentioned in Section 7.2.2, and the microstructures which may account for it. Evidence for the asymmetry is widespread. Kemsley and Paterson (1960), and Wadsworth (1963) report the effect for copper single crystals of various orientation. Grosskreutz (1970) observed similar behavior for polycrystalline copper. Coffin (1963) found that cyclic creep occurred in polycrystalline aluminum even though the mean stress was zero, and Kawamoto and Shibata (1971) found that a small compressive mean stress is required to prevent cyclic creep in mild steel and 60/40 brass. All of these authors consider the phenomenon to be a real material property and not an artifact of the measuring technique.*

No attempt has yet been made to account for this phenomenon either in terms of the cyclic work-hardened structures or the mechanisms of cyclic strain accommodation discussed in Section 7.5.1. We do not attempt to develop such a rationale here, but it does seem reasonable that the effect have its origin in the metastability of the cyclic microstructure, which is common to all the materials cited in the foregoing paragraph. Using copper as an example, one can argue that the decomposition of dislocation veins and walls (as in the case of strain bursts) is more likely when there is a tensile stress <u>normal</u> to the slip plane than when there is a compressive stress. Therefore, a small compressive mean stress is required to equalize the probability of decomposition during each half cycle. That decomposition of multiple bundles occurs regularly during each reversal of stress is an assumption which has not been verified, but it seems reasonable that it may happen on a random microscale, contributing to the overall strain at any given flow stress. The effects of stress components normal to the glide plane have not received much attention in the case of FCC metals. In BCC metals, stress asymmetry (and the nonvalidity of Schmid's law of resolved shear stresses) are important and well-documented effects exist (cf. Hirsch 1968,

---

* The slightly reduced specimen cross-section in tension as compared to the increased cross-section in compression can give rise to asymmetric tensile and compressive loads in controlled strain cycling. The contribution of this effect can be calculated and, depending on strain amplitude, amounts to load asymmetries of the order of 1%, which is too low by a factor of 5 to account for the observed asymmetry.

Christian 1970, Šesták 1970) which are related to the complex structure of the screw dislocation cores. The latter may be modified by stress components normal to the glide plane.

7.7 BASIC DIFFERENCES BETWEEN CYCLIC AND UNIDIRECTIONAL WORK-HARDENED STRUCTURES

The introductory sentence to this Chapter stated that the cyclic work-hardened structure, while containing much that resembles the structure induced by unidirectional deformation, has unique properties of its own. In this Section, we summarize briefly the basic differences between these two structures.

7.7.1 *Qualitative Differences*

Repeated cyclic glide, always partly reversible, produces smaller structural changes and hardening increments per unit strain than glide in one direction*. The latter is characterized by irreversible long range cooperative slip processes (Seeger and Wilkens 1967) that are much less important in cyclic deformation. The more gradual structural changes in cyclic hardening result in regular and "tidy" low energy arrangements of dislocations (multipole bundles and walls, low-angle grain boundaries), culminating in a saturation or equilibrium structure which is unique to cyclic deformation. By comparison the veins and walls formed in unidirectional deformation are less dense and more ragged (fig. 7.25); disordered tangles are more characteristic of tensile deformation.

From the point of view of internal stresses and secondary dislocation activity, low and high amplitude cyclic deformation can be compared qualitatively with stage I and stage II unidirectional behavior (cf. Avery and Backofen 1963, Feltner and Laird 1967b).

Secondary glide activity (which "pins down" sources of long range stress such as dislocation pile-ups in unidirectional deformation, cf. Chapter 6) is less pronounced in cyclic hardening for three reasons. Orientation changes are small. This favors persisting single slip. Activation of secondary slip by internal stresses is reduced, especially at low amplitudes. The high dislocation density of the low energy dislocation

* Some interesting differences between prescribed stress and prescribed strain cycling, also with respect to comparison with unidirectional deformation, have been pointed out by Kettunen and Kocks (1972).

arrangements causes stronger latent hardening than after tensile deformation (Basinski and Saimoto 1967).

The temperature-dependence of the flow stress is larger after cyclic deformation than after unidirectional deformation (in spite of the generally lower forest dislocation density) because of the high density of point defect clusters and dislocation debris. The (Frank-Read) bowing stress makes a larger contribution in cyclic than in unidirectional deformation, where the athermal component of the flow stress is almost entirely due to the elastic interaction of dislocations (cf. Chapter 6).

The role of cross slip in fatigue has been emphasized repeatedly (cf. Alden 1962) and has been compared to unidirectional stage III behavior (Feltham 1961; Rudolph, Haasen, Mordike and Neumann 1965, Kettunen and Kocks 1969, 1972). Perhaps a distinction should be made between the (possibly largely athermal) stress induced cross slip permitting annihilation of near screw segments of opposite sign at almost any stress level and the strongly temperature-dependent cross slip of a number of dislocations of the same sign at the head of a pile-up at higher stresses (Seeger, Diehl, Mader and Rebstock 1957). It appears doubtful at the present time whether the detailed mechanisms of these two cross slip processes are fundamentally equivalent. In fact, this may be the key to the conflicting standpoints on the role of cross slip in the formation of fatigue striations (cf. Nine and Kuhlmann-Wilsdorf 1967).

7.7.2 *Quantitative Differences*

The dislocation densities attained in cyclic deformation are significantly higher than in unidirectional deformation (Table 7.3). A further striking feature of the cyclically-hardened state is the high density of point defect clusters ($\sim 10^{15}$ cm^{-3}) and the associated large friction stress which can be as high as 25% of the peak stress at room temperature. The mean lengths of free segments of dislocations in dense structures are shorter by at least a factor of two at comparable stresses after cyclic deformation at high amplitudes than after tensile deformation (cf. Chapter 6). Thus activation of dislocation sources requires the aid of the internal stresses in high amplitude fatigue but not in tension. The amplitudes of the internal stresses are of the order of the flow stress in both cases, whereas internal stresses are probably negligible at low amplitudes.

7.7.3 *Stability*

Questions of stability in unidirectional work-hardened pure metals have not been given too much attention - an indication of the fact that instabilities of the mechanical state play a minor role, at least on a macroscopic scale. On a microscopic scale, plastic deformation is per se unstable, typical examples being bowing out of dislocation segments, dissociation of dipoles, intersection processes and cross slip - slip band formation. The propagation of these instabilities is limited by the efficient immobilization of dislocations at the end of their glide paths by secondary slip and sessile reaction products which form strong obstacles. This is fundamental to unidirectional work-hardening.

Cyclic deformation does not favor long range cooperative slip on different glide systems. As a result, stabilization of the structure by secondary dislocations and by internal stresses is reduced drastically. Cyclic slip lines do not (usually) reflect the distances travelled by dislocations before being trapped, as is evident by the high degree of reversibility. Structures that evolve from basically non-hardening dislocation glide are inherently metastable. The precarious mechanical stability of the cyclic work-hardened state manifests itself in the catastrophic large-scale breakdown of the originally formed dislocation structures and the formation of energetically more favorable ones (PSB formation, strain bursts, coarse slip in subsequent tensile deformation).

As Basinski et. al. (1969) put it, the specimen is "trained" to a particular fatigue cycle. When this is exceeded with regard to stress or strain level, the mode of deformation changes drastically, accompanied by structural changes. On a microscopic scale, the structural reversibility of dislocation motion is confined to "safe" regions, which we have characterized by their range of coherency.

Systematic investigations on the stability of cyclic work-hardened structures, especially in saturation, have not been performed. A better understanding of the extreme situations at low amplitudes and in the approach to the conditions of unidirectional deformation at high amplitudes would be most desirable.

ACKNOWLEDGEMENTS

One of the authors (H. M.) would like to express his sincere thanks to Dipl. Phys. K. Herz, Dipl. Phys. R. Kütterer and Dipl. Phys. X. Stark for their cooperation and their permission to quote unpublished results. Helpful discussions with Dr. W. Frank and Dr. K. Urban's aid in the high voltage electron microscopy (fig. 7.23) are acknowledged. H. M. is grateful to Dr. M. Wilkens and Prof. H. Kronmüller for their stimulating interest. The kind support of Prof. A. Seeger is deeply appreciated. Some of this work received financial support from the Deutsche Forschungsgemeinschaft. J. C. G. acknowledges with thanks the discussions held with Dr. L. M. Brown and Dr. P. J. Woods concerning deformation homogeneity.

REFERENCES

Abdel-Raouf, H., Benham, P. P. and Plumtree, A., (1971) Canad. Metal. Quart., 10, 87.

Abdel-Raouf, H. and Plumtree, A., (1971) Metal. Trans., 2, 1863.

Abdel-Raouf, H., Plumtree, A., and Topper, T. H., (1973) ASTM, STP 519, (Philadelphia: ASTM) p. 28.

Adams, M. A., and Cottrell, A. H., (1955) Phil. Mag., 46, 1187.

Alden, T. H., (1962) J. Metals, 14, 828.

Argon, A. S., and East, G., (1968) in Proc. Int. Conf. on Strength of Metals and Alloys, Trans. Japan Inst. Metals, Suppl., 9, 756.

Atkinson, J. D., Brown, L. M., Kwadjo, R., Stobbs, W. M., Winter, A. T., and Woods, P. J., (1973) Proc. 3rd Int. Conf. on Strength of Metals and Alloys, Cambridge, p. 402.

Avery, D. H., and Backofen, W. A., (1963) Acta Met., 11, 653.

Basinski, S. J., Basinski, Z. S., and Howie, A., (1969) Phil. Mag., 19, 899.

Basinski, Z. S., and Saimoto, S., (1967) Canad. J. Phys., 45, 1161.

Bergstrom, Y., Vingsbo, O., and Lagerberg, G., (1969/70) Mat. Sci. Engng., 5, 153.

Broom, T., and Ham, R. K., (1959) Proc. R. Soc., (London) A 251, 186.

Christian, J. W., (1970) Proc. 2nd Int. Conf. on Strength of Metals and Alloys, Asilomar, (Metals Park, Ohio: ASM) p. 31.

Clarebrough, L. M., Hargreaves, M. E., West, G. W. and Head, A. K., (1957) Proc. R. Soc., (London) A 242, 160.

Coffin, L. F., (1963) Trans. ASME, paper no. 63-WA-109.

Cottrell, A. H., (1953) Dislocations and Plastic Flow in Crystals, (Oxford: Clarendon Press) p. 111.

den Buurman, R., and Snoep, A. P., (1972) Acta Met., 20, 407.

Edington, J. W., (1969) Phil. Mag., 19, 1189.

Essmann, U., (1964) Acta Met., 12, 1468.

Essmann, U., (1966) phys. stat. sol., 17, 725.

Essmann, U., and Rapp, M., (1973) Acta Met., 21, 1305.

Essmann, U., and Strunk, H., (1970) Z. Metallk., 61, 667.

Feltham, P., (1961) Phil. Mag., 6, 1479.

Feltner, C. E., (1963) Phil. Mag., 8, 2121.

Feltner, C. E., (1965) Phil. Mag., 12, 1229.

Feltner, C. E., and Laird, C., (1967) Acta Met., 15 (a), 1621, and (b) 1633.

Feltner, C. E., and Laird, C., (1968) Trans. Metall. Soc. AIME, 242, 1253.

Feltner, C. E., and Laird, C., (1969) Trans. Metall. Soc. AIME., 245, 1372.

Frank, W., Rühle, M., and Saxlová, M., (1968) phys. stat. sol., 26, 671.

Gilman, J. J., and Johnston, W. G., (1962) Solid State Physics, 13, 147.

Gostelow, C. R., (1971) Metal Sci. Journ., 5, 177.

Greenfield, I. G., (1972) in Corrosion Fatigue, NACE-2, National Association of Corrosion Engineers, edited by R. Staehle, et. al. (Houston, Texas: NACE), p. 133.

Grosskreutz, J. C., (1962) J. Appl. Phys., 34, 372.

Grosskreutz, J. C., (1970) (Unpublished Work).

Grosskreutz, J. C., (1971) phys. stat. sol., 47, 11.

Grosskreutz, J. C., Reimann, W. H., and Wood, W. A., (1966) Acta Met., 14, 1549.

Grosskreutz, J. C., and Shaw, G. G., (1964) Phil. Mag., 10, 961.

Grosskreutz, J. C., and Waldow, P., (1963) Acta Met., 11, 717.

Hancock, J. R., and Grosskreutz, J. C., (1969) Acta Met., 17, 77.

Hartmann, R. J., and Macherauch, E., (1963) Z. Metallk., 54, 197.

Helgeland, O., (1967) Trans. Met. Soc. AIME, 239, 2001.

Herz, K., (1973) Diploma Thesis, University of Stuttgart.

Hirsch, P. B., (1968) in Proc. Int. Conf. Strength of Metals and Alloys, Trans. Japan Inst. Metals, Suppl. 9, XXX.

Ivanova, V. S., Goritskiy, V. M., Orlov, L. G. and Terent'Yev, V. F., (1972) Fiz, Metal. Metalloved., 34, No. 3, 456.

Johnson, E. W., and Johnson, H. H., (1965) Trans. Met. Soc. AIME, 233, 1333.

Kawamoto, M., and Shibata, T., (1971) Dept. of Mech. Engrg., Kyoto University, Kyoto, Japan (private communication).

Keh, A. S., (1965) Phil. Mag., 12, 9.

Kemsley, D. S., and Paterson, M. S., (1960) Acta Met., 8, 453.

Kettunen, P. O., (1967) Acta Met., 15, 1275.

Kettunen, P. O., and Kocks, U. F., (1969) Czech. J. Phys., B19, 299.

Kettunen, P. O., and Kocks, U. F., (1972) Acta Met., 20, 95.

Klesnil, M., Holzman, M., Lukáš, P., and Rys, R., (1965) J. Iron Steel Inst., 203, 47.

Klesnil, M., and Lukáš, P., (1965) J. Iron Steel Inst., 203, 1043.

Kralik, G., and Mughrabi, H., (1973) in Proc. 3rd Int. Conf. on Strength of Metals and Alloys, (Cambridge: Inst. Metals) p. 410.

Krejci, J., and Lukáš, P., (1971) phys. stat. sol., 5, 315.

Kronmüller, H., (1967) Canad. J. Phys., 45, 631.

Kütterer, R., (1973) Diploma Thesis, University of Stuttgart.

Laufer, E. C., (1969) Czech. J. Phys., B19, 333.

Laufer, E. C., and Roberts, W. N., (1964) Phil. Mag., 10, 993.

Laufer, E. C., and Roberts, W. N., (1966) Phil. Mag., 14, 65.

Levine, E., and Weissman, S., (1968) ASM Trans. Quart., 61, 128.

Lukáš, P., and Klesnil, M., (1964) Czech. J. Phys., B14, 600.

Lukáš, P., and Klesnil, M., (1967) phys. stat. sol., 21, 717.

Lukáš, P., and Klesnil, M., (1970) phys. stat. sol., 37, 833.

Lukáš, P., and Klesnil, M., (1971) phys. stat. sol., 5, 247.

Lukáš, P., and Klesnil, M., (1972) Corrosion Fatigue, NACE-2, National Association of Corrosion Engineers, edited by R. Staehle et. al. (Houston, Texas: NACE), p. 118.

Lukáš, P., and Klesnil, M., (1973) Mater. Sci. Eng., 11, 345.

Lukáš, P., Klesnil., M., and Krejci, J., (1968) phys. stat. sol., 27, 545.

Lukáš, P., Klesnil, M., Krejci, J., and Rys, P., (1966) phys. stat. sol., 15, 71.

Lukáš, P., Klesnil, M., and Rys, P., (1965) Z. Metallk., 56, 109.

Mandigo, F. N., (1972) Ph.D. Thesis, Cornell University, Ithaca, N.Y.

McEvily, A. J., and Johnson, T. L., (1967) Int. J. Fract. Mech., 3, 45.

McGrath, J. T., and Bratina, W. J., (1965) Phil. Mag., 11, 429, (a); 12, 1293, (b).

Mitchell, A. B., and Teer, D. G., (1969) Phil. Mag., 19, 609.

Mitchell, A. B., and Teer, D. G., (1970) Phil. Mag., 22, 399.

Mitchell, T. E., Foxall, R. A., and Hirsch, P. B., (1963) Phil. Mag., 8, 1895.

Mughrabi, H., (1971) Phil. Mag., 23, 931.

Mughrabi, H., (1973) in Proc. 3rd Int. Conf. on Strength of Metals and Alloys, Cambridge, (Cambridge: Inst. Metals) p. 407.

Nabarro, F. R. N., Basinski, Z. S., and Holt, D. B., (1964) Adv. Phys., 13, 193.

Neumann, P., (1968) Z. Metallk., 59, 927.

Neumann, R., and Neumann P., (1970) Scripta Met., 4, 645.

Nine, H. D., and Bendler, H. M., (1964) Acta Met., 12, 895.

Nine, H. D., and Kuhlmann-Wilsdorf, D., (1967) Canad. J. Phys., 45, 865.

Piqueras, J., Grosskreutz, J. C., and Frank, W., (1972) phys. stat. sol., 11(a), 567.

Plumbridge, W. J., and Ryder, D. A., (1969) Metal., Reviews, 14, 119.

Polák, J., (1969) Czech. J. Phys., B19, 315.

Pratt, J. E., (1967) Acta Met., 15, 319.

Roberts, W. N., (1969) Phil. Mag., 20, 675.

Rudolph, G., Haasen, P., Mordike, B. L., and Neumann, P., (1965) in Proc. 1st Int. Conf. on Fracture, Sendai, Vol. 2, p. 501.

Schoeck, G., and Seeger, A., (1955) Rep. Conf. Defects in Solids (London: Physical Society), p. 340.

Seeger, A., (1957) Dislocations and Mechanical Properties of Crystals, edited by J. C. Fisher, (New York: John Wiley), p. 243.

Seeger, A., (1963) in Proc. of Conf. on Relation between Structure and Mechanical Properties of Metals, Vol. I (London: H.M.S.O.), p. 4.

Seeger, A., and Wilkens, M., (1967) Realstruktur und Eigenschaften von Reinststoffen, edited by E. Rexer (Berlin: Akademie-Verlag), part III, p. 29.

Segall, R. L., (1968) Adv. Mat. Res., 3, 109.

Segall, R. L., Partridge, P. G., and Hirsch, P. B., (1961) Phil. Mag., 6, 1493.

Šesták, B., (1970) 3rd Int. Conf. Reinststoffe in Wissenschaft und Technik, Dresden, p. 222.

Šesták, B., and Seeger, A., (1971) phys. stat. sol., (b) 43, 433.

Solomon, H. D., and McMahon, C. J., (1968) Work Hardening, edited by J. P. Hirth and J. Weertman (New York: Gordon and Breach), p. 311.

Staker, M. R., and Holt, D. L., (1972) Acta Met., 20, 569.

Stark, X., and Mughrabi, H., (1973) unpublished.

Steeds, J. W., (1966) Proc. R. Soc. (London) A 292, 343.

Tetelman, A. S., (1962) Acta Met., 10, 813.

Veith, H., (1971) Doctorate Thesis, Technische Universitat Berlin.

Vingsbo, O., (1970) University of Uppsala Report U.U.I.P.-713.

Vingsbo, O., Bergström, Y., and Lagerberg, G., (1969) Phil. Mag., 20, 1271.

Wadsworth, N. J., (1963) Acta Met., 11, 663.

Watt, D. F., (1967) Ph.D. Thesis, McMaster University.

Watt, D. F., and Ham, R. K., (1966) Nature, 211, 734.

Watt, D. F., Embury, and Ham, R. K., (1968) Phil. Mag., 10, 195.

Wei, R. P., and Baker, A. J., (1965) Phil. Mag., (a) 11, 1087, (b) 12, 1005.

Wells, C. H., (1969) Acta Met., 17, 443.

Wilkens, M., (1970) phys. stat. sol. (a) 2, 359.

Wilkens, M., (1973) unpublished.

Wilson, D. V., (1973) Acta Met., 21, 673.

Winter, A., (1973) private communication.

Wood, W. A., Reimann, W. H., and Sargant, K. R., (1964) Trans. Metall. Soc. AIME, 230, 511.

Wood, W. A., and Segall, R. L., (1957) Proc. R. Soc., (London) A 242, 180.

Woods, P. J., (1973) Phil. Mag., 28, 155; (1973 a) (private communication).

Yoshikawa, A., and Okamoto, M., (1968) Proc. Int. Conf. on Strength of Metals and Alloys, Trans. Japan Inst. Metals, Suppl., 9, 471.

LIST OF SYMBOLS

$\underline{b}_p$	primary Burgers vector
d	cell size (average linear dimension)
$H(\ell)$	frequency distribution of dislocation segment lengths
ℓ	distance between dislocation pinning points
L	glide path distance
n	number of mobile dislocations per unit length
n'	cyclic strain hardening exponent
N	number of cycles
N_f	number of cycles to failure
$r_{\odot}$	radius of curvature of screw dislocations
T_m	melting temperature
y	dislocation dipole spacing
γ_p	plastic shear strain amplitude
$\dot{\gamma}_p$	plastic shear strain rate
γ_t	total shear strain amplitude (elastic + plastic)
μ	shear modulus
ν	Poisson's number
ρ	average dislocation density
ρ_{dip}	dislocation dipole density
ρ_{loc}	local dislocation density
ρ_m	mobile dislocation density
τ	resolved shear stress

$\Delta\tau$	shear stress increment per unit cumulative resolved shear strain
τ^*	friction stress (thermal component of τ_s)
τ_a	shear stress amplitude
τ_{eff}	local effective shear stress
$\tau_{F.R.}$	Frank-Read bowing stress
τ_i	internal shear stress
τ_o	cyclic stress coefficient
τ_s	saturation shear stress amplitude

The editor notes with great regret the recent death of Professor H.M. Miekk-oja. Professor Miekk-oja, who died shortly before this contribution was invited had been Professor of Physical Metallurgy at the Helsinki University of Technology for the past 22 years. He was formally educated as a physicist, however, through his tremendous initiative and energy he made himself into a modern physical metallurgist whose contributions in the fields of dislocation theory and transmission electron microscopy are well known. His work and that of his students has formed the basis of physical metallurgy in Finland and it is certain that his example will serve well for its further development.

ABSTRACT. *Quantitative methods are presented for the description and analysis of dislocation structures developed during elevated temperature deformation processes such as dynamic recovery and creep. Furthermore, the basic characteristics of the formation of dislocation network through knitting processes is described and it is shown how these processes enable dislocation sub-structures to be held in a steady-state condition under an applied stress; i.e. the condition of dynamic recovery during elevated temperature deformation.*

8. DISLOCATION STRUCTURES IN DEFORMATION AT ELEVATED TEMPERATURES

H.M. Miekk-oja and V.K. Lindroos

8.1 INTRODUCTION

When metals are plastically deformed at different temperatures a particular transition temperature can be found above which the flow stress adopts a steady value. This value is then maintained up to high strains in contrast with the behavior below the transition temperature where work-hardening occurs instead. This can be seen by considering fig. 8.1 (a) which illustrates the stress-strain curves for Al-2% Mg alloy specimens deformed at different temperatures between room temperature and 500°C (Lindroos and Miekk-oja, 1968). Based on the shape of these curves they can be classified into two general categories: a high temperature region (400 and 493°C) where after initial rapid work-hardening the flow stress maintains a rather steady value up to high strains; and a low temperature region (25, 100 and 200°C) where the test specimens begin to neck as soon as work-hardening has stopped. At intermediate

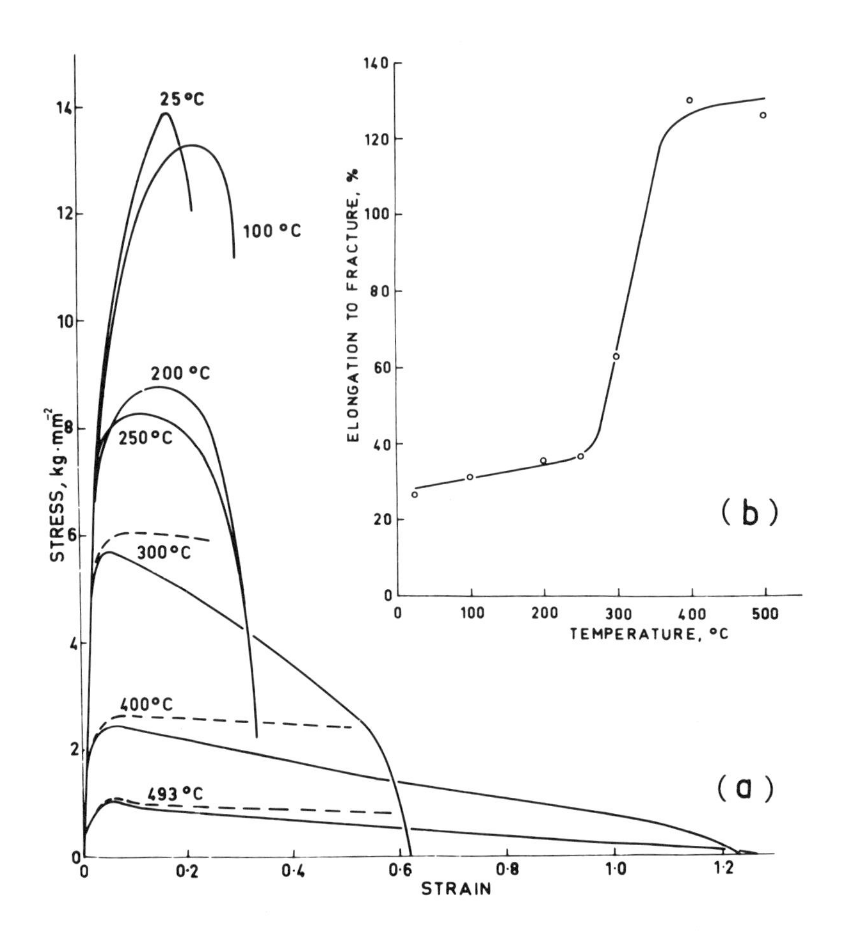

Fig. 8.1 (a) Stress-strain curves for a rapidly solidified Al-Mg alloy tested at temperatures between 25°C and 493°C. (b) Elongation to fracture at different temperatures corresponding to the curves in fig. 8.1 (a) (from Lindroos and Miekk-oja 1968, courtesy of Taylor and Francis).

temperatures (250 and 300°C) the curves adopt a transitional shape.

The difference in the mode of plastic deformation in the high and low temperature regions is also apparent in fig. 8.1 (b), which shows the elongation to fracture at different temperatures. The values of elongation are small, approximately 30%, at low temperatures and quite large, approximately 130%, at high temperatures. The fact that both the elongation and the nature of the stress-strain curves change abruptly within a narrow temperature range near 300°C gives reason to believe that the mechanism of plastic deformation also changes.

The dislocation structure developed during straining below the transition temperature of approximately 300°C is shown in fig. 8.2. As expected, the plastic deformation has been quite

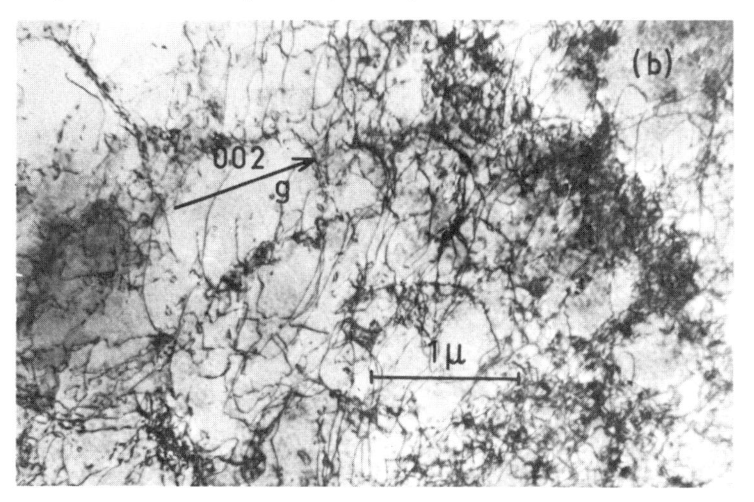

Fig. 8.2 Dislocation arrays generated in the rapidly solidified alloy as a result of deformation below the 'transition' temperature, 35% strain at 250°C (from Lindroos and Miekka-oja, 1968, courtesy of Taylor and Francis).

uniform resulting in a structure in which the dislocations are heavily tangled. The overall dislocation density increases with increasing strain, as is generally the case during plastic deformation at low temperatures. This explains the continuous increases in the flow stress with straining, revealed by the corresponding stress-strain curves (fig. 8.1).

As can be seen from fig. 8.3 (and also fig. 8.17) the dislocation structures in samples strained above the 300°C transition temperature differ completely from those developed at low temperatures. Instead of dislocation tangles, a regular cell structure is formed, where the number of dislocations inside the cells is remarkably small, but the networks themselves consist of quite densely knitted dislocation arrays.

Since the topic of this contribution is the description of dislocation structures associated with elevated temperature

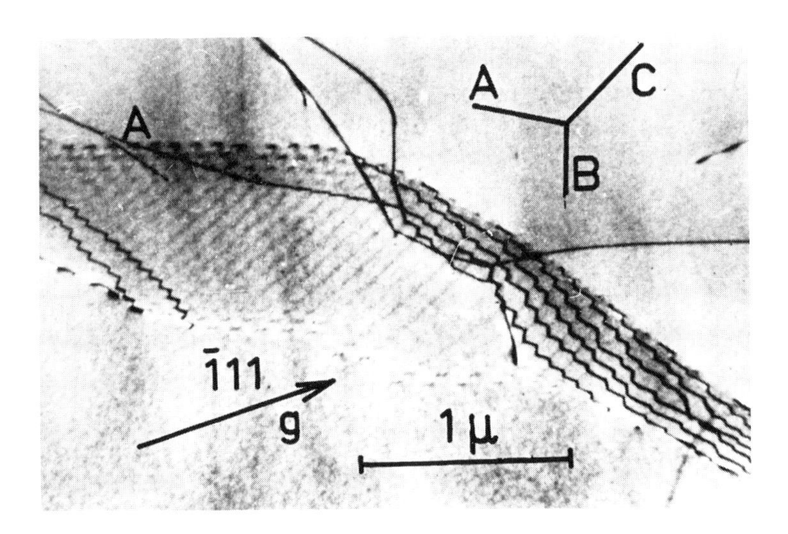

Fig. 8.3 Dislocation arrays generated by straining the rapidly solidified alloy above the 'transition' temperature, 2.5% strain at 400°C (from Lindroos and Miekk-oja, 1968, courtesy of Taylor and Francis).

deformation, we shall not consider further the temperature region resulting in cold working. In concentrating our attention on the details of the mechanisms of plastic deformation at elevated temperatures we will discuss: (1) the accurate description of dislocation structures; (2) the detailed characteristics of forming dislocation networks through the knitting process; and (3) the various basic aspects of dynamic recovery. General discussions of the sub-structures developed during high temperature deformation (e.g. Jonas, Sellars, and Tegart, 1969) and diffusional creep (e.g. Ashby, 1972) have been presented elsewhere.

8.2 ACCURATE DESCRIPTION OF DISLOCATION STRUCTURES

Information about structures developed during plastic deformation at various temperatures can be gained by X-ray diffraction and by light optical as well as electron optical microscopy. In this paper, after discussing some aspects of producing dislocations for fundamental studies, we shall deal only with observations by transmission electron microscopy, which is no doubt the most useful method for studying dislocation networks and dislocation interactions.

8.2.1 *The Experimental Production of Dislocations*

For the purposes of examining dislocation behavior we have utilized a variety of methods to produce dislocation arrangements for detailed study; (1) rapid solidification, (2) tensile testing at elevated temperature, (3) hot torsion testing, (4) creep testing, and (5) different thermomechanical treatments. Among these, we would like to emphasize the usefulness of rapid solidification in producing dislocation arrangements for studies of this type (Lindroos and Miekk-oja, 1967). The two-dimensional dislocation networks developed during solidification offer particularly favorable conditions for the study of dislocation reactions for at least three reasons: (1) during solidification, all possible kinds of forest dislocations are generated in connection with the boundaries formed when growing cells meet; this is in contrast to plastic deformation which produces dislocations belonging predominantly to the primary and conjugate systems; (2) the resulting networks are quite stable since once formed they do not easily disappear during the preparation of thin foils; and (3) the dislocation reactions resulting in the network formation are abundantly repetitive and therefore represent real processes in metals. Finally, it should be noted that the fundamental aspects of dislocation behavior observed in this way can then be applied to the different stages of plastic deformation.

8.2.2 *Preparation of Thin Foils*

The proper preparation of thin foil specimens for study, after a dislocation structure has been introduced into a bulk specimen, has often presented problems. In this regard, we would like to emphasize the advantages of electropolishing at low temperatures. This is briefly illustrated in fig. 8.4 which shows schematically how the polishing plateau associated with the current density-anode potential curves becomes more horizontal and longer as the temperature of the electrolyte is decreased; the details of these effects have been reported elsewhere (Räty, Lindroos, Saarinen, Forstén and Miekk-oja, 1966). Therefore, it is common practice in our laboratory to electropolish thin foil specimens at temperatures near -35 to -40°C.

8.2.3 *Analysis of Burgers Vectors*

The first task when beginning transmission electron microscopy work is to fix the orientation of the thin foil with the aid of Kikuchi maps, examples of which are shown in fig. 8.5 for

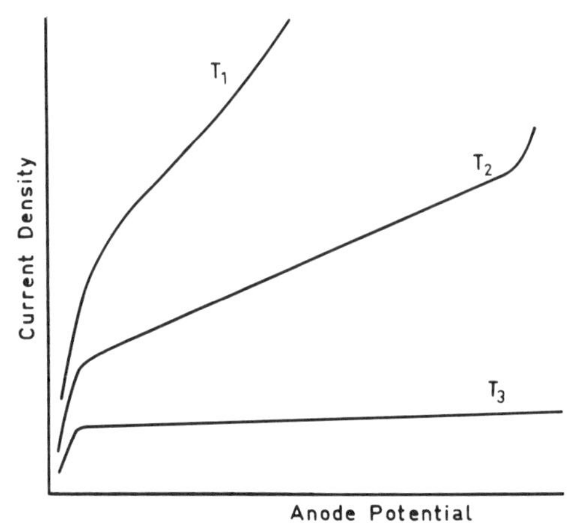

Fig. 8.4 Schematic current density-anode potential curves at different electrolyte temperatures where T_1, T_2, and T_3 correspond approximately to room temperature, zero, and sub-zero temperatures, respectively.

FCC and BCC crystals. This is done by searching for the closest low index direction of the thin foil with aid of a diffraction pattern and then locating that particular direction with respect to the appropriate Kikuchi map. This treatment assures the exact indexing of any reflecting vector as well as any crystallographic direction in the thin foil; i.e. we are then no longer considering the general type of reflecting vector or the general type of habit plane, etc.

Once the process described above is completed, a Burgers vector analysis can be accomplished with the conventional dislocation contrast analysis (Hirsch, Howie, Nicholson, Pashley and Whelan, 1965), according to which the criteria for invisibility of a dislocation is $\underline{g}\cdot\underline{b} = 0$ where $\underline{g}$ is the operating reflecting vector and $\underline{b}$ is the Burgers vector of the dislocation. For certain difficult cases, the so-called m-contrast effect can also be used (Hirsch et al., 1965), $m = 1/8\ (\underline{g}\cdot\underline{b}\mathrm{x}\underline{u})$, where $\underline{u}$ is the positive direction of the dislocation; the invisibility criterion for a dislocation in this case is that m is less than 0.08 - 0.10. Examples of analysis of the Burgers vectors of dislocations forming typical FCC and BCC networks are shown in figs. 8.6 and 8.7.

For FCC crystals, such as shown in fig. 8.6, it is sufficient to use three different 111 type reflections to analyze all possible Burgers vectors of the type a/2<110>; the Hirth dislocations and partial dislocations which require a slightly different analysis will not be discussed here. Initially, we have the foil in the 111 orientation. Then by using a high angle tilting stage we examine three different 111 reflections, i.e. $\bar{1}\bar{1}1$, $\bar{1}1\bar{1}$ and $1\bar{1}\bar{1}$ sequentially, by 'driving' along Kikuchi

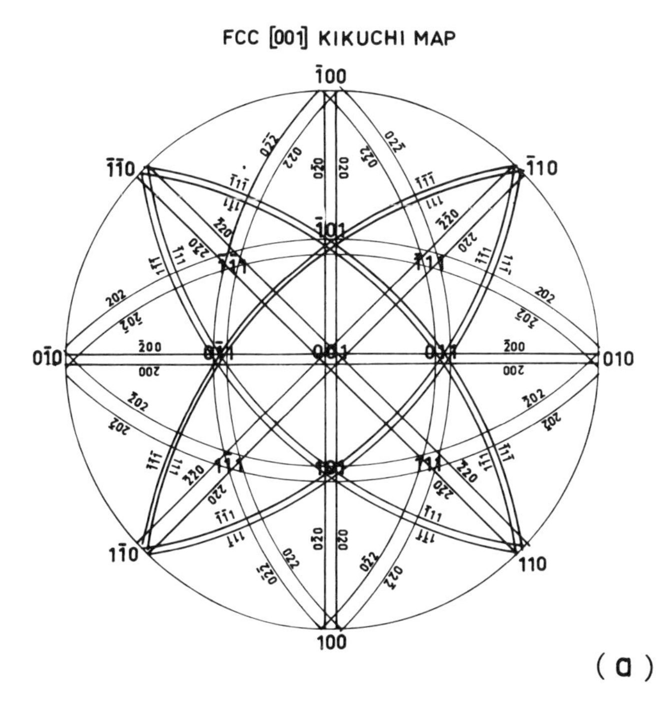
FCC [001] KIKUCHI MAP
$\bar{1}00$
$\bar{1}\bar{1}0$
$\bar{1}10$
$0\bar{1}0$
010
$1\bar{1}0$
110
100
(a)

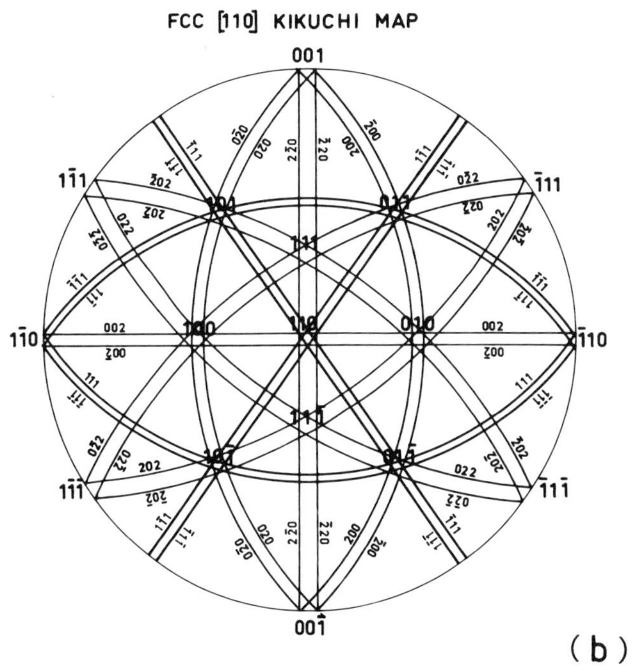
FCC [110] KIKUCHI MAP
001
$1\bar{1}1$
$\bar{1}11$
$1\bar{1}0$
$\bar{1}10$
$1\bar{1}\bar{1}$
$\bar{1}1\bar{1}$
$00\bar{1}$
(b)

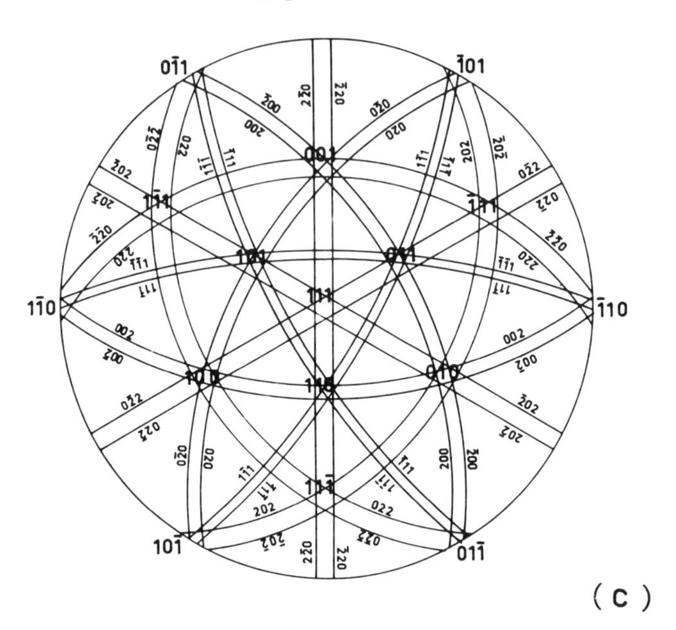
FCC [111] KIKUCHI MAP
0$\bar{1}$1
$\bar{1}$01
1$\bar{1}$0
$\bar{1}$10
10$\bar{1}$
01$\bar{1}$

(c)

BCC [001] KIKUCHI MAP

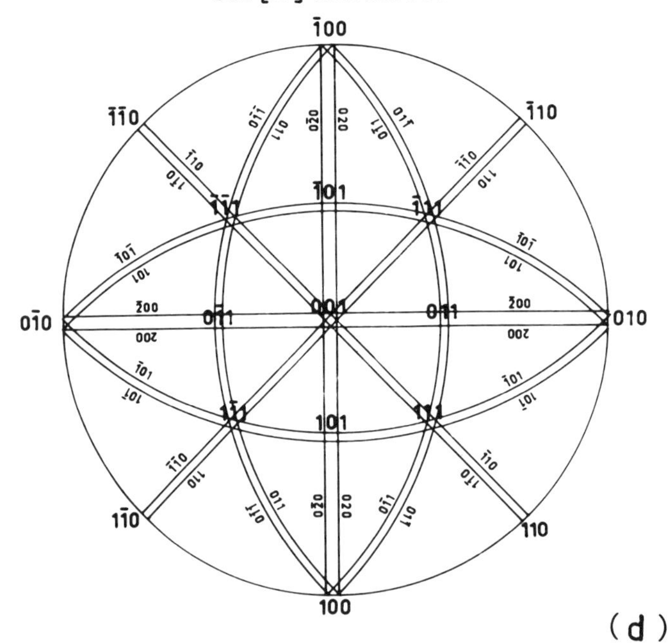
$\bar{1}$00
$\bar{1}\bar{1}$0
$\bar{1}$10
0$\bar{1}$0
010
1$\bar{1}$0
110
100

(d)

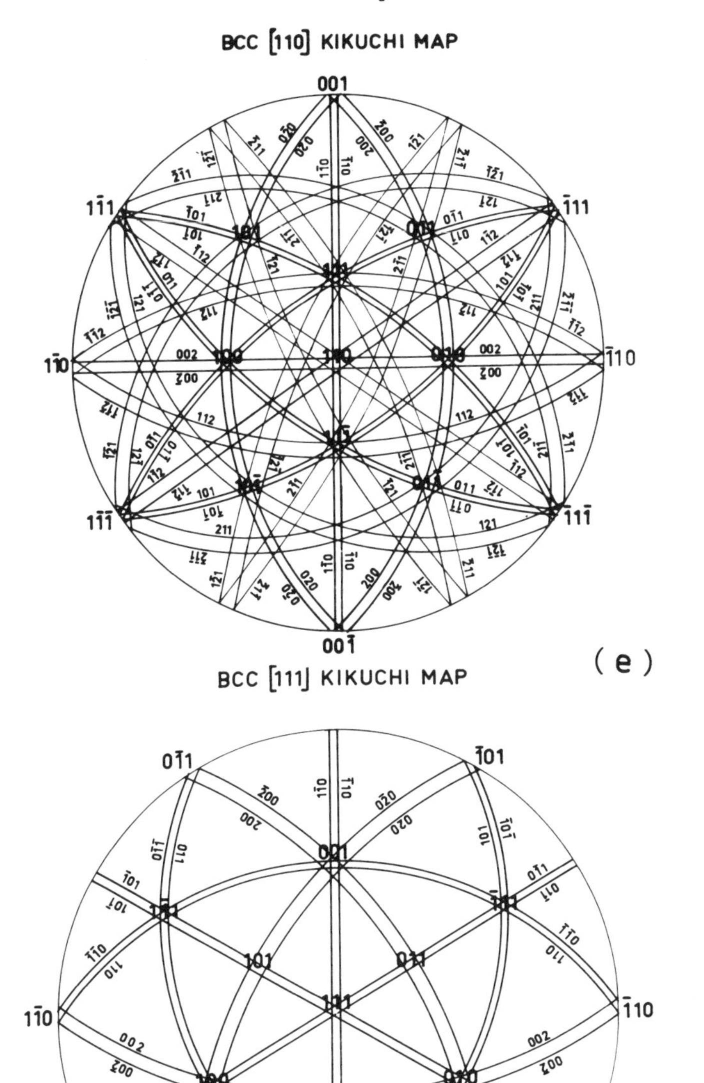

Fig. 8.5 Low index Kikuchi maps for (a) - (c) FCC and (d) - (f) BCC crystals (Lindroos, 1968, unpublished research).

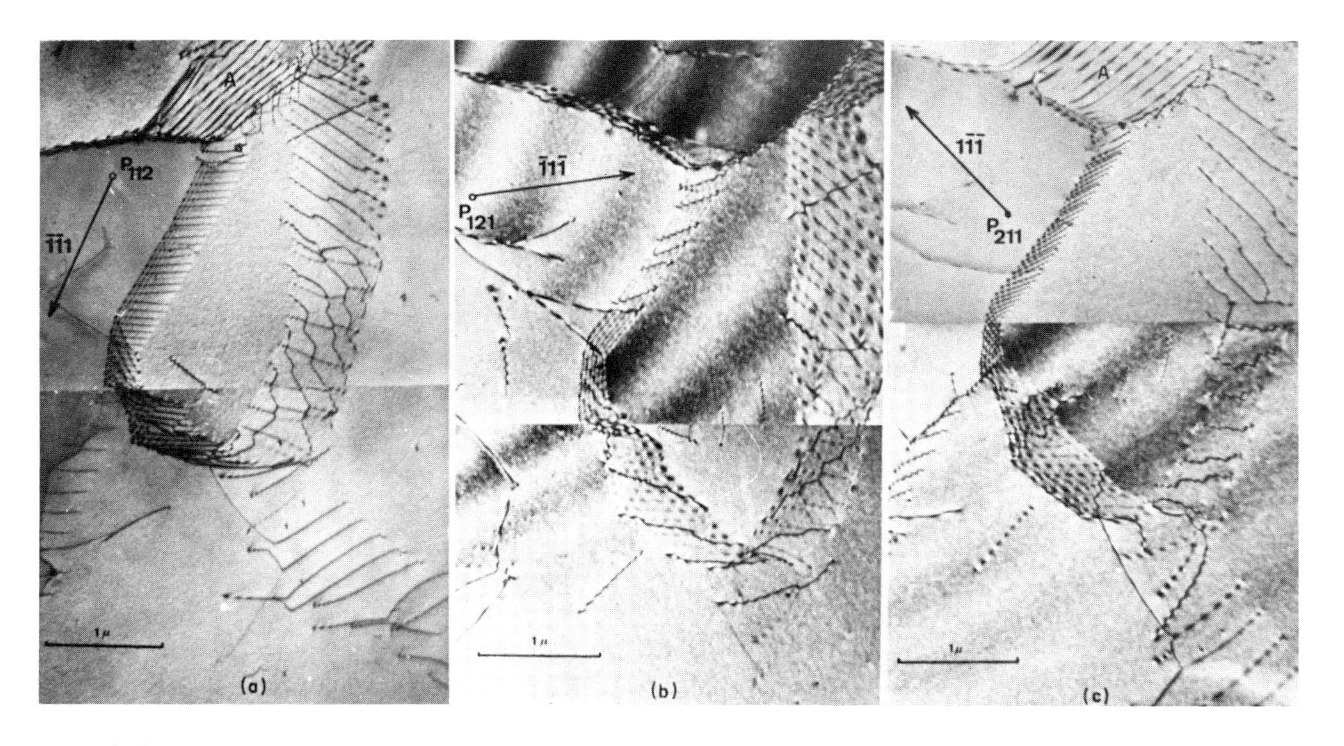

Fig. 8.6 Small-angle boundaries making up a complete subgrain section in a FCC structure (Al-Mg alloy). As an example of Burgers vector analysis, the $\tilde{g} \cdot \tilde{b} = 0$ criterion results in the A dislocations with Burgers vector $a/2[10\bar{1}]$ being visible in the $\bar{1}\bar{1}1$ and $1\bar{1}\bar{1}$ reflections, but invisible in the $\bar{1}1\bar{1}$ reflection (Kivilahti and Lindroos, 1973, unpublished research).

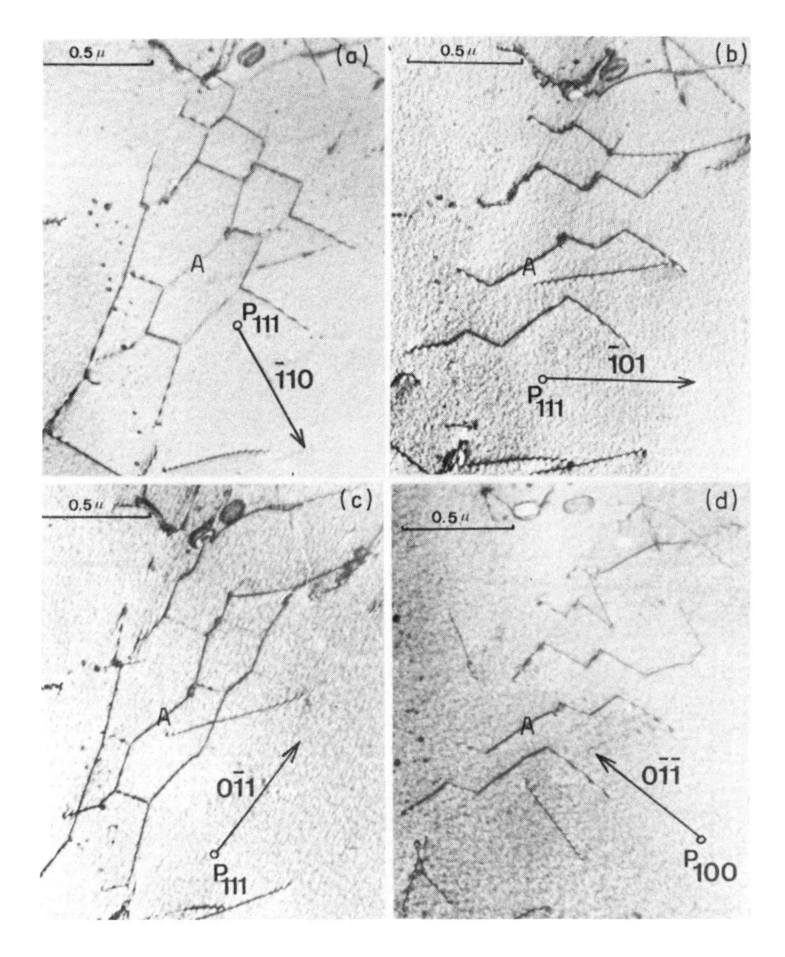

Fig. 8.7 Dislocation network in a BCC structure (low-carbon steel). As an example of Burgers vector analysis, the $\underset{\sim}{g}\cdot\underset{\sim}{b} = 0$ criterion results in the a[001] dislocations, denoted by letter A, being visible in the $\bar{1}01$, $0\bar{1}1$ and $0\bar{1}\bar{1}$ reflections and invisible in the $\bar{1}10$ reflection (Havola, Kivilahti and Lindroos, 1973, unpublished research).

bands $\bar{2}20$, $\bar{2}02$ and $0\bar{2}2$, respectively; as can be seen from fig. 8.5 (c).

For BCC crystals, fig. 8.7 shows that in addition to the three 110 type reflections available around a 111 pole, the fourth 110 reflection is needed for a complete Burgers vector analysis, i.e. to identify the four a/2<111> and three a<100> type Burgers vectors. The fourth 110 reflection can be reached by 'driving' along 011 type Kikuchi bands to any of the three 100 type poles, as shown in fig. 8.5 (f).

Final Burgers vector analysis can be carried out with either (1) the Kikuchi map method, or (2) the Thompson tetrahedron for FCC and the Burgers vector pentahedron method for BCC crystals.

8.2.4 *Trace Analysis of Habit Planes and Directions*

In order to further characterize the sub-structure, the habit planes and crystallographic directions of lattice defects can be determined. However, the common trace analysis technique involving the combination of information from a selected area diffraction pattern together with the geometry available from the electron micrograph, generally requires a determination of both the thickness and crystallographic orientation of the specimen (e.g. Hirsch et al., 1965). In order to eliminate these tedious determinations, which are also the main sources of inaccuracy, a new trace analysis method has been developed (Kivilahti and Lindroos, 1972) in which neither the thickness nor the exact crystallographic orientation of the specimen are necessary.

For example, in order to determine the orientation indices [uvw] of the line defect in the lattice co-ordinate system shown in fig. 8.8, the electron optical image projections OQ_1

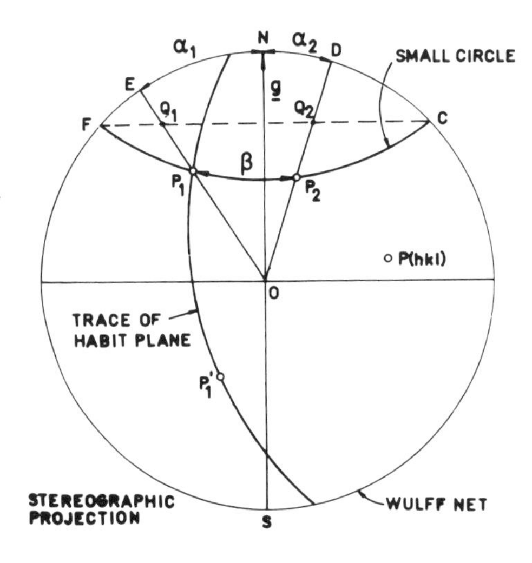

Fig. 8.8 Trace analysis method (Kivilahti and Lindroos, 1972).

and OQ_2 are drawn with the aid of the angles α_1 and α_2 derived from two micrographs containing the image projections of the defect. A small circle of the superimposed Wulff net is then determined which intersects the projections OQ_1 and OQ_2 at the poles P_1 and P_2, these being an angular distance β apart, where β is the goniometer tilt angle of the specimen. Finally, the orientation indices [uvw] of the line defect are found by determining the indices of the pole P_1 from the stereogram.

For planar defects, the poles P_1 and P_1' of two directions, which may be, for example, dislocations or intentionally drawn reproducible lines lying along the defect plane, are determined in the manner described above. The indices of the habit plane are then given by determining that pole P(hkl) which represents the great circle (i.e. trace of the habit plane) intersecting the two poles P_1 and P_1' as shown in fig. 8.8.

The same method can also be used for the rapid determination of the top and bottom surface intersections of a line defect in thin foils (Kivilahti and Lindroos, 1974).

8.2.5 *Devices for the Study of Dislocation Interactions*

After acquiring information about Burgers vectors as well as dislocation directions and habit planes from the methods described above, detailed dislocation interactions can be considered. For FCC crystals, the Thompson tetrahedron (Thompson, 1953) is at the present time a standard tool for analyzing dislocation reactions. However, such a device reveals only a limited portion of Burgers vectors of the FCC dislocations, these being perfect dislocations, a/2<110>, Shockley and Frank partial dislocations, a/6<112> and a/3<111>, and Lomer-Cottrell lock dislocations, a/6<110>. Furthermore, as some dislocations, such as the Hirth dislocation (Hirth 1961, Nabarro, Basinski and Holt, 1964) were not known at the time of the presentation of the Thompson tetrahedron, the need for an additional device called the Burgers vector octahedron (Lindroos, 1974a), fig. 8.9, became obvious.

In such an octahedron the Burgers vectors of perfect and partial Hirth dislocations, a<100> and a/3<100>, as well as the stair-rod dislocation at an acute bend, a/3<110>, are illustrated. Finally, the combination of the Thompson tetrahedron and the octahedron, which can reveal the Burgers vector of the stair-rod dislocation at an obtuse bend, a/6<301>, completes the representation of the Burgers vector notations in FCC crystals.

For BCC crystals an analog to the Thompson tetrahedron, the so-called Burgers vector pentahedron shown in fig. 8.10, was developed (Lindroos, 1970). This pentahedron reveals seven different Burgers vectors, viz. four a/2<111> type and three a<100> type, and six different slip planes of the type {110}; therefore all possible BCC slip system combinations of the types {110} <111> and {110} <100> are shown. Because of the high stacking-fault energy, dislocations do not extend in most BCC structures even though core dissociation may occur, as suggested by Vitek (1968). Nevertheless, if partial dislocations with Burgers vectors on {110} planes, such as a/6<111>, a/3<111>, a/3<112>, and a/6<115>, are also to be considered it may be conveniently done by adding these vectors to the pentahedron described above.

8.3 BASIC CHARACTERISTICS OF THE DISLOCATION KNITTING PROCESS

This section will describe the basic characteristics of the process by which dislocation networks are formed during deformation at elevated temperatures; this process we have termed as the 'knitting of dislocation networks'.

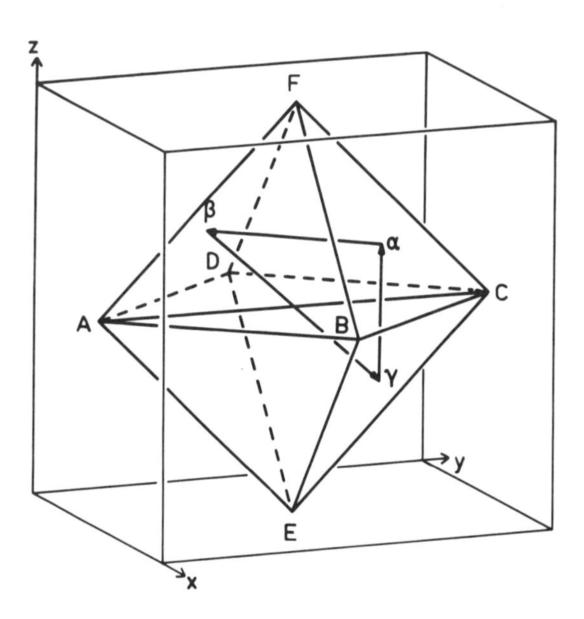

Fig. 8.9 Burgers vector octahedron in the FCC lattice unit cell, where $\underline{AC}$, $\underline{\gamma\alpha}$ and $\underline{\alpha\beta}$ represent Burgers vectors of a perfect Hirth dislocation, a partial Hirth dislocation, and a stair-rod dislocation at an acute bend, respectively. The Burgers vector $\underline{\beta\gamma}$ is considered elsewhere.

8.3.1 *Forces on a Dislocation*

Dislocation interactions arise from the forces on dislocations which can be in principle either of elastic or thermodynamic origin. Elastic forces arise from applied stresses, external or internal, whereas thermodynamic forces are generated by a disturbance in the thermodynamic equilibrium. Forces of both types can affect plastic deformation at elevated temperatures in an important way.

As an example, the elastic force produced by the tensile stress σ in fig. 8.11 drives the edge dislocation to climb in such a direction that vacancies are created; the more detailed distribution of these vacancies is shown in fig. 8.12, where the vacancy concentration contours around a climbing edge dislocation are shown (Turunen and Lindroos, 1973). As

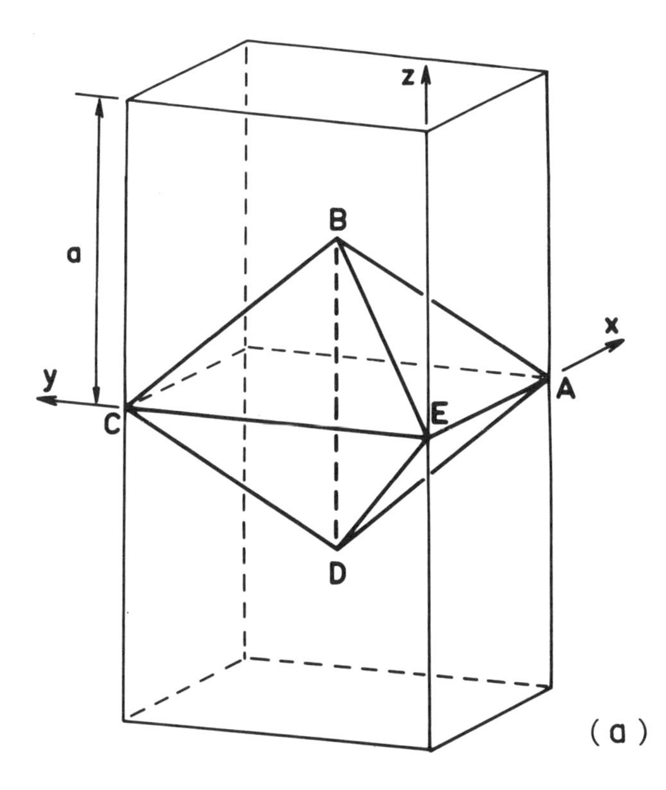

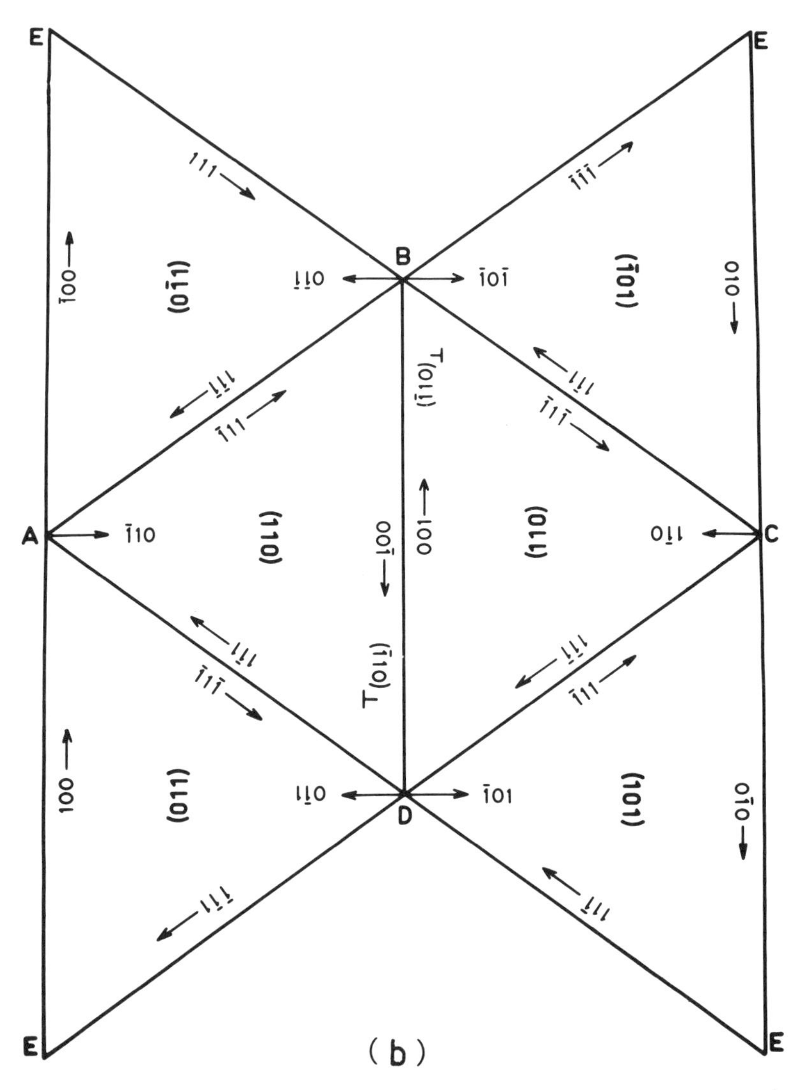

Fig. 8.10 (a) Representation of the Burgers vectors in the BCC lattice unit cell by means of the pentahedron; (b) indexing of the Burgers vectors and slip planes associated with the pentahedron (from Lindroos, 1970, courtesy of Taylor and Francis).

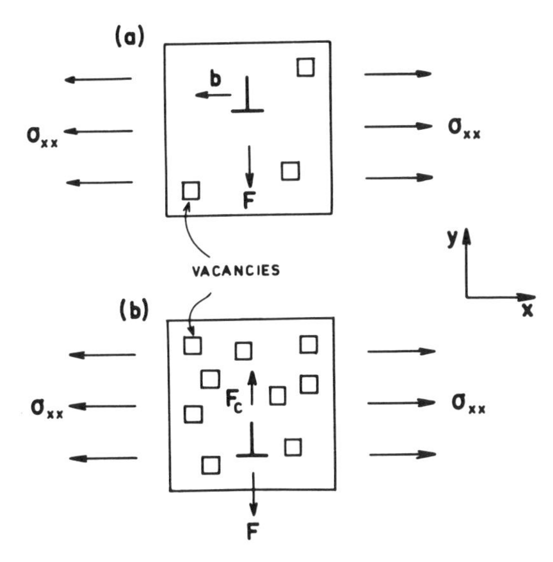

Fig. 8.11 (a) Elastic force F acting on an edge dislocation as a result of an external tensile stress σ_{xx}, (b) chemical force F_c produced by stress-induced climb motion of an edge dislocation (from Weertman and Weertman, 1964, courtesy of The Macmillan Company).

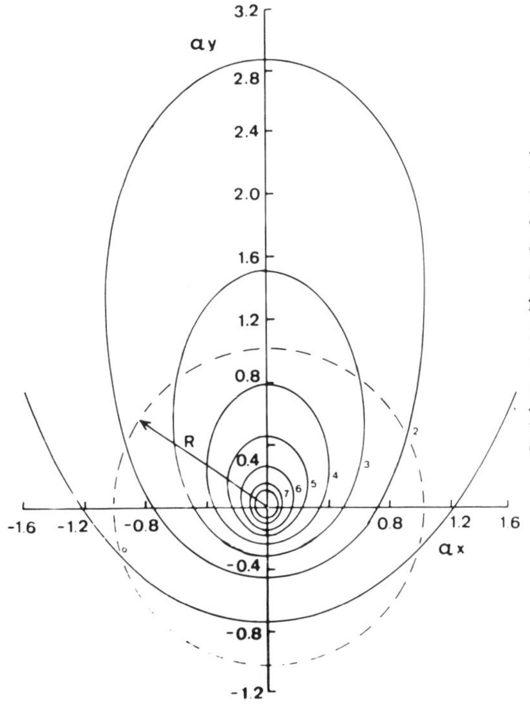

Fig. 8.12 Vacancy concentration contours around a positive edge dislocation climbing downward. The numbers marked on the contour lines indicate the relative vacancy concentrations (from Turunen and Lindroos, 1973, courtesy of Taylor and Francis).

climb occurs and the vacancy concentration increases and exceeds that associated with thermodynamic equilibrium in the crystal, it is reasonable to expect that it will become more difficult for the dislocation to continue its climb. In other words, it is reasonable to believe that the excess vacancy concentration gives rise to a force, a 'chemical force', F_c, which opposes the elastic force F. This idea was first proposed by Bardeen and Herring (1952) and the chemical force acting on a unit length of the edge dislocation is commonly expressed as (e.g. Weertman and Weertman, 1964, Hirth and Lothe, 1968)

$$F_c = \frac{kTb}{V} \ln \frac{C_d}{C_0} \tag{8.1}$$

where b is the magnitude of Burgers vector, V is the atomic volume, C_d is the vacancy concentration near the dislocation core, and C_0 is the thermodynamic equilibrium concentration of vacancies in the crystal.

In summary, the chemical force which develops to oppose the elastic force produced by an applied stress, results in the following: (1) At low temperatures, the chemical force, F_c, is able to eliminate dislocation climb with the result that glide is virtually the only mechanism for plastic deformation. This occurs because the equilibrium concentration of vacancies is extremely small at low temperatures, and also because the mobility of vacancies is small at low temperatures, resulting in a high local concentration of vacancies in the vicinity of the dislocation core during climb. (2) On the other hand, at elevated temperatures (e.g. above 300°C in the aluminum alloy considered here) the chemical force F_c does not increase sufficiently to compensate for the elastic force with the result that F_c is not able to prevent dislocation climb. This is because of the high equilibrium concentration of vacancies, and furthermore, the high mobility of these vacancies which permits their annihilation in sinks of various kinds.

Consequently, the climb of dislocations is an important aspect of plastic deformation at elevated temperatures, particularly in such processes as dynamic recovery and creep where there is no increase in flow stress. We have studied these deformation processes in Al-alloys, iron, austenitic stainless steels, and in copper and its alloys, and the results of these studies will form the main topics of the remainder of this contribution.

8.3.2 *The Basic Knitting Mechanism*

As a basic example of the knitting process, the mechanism for the formation of the simple dislocation network making up a symmetrical tilt boundary, as shown in fig. 8.13, will be con-

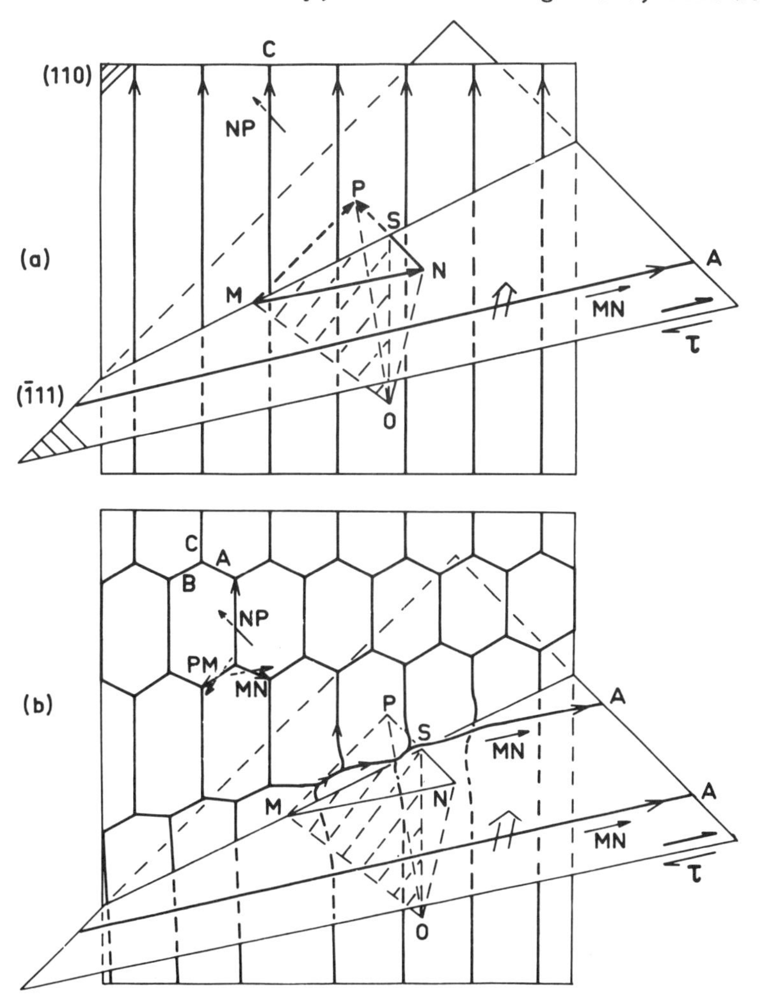

Fig. 8.13 The dislocation knitting mechanism for the formation of a dislocation network on a symmetrical tilt boundary (from Lindroos and Miekk-oja, 1967, courtesy of Taylor and Francis).

sidered. According to the knitting mechanism, a screw dislocation A with Burgers vector MN = a/2[$\bar{1}0\bar{1}$] glides under a shear stress in the ($\bar{1}$11) plane and interacts with a forest consisting of dislocations C with Burgers vector NP = a/2[110] on the (110) plane. Due to their mutual attraction A and C align themselves by climb, and as can be seen from the Thompson tetrahedron:

$$\underline{\underline{MN}} + \underline{\underline{NP}} \rightarrow \underline{\underline{MP}} \; , \tag{8.2}$$

i.e.

$$\frac{a}{2}[\bar{1}0\bar{1}] + \frac{a}{2}[110] \rightarrow \frac{a}{2}[01\bar{1}] \quad (A + C \rightarrow B) \; . \tag{8.3}$$

This produces a junction dislocation B with Burgers vector MP = a/2[01$\bar{1}$] lying in the plane of the forest and, not being located in its normal glide plane, it is driven by elastic and/or chemical forces to climb onto the forest plane (110). At the same time its line tension pulls the dislocation A onto the same plane, which is a plane of easy climb for all three mesh dislocations A, B and C. Finally, as will be shown in Section 8.4.1, the vacancies generated during stress-induced climb of one dislocation are annihilated by opposite climb of another dislocation with the necessary vacancy transport occurring mainly by means of pipe diffusion.

In a similar manner, junction dislocations are formed at all forest dislocations, and when agreat number of A dislocations are pushed against the forest C , the final network with the three-fold nodes is knitted. During the whole knitting process, the network conserves its original plane, i.e. that of the forest, since all mesh dislocations climb onto this plane and their glide is prevented because they do not lie on their glide planes. Experimental evidence for the knitting process described above is shown in fig. 8.14.

Obviously, a necessary condition for the operation of the knitting procedure is the alignment of dislocations associated with the formation of junction dislocations in the plane of the forest; this enables the dislocations to change their mode of motion from glide to climb. In those cases where alignment cannot occur, for example at low temperatures, the glide dislocations are trapped in the forest where they may bow out or penetrate the forest resulting in the more or less tangled dislocation structures of the type shown in fig. 8.15. Examination of fig. 8.15 shows that although the glide dislocations have penetrated the forest, the mesh dislocations have not been involved, obviously because they do not lie in their

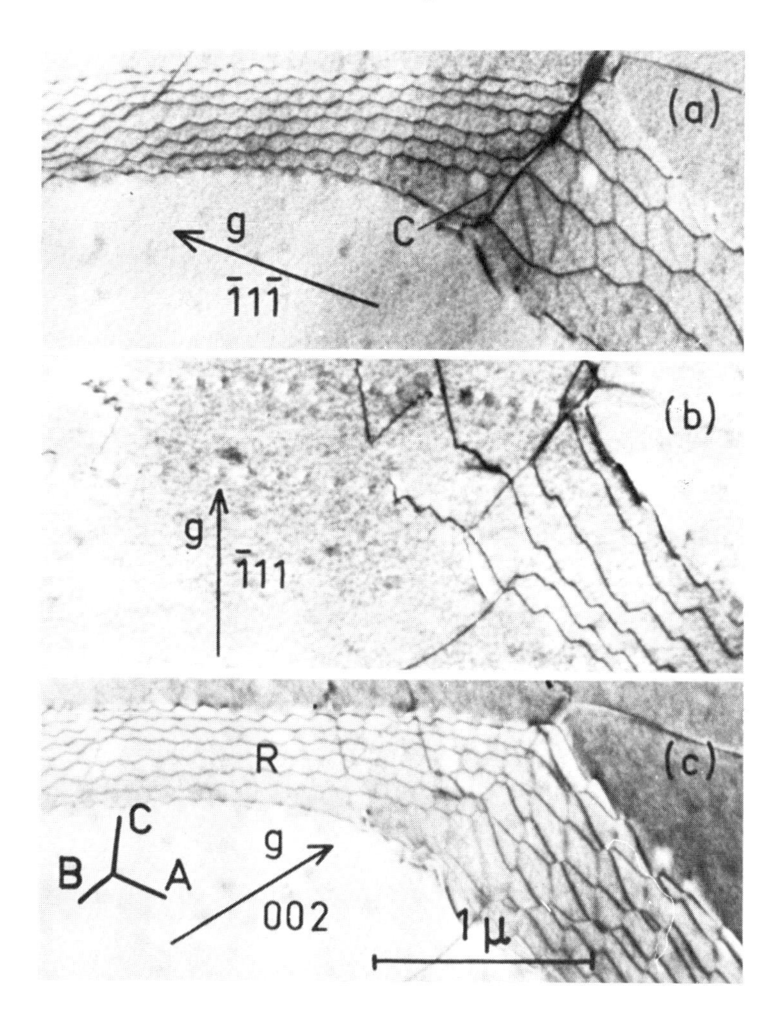

Fig. 8.14 Dislocation network knitted on a symmetrical tilt boundary in the Al-Mg alloy (from Lindroos and Miekk-oja, 1967, courtesy of Taylor and Francis).

glide planes.

In addition to this basic example we have presented a great number of other mechanisms (Miekk-oja, 1966, Lindroos and Miekk-oja, 1967, 1968, 1969, Lindroos, 1971, Miekk-oja and Lindroos, 1972, Lindroos and Kivilahti, 1972) where it is shown that the various combinations of both forest and glide source dislocations can result in quite complex networks. Nevertheless, in all cases these mechanisms have the important

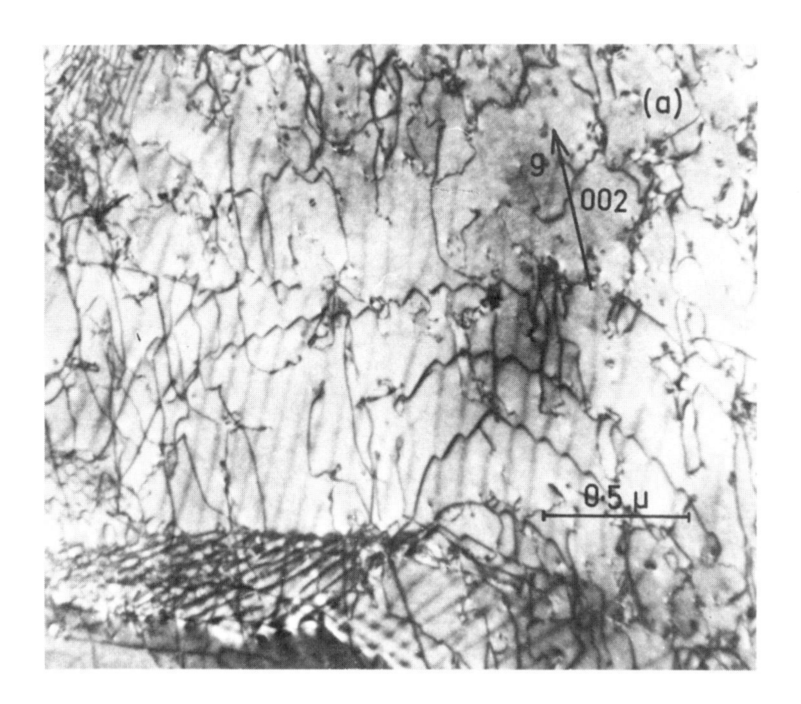

Fig. 8.15 Dislocation arrays generated in the rapidly solidified Al-Mg alloy deformed below the 'transition' temperature, 2% strain at 25°C (from Lindroos and Miekk-oja, 1968, courtesy of Taylor and Francis).

common feature, that when any two dislocations react during deformation at elevated temperatures there will be no real repulsive junctions, but instead all interactions result finally in attractive reactions. In order to examine this statement in more detail some schematic configurations will be considered next.

Firstly, as shown in case (a), fig. 8.16, the forest dislocation A-A' lying in the plane MRO of the type {110} will attract the dislocation B-B', which meets the forest gliding on the plane MPN of the type {111}; this is because the energy of the dislocation configuration decreases as a result of the

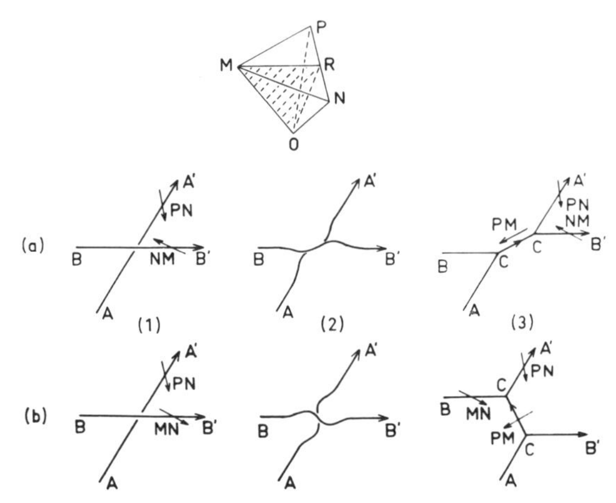

Fig. 8.16 Alignment preceding the formation of a junction dislocation under conditions where the two reacting dislocations are initially (a) attractive and (b) repulsive (from Miekk-oja and Lindroos, 1972, courtesy of North Holland).

reaction:

$$\underline{\underline{PN}} + \underline{\underline{NM}} \rightarrow \underline{\underline{PM}} \, . \qquad (8.4)$$

Consequently, when coming together, the dislocations which are both able to climb on the MRO plane align themselves as shown in (a2), with the result that a junction dislocation C-C' is created as shown in (a3).

On the other hand in case (b), the dislocations A-A' and B-B' with Burgers vectors $\underline{\underline{PN}}$ and $\underline{\underline{MN}}$ repel each other, however, this repulsion changes into attraction when the dislocations align themselves in the way shown in fig. 8.16 (b2). Because the dislocation segments coming together now have opposite directions, they react:

$$\underline{\underline{PN}} + (-\underline{\underline{MN}}) \rightarrow \underline{\underline{PM}} \qquad (8.5)$$

with the result that a junction dislocation, C-C', is also created in this case, fig. 8.16 (b3); an example of junctions generated between dislocations originally repelling each other is shown at J in fig. 8.17. The work required to push the dislocations together against the repulsive force and to align them locally is presumably performed by the applied stress.

Secondly, a more detailed picture of the climb alignment of two dislocations without lattice diffusion can be seen in fig. 8.18. Here the neighboring segments of each dislocation climb in opposite directions in such a way that the vacancies produced by one segment are consumed by the other, with the vacancy transport occurring by pipe diffusion along the disloca-

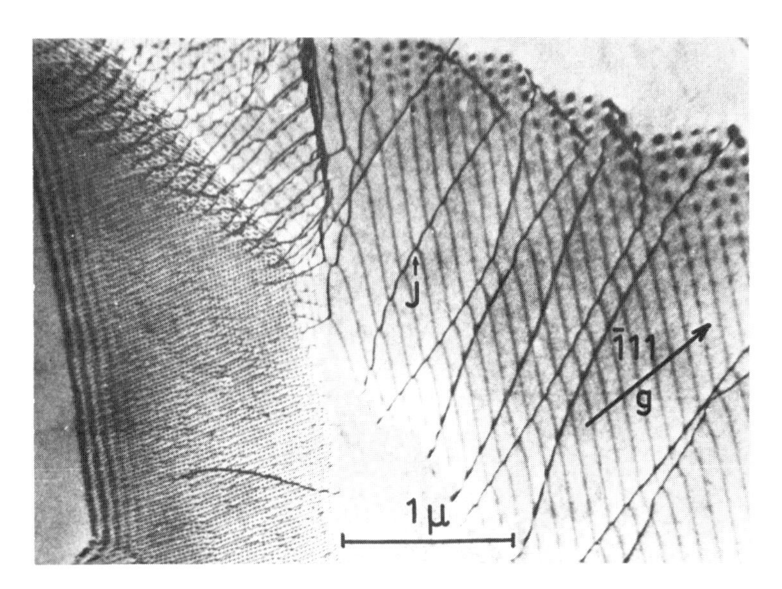

Fig. 8.17 Junction J formed by the reaction of two originally repulsive dislocations; Al-2% Mg alloy strained 10% at 500°C (from Miekk-oja and Lindroos, 1972, courtesy of North Holland).

tion cores so that no transport of vacancies away from the dislocations is required. The details of the vacancy pipe diffusion mechanism have been given elsewhere (Turunen and Lindroos, 1974).

8.4 REQUIREMENTS FOR STEADY-STATE DYNAMIC RECOVERY

The primary feature of the steady state situation occurring during dynamic recovery is that plastic deformation proceeds with no increase in flow stress; i.e. no work-hardening occurs during straining. This characteristic feature gives rise to two necessary conditions: (1) the number of vacancies created must be equal to the number destroyed, and (2) dislocations produced by sources within the cells must be annihilated during the process, i.e. the dislocation density has to remain unchanged. These two requirements will now be considered in detail.

8.4.1 *Dynamic Equilibrium of Vacancies*

A discussion of the vacancy balance during the knitting pro-

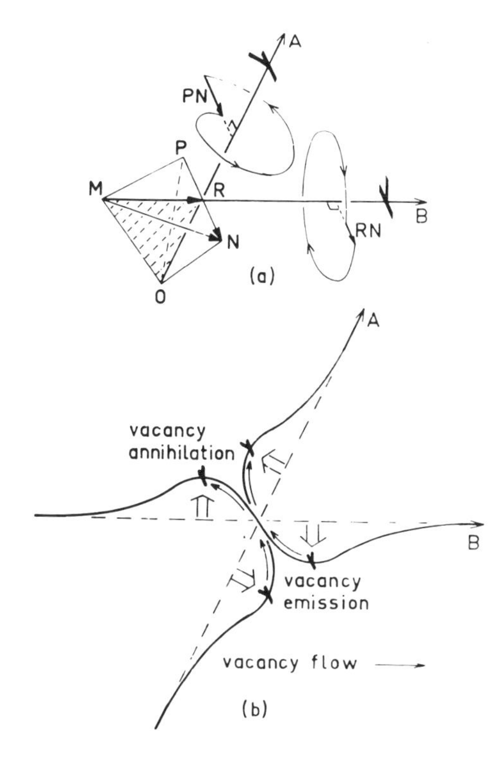

Fig. 8.18 Opposite climb of neighboring dislocation segments resulting in alignment thereby facilitating the attractive interaction of the two dislocations (from Miekk-oja and Lindroos 1972, courtesy of North Holland).

cess has been presented recently (Lindroos, 1974b). The basic idea can be seen from the fig. 8.19 which represents the simplest case, i.e. a dislocation network knitted on a symmetrical tilt boundary (Lindroos and Miekk-oja, 1967). Here the forest dislocations, C, which form the symmetrical tilt boundary are taken to lie on the (110) plane. By using the Thompson tetrahedron, MNOP, it can be seen that both the knitting dislocation, A, and the reaction dislocation, B, have equal edge components $\underline{\underline{RN}}$ and $\underline{\underline{PR}}$ in their Burgers vectors $\underline{\underline{MN}}$ and $\underline{\underline{RM}}$. By applying the FS/RH Burgers circuit around the A and B dislocations, it can be seen that their components of extra half plane are of opposite sign. During the knitting of the A and B dislocations along the forest dislocations, C, in the direction shown by arrows the segments of the A dislocations create exactly as many vacancies as the B dislocations annihilate. Consequently, the knitting process, and therefore the plastic deformation, can proceed through the diffusion of vacancies along the dislocation cores, from the A dislocations to the B dislocations. Since this can occur at all the nodes in the dislocation network, no lattice dif-

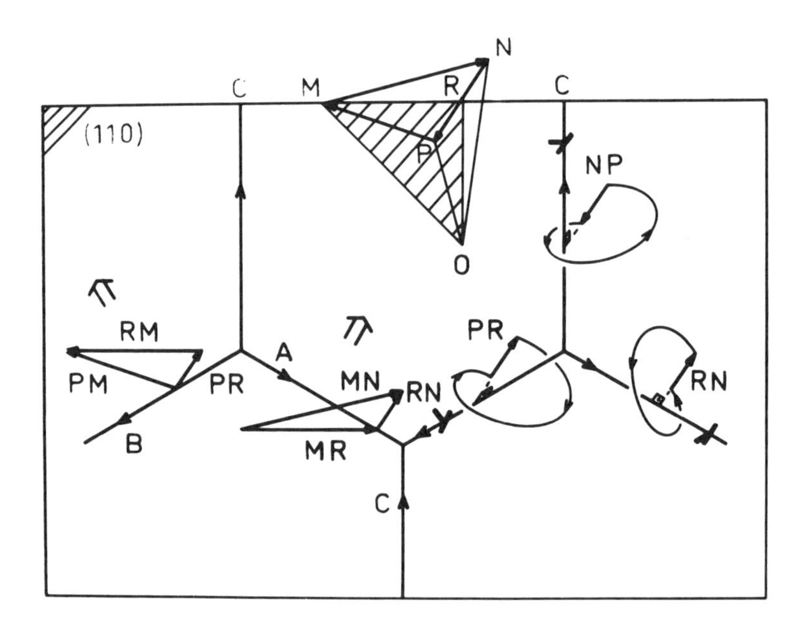

Fig. 8.19 Dislocation node configuration associated with a network knitted on a symmetrical tilt boundary in a FCC structure.

fusion of vacancies is required.

Although in the general case the situation is more complicated because of the many possible kinds of forest, knitting, and reaction dislocations (Lindroos and Miekk-oja, 1967, 1968, 1969, Miekk-oja and Lindroos, 1972), and the fact that the knitting plane may deviate from (110) or even be curved, etc., the basic principle of the process is the same. Therefore, in the general case the edge components of all mesh dislocations entering the node must be considered. For example, two mesh dislocations of a three-fold node can create vacancies which are annihilated by the third dislocation, or vice versa. Furthermore, the vacancy balance condition can determine the length of mesh dislocations such that a dislocation segment having a small edge component in its Burgers vector will be long compared with a dislocation segment having a large proportion of edge component. Therefore, if the forest plane in fig. 8.19 deviates from its (110) orientation in such a way, for example, that for the Burgers vectors RN and PR we have the case RN = 2PR, then the lengths of the B dislocation segments will be half those of the A dislocation segments; for simplicity the role of the C dislocations has been omit-

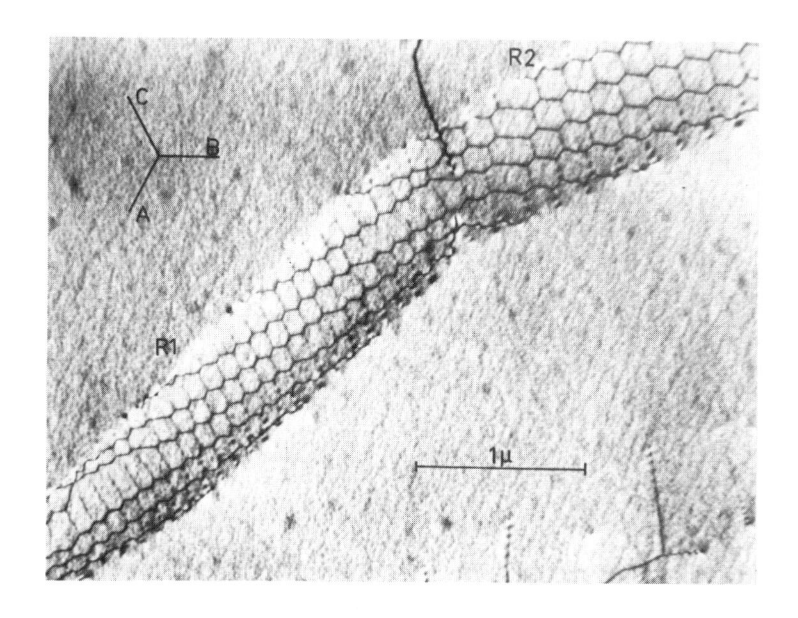

Fig. 8.20 Dislocation network in the Al-Mg alloy where the mesh dislocations A and B have different lengths at different parts R1 and R2 of the network.

ted here. Such a length proportioning has been verified experimentally in cases where the orientation of the forest boundary changes, i.e. the edge components of the dislocations have been found to have different lengths at different parts of the boundary; this is illustrated in fig. 8.20.

8.4.2 *Dynamic Equilibrium of Dislocations*

Because no increase in flow stress is observed during the steady-state stage of the knitting process in Al-Mg alloys, we have concluded that the dislocations continually generated within the polygons must be annihilated during deformation. We further suggest that the annihilation mechanism operates on the small angle boundaries which, as mentioned above, can absorb arbitrary glide dislocations pushed against them by a shear stress. A detailed mechanism has been proposed for this previously (Lindroos and Miekk-oja, 1968) and is illustrated in fig. 8.21.

Here, a dislocation forest, C, in the form of a symmetrical tilt boundary separates two adjacent polygons (1) and (2). When the crystal is subjected to stress, primary slip systems begin to operate in both polygons, and the dislocations A1 and

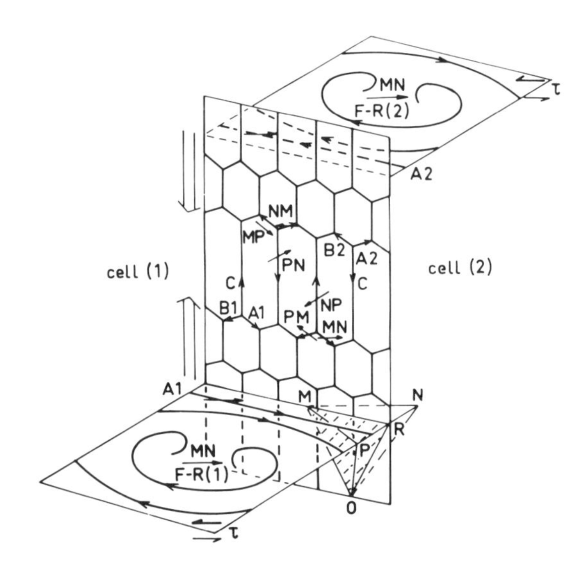

Fig. 8.21 A model for the annihilation of dislocations by means of stress-induced knitting (from Lindroos and Miekk-oja, 1968, courtesy of Taylor and Francis)

A2 produced by Frank-Read sources F-R(1) and F-R(2), respectively, interact with the forest. These two types of dislocations coming from opposite sides react with the forest dislocations, C, to form junction dislocations, B1 and B2, and thereby knit separate networks by means of stress-induced climb. However, because A1 and A2 are of opposite sign, they knit in opposite directions and when the leading dislocations of both networks come together they will annihilate. Thus the dislocation forest, which in this case is the symmetrical tilt boundary, acts as a kind of collector where dislocations meet and annihilate each other. During the actual process dislocation pile-ups may form within the polygons, but these may withdraw or relax when the applied stress is removed. Therefore, only a few free dislocations are observed within the polygons. It should be noted for completeness that the same annihilation mechanism can operate regardless of the nature of the dislocation sources, i.e. one can also consider sources other than the Frank-Read type.

Summarizing, we suggest that a steady-state stage occurs during plastic deformation at elevated temperatures because of (1) close collaboration between the glide process by which new dislocations are continually created, and (2) stress-induced climb which makes possible the annihilation of dislocations through the knitting process, thus preventing an increase in flow stress.

8.5 SUMMARY

Various detailed aspects of dislocation structures developed during elevated temperature deformation have been considered. In particular, the accurate description of such structures by means of the analysis of Burgers vectors and trace analysis of habit planes and directions, together with the devices available for the study of dislocation interactions, has been discussed in detail. The basic features of dislocation knitting processes for forming dislocation networks have been described, with particular emphasis on the importance of chemical forces in these processes. Finally, the knitting process, characteristic of elevated temperature deformation, has been applied to steady-state dynamic recovery, where it is shown that there at any instant is a dynamic balance between the behavior of vacancies and the existing dislocation sub-structure. The diffusion of vacancies along dislocation cores, i.e. 'pipe diffusion', is of primary importance in reaching this balance.

The mechanisms reviewed here can be applied to a wide range of elevated temperature processes, such as hot working, creep, and dynamic recovery as well as recovery annealing and substructure relaxation.

ACKNOWLEDGEMENTS

The authors gratefully acknowledge the support of the Finnish Academy in this work. Discussions with Professors M.F. Ashby, J.P. Hirth, J.J. Jonas, and Dr. U.F. Kocks are greatly appreciated. During preparation of the manuscript the valuable comments and help of Professor D.E. Mikkola, who was visiting us during a sabbatical leave, and the able technical assistance of our laboratory staff, in particular Mr. E. Poutiainen, H. Räikkönen, and Mrs. P. Korpiala, V. Klemetti and M. Salo, are gratefully acknowledged.

REFERENCES

Ashby, M. F., (1972) Surface Sci., 31, 498.

Bardeen, J., and Herring, C., (1952) Imperfections in Nearly Perfect Crystals, edited by W. Shockley et al. (New York: Wiley), p. 261.

Hirsch, P. B., Howie, A., Nicholson, R. B., Pashley, D. W., and Whelan, M. J., (1965) Electron Microscopy of Thin Crystals (London: Butterworths).

Hirth, J. P., (1966) Acta Met., 14, 1394

Hirth, J. P., and Lothe, J., (1968) Theory of Dislocations (New York: McGraw-Hill), p. 507.

Jonas, J. J., Sellars, C. M., and Tegart, W. J. Mc G., (1969) Metals and Materials, 3, 33.

Kivilahti, J. K., and Lindroos, V. K., (1972) Proc. Fifth European Congress on Electron Microscopy, University of Manchester, p. 368.

Kivilahti, J. K., and Lindroos, V. K., (1973) Metallography, in press.

Lindroos, V. K., (1970) Phil. Mag., 22, 637.

Lindroos, V. K., (1971) Phil. Mag., 24, 709.

Lindroos, V. K., (1974a) Metallography, in press.

Lindroos, V. K., (1974b) to be published.

Lindroos, V. K., and Kivilahti, J. K., (1972) Phil. Mag., 26, 833.

Lindroos, V. K., and Miekk-oja, H. M., (1967) Phil. Mag., 16, 593.

Lindroos, V. K., and Miekk-oja, H. M., (1968) Phil. Mag., 17, 119.

Lindroos, V. K., and Miekk-oja, H. M., (1969) Phil. Mag., 20, 329.

Miekk-oja, H. M., (1966) Phil. Mag., 13, 367.

Miekk-oja, H. M., and Lindroos, V. K., (1972) Surface Sci., 31, 422.

Nabarro, F. R. N., Basinski, Z. S., and Holt, D. B., (1964) Adv. Phys., 13, 193

Räty, R., Lindroos, V., Saarinen, A., Forstén, J., and Miekk-oja, H. M., (1966) J. Scient. Instrum., 43, 367.

Thompson, N., (1953) Proc. Phys. Soc., B., 66, 481.

Turunen, M. J., and Lindroos, V. K., (1973) Phil. Mag., 27, 81.

Turunen, M. J., and Lindroos, V. K., (1974) Phil. Mag., 29, 701.

Weertman, J., and Weertman, J. R., (1964) Elementary Dislocation Theory (New York: Macmillan), p. 72.

Vitek, V., (1968) Phil. Mag., 18, 773.

LIST OF SYMBOLS

a	lattice constant
$\underset{\sim}{b}$	Burgers vector
C_d	vacancy concentration near the dislocation core
C_o	thermodynamic equilibrium concentration of vacancies
F	elastic force
F_c	chemical force
$\underset{\sim}{g}$	operating reflection vector
h, k, l	Miller indices
k	Boltzmann constant
m	factor due to the contrast effect
T	absolute temperature
$\underset{\sim}{u}$	positive direction of dislocation
u, v, w	orientation indices
V	atomic volume
α_1, α_2	angles in micrographs
σ, σ_{xx}	tensile stress

ABSTRACT. *A mathematical framework is given for a single crystal under complex loading and undergoing large strains. A physical model based on the known behavior of single crystals is used in the mathematical framework. The framework allows the quantitative description of the changes in the dislocations structure during for example a tensile experiment.*

9. MODELLING OF CHANGES OF DISLOCATION STRUCTURES IN MONOTONICALLY DEFORMED SINGLE PHASE CRYSTALS

J. Zarka

9.1 INTRODUCTION

Several theories of plasticity and viscoplasticity for metals have been given during the last few years either from an engineer's or a metallurgist's point of view. Theories of the first type have tried to describe the behavior with a phenomenological approach using a functional representation or invoking internal parameters. The utility of these representations have been limited since numerous experiments are required to fix the internal parameters, and functionals are often not well defined. Theories of the second type have tried to describe the material, in the form of a single crystal by a microscopical, physical approach. In such theories the physical phenomena which occur during plastic straining are usually well explained. But, most often, only one-dimensional models are proposed which do not allow generalization. In an earlier theory (Zarka, 1968) we have tried to furnish a link between these two groups of theories by introducing the physical phenomenae in a general mathematical framework, with the object to give to the engineer readily usable explicit constitutive relations for a real medium, and to the metallurgist, a framework for including any elementary physical deformation process, under complex loading and capable of going to large strains. This book, being intended for the engineer as well as materials scientist, our purpose will be to show, with a particular example, how this generalization is possible. The theory will allow us to follow in detail the changes of dislocation structure during a tensile deformation experiment on a single crystal. How the behavior of the single crystal is related to that of a polycrystal has been discussed by Rice in Chapter 2 (see also Zarka 1973).

9.2 MODELLING THE SINGLE CRYSTAL BEHAVIOR

9.2.1 *Mathematical Framework*

We begin by discussing the behavior of a single crystal. We limit ourselves to a summary of earlier work to which the reader is referred for further details (Zarka, 1968, 1973, it is to be noted that the symbols used here differ somewhat from those used in the earlier work).

Around a geometrical point, we choose a volume element of a large enough size as to contain a large number of well distributed defects. We define by $\{\xi\}$ the family of parameters which describes this distribution. In order to characterize the elastic and plastic strains, relative to a fixed frame of reference $\vec{g}_i$, we define two families of vectors $\vec{F}_K$ and $\vec{e}_\alpha$ with the following properties: The family $\vec{F}_K$ represents 3 macroscopic vectors, that give the total transformation

$$\vec{F}_K = \frac{\partial x_i}{\partial X_K} \vec{g}_i \tag{9.1}$$

where $F = \left(\frac{\partial x_i}{\partial X_K}\right)$ is the classical displacement gradient. The family $\vec{e}_\alpha$ represents 3 microscopic vectors linked to the <u>mean lattice</u> of the volume element, that give the elastic transformation

$$\vec{e}_\alpha = x^i_\alpha \vec{g}_i \ . \tag{9.2}$$

Generally, the single crystal is subjected to both large strains and large rotations - the elastic strains, however, are always very small.

During the deformation, we follow step by step the same volume element (or the same material point) and divide the Eulerian velocity gradient

$$\nu_{ij} = \frac{\partial \nu_i}{\partial x_j} \tag{9.3}$$

into elastic and viscoplastic parts:

$$\nu_{ij} = \nu'_{ij} + \nu''_{ij} \tag{9.4}$$

In this way we have at once the variations of $\vec{F}_K$ and $\vec{e}_\alpha$:

$$\left(\frac{\partial \dot{x}_i}{\partial X_K}\right) = \nu_{i\ell} \frac{\partial x}{\partial X_K} \tag{9.5}$$

$$\dot{x}^i_\alpha = \nu'_{i\ell} x^\ell_\alpha . \tag{9.6}$$

(The dot above a quantity marks its material derivative).

Macroscopically, the plastic strains are the result of displacements on some crystallographic planes and along some crystallographic directions. These planes and directions have always the same components in the frame $\vec{e}_\alpha$. The unit vectors along them, $\vec{n}$ and $\vec{h}$, are thus easily defined in the fixed frame $\vec{g}_i$. By definition the pair of quantities $(\vec{n}^{(R)}, \vec{h}^{(R)})$ is associated with a <u>displacement system</u> (R). ($h_i^{(R)} n_i^{(R)}$ is not necessarily equal to 0).

In Eulerian coordinates, if $\gamma^{(R)}$ represents the relative <u>velocity</u> along $\vec{h}^{(R)}$ between two planes $\vec{n}^{(R)}$ which are unit distance apart, then the velocity gradient is equal to:

$$\nu''_{ij} = \sum_{\text{active (R)}} h_i^{(R)} n_j^{(R)} \gamma^{(R)} , \tag{9.7}$$

where the summation is extended over all active systems. We assume that a system is active when the inequality:

$$f^{(R)} \equiv h_i^{(R)} n_j^{(R)} \sigma_{ij} - \tau^{P(R)} (T , \{\xi\}) = \sigma^{(R)} - \tau^{P(R)} > 0 \tag{9.8}$$

is satisfied. In eqn. (9.8) σ_{ij} is the classical Cauchy stress tensor, and

$$\sigma^{(R)} = h_i^{(R)} n_j^{(R)} \sigma_{ij} \tag{9.9}$$

is a scalar, and represents the applied resolved stress; $\tau^{P(R)}$ is the plastic resistance for this system assumed to be a function of the temperature T and $\{\xi\}$, the distribution of internal parameters. Considering that in metals, displacements result from time dependent motions of defects (viscoplasticity), we write:

$$\gamma^{(R)} \equiv \gamma^{(R)} (\sigma^{(K)} , T , \{\xi\}) < f^{(R)} > \tag{9.10}$$

We expect that some parameters among the family $\{\xi\}$ may change. Toward this end we assume that:

$$\left[<x> \text{ is the positive part of } x = \begin{cases} 0 \text{ if } x \leq 0 \\ x \text{ if } x \geq 0 \end{cases} \right] . \tag{9.11}$$

During the elastic transformation, the defects stay immobile. The symmetrical part of ν'_{ij} may be locally known as:

$$d'_{ij} = \nu'_{(ij)} = S_{ijhk}(\sigma_{\ell m},T,\{\xi\}) \check{\sigma}_{hk} + \alpha_{ij}(\sigma_{\ell m},T,\{\xi\}) \dot{T} \tag{9.12}$$

$\check{\sigma}_{ij}$ is an objective derivative of σ_{ij} ; S_{ijhk} and α_{ij} are the elastic and thermal expansion coefficients, which are generally functions of $\sigma_{\ell m}$, T and $\{\xi\}$. The elastic rotation or antisymmetrical part $\nu'_{[ij]}$ can only be determined after solving the velocity field in the whole crystal.

9.2.2 *The Particular Physical Model*

Here, we consider a pure FCC single crystal, maintained at a constant temperature $T_0 \simeq 300^{\circ}K$, and loaded progressively to a rather low stress level where cross-slip and climb of dislocations can be ignored (there is no work-softening). In this case, it is known that slip occurs only on the planes $\vec{n} = (1/\sqrt{3})\{111\}$ and in the directions $\vec{h} = (1/\sqrt{2})\langle 110\rangle$. We have 24 such slip systems (we distinguish systems of different direction with opposite $\vec{h}$ vectors). We assume that the dislocations are small segments which are parallel to the <110> directions and that in each plane these orientations are equally probable. Let $\rho^{(R)}$ be the mean number per unit volume of small segments (R) which are in parallel planes, $\vec{n}^{(R)}$, and have the same Burgers vector $\vec{b}^{(R)}$ ($\vec{b}^{(R)} = b\,\vec{h}^{(R)}$); they thus belong to the slip system (R). Furthermore, let $\ell^{(R)}$ be the mean length of these dislocations segments; the important quantity $\rho^{(R)}\,\ell^{(R)}$ is then the total length per unit volume of these segments. Then, let $N^{(R)}$ be the mean number per unit surface of all dislocations which pierce a slip plane (R); this is linked to the $\rho^{(K)}\,\ell^{(K)}$ by the geometrical relation:

$$N^{(R)} = 0.54 \sum_{(K)} \rho^{(K)}\,\ell^{(K)} \tag{9.13}$$

where the summation is only extended over all systems (K) which have slip planes other than (R). Now, the plastic resistance for the system (R) is given by

$$\tau^{P(R)} \simeq \mu\, b \sum_{\text{over all }(K)} D^{(R)}_{(K)} \left(\rho^{(K)}\,\ell^{(K)}\right)^{1/2} \tag{9.14}$$

where μ is the shear modulus

and $D^{(R)}_{(K)} \simeq 1/8$, when (R) and (K) have the same slip plane and parallel burgers vectors (no Bauschinger's effect)

$D^{(R)}_{(K)} \simeq 1/16$, when (R) and (K) have only the same slip plane.

$D^{(R)}_{(K)}$ 1/12 , when (R) and (K) do not have the same slip plane nor parallel Burgers vectors.

$D^{(R)}_{(K)}$ 1/20 , when (R) and (K) do not have the same slip plane but have parallel Burgers vectors.

The calculations to be discussed below were carried out based on the classical hypothesis of regularly distributed straight and infinitely long dislocation lines in an infinite isotropic elastic medium, but taking account of their finite length and their random distribution.

Equation (9.14), which gives the interactions between slip systems, may be considered as a rational generalization of the classical one-dimensional formula $\tau^P = \mu\ b\ N^{1/2}/\beta$, where β is a constant of order unity. A slip system (R) is active when the applied resolved stress $\sigma^{(R)} = h_i^{(R)}\ n_j^{(R)}\ \sigma_{ij}$ is higher than $\tau^{P(R)}$,

$$\sigma^{(R)} > \tau^{P(R)} \tag{9.15}$$

As in the Saada's Model, we assume that, globally, for each active system (R) , during a small interval of time dt , some dislocations act as sources and emit $d\rho^{(R)}$ dislocation loops (per unit volume). These new loops, of mean length $2\ \pi\ L^{(R)}$ sweep a mean area $\pi\ L^{(R)^2}$ by glide before being stopped on the dislocations that pierce their plane: Thus,

$$\left.\begin{aligned} \gamma^{(R)}\ dt &= b\ d\rho^{(R)}\ \pi\ L^{(R)^2} \\ d(\rho^{(R)}\ \ell^{(R)}) &= d\rho^{(R)}\ 2\ \pi\ L^{(R)} = 2\ \frac{\gamma^{(R)} dt}{b\ L^{(R)}} \end{aligned}\right\} \tag{9.16}$$

We make the substitution

$$\pi\ L^{(R)^2} \equiv \frac{x_0}{N^{(R)}} \tag{9.17}$$

and assume that x_0, the mean number of jogs on a new loop at the time when it stops expanding (or the mean number of intersections with other forest dislocations) is constant. If then, $t_{GD}^{(R)}$ is the mean elapsed time for accomplishing one intersection, ($t_{GD}^{(R)} = 1/\nu_{GD}^{(R)}$, $\nu_{GD}^{(R)}$ is the frequency factor of the intersection) given by:

$$t_{GD}^{(R)} \simeq \frac{c_0}{\langle \sigma^{(R)} - \tau^{P(R)} \rangle} \tag{9.18}$$

where c_0 is a physical constant, we may write,

$$d\rho^{(R)} \simeq w_0 \, \rho^{(R)} \frac{dt}{x_0 \, t_{GD}^{(R)}} = \frac{w_0 \, dt}{c_0 \, x_0} \rho^{(R)} \langle \sigma^{(R)} - \tau^{P(R)} \rangle . \tag{9.19}$$

In eqn. (9.19) w_0 is the fraction of dislocations which may act as sources which is assumed here to be constant. When the applied stress is very slow, we are in the plasticity framework and the eqn. (9.19) does not matter. We then have for the active systems:

$$\sigma^{(R)} \simeq \tau^{P(R)} \quad \text{and} \quad d\sigma^{(R)} \simeq d\tau^{P(R)} \tag{9.20}$$

and the the latent systems:

$$\left.\begin{array}{l} \sigma^{(R)} \simeq \tau^{P(R)} \quad \text{and} \quad d\sigma^{(R)} < d\tau^{P(R)} , \quad \text{or} \\ \sigma^{(R)} < \tau^{P(R)} . \end{array}\right\} \tag{9.21}$$

We may easily solve the eqns. (9.20) and (9.21) to get $d(\rho^{(R)}\ell^{(R)})/2(\rho^{(R)}\ell^{(R)})^{1/2}$ as functions of $d\sigma^{(K)}$ by defining quadratic minimization problem with linear constraints. With this elementary model we will see how to give explicit forms to eqns. (9.8), (9.10) and (9.11). It is a relatively simple matter to introduce into the model cross-slip, climb, (work-softening), impurities, etc. (see Zarka, 1968, or 1973).

9.3 CHANGES IN THE DISLOCATION STRUCTURE DURING A MONOTONIC TENSILE EXPERIMENT

Single crystals are most commonly deformed in tension. Such

experiments are generally very difficult to perform as well as to interpret, chiefly because of the large shape changes which they undergo. During the straining, one commonly measures only the applied force F and the crystal length D. Hence, as a first task we verify the validity of the model on these two quantities. Furthermore, there exist detailed observations of the dislocation structures on such deformed crystals which we can use with the theoretical dislocation structures.

9.3.1 *The Basic Equations of the Problem*

We assume that, at any moment, we have a simple uniaxial force F acting on the crystal. Since the elastic strains are very small, we assume that the frame of reference $\vec{g}_i$ is indistinguishable from the frame $\vec{e}_\alpha$. In addition there is no elastic rotation. The axes Ox, Oy, Oz are respectively parallel to the directions [100], [010] and [001]. We then assume that the elastic coefficients S_{ijhk} are constant, and adopt the particular physical model given in Section 9.2.2. Let us now denote the actual normal cross section of the crystal by S. The Cauchy stress tensor is given by:

$$\sigma_{ij} = \Sigma\, d_i\, d_j \quad , \quad \text{where ,} \tag{9.22}$$

$$\Sigma = \frac{F}{S} \tag{9.23}$$

and $\vec{d} = d_i\, \vec{g}_i$ is the actual unit vector along the tensile axis OZ. This vector may be defined by its Euler angles Θ, Ψ, giving components for the vector

$$\left.\begin{aligned} d_x &= \sin\theta \sin\Psi \\ d_y &= -\sin\theta\cos\Psi \\ d_z &= \cos\theta \end{aligned}\right\} \tag{9.24}$$

On the other hand the vector $\vec{D} = D\,\vec{d}$ is parallel to the tensile axis and is proportional to the crystal length. Since elastic strains are small, and plastic strains, resulting from slip, do not produce a change of volume, we have,

$$S \simeq \frac{S_o D_o}{D} \tag{9.25}$$

where S_o and D_o are the initial values of S and D. Thus by eqn. (9.23) we deduce the simple relation:

$$\Sigma \simeq F \frac{D}{S_o D_o} \tag{9.26}$$

The applied resolved stress on the slip system (R), geometrically defined by $(\vec{n}^{(R)}, \vec{h}^{(R)})$ and physically characterized by $(\rho^{(R)}, \rho^{(R)} \ell^{(R)})$, is given by,

$$\sigma^{(R)} = h_i^{(R)} n_j^{(R)} \sigma_{ij} = (h_i^{(R)} d_i)(n_j^{(R)} d_j) \Sigma \tag{9.27}$$

and after making the substitution

$$m^{(R)} = (h_i^{(R)} d_i)(n_j^{(R)} d_j) , \tag{9.28}$$

$$\sigma^{(R)} = m^{(R)} \Sigma = m^{(R)} F \frac{D}{S_o D_o} , \tag{9.29}$$

where $m^{(R)}$ is the classical geometrical stress resolution coefficient. The plastic resistance for this system is then equal to,

$$\tau^{P(R)} = \mu b \sum_{\text{at all (K)}} D_{(K)}^{(R)} (\rho^{(K)} \ell^{(K)})^{1/2} \tag{9.30}$$

We have now,

$$\left.\begin{aligned} \dot{\gamma}^{(R)} &\simeq b \frac{w_o}{c_o} \frac{\rho^{(R)}}{N^{(R)}} < \sigma^{(R)} - \tau^{P(R)} > \\ \dot{\rho}^{(R)} &= \frac{N^{(R)}}{b x_o} \dot{\gamma}^{(R)} \\ (\dot{\rho^{(R)} \ell^{(R)}}) &= \frac{2}{b} \sqrt{\frac{\pi N^{(R)}}{x_o}} \dot{\gamma}^{(R)} \end{aligned}\right\} \tag{9.31}$$

The Eulerian velocity gradient is thus given by,

$$\nu_{ij} = s_{ijhk}\,\dot{\sigma}_{hk} + \sum_{\text{active (R)}} h_i^{(R)}\, n_j^{(R)}\, \gamma^{(R)} \qquad (9.32)$$

$$\nu_{ij} = \nu'_{ij} + \nu''_{ij} \quad \text{(elastic + plastic parts)}$$

For a cubic crystal in this particular frame $\vec{g}_i$, the elastic part can be explicitly given in the form:

$$\left.\begin{aligned} \nu'_{xx} &= s_{11}\,\dot{\sigma}_{xx} + s_{12}\,(\dot{\sigma}_{yy} + \dot{\sigma}_{zz}) \\ 2\nu'_{xy} &= s_{44}\,\dot{\sigma}_{xy} \end{aligned}\right\} \qquad (9.33)$$

with similar relations for the other components.

The relation giving the change in the macroscopical vector $\vec{D}$ is (see eqn. 9.5),

$$\begin{vmatrix} \dot{D}_x \\ \dot{D}_y \\ \dot{D}_z \end{vmatrix} = \begin{vmatrix} \nu_{xx} & \nu_{xy} & \nu_{xz} \\ \nu_{yx} & \nu_{yy} & \nu_{yz} \\ \nu_{zx} & \nu_{zy} & \nu_{zz} \end{vmatrix} \begin{vmatrix} D_x \\ D_y \\ D_z \end{vmatrix} \qquad (9.34)$$

From here, we deduce the variation of the length of the crystal as,

$$\dot{D} \equiv S_Z\,\dot{\Sigma} + \dot{S}_Z\,\Sigma + \sum_{\text{active (R)}} m^{(R)}\,\gamma^{(R)}\, D \qquad (9.35)$$

where S_Z , the inverse of the Young's modulus in the direction OZ of the tensile axis is given by:

$$S_Z = s_{11} - \{2\,s_{11} - 2\,s_{12} - s_{44}\}\{(d_x\,d_y)^2 + (d_x\,d_z)^2 + (d_y\,d_z)^2\} \qquad (9.36)$$

Generally, tensile experiments are performed on machines where a constant cross-head velocity C, is prescribed, i.e.,

$$\dot{D} = C \tag{9.37}$$

Hence, we need to solve for this problem a system of differential equations with: Σ (or F), Ψ, θ, and all the $(\rho^{(R)}, \rho^{(R)}\ell^{(R)})$ as unknown functions, and with the initial conditions for example:

$$\left.\begin{array}{l} \Sigma(0) = 0, \ \Psi(0) = \Psi_o, \ \theta(0) = \theta_o \ \text{and} \\ \text{for all (R)} \\ \rho^{(R)}(0) = \rho_o, \ \rho^{(R)}(0)\ell^{(R)}(0) = \rho_o\ell_o \end{array}\right\} \tag{9.38}$$

(The slip systems are initially equivalent).

9.3.2 *Integration of the Equations*

When only the primary slip system is active, the integration can be performed explicitly. As eqn. (9.14) or (9.30) shows, the work-hardening in the latent slip systems is lower than that in the primary one. When several slip systems are active, the integration was performed numerically. It is possible to write two different kinds of program for this purpose. The first one is based on a framework of viscoplasticity where we do the direct explicit integration of the system of differential equations. There are several numerical algorithms among which the best seems to be the method of Treanor having a variable step. In this the velocity C must be introduced and we have to fit the parameters x_o, w_o/c_o and ρ_o. The second one is based on the classical framework of plasticity where we use an implicit scheme of integration. At time between t and $t + \Delta t$ we must have for any (R) :

$$\sigma^{(R)}(t) \leq \tau^{P(R)}(t) \quad \text{and} \quad \sigma^{(R)}(t + \Delta t) \leq \tau^{P(R)}(t + \Delta t) \tag{9.39}$$

together with linear constraints $\Delta(\rho^{(R)}\ell^{(R)}) \geq 0$ which result from the condition of no work-softening.

By means of a linearly constrained algorithm for quadratic minimization eqn. (9.39) gives the quantities: $\Delta(\rho^{(R)}\ell^{(R)})/2(\rho^{(R)}\ell^{(R)})^{1/2}$, and $(\rho^{(R)}\ell^{(R)})$, as functions of all of the $\Delta\sigma^{(K)}$ (or $\Delta\Sigma$). The eqns. (9.31) are modified into:

$$\gamma^{(R)} \Delta t = \Delta(\rho^{(R)} \ell^{(R)}) \frac{b}{2} \sqrt{\frac{x_o}{\pi N^{(R)}}} \qquad (9.40)$$

The velocity C , the dislocation segment density $\rho^{(R)}$, and the parameter w_o/c_o do not appear in this solution(*).

9.3.3 *Theoretical Results*

Using the program described above, the tensile behavior of FCC crystals with seven different starting orientations shown in Fig. 9.1 and having Euler angles given in Table 9.1 were computed. Two different and relatively representative initial densities of dislocation $\rho_o \ell_o$, and the terminal number of jogs x_o , also given in Table 9.1, were taken for the computations. The computed tensile stress-extension ratio curves are given in Figs. 9.2 and 9.3. In addition to these curves, the activity on all possible slip planes given in Table 9.2 was also computed. Figures 9.4-9.10 show as a function of tensile stress for each of the seven orientations: a) the total integrated shear strains on the most active slip systems, identified by the numbering scheme of Table 9.2; and b) the increases in forest dislocation density piercing each of the four slip planes identified in Table 9.2.

Examination of the tensile stress-extension ratio curves show, except for the high symmetry orientation (4), an initial region of very low work hardening which is followed by a region of much higher work hardening. Thus, we obtain directly the familiar stages I and II of work-hardening. Since no provision was made in the computation for cross-slip and climb of dislocations, stage III does not appear. In this sense our curves represent low temperature behavior. For the high symmetry orientation (4), the region of low work hardening is absent. Examination of the curves on increases in forest dislocation density shows clearly that only one slip system is substantially active (in some cases there may be two coplanar slip systems active). It is clear also that at the transition from stage I to stage II new slip systems become

(*)The detailed listing of the computational programs described above can be obtained from the author upon request.

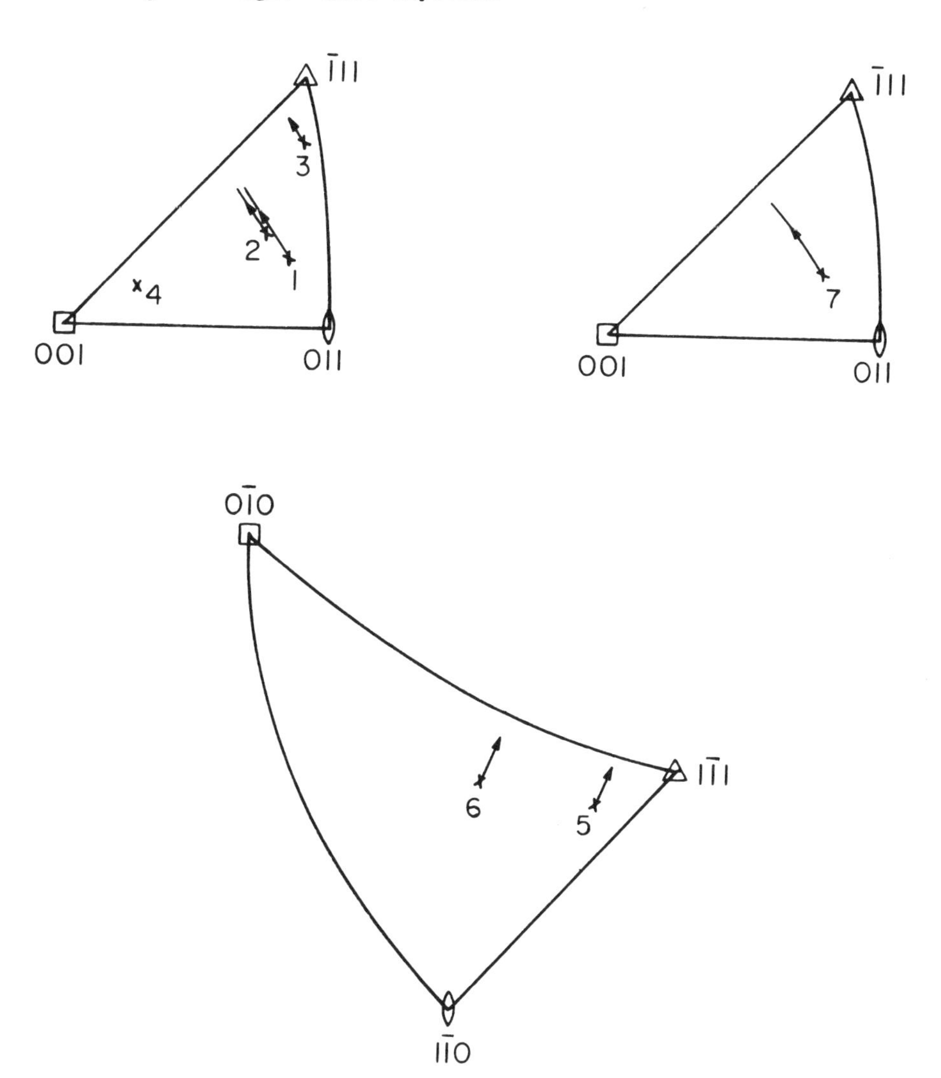

Fig. 9.1 Orientations of seven crystals considered in the computation.

activated which cut through the primary slip plane. In addition we see also the familiar effect that the extent of the region of low work-hardening is orientation sensitive.

Since the crystallographic rotations resulting from slip are of opposite sense in compression from those in tension, the stress-extension ratio curves are anti-symmetrical about the

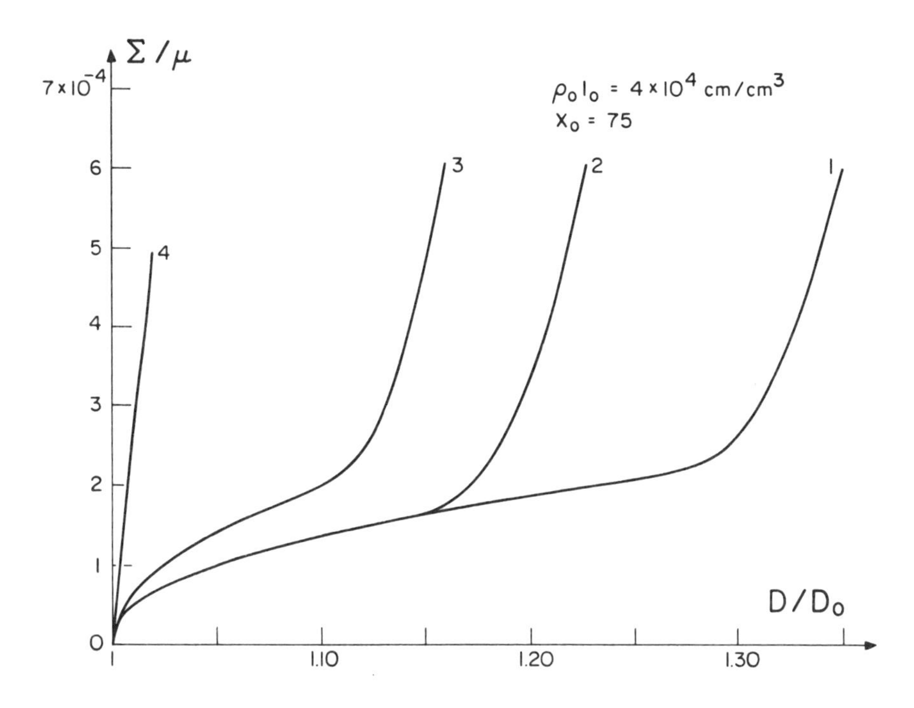

Fig. 9.2 Curves of normalized tensile stress-extension ratio for crystals with orientations (1)-(4).

origin and for stress reversal at any stage. Figures 9.11-9.13 show the behavior of crystals with orientations (5), (6), and (7) in tension and compression and for reversal of stress at two points along the tensile extension curve. It must be noted that since the theoretical model did not specifically model it the Bauschinger effect is absent in these curves.

Needless to say the results just presented for the FCC crystals furnish only an example, other curves for other FCC and BCC crystals with different orientations or different cases of loading can be calculated with similar ease.

9.4 CONCLUSIONS

The computed curves give results closely resembling the known stress strain curves of single crystals. At the present there are no quantitative comparisons between the predictions of the theory and experimental results; especially in the dislocation structures resulting from strain. It is hoped that such comparisons can be furnished by experimental metallurgists.

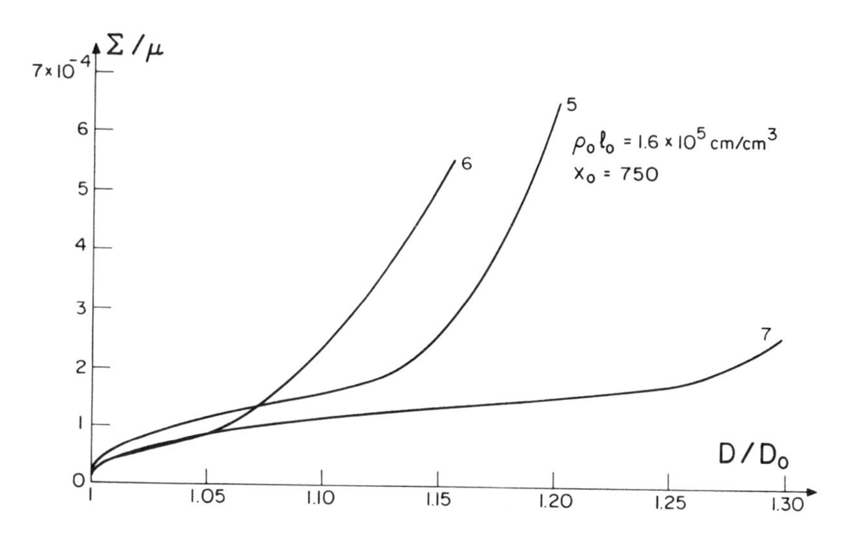

Fig. 9.3 Curves of normalized tensile stress-extension ratio for crystals with orientations (5)-(7).

The theory too could be perfected further to deal with cross slip and climb and with twinning wherever that is an important mechanism.

ACKNOWLEDGEMENT

The author is indebted to the C.N.R.S. for financial support of this work through Contract A.T.P. Plasticité 1971.

Table 9.1 Angles of Orientation and Initial Structure Parameters for Crystals Considered

No. of Crystal	θ_o	Ψ_o	$\rho_o \ell_o$	x_o
(1)	-38.86^o	16.36^o		
(2)	-37.15^o	23.99^o	4×10^4	75
(3)	-48.00^o	37.00^o	cm/cm^3	
(4)	-15.00^o	26.00^o		
(5)	65.00^o	40.00^o		
(6)	74.25^o	29.50^o	1.6×10^5	750
(7)	-37.15^o	16.36^o	cm/cm^3	

Table 9.2 Symbols for Slip Planes and Directions

Symbol	Slip Plane	Symbol	Slip Direction
A	$(11\bar{1})$	①	[101]
		②	[011]
		③	$[1\bar{1}0]$
B	$(1\bar{1}1)$	④	$[\bar{1}01]$
		⑤	[011]
		⑥	[110]
C	$(\bar{1}11)$	⑦	[110]
		⑧	$[0\bar{1}1]$
		⑨	[101]
D	(111)	⑩	$[\bar{1}10]$
		⑪	$[01\bar{1}]$
		⑫	$[10\bar{1}]$

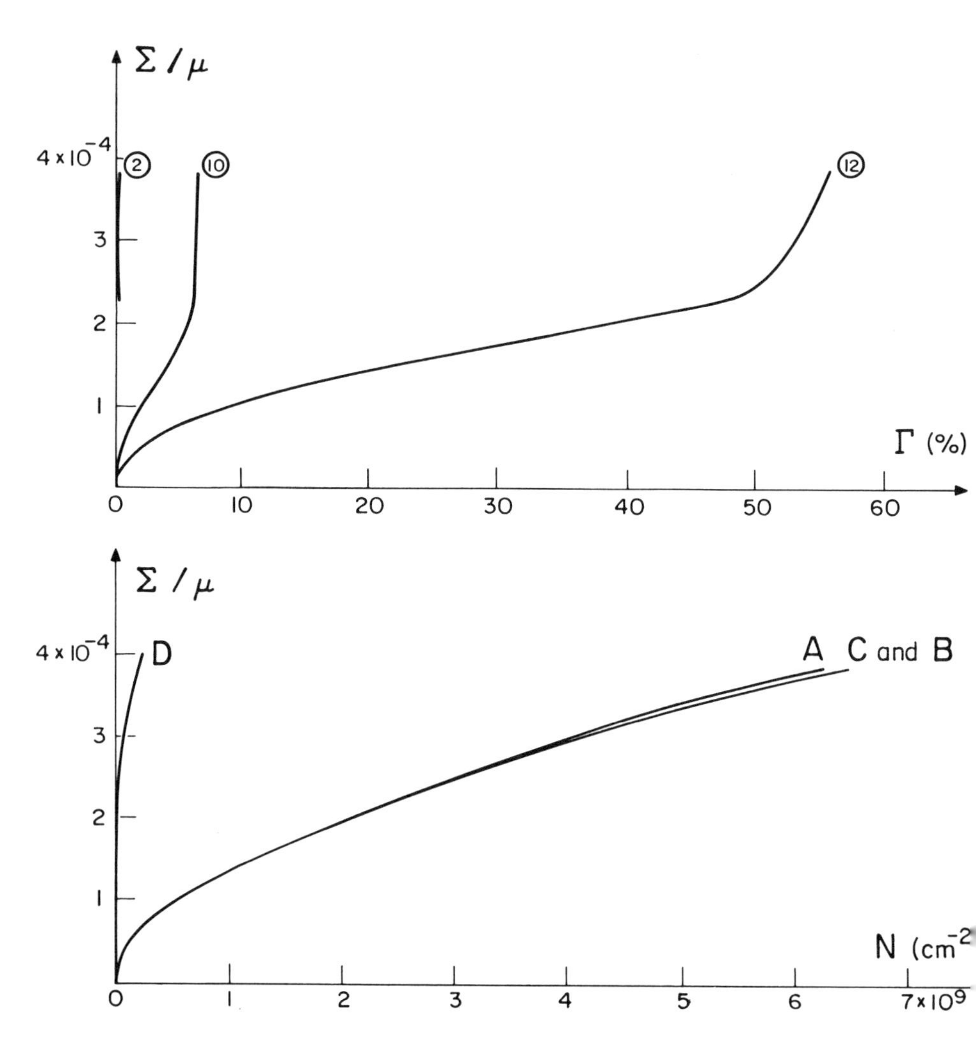

Fig. 9.4 Curves of integrated shear strain on the most active systems vs. the normalized tensile stress, and increase in forest dislocation density across the four slip planes vs. the normalized tensile stress for the crystal with orientation (1).

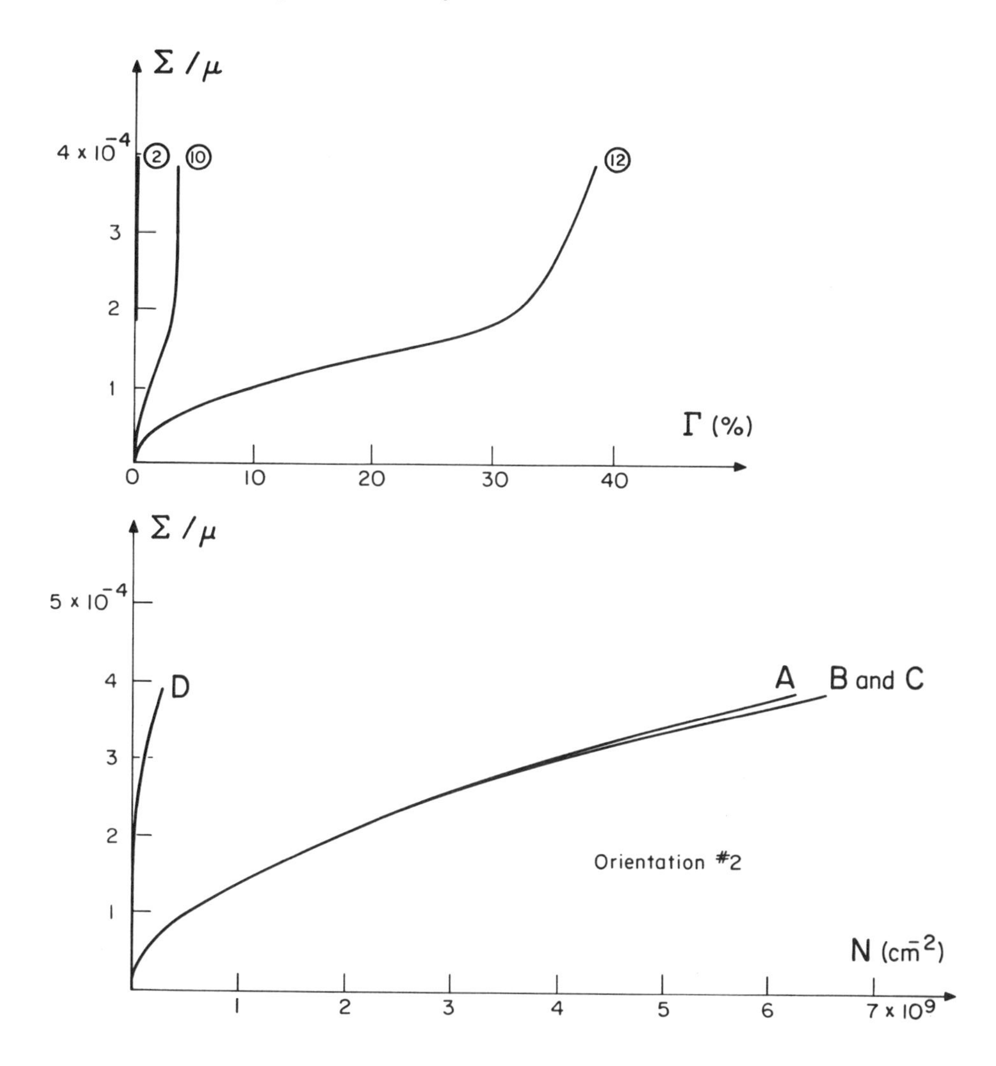

Fig. 9.5 Curves of integrated shear strain on the most active systems vs. the normalized tensile stress, and increase in forest dislocation density across the four slip planes vs. the normalized tensile stress for the crystal with orientation (2).

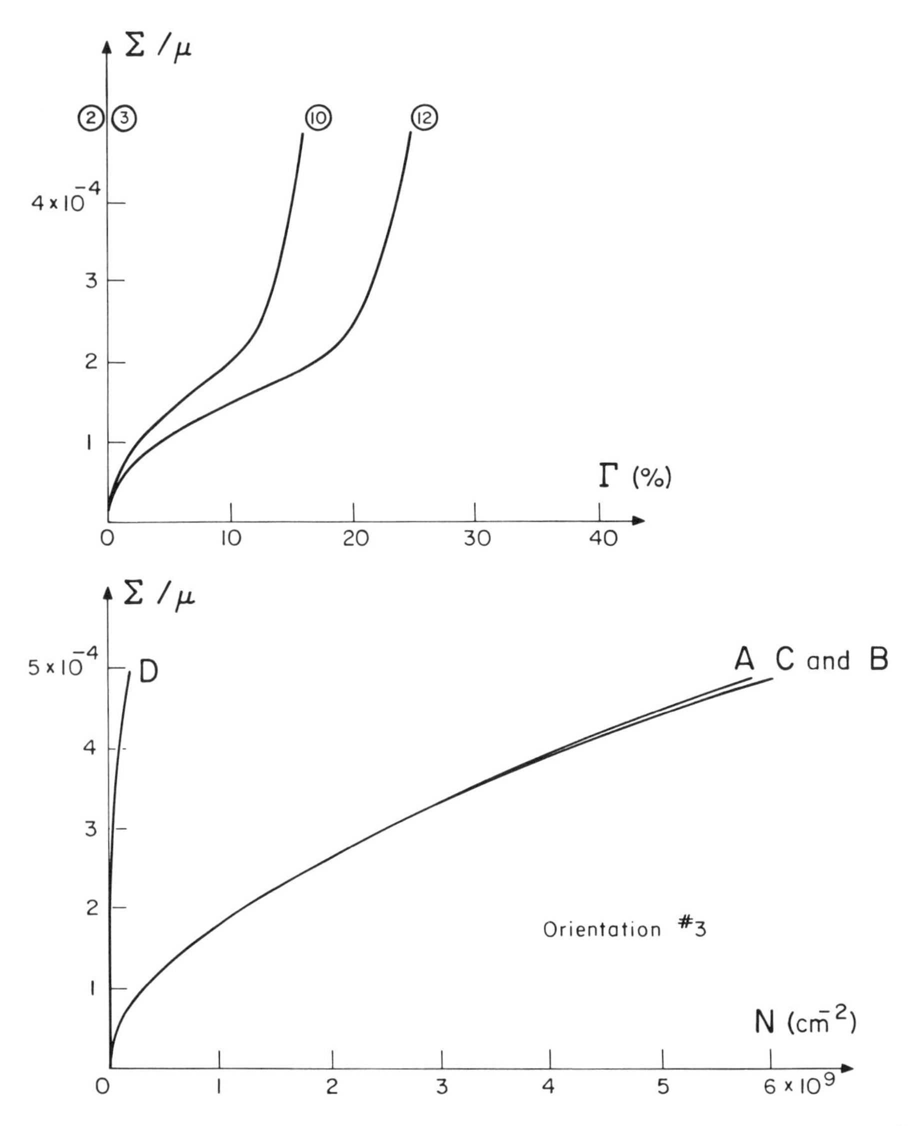

Fig. 9.6 Curves of integrated shear strain on the most active systems vs. the normalized tensile stress, and increase in forest dislocation density across the four slip planes vs. the normalized tensile stress for the crystal with orientation (3).

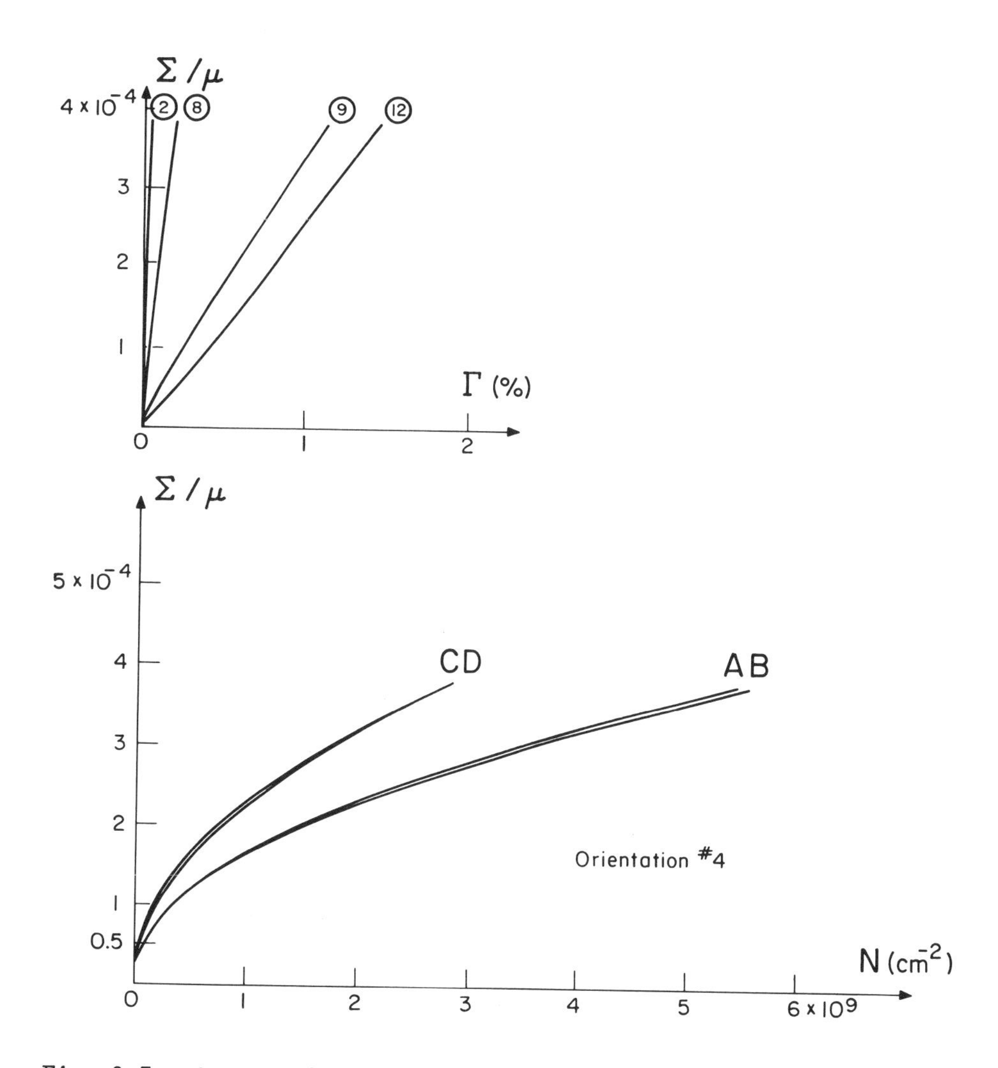

Fig. 9.7 Curves of integrated shear strain on the most active systems vs. the normalized tensile stress, and increase in forest dislocation density across the four slip planes vs. the normalized tensile stress for the crystal with orientation (4).

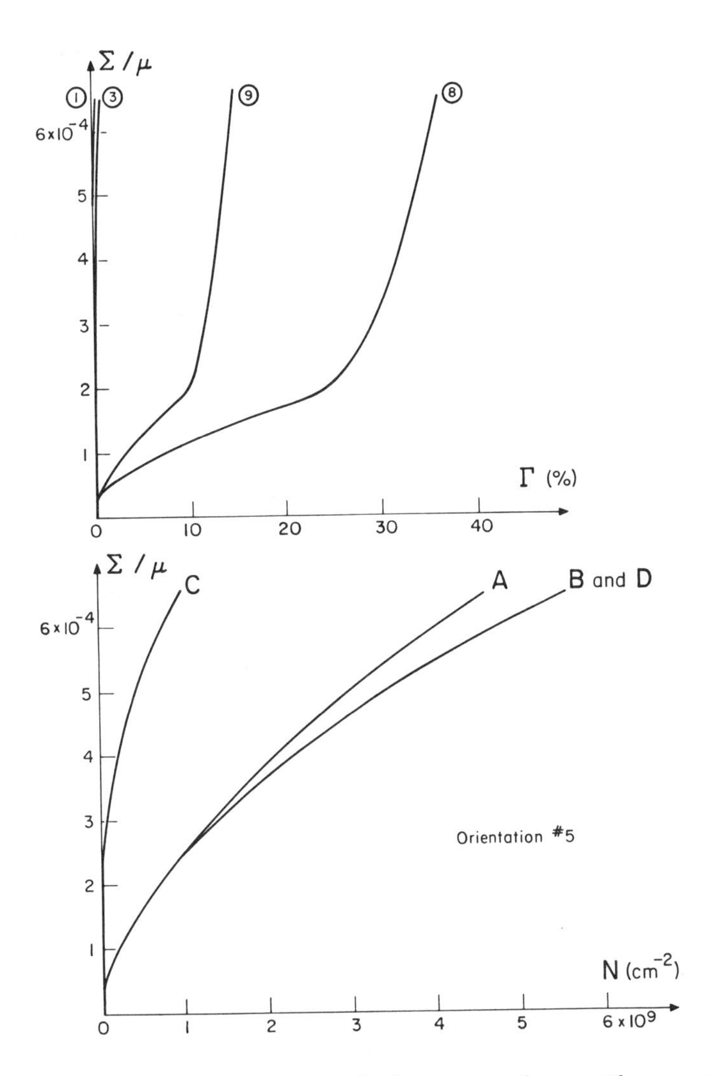

Fig. 9.8 Curves of integrated shear strain on the most active systems vs. the normalized tensile stress, and increase in forest dislocation density across the four slip planes vs. the normalized tensile stress for the crystal with orientation (5).

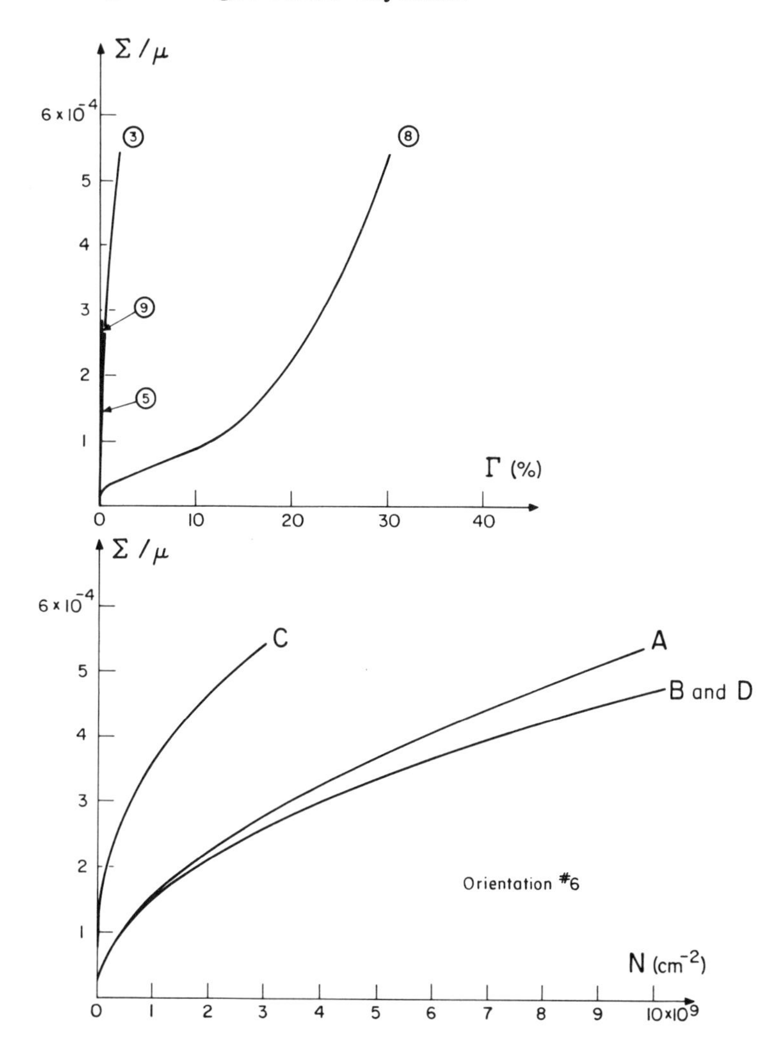

Fig. 9.9 Curves of integrated shear strain on the most active systems vs. the normalized tensile stress, and increase in forest dislocation density across the four slip planes vs. the normalized tensile stress for the crystal with orientation (6).

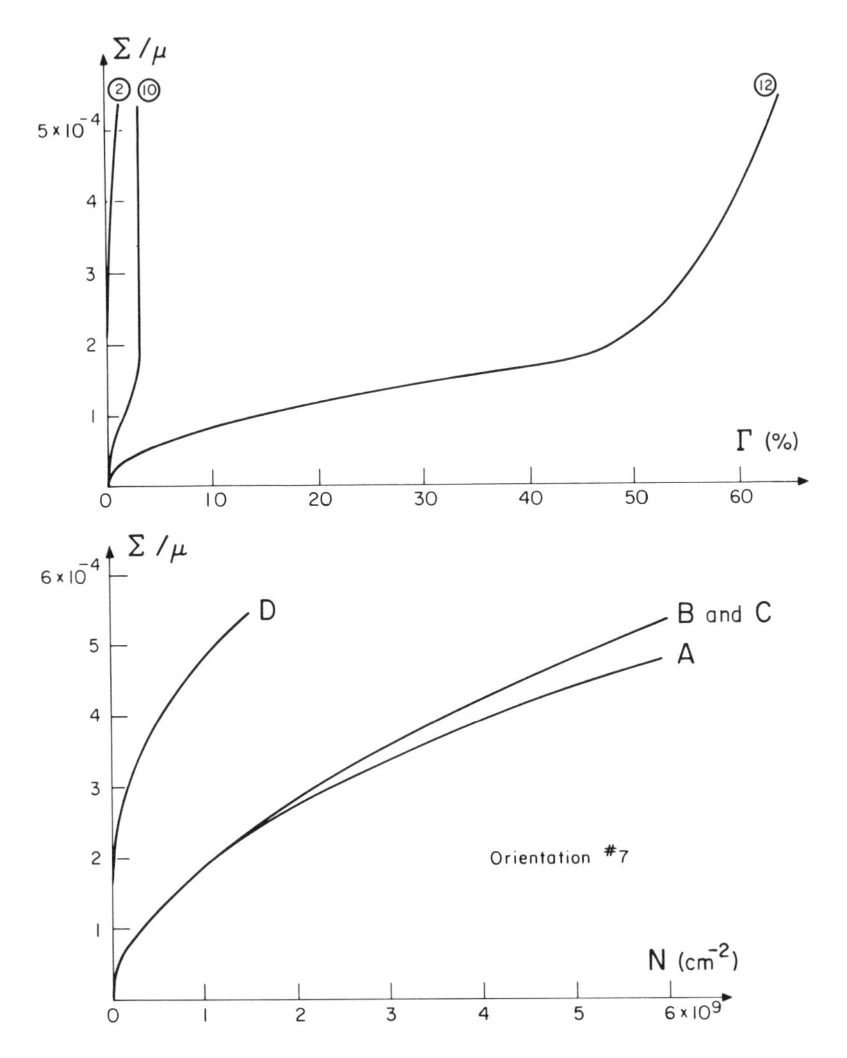

Fig. 9.10 Curves of integrated shear strain on the most active systems vs. the normalized tensile stress, and increase in forest dislocation density across the four slip planes vs. the normalized tensile stress for the crystal with orientation (7).

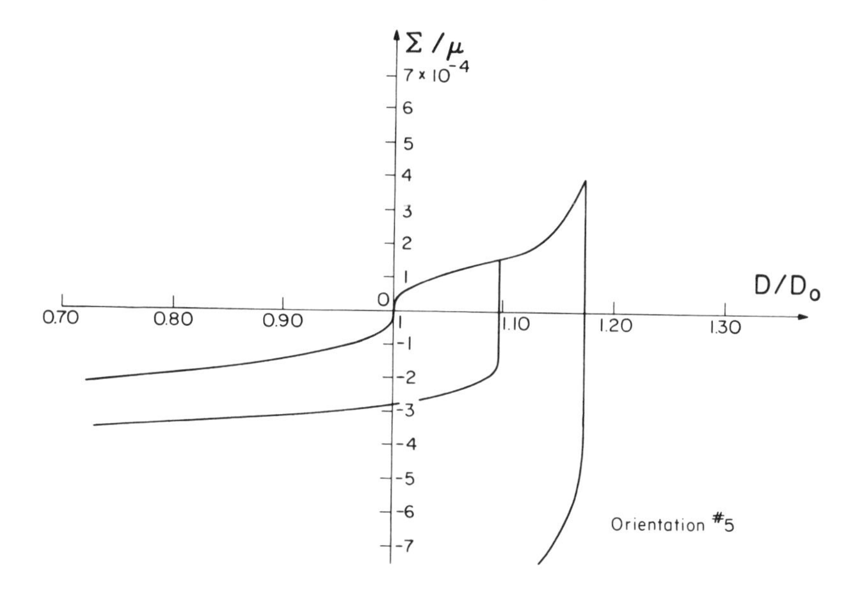

Fig. 9.11 Computed tension-compression curves with stress reversal for orientation (5).

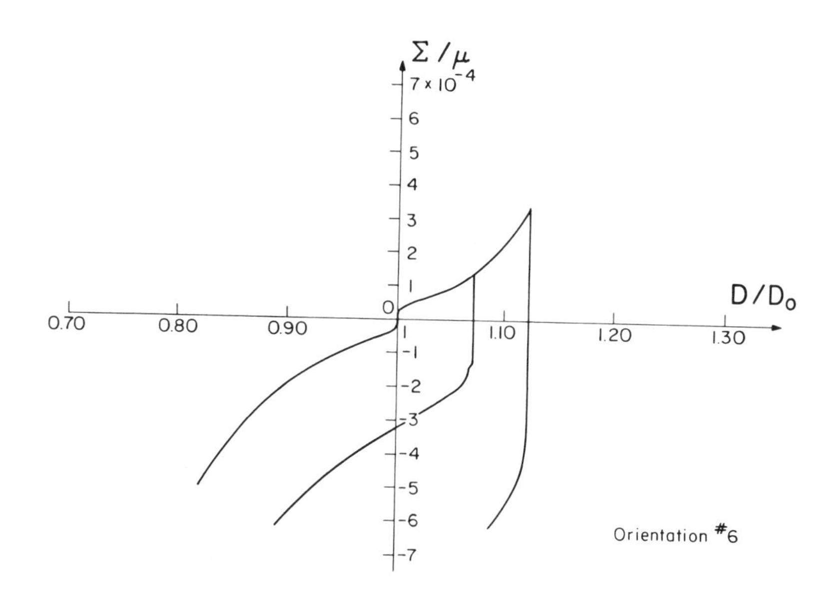

Fig. 9.12 Computed tension-compression curves with stress reversal for orientation (6).

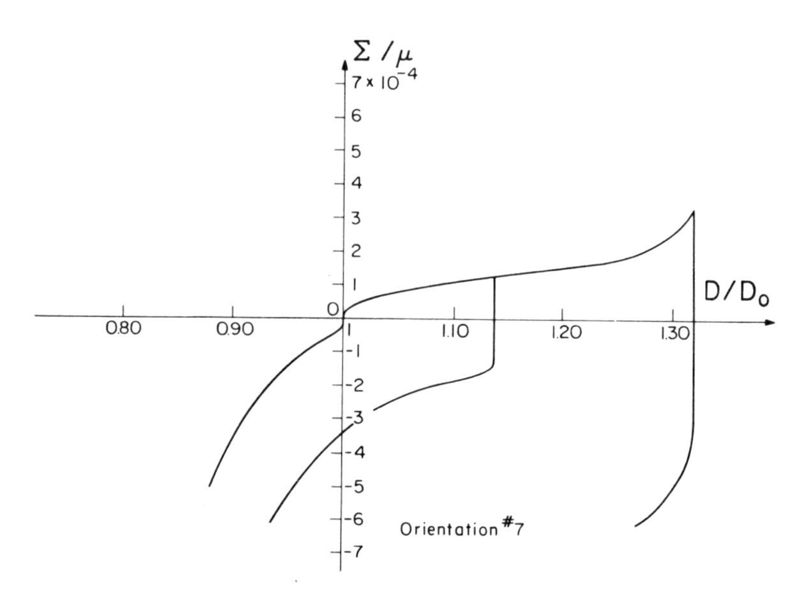

Fig. 9.13 Computed tension-compression curves with stress reversal for orientation (7).

REFERENCES

Saada, G., (1961) Publication de l'I.R.S.I.D., Series A, No. 251.

Zarka, J., (1968) Ph.D Thesis, Paris, Mem. Art. Franc., 2ème Fascicule, 223 (1970).

Zarka, J., (1973) J. de Mécanique, 12, 275.

LIST OF SYMBOLS

$\{\xi\}$	family of internal parameters
$\vec{g}_i$	fixed frame of reference
$\vec{F}_K$	macroscopic vectors
$\vec{e}_\alpha$	microscopic vectors linked to the mean lattice
ν_{ij}	Eulerian velocity gradient
ν'_{ij} and ν''_{ij}	elastic part and plastic part of this gradient
$\vec{n}$	unit vector normal to a crystalline plane
$\vec{h}$	unit vector along a crystalline direction
γ	relative velocity
σ_{ij}	Eulerian stress tensor
σ	applied resolved stress
τ^P	plastic resistance
T	absolute temperature
d'_{ij}	elastic strain rate tensor
$\vec{b}$	Burgers vector
ρ	mean number per unit volume of dislocations
N	mean number per unit surface of dislocations
ℓ	mean length of the dislocations
μ	shear modulus
$D^{(R)}_{(K)}$	interaction matrix between slip systems
L	mean radius of a newly created dislocation
x_o	mean number of jogs on a new dislocation
m	classical Schmid geometrical coefficient
$\vec{D}$	macroscopical vector linked to the axis of the crystal

ABSTRACT. *A description is given of the continuum model for the plastic deformation of a dispersion strengthened alloy in which no stress relaxation occurs. The consequences of the resulting concept of a "Perfect Memory Solid" are then briefly reviewed. Particular emphasis is given to the importance of the uniform stress in the matrix opposing further forward deformation and aiding reverse flow. The various ways by which the mean matrix strain is reduced by stress relaxation at the particles are then described as a function of the dispersion characteristics, testing temperature and applied strain. A quantitative interpretation of the Bauschinger effect is then given so that the degree to which the material deviates from a perfect memory solid, under the given testing conditions, can be assessed. Finally, constitutive equations are given for the various classes of mechanical behavior which are distinguished by the approach described in this paper.*

10. MODELLING STRUCTURAL CHANGES IN DEFORMED DISPERSION STRENGTHENED CRYSTALS

L.M. Brown and W.M. Stobbs

10.1 INTRODUCTION

A dispersion-hardened system is one in which strength is achieved by a dispersion of non-deforming particles. Many commercial materials are prepared with an element of dispersion strengthening, either by design or because oxide or other inclusions have become incorporated into them accidentally in the process of fabrication. Most over-aged precipitation-hardening alloys can be regarded as dispersion-hardened. Such alloys have a strength which is controlled by the stress required to bow a dislocation between two particles - the so-called Orowan strength. Recent work has established this fact quite clearly, and the reader is referred to recent reviews for details (Ashby 1969 and 1971, Hirsch and Humphreys 1969, Brown and Ham 1971).

In this paper, we shall concentrate on systems with equiaxed particles - indeed, most of our recent work has been on the system Cu-SiO_2, in which a small volume fraction of silica spheres are embedded in a matrix of pure copper. We shall not attempt to write constitutive equations which will be more accurate than $\pm$ 25% or so, and we shall not concern ourselves with recent theoretical refinements which aim at greater accuracy. Thus we shall take the Orowan strength σ_o to be given by

$$\sigma_o = \frac{\mu b}{\ell} = \mu b N^{1/2} \tag{10.1}$$

rather than by modified forms for the equation such as that given by Bacon, Kocks and Scattergood (1973). In adopting this approach we are governed firstly by the need to clarify the principles of the models which we use to describe the different regimes of mechanical behavior and, secondly, by various uncertainties in experiment and theory, of order of 20%.

When a dislocation passes through a dispersion-hardened material, it leaves "Orowan loops" around every particle in its slip plane. It has been recognized since the work of Fisher, Hart and Pry (1953) that these loops will oppose further deformation of the material. Work-hardening models in these materials concern themselves with the fate of the Orowan loops, and with their effect on strength.

Actually there are two basic methods of describing what happens during plastic deformation. These are the continuum model, introduced by Ashby (1966, 1970a) and the dislocation model, which is of course not really a model but in principle a description of real atomic movements. The trouble with the dislocation description is its complexity; it has been remarked that using dislocation mechanics and high-resolution microscopy to model plastic deformation is like modelling the behavior of gases without the simplifications of statistical mechanics, but by studying individual molecular trajectories. It is quite widely agreed that the most useful model will be a hybrid.

We examine the continuum model in Section 10.2 and review the consequences of its adoption in Section 10.3. We give particular emphasis to the concept of a "Perfect Memory Solid", or PMS, as a means of classifying all solids which contain elastic strains proportional to the plastic shape change applied to them. In Section 10.4 we describe briefly the microstructural evidence, obtained by electron microscopy, for various different modes of stress relief. We then show in Section 10.5 that the Bauschinger effect can be used not only to give a quantitative assessment of the degree of stress relief which has occurred during a given test but also to assess the effect of different dislocation microstructures in dynamic and static softening. We are thus able to give, in Section 10.6, realistic models for the various types of observed behavior, ranging from that of the PMS, in which no stress relaxation occurs, to that in which full stress relaxation occurs by diffusion.

10.2 THE ELASTIC CONTINUUM MODEL OF INTERNAL STRESSES

In this model the matrix undergoes a uniform plastic distortion, which consists of pure shear on the primary slip system. We shall call the symmetrical shear distortion (half the glide strain) ε_p . The particles, on the other hand, can undergo only elastic distortions. The plastic distortion of the matrix is a 'stress-free strain' - it changes the shape of the holes in which the particles sit, but in the absence of the particles no stress would build up in the matrix. The theory of elastic inclusions developed by Eshelby (1956, 1961) enables one to calculate the elastic stresses and strains in the neighborhood of the particles. These stresses are as follows:

10.2.1 *The Stress Inside the Particle*

Provided the particle is elliosoidal in shape, this stress is uniform. The order-of-magnitude of the stress is $\mu\varepsilon_p$. Eshelby's theory enables one to calculate the stress exactly. If the particle is spherical in shape, with shear modulus μ^*, and is embedded in a matrix of shear modulus μ and Poisson's ratio ν , the stress inside the particle is pure shear, and is given by

$$\sigma^I = -2\gamma\mu^*\varepsilon_p \frac{\mu}{\mu^* - \gamma(\mu^* - \mu)} \qquad (10.2)$$

where γ , the accommodation factor, is given by

$$\gamma = \frac{7 - 5\nu}{15(1-\nu)} \qquad (10.3)$$

10.2.2 *The Stress in the Neighborhood of the Particle*

The particle produces a local stress field which, at large distances from it, varies inversely with the cube of the distance from the particle. The field has a complicated angular variation, with regions both of large positive and of large negative stresses. This field is called the 'constrained' field by Eshelby, because it results from the elastic reaction of the matrix which constrains the shape change of the particle. If the particle has radius r_o, and if r, θ, ϕ are spherical polar coordinates,

$$\sigma^C(r,\theta,\phi) = \frac{\mu r_o^3 \varepsilon_p}{r^3} f(\theta,\phi) + \text{terms of order } r^{-5} \quad (10.4)$$

The constrained stress has the property that its mean value is zero when it is averaged over any surface parallel to the surface of the inclusion. Thus for a rigid dislocation the constrained stress on average aids slip as much as it hinders it; of course the precise effect of the local fluctuating stress on a flexible dislocation depends very much on how the dislocation samples the stress field.

10.2.3 *The Image Stress, and the Mean Stress in the Matrix*

Because the body has traction-free surfaces (when it is unloaded) each particle which is the center of internal stresses causes an image stress - a stress-field which is divergence-free inside the matrix, but cancels the tractions resulting from the constrained stress at the surface. This stress is a long-range, quasi-uniform stress which is difficult to calculate explicitly except in certain simple cases. However, the mean stress in the matrix can be easily calculated. If we think of a plane which samples both matrix and inclusions, the mean stress over the entire plane must be zero. But the area fraction of the plane inside the particles is just f, the volume fraction of the particles. It follows that the mean stress in the matrix is the negative of the volume fraction times the mean stress inside the particles: and for the case of spherical particles considered here, we have

$$\langle\sigma^F\rangle_M = + 2\gamma f \mu \varepsilon_p \frac{\mu^*}{\mu^* - \gamma(\mu^* - \mu)} \quad (10.5)$$

where the notation $\langle\sigma^F\rangle_M$ means 'the mean stress in the matrix of a finite body'. The sense of this mean stress is such that it opposes further deformation: it is a "back stress".

The theory outlined here is appropriate to the case of spherical particles embedded in a matrix which is subject to a stress-free shape change of pure shear. In general, the stresses and strains involved are tensors, and the particle may not be spherical. For these more complicated cases, the theory has been taken to a point where explicit formulae exist for many cases of practical interest, and we refer the reader to the following papers: Tanaka and Mori (1970), Brown (1973), Tanaka, Narita and Mori (1972). Also, it has been shown that the mean stress in the matrix is linear in f to second order, i.e. there are no terms in f^2 (Brown 1973, Tanaka 1974). Our purpose in this review is to stress the principles involved in the deformation of dispersion-hardened materials, and we do not have space to give a detailed review of the mathematical theory. The description given here follows closely the paper by Brown and Stobbs (1971a). The accommodation factor γ can vary from nearly zero (for thin plates in the slip plane) to nearly unity (for fibres parallel to the tensile axis).

10.3 ANALOGUES AND INSIGHTS INTO THE ELASTIC CONTINUUM MODEL

The first and most important point is that the mean stress in the matrix can be derived from the rate of elastic energy storage. If we call the total elastic energy/unit volume resulting from plastic deformation $E(\varepsilon_p)$ then

$$(1 - f) \langle\sigma^F\rangle_M = + \frac{dE}{d\varepsilon_p} \tag{10.6}$$

Tanaka and Mori (1970) first used this equation to derive the back stress, and it has been further discussed by Ashby (1970b), Hart (1972) and Brown (1973). If all the work done on the specimen in an incremental deformation were converted into elastic energy, then the stress required to produce the deformation would be $\langle\sigma^F\rangle_M$. The true flow stress must thus be greater than this, because some of the work done goes into heat. In dislocation terms, if the dislocation producing the plastic deformation of the matrix is rigid, then the stress it feels is $\langle\sigma^F\rangle_M$ and the applied stress must be equal to (or infinitesimally greater than) this to produce deformation. If the dislocation is not rigid, it is bent by the fluctuating internal stresses and it will require a finite additional stress to force the dislocation into a critical bowed configuration which will lead to movement through large distances in the stress field. This aspect of the theory of strength is not very well understood at present; it is closely related to the problem of calculating the strength due to coherency strains. For the coherency strain problem, there is no average shear stress - the mean stress in the matrix is purely hydrostatic and cannot affect the strength. But the local fluctuating shear stresses act to strengthen the matrix, and the problem of calculating their effect (first tackled by Mott and Nabarro 1948) is only partially solved (Brown and Ham 1971).

A further insight into the stresses quoted in Section 2 may be gained from considering the electrostatic fields in a capacitor containing a gas. The applied field (analogous to ε_p) polarizes each molecule of the gas. Inside each molecule there is an electrostatic field, and if the molecule is modelled by a dielectric ellipsoid, this field is uniform - its analogue is $-\gamma\varepsilon_p$, the field inside the inclusion. Surrounding each molecule is a local dipole field, falling off inversely with the cube of the distance from the molecule, and possessing the property that its mean value (suitable defined) is zero - its analogue is the constrained strain. In addition, there is the polarization field which results from the surface of the gas; its value is proportional to the number of molecules per unit volume and to the dipole moment of each of them. This field is analogous to $\langle\sigma^F\rangle_M$. It is therefore possible to speak of 'plastic polarization' and it sometimes helps to think in this way. However, it must always be borne in mind that the elastic continuum model for plastic deformation is dissipative in a way that the electrically polarized gas is not. In any practical case, the work done by plastic deformation is far greater than the stored elastic energy given by eqn. (10.6). In the theory of plastic polarization, the quantity γ in eqns. (10.2, 10.3 and 10.5), the so-called 'accommodation factor' plays a role analagous to the 'demagnetization factor' or 'shape factor' in electrostatics or magnetostatics.

An important and difficult question concerns the relationship between the continuum model and the dislocation model. It seems clear that if the slip plane spacing is small compared to the diameter of a particle, so that one can imagine the Orowan loops to be smeared into a continuum in the interface, then the continuum model will be accurate. On the other hand, if the slip plane spacing is coarse, so that the loops around a given particle lie only in one or two slip planes, then the continuum model cannot apply. It is easy to see that the continuum model provides an estimate of the back stress which is lower than the value appropriate to the coarse slip model. This is because the net area of loops in a pile-up will be greater than their area when the loops are uniformly distributed in the interface. Using eqn. (10.6), it can be seen that coarse slip will produce a larger elastic energy than fine slip, so if the crystal contains sufficient sources of dislocations it will produce slip on a scale which is fine compared to the particle size. However, at the time of writing it does not seem possible to give a convincing account of the slip plane spacing, and it is very difficult to estimate the error involved in using the continuum model.

Before leaving this theoretical discussion, it is worth pointing out the connection between these ideas and mechanical memory. The matrix in the dispersion-hardened material is subject to a uniform strain, on which is superposed local fluctuating strains. The uniform strain, if it could be measured by X-rays for instance, enables one to deduce the original shape of the composite after 'plastic' deformation. Furthermore, the stress associated with the uniform strain are in principle capable of driving viscous or other plastic processes and so are capable of returning the material to its original shape. Thus the continuum model of plastic deformation in a two-component system is closely related to viscoelasticity. To illustrate this further, let us briefly consider a shear stress σ applied to a two component system with equal shear moduli μ, but different coefficients of viscosity, η_1 and η_2. By coefficient of viscosity we mean that the component is capable of linear Newtonian creep in response to an applied stress. This flow is essentially irreversible, and the creep strain introduced in this way is not accompanied by an increase in stress - such a strain is similar to a 'plastic' strain in that it is a stress-free or transformation strain. If the volume fraction of component 2 is f, the equations governing flow in this system are as follows:

$$\begin{aligned}
\varepsilon_1 &= (1/2\mu)\,(\sigma - 2\,\gamma\,f\,\mu\,(\varepsilon_1^p - \varepsilon_2^p)) \\
\dot{\varepsilon}_1^p &= (1/2\eta_1)\,(\sigma - 2\,\gamma\,f\,\mu\,(\varepsilon_1^p - \varepsilon_2^p)) \\
\varepsilon_2 &= (1/2\mu)\,(\sigma + 2\,\gamma\,\mu\,(\varepsilon_1^p - \varepsilon_2^p)) \\
\dot{\varepsilon}_2^p &= (1/2\eta_2)\,(\sigma + 2\,\gamma\,\mu\,(\varepsilon_1^p - \varepsilon_2^p))
\end{aligned} \tag{10.7}$$

Here the irreversible strains are written with the superscript p for plastic. Now what is measured is the <u>total</u> strain,

$$\varepsilon = (1 - f)\,\varepsilon_1 + f\,\varepsilon_2 + (1 - f)\,\varepsilon_1^p + f\,\varepsilon_2^p \tag{10.8}$$

Let us imagine that the matrix is nearly perfectly elastic, and the inclusions nearly perfectly viscous: that is, let us take the limit as $\eta_1 \to \infty$ and $\varepsilon_1^p = 0$. Then some algebra shows that the composite behaves exactly as a standard linear solid (Zener 1948) and satisfies the linear differential equation

$$(\sigma + \tau_\varepsilon \dot{\sigma}) = M_R (\varepsilon + \tau_\sigma \dot{\varepsilon}) \qquad (10.9)$$

where the unrelaxed modulus, $M_u = 2\mu$; the relaxed modulus, $M_R = 2\mu \, \gamma/(f + \gamma)$; and the constant strain relaxation time, $\tau_\varepsilon = \eta_2/\mu(f + \gamma)$; and the constant stress relaxation time, $\tau_\sigma = \eta_2/\mu\gamma$. The physical meaning of this model is very clear: the elastic matrix retains a memory of the shape the composite has in the unstressed state, and it is capable of driving the viscous elements back to their original configuration. It is useful to classify all solids which on deformation contain elastic strains which are proportional to the 'plastic' shape change as 'Perfect Memory Solids' or PMS for short. In the case of the standard linear solid the fact that it has a perfect memory is demonstrated by observing that after a long enough time it will return to its original shape - all the plastic strain is recoverable. However, there are many instances where a solid has a perfect memory in the sense that an X-ray examination of internal stresses will enable one to deduce the original shape, but subsequent deformation of the solid allows plastic relaxation and prevents all the strain from being recoverable. Thus we shall use PMS to stand for a solid for which the plastic strain is a state variable - the solid contains elastic energy which is zero only in the undeformed state, and whose value is uniquely related to the external shape change.

In dispersion-hardened systems, many mechanisms exist which can destroy the memory. The particles may yield, twin or fracture; diffusion may occur to relieve the elastic stresses; the matrix in the neighborhood of the particles may yield and the consequent local plastic flow relieve the stresses. In the next section we shall briefly review the state of knowledge of these mechanisms gained by electron microscopy.

10.4 RELAXATION PROCESSES

In this section, we shall review briefly the electron microscopy of deformed dispersion-hardened single crystals. It is important to bear in mind that the foil thickness used in this work is often comparable with the particle size, and in experiments so far is at most ten times the particle diameter, so dislocation structures must often suffer from the escape of dislocations to the foil surfaces given the large internal stresses present. Nevertheless, a consistent picture emerges which one might hope to correlate with the mechanical properties. The structures which have been observed are as follows:

10.4.1 *Orowan Loops*

In most systems, Orowan loops around particles are conspicuous by their absence and one must conclude that in the thin foil relaxation is very nearly complete. However, in some alloy systems Orowan loops are clearly observed. Figure 10.1 shows

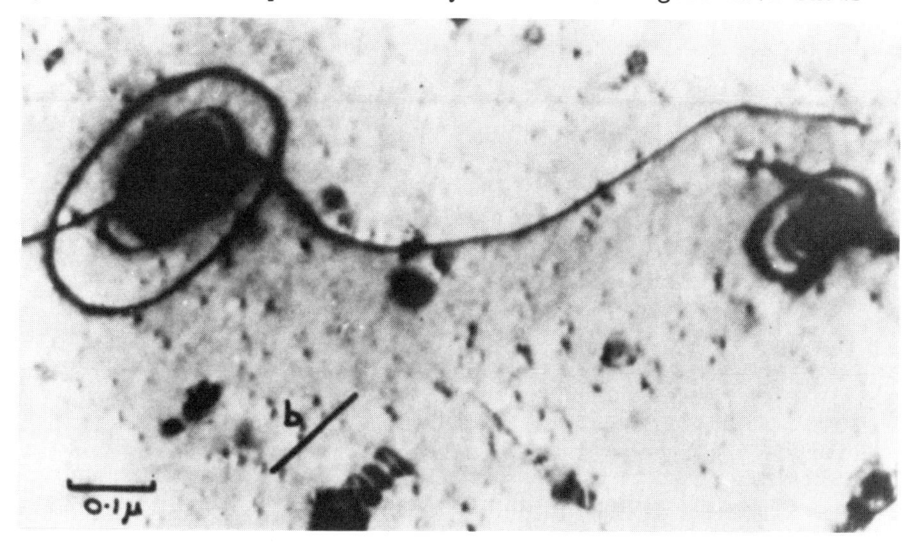

Fig. 10.1 Orowan loops as viewed on the primary slip plane. They surround Al_2O_3 particles in a copper alloy containing 30% Zn and 0.049% Al, deformed at 77°K to a resolved shear strain of 0.06 (Humphreys and Hirsch 1970, Courtesy of the Royal Society)

an example taken from the work of Humphreys and Hirsch (1970) on Cu-Zn-Al_2O_3. Orowan loops have also been seen by Merrick (1962) around Ni_3Al particles in a Ni-Cr-Ti-Al alloy. Several factors may operate to stabilize the Orowan loops: the solution hardening may provide a high friction stress, and so prevent their loss to the foil surface; the solute may lower the stacking-fault energy and so prevent cross-slip; and at least in the brass it is clear that the slip-plane spacing is large, so that loops from a single slip plane can be seen to stand off around the particles, and the visibility of the loops is enhanced.

10.4.2 *Stacks of Prismatic Loops on the Primary Slip Plane (A Structure)*

It was originally proposed by Hirsch (1957) that dislocations could by-pass particles by cross-slip, leaving behind prismatic loops. The modern version of this mechanism recognizes that it would be energetically unfavorable for a single dislocation to cross-slip, but that when Orowan loops are stacked around a particle it is energetically favorable to remove all but one or two of them by cross-slip and conversion to prismatic loops. The reader unfamiliar with these ideas should refer to the paper by Hirsch and Humphreys (1970) or to the review by Brown and Ham (1971). A schematic version of this mechanism is shown in fig. 10.2A. A single shear loop (the Orowan loop) is converted into a dipole of prismatic loops - one of vacancy character and one of interstitial character. It is clear from the discussion of Section 10.2 that both the stress inside the particle and the image stress due to the Orowan loop are removed by this process.

Stacks of loops of primary Burgers vector have been observed by Humphreys and Hirsch (1970) and earlier by many other authors. For small particles, and particularly at low temperatures, the number of these loops approaches that which one would expect if every Orowan loop were converted into one prismatic loop. Figure 10.3 shows micrographs of the primary slip plane of a deformed Cu-SiO_2 specimen (Brown and Stobbs 1971b). At almost every particle a stack of loops can be seen, and the number of these increases with strain. Hirsch and Humphreys (1970) present loop counts which are used in Section 10.6 of this review. For brevity, we shall call the stacks of prismatic loops of primary Burgers vector A structure.

In the system Cu-Zn-Al_2O_3, particularly clear micrographs can be obtained, and it is abundantly clear that the loops are interstitial in nature. The vacancy loops have evidently been carried off as jogs on glide dislocations, or possibly (in view of later work, see Section 10.4.4 below) they have condensed to form a void in the particle-matrix interface. Hirsch and Humphreys (1970) and MacEwen, Hirsch and Vitek (1973) present arguments which suggest that residual thermal stresses

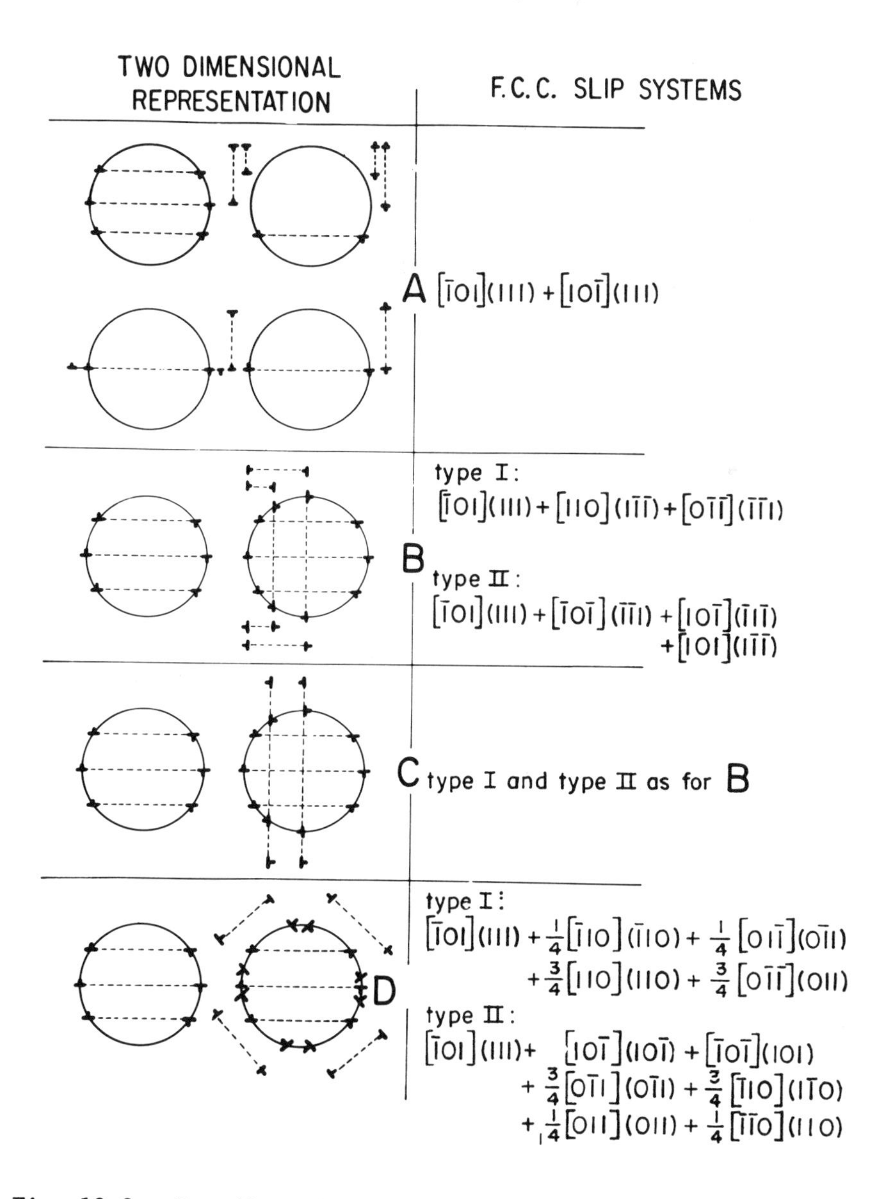

Fig. 10.2 Two dimensional representations of the various possible modes of plastic relaxation. The equivalent FCC slip systems necessary for the given types of accommodation are given with each mode (Brown and Stobbs 1971b, Courtesy of Taylor & Francis).

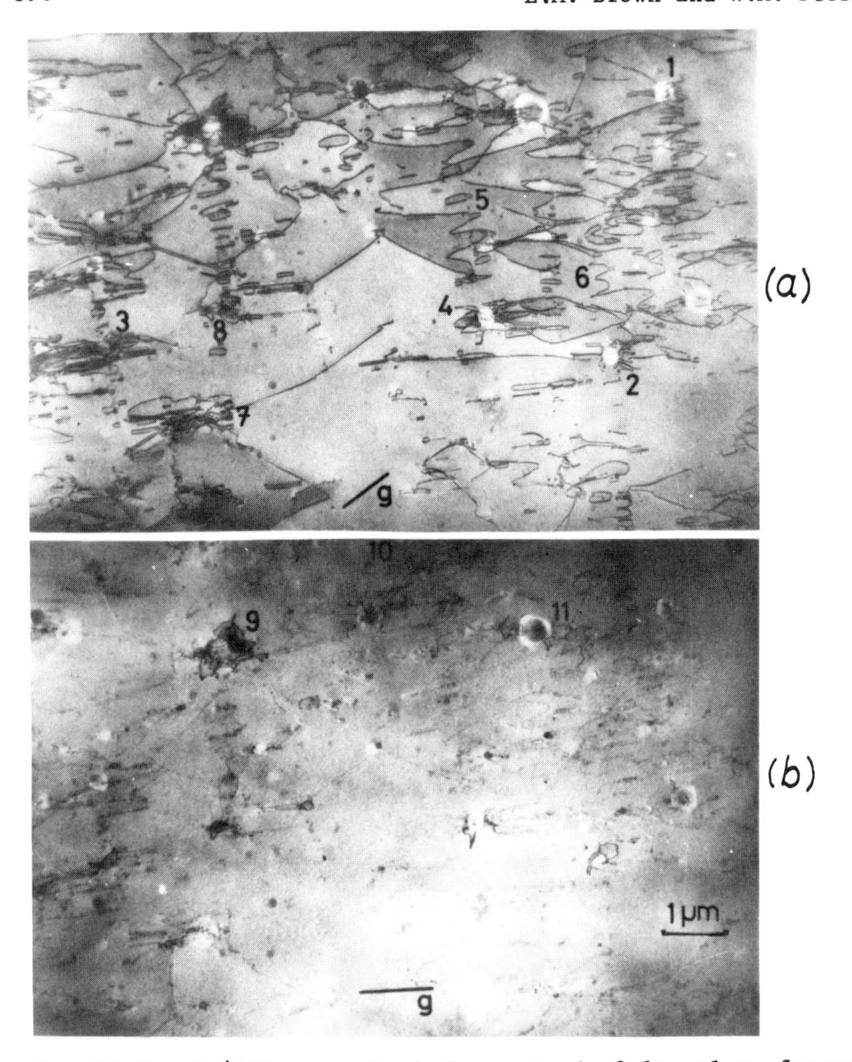

Fig. 10.3 Cu/SiO_2; resolved shear strain 0.1; volume fraction 0.3%; the primary slip plane. The A structure, prismatic loops of primary Burgers vector aligned in the slip direction, is in contrast in a) and invisible in b). In the latter micrograph the B structures are seen at the larger particles 9 and 10. (The particle at 11 is partially out of the foil.) In a) interaction of the loops with primary dislocations to form helices may be seen at 5 and 6, while the characteristic interaction of the loops with the remaining internal stresses at the particles may be seen at 1,2,3,4,7 and 8. (Brown and Stobbs 1971b, Courtesy of Taylor & Francis)

around the particles attract vacancy loops and repel interstitial loops and so account for the observations. It seems unlikely to us that thermal stresses can account for the very large number of vacancy loops apparently absorbed at the particle, although they certainly can account for one or two of them. This aspect of the observations remains somewhat ill-explained.

An observation which it might be argued could give some indication of the active slip plane spacing is that some of the smaller particles have no dislocation structure at all associated with them. (A few examples of such particles, which are apparently contained within the foil, may be seen in fig. 10.3.) Furthermore, Hirsch and Humphreys' (1970) exhaustive loop counts appear to imply that, at room temperature in copper containing $A\ell_2O_3$ particles of mean radius about 250 Å, the fraction of particles with no A structure is reduced from two thirds to about a half on increasing the resolved shear strain from 0.04 to 0.10. It is, however, difficult to interpret the results unambiguously because of the presence of a number of loops not associated with any particular particle. In any case the conclusion that the slip plane spacing decreases with strain can only be made if the prismatic loops are produced during deformation. In view of the results described in Section 6 this seems unlikely.

10.4.3 *Secondary Dislocation Structures of Shear Character (B Structures)*

Extensive observations of dislocation tangles of secondary dislocations have been made; the observations are reviewed by Brown and Stobbs (1971b). These structures extend perpendicular to the primary slip plane and they contain certain characteristic Burgers vectors, in particular the two vectors on the cross-slip plane are prominent in a structure called BI by Chapman and Stobbs (1969) and the vector perpendicular to the primary Burgers vector in a structure called BII. These dislocation structures show very characteristic small rotations of the lattice inside the tangle (Chapman and Stobbs 1969). At a shear strain of 0.08 the rotation is of a rather variable magnitude $\lesssim 10^{-2}$ radians and is of a definite sense - its main component is in the same sense as the sense of rotation of a particle caused by Orowan loops. Brown and Stobbs concluded on the basis of rather indirect evidence that these structures must result from the relief of stresses by shear on secondary systems in such a sense as to cancel the shear caused by the

Orowan loops. Their analysis is summarized in fig. 10.2B, which shows the Burgers vectors expected for structures BI and BII; these are precisely the ones observed. At present, however, it is not clear whether the rotations associated with the structures show that the local forest is an active obstacle or that some further softening process occurs on unloading. The structures can be seen in fig. 10.3 although they are best viewed in foils of other normals. They appear around larger particles, so that particles smaller than about 3000 Å tend to have a well-developed A structure, whereas particles larger than this have well-developed B structure. However, the structures are very variable in appearance, and although part of this variation may be due to the effect of foil surfaces, one forms an impression of a structure which is intrinsically rather variable. Figure 10.4 shows a section with the cross-slip plane normal. This aspect is very nearly the same as is drawn schematically in fig. 10.2. The structure can be seen to extend perpendicular to the trace of the primary slip plane and to contain roughly equal numbers of the non primary Burgers vectors on the cross-slip plane. Other views of this structure are to be found in the paper by Brown and Stobbs (1971b). Here we wish to present a piece of new evidence from some recent experiments. As fig. 10.2 clearly shows, the particle is rotated by the action of the secondary shear. This rotation can be seen when the particles are crystalline, although not then they are amorphous as are the silica particles previously studied. Figure 10.5 shows a diffraction pattern taken from a deformed specimen of Cu-$A\ell_2O_3$. The normal to the pattern is near to that of the cross-slip plane, so the aspect of the particle is similar to that in fig. 10.4. It can be clearly seen that the $A\ell_2O_3$ particle is twinned and rotated with respect to the copper matrix; furthermore the copper spots are streaked in a characteristically asymmetrical way indicating local rotations of the same sense inside the plastic zone. These rotations are of a sense opposite to the sense of rotation of the tensile axis in a tensile test: the particles tend to stay stationary while the crystal axes rotate. The magnitude of the rotation is $\sim 6.10^{-2}$ radians which is about half what would be expected for the applied strain, assuming that the B structure accounts for all the stress-relief and neglecting whatever is due to twinning. Neither the twins nor the rotations are observed in undeformed specimens.

It is of some interest to ask why the larger particles produce B structure. One obvious reason which clearly works in

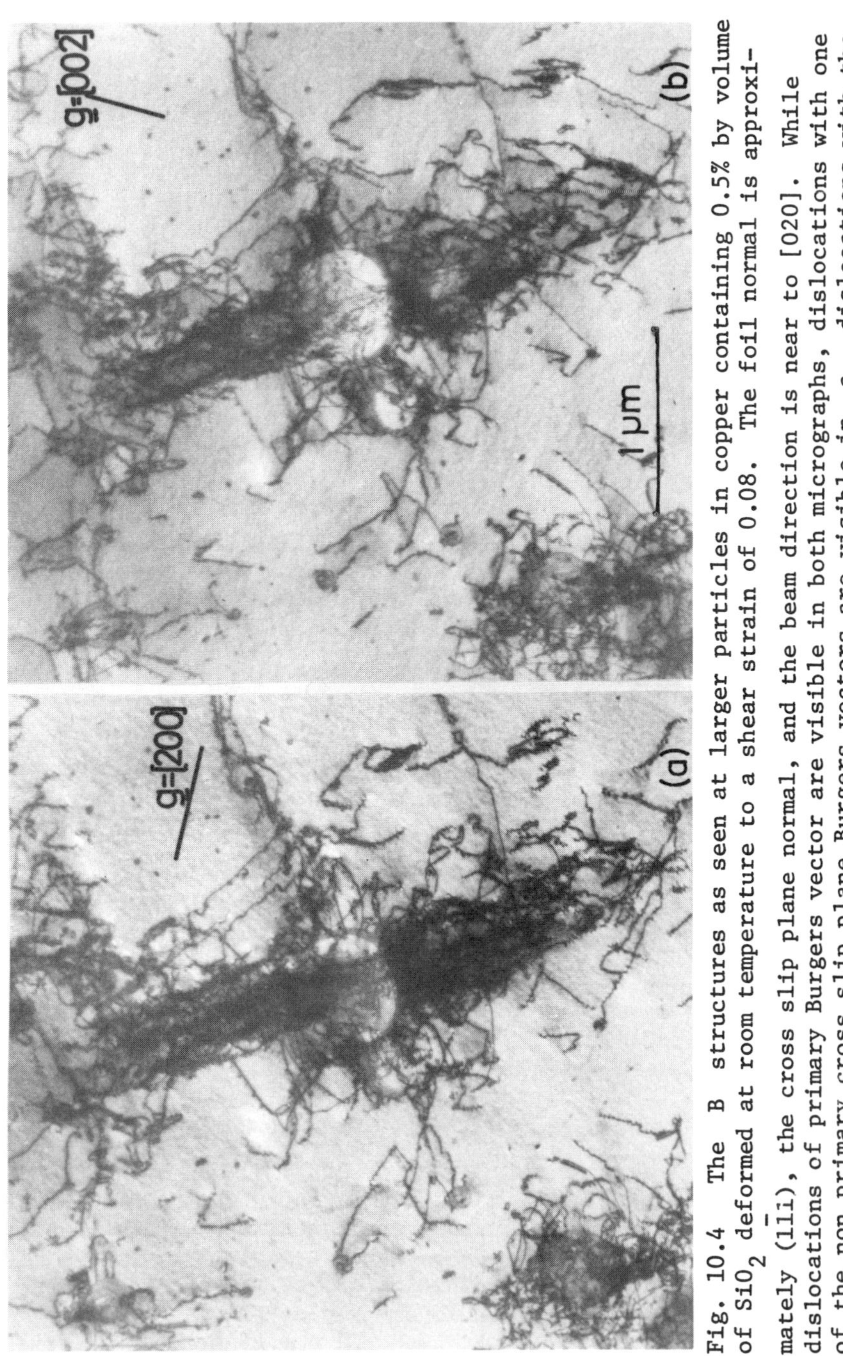

Fig. 10.4 The B structures as seen at larger particles in copper containing 0.5% by volume of SiO_2 deformed at room temperature to a shear strain of 0.08. The foil normal is approximately $(1\bar{1}1)$, the cross slip plane normal, and the beam direction is near to [020]. While dislocations of primary Burgers vector are visible in both micrographs, dislocations with one of the non primary cross slip plane Burgers vectors are visible in a , dislocations with the other in b . The BII structure is in contrast in both a and b . Some radiation damage may be seen since the foil was observed at 500 kV to facilitate the observation of the structures at large particles.

Fig. 10.5 Diffraction pattern near the cross slip plane normal of a copper specimen containing $A\ell_2O_3$ particles deformed to a resolved shear strain of 0.25. The sense of rotation of the $(1\bar{1}1)$ copper normal may be seen by the asterism to be into the paper. The copper spots are marked. The unmarked spots are the $A\ell_2O_3$ pattern which may be seen to be twinned or heavily faulted. The relative rotation of the two patterns is about 4° compared with an 'expected value', assuming that all stress relief is by the B structures, of about 13°. Other examples show rotation ranging from 4° to 7°. No relative rotation of the diffraction patterns nor twinning, is seen in undeformed specimens.

many cases is that larger particles in any population of slowly coarsening particles tend to be associated with 'grown-in' dislocations and so have sources of secondary slip near them. This seems to be an inevitable result of competitive growth where pipe diffusion along dislocations may play a role, but there is also a statistical aspect which should not be neglected: large particles have a large surface area and will thus tend to have a dislocation threading them. If the grown-in dislocation density is $10^8 cm^{-2}$, then a particle 3000 Å in radius will have on average one dislocation threading its interface.

The B-structures remain more-or-less constant in size over the strain range in which they have been studied (0 to 0.5 shear strain) and this distinguishes them in a fundamental way from the A structures, which get longer linearly with strain. However, as Brown and Stobbs (1971b) point out, it seems difficult to present a clear model for the dislocation arrangement in the B-structure. Although the type of dislocation is understood, and its approximate disposition in the matrix around the particle, it has not been possible to draw a model which can account for the detailed appearance of the micrographs. One observation is clear, however: the dislocation structure outside the particle contains numerous primary dipoles which begin and end on secondary dislocations, indicating that the B-structure must be cut by the primary glide dislocations and so it must produce forest hardening. One other fact is clear: the dislocations outside the particle are <u>not</u> in the form of shear loops, as shown in fig. 10.2C. Shear loops produce the wrong sense of rotation.

10.4.4 *Voiding and Punching Mechanisms*

Another type of dislocation structure has been observed by Humphreys and Stewart (1972). They worked with Cu-Zn-SiO_2 and found very clear secondary structures, with prismatic loop character. Figure 10.6 shows the structure viewed in the cross-slip plane; one can see a stack of prismatic loops whose Burgers vector is one of the vectors in the cross-slip plane. At the same time, one can see a void opening in the particle-matrix interface. The complete array of dislocations at a shear strain of approximately 0.04 has been analysed by Humphreys and Stewart, who are able to present the detailed model shown in fig. 10.7. As well as loops in the cross-slip plane, some loops in the primary slip plane of non-primary Burgers vectors are observed. The reason for the operation of this stress-relief mechanism is most clearly appreciated from a drawing due to Ashby (1966) reproduced in fig. 10.8. The action of the tensile stress tends to separate the particle-matrix interface where it is normal to the tensile axis, but it tends to produce prismatic punching of interstitial loops perpendicular to the tensile axis. Clearly the FCC geometry changes the angles somewhat, but in general this very simple prediction of Ashby is well borne out. In fig. 10.2 a closely

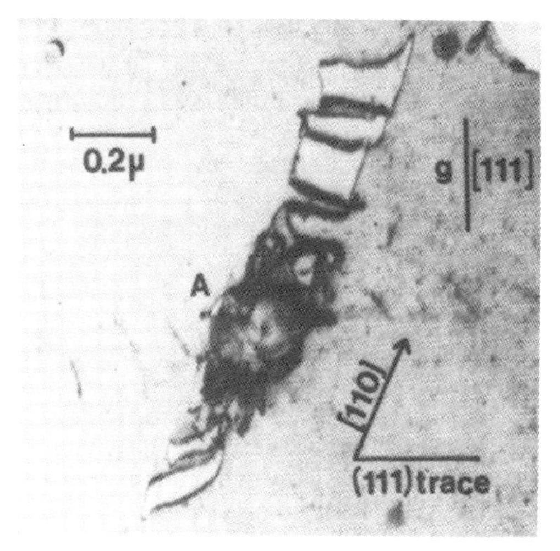

Fig. 10.6 The secondary prismatic loop array in a cross-slip plane section described as the 'δ' structure by Humphreys and Stewart (1972). The specimen, containing 20wt%Zn and a dispersion of SiO_2 particles, was deformed to a shear strain of 0.04. The [110] loops have twisted on their glide prism and thus have some shear character. A void may be seen at A . (Humphreys and Stewart 1972, Courtesy of North Holland)

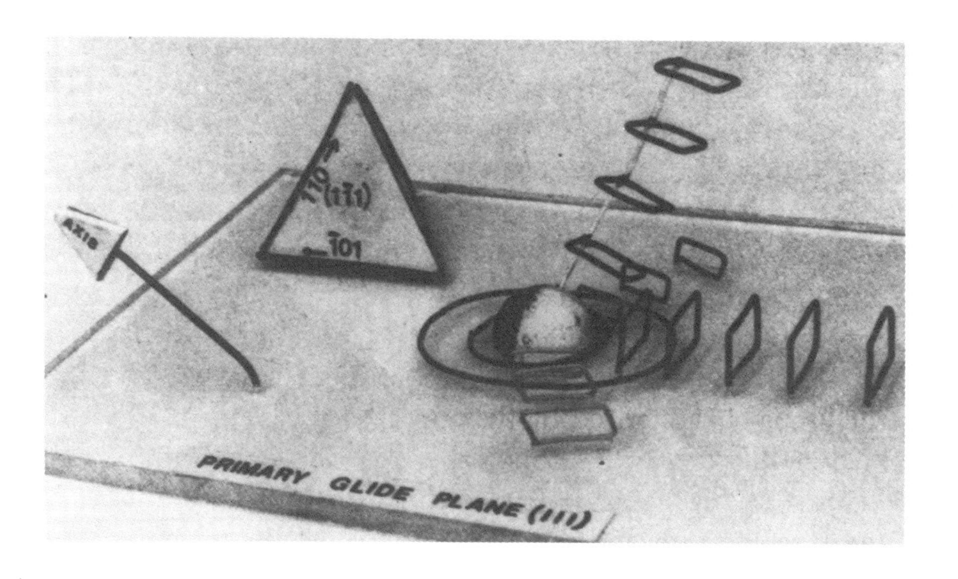

Fig. 10.7 A wire model of the full prismatic array and void structure for the top half of a particle as seen in Cu-Zn-SiO_2, prior to the secondary loops attaining shear character. A void is indicated by the darkened surface of the particle and the direction of the tensile axis is shown (Humphreys and Stewart 1972, Courtesy of North Holland).

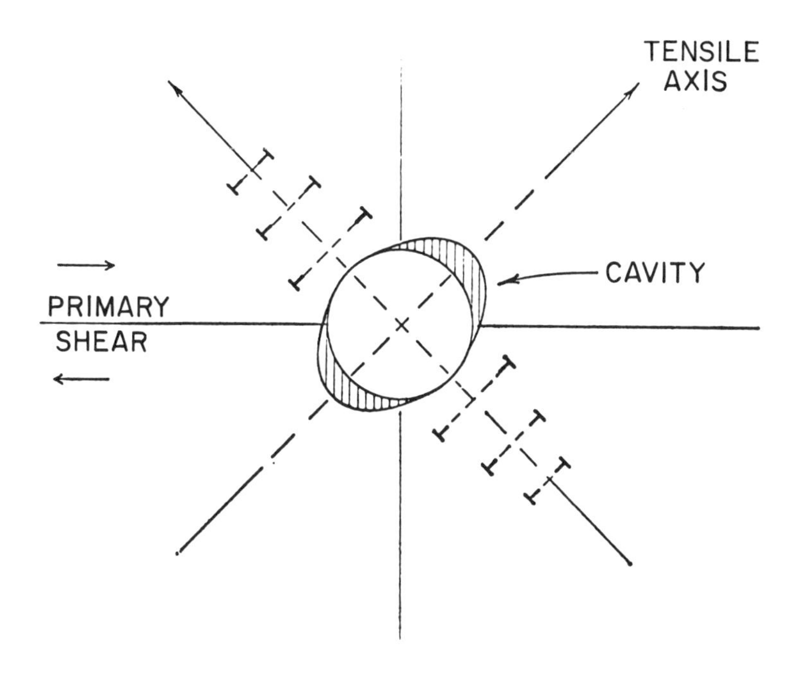

Fig. 10.8 As deformation proceeds, the tensile stress in the particle matrix interface increases until fracture occurs, and a cavity opens up. The cavity increases in volume as straining continues and interstitial prismatic loops continue to be punched out perpendicular to the tensile axis direction (Ashby 1966, Courtesy of Taylor & Francis).

related mechanism is indicated - the D mechanism. The Burgers vectors observed by Humphreys and Stewart fit quite accurately with those predicted by mechanism DI , except that the vacancy loops are replaced by voids - a possibility not considered by Brown and Stobbs. Brown and Stobbs point out that the punching mechanism does not relieve the long-range stresses, although the stresses in the neighborhood of the interface are lowered. This provides a partial explanation for the tendency of the secondary prismatic loops to assume some shear character. At larger strains, Humphreys and Stewart observe that the punching mechanism changes to something more like the B mechanism, and it may well be that even with voiding the punching mechanism is incapable of full stress relief.

The operation of the punching and voiding mechanism seems to be bound up with the addition of Zn to the matrix. This may perhaps lower the strength of the interface, and make cross-slip more difficult, so that other mechanisms are suppressed. It is clear from fig. 10.6 that the spacing of the slip planes in the Cu-Zn-SiO_2 system is much larger than the spacing in Cu-SiO_2: there appear to be few if any primary dipoles tangled up in the secondary structure.

Finally, it should be mentioned that it is difficult to see very small voids in the electron microscope, unless the foil is examined in a heavily absorbing orientation. Unmistakable evidence for the existence of voids seems to be the asymmetrical stack of prismatic loops, all interstitial, in the cross-slip plane. This observation is very different from the appearance of the B structure (compare fig. 10.4 with fig. 10.6) and is one way of distinguishing the two mechanisms. Other distinguishing features are the secondary Burgers vectors in the primary slip plane, which play an important role in the D structures but are theoretically absent in the B structures. However, in a given dislocation structure reactions can take place, so that the B structures as observed in Cu/siO_2 always contain a small fraction of dislocations of these Burgers vectors, though they never look like the prismatic loop stacks as seen in Cu-Zn-Sio_2. It is interesting to note that the A structure also shows a stack of interstitial prismatic loops, so that one wonders if a small void exists in that case as well. No direct evidence supports this idea.

10.4.5 *Void Formation*

It is clear from the examples quoted so far that the strength of the interface must be an important variable in deciding the mechanism of the stress-relief. Even in the Cu-SiO_2 system voids appear after a certain critical shear strain, which is estimated to be between 0.2 and 0.25 (Atkinson 1973). These voids were first described by Palmer and Smith (1968) and their appearance is very different from the angular void of fig. 10.6 - these voids have a rounded appearance and appear to be approximately radially symmetrical about the tensile axis. It seems clear that any dislocation mechanism can afford only limited stress relief, and at a high enough strain gives way to voiding. This void formation is, of course, the beginning of the fracture process, but the total strain to fracture is much larger than the strain to nucleate a void, because the voids have to grow and link up to produce failure. However, the fact that voids are nucleated at all after extensive stress relief by dislocation motion shows that the material inside the dislocation tangles is work-hardened and can support quite large elastic stresses - large enough to break the interface.

10.4.6 *Diffusional Relaxation*

Stobbs (1973) has used the electron microscope to measure the rate of spheroidization of SiO_2 particles in copper and thus to estimate an interface diffusion coefficient. It appears that the diffusion is controlled by an activation energy indistinguishable from that for surface diffusion in copper, as is the pre-exponential factor. Using this diffusion coefficient, it is possible to calculate the range of temperature at which diffusional relaxation will occur: this is about 520°K for particles of 3000 Å diameter, and about 470°K for particles of 300 Å diameter. Experimentally, it is known that for temperature rising above this range the flow stress starts to drop below the Orowan stress (eqn. 10.1) and in electron micrographs stress-relief structures become less prominent (Shewfelt 1972). There seems to be no doubt that diffusional relaxation is the major stress relief mechanism at higher temperatures than those just quoted. At room temperature, this process is very slow: Stobbs (1973) estimates that for a 100 Å particle diffusional stress relief would take several years at 300°K.

10.4.7 *Other Mechanisms*

It is clear that a wide variety of stress-relief mechanisms can operate, even if the particle is relatively inert. However, particle twinning has been observed in Cu-$A\ell_2O_3$ (see Section 10.4.3 above) and the fracture of particles is commonly observed in fibre hardened systems. The latter phenomenon has been extensively studied in the Fe-Fe_3C system (Lindley, Oates and Richard 1970) and in fibre composites. Clearly plastic flow around a needle-like particle is difficult - a fact which forms the basis for strengthening ductile materials by strong fibres (Kelly 1966). It is thus a mistake to imagine that there is any one 'universal' or 'principal' stress relief mechanism. There are many choices, and the decision as to which path a particle will follow must be affected by many microscopic variables: stacking-fault energy, interface adhesion, dislocations in the vicinity, surface-steps or other microscopic stress-raisers in the interface. Clearly also the macroscopic variables of gross particle geometry, temperature, deformation modes available, and friction stress also play a role. It is most unlikely that theoretical attacks on this problem will be very profitable, because it is well-known that nucleation problems are the most intractable of atomic problems, involving detailed models of dislocation cores, interatomic potentials, and so on. It seems wiser at the present time to rely on empirical observations and to try to relate them to macroscopic behavior.

10.5 THE BAUSCHINGER EFFECT AND PLASTIC RELAXATION

For very small strains, small enough that the stress in the neighborhood of the particle is insufficient to actuate a stress relief mechanism, the material will behave like a PMS. At larger strains, plastic relaxation will occur. This point has been recognized by virtually every study of work-hardening in dispersion hardened materials, starting with Fisher, Hart and Pry (1953). However, as Ashby (1966) was the first to point out, the mechanical behavior of these systems can be most easily understood if the stresses around a particle are completely relaxed, and every Orowan loop has been converted into a dislocation loop which acts as a forest obstacle to further slip. Similarly, Hirsch and Humphreys (1970a) have proposed a model in which Orowan loops play no part in the hardening. Good agreement with experimental data on the work-hardening characteristics has been claimed by all authors, so that a rather confusing state of affairs has arisen. The origin of this confusion is the fact that the work-hardening curve is expected to be rather insensitive to the mechanism of hardening (Brown 1973); we shall see more evidence of this in the next section. Additional data are required to distinguish the various mechanisms. Our view (Atkinson, Brown and Stobbs 1974) has been that the main additional evidence required is a measure of the mean stress or strain in the matrix. Two very important papers (Wilson and Konnan 1964, Wilson 1965) showed very clearly that the matrix after substantial deformation contains very large internal stresses, capable of accounting for a significant portion of the total hardening. Furthermore, these papers showed that the internal stresses were closely bound up with the very large Bauschinger effect characteristic of work-hardening in dispersion hardened materials.

Let us suppose that the hardening consists of two components: a 'frictional' component, which may include forest hardening, or the Orowan stress; and the mean stress in the matrix which is directional in its effect. The flow stress in the forward direction is equal to the sum of the friction stress plus the internal stress. On unloading and reverse straining, dislocation flow will not begin until the stress felt by a dislocation has been lowered by twice the friction stress: thus the flow stress on unloading is equal to the internal stress minus the friction stress. This simple model suggests that the difference between the forward flow stress and the modulus of the backward flow stress should be equal to twice the mean internal stress: $|\sigma_{forward}| - |\sigma_{backward}| = 2\langle\sigma^F\rangle_M$.

It has been known for some time that actual Bauschinger tests produce stress strain curves more complicated in appearance than the above model can account for. A good discussion is given by Orowan (1959) who divides the stress-strain curve on reverse straining into two regions: a 'microstrain' region and a permanent softening. Figure 10.9 shows the shape of the stress-strain curves. In the permanent softening region, the forward and backward stress-strain curves become parallel when plotted against cumulative strain: i.e. if $|\varepsilon| = |\varepsilon_{forward}| + |\varepsilon_{backward}|$,

$$\Delta\sigma_p = \lim_{|\varepsilon|\,\mathrm{large}} \left\{ \left|\sigma_{forward}(|\varepsilon|)\right| - \left|\sigma_{backward}(|\varepsilon|)\right| \right\} \tag{10.10}$$

where $\Delta\sigma_p$ is a well-defined stress difference independent of $|\varepsilon|$ for a given $|\varepsilon_{forward}|$. The 'microstrain' region may by typically observed over a reverse strain interval equal to about 1/3 the forward strain, so $|\varepsilon|$ has to be greater than about $4\,\varepsilon_{forward}/3$ to achieve permanent softening. From the arguments given above, we must expect the permanent softening to be a measure of the internal stress in the matrix. Recent studies by Atkinson (1973) on dispersion-hardened copper and by Lilholt (unpublished) on Cu-tungsten fibre composites show that, when the permanent softening is well-defined, it can be correlated with the mean stress in the matrix, as calculated from eqn. (10.5), if it is simply assumed that

$$\Delta\sigma_p = 2\langle\sigma^F\rangle_M \tag{10.11}$$

In eqn. (10.11), the mean stress in the matrix is that existing at the strain where reverse straining starts. Detailed empirical arguments for the correctness of eqn. (10.11) have been

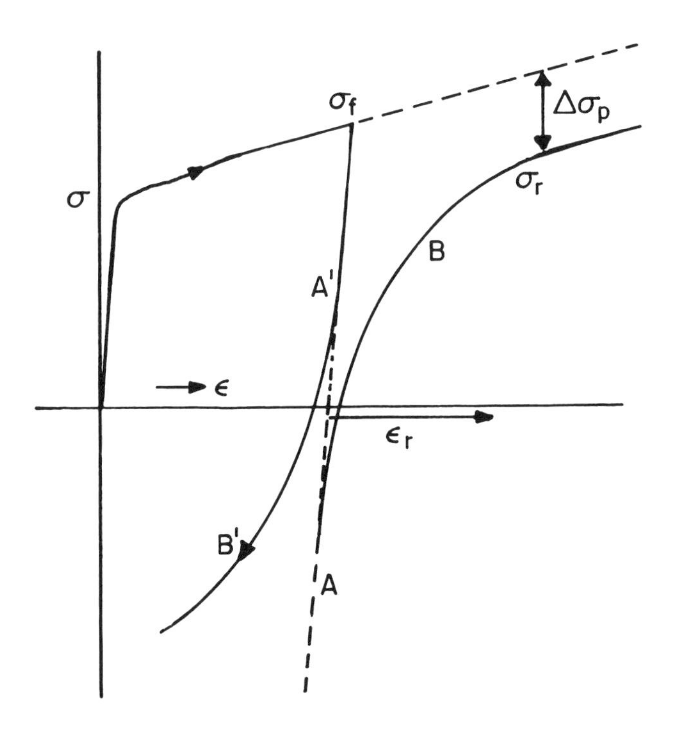

Fig. 10.9 The Shape of the Stress-strain Curves (Schematic). The backward curve, A'B', is replotted, A B , as a function of the cumulative strain, with the backward flow stress taken to be positive, to show the well-defined difference between this and the extrapolated forward curve.

presented by Atkinson, Brown and Stobbs (1973, 1974), who show that the equation can be regarded as a 'calibration' of Bauschinger tests in terms of Wilson's (1965) absolute measurements of internal stress by X-rays. Evidence for the correctness of eqn. (10.11) will be given below, but the detailed arguments cannot be presented here. The sceptical reader can interpret the rest of this review in a qualitative way, without accepting the quantitative statements, if he feels that eqn. (10.11) is not sufficiently justified. There is no doubt that a detailed understanding of the equation, in terms of dislocation mechanics, does not exist at present. In what follows, we shall assume that the mean internal stress, calculated as above, is not merely a qualitative measure of the Bauschinger effect (which it obviously is) but that it is a quantitative measure of the internal stress.

The Bauschinger effect distinguishes between the various models that have been proposed for the work-hardening. Let us call the proportion of plastic strain which is unrelaxed ε_p^*. In terms of ε_p^*, and the continuum model, the number of Orowan loops encircling a particle of radius r_o is given by

$$n = 4\,\varepsilon_p^*\, r_o/b \qquad (10.12)$$

For the PMS, $\varepsilon_p^* = \varepsilon_p$: all the Orowan loops are present, and the external shape change, ε_p, is a measure of the internal stress. When plastic relaxation occurs, ε_p^* becomes less than ε_p. For a pure forest model, $\varepsilon_p^* = 0$. But, combining eqn. (10.11) and eqn. (10.5), ε_p^* can be measured from the permanent softening:

$$\varepsilon_p^* = \frac{\Delta\,\sigma_p}{\gamma f \mu \left(\dfrac{\mu^*}{\mu^* - \gamma\,(\mu^* - \mu)} \right)} \qquad (10.13)$$

Systematic measurements of the Bauschinger effect in dispersion-hardened single crystals of copper have been made by Atkinson (1973) and by Gould, Hazzledine, Hirsch and Humphreys (1973). The two sets of experimental results show substantial agreement as to magnitudes and qualitative behavior. Here we present Atkinson's results, as analyzed in terms of eqn. (10.13), by Atkinson, Brown and Stobbs (1973, 1974). Figure 10.10 shows ε_p^* as a function of ε_p for strains from 0.01 to 0.10, and for four temperatures: 77^oK, 300^oK, 400^oK and 500^oK. It is clear that there is a strain range at 77^oK and a smaller one at 300^oK where $\varepsilon_p^* = \varepsilon_p$, as expected. This range is the PMS range. The reader should consult the original papers for details, but the authors demonstrate that within this range of strain ε_p^* depends only on volume fraction (linearly) and not on particle radius; furthermore, the elastic constants of the particle enter eqn (10.13) in such a way that all the particles examined ($A\ell_2O_3$, BeO and SiO_2) produce results which lie on the

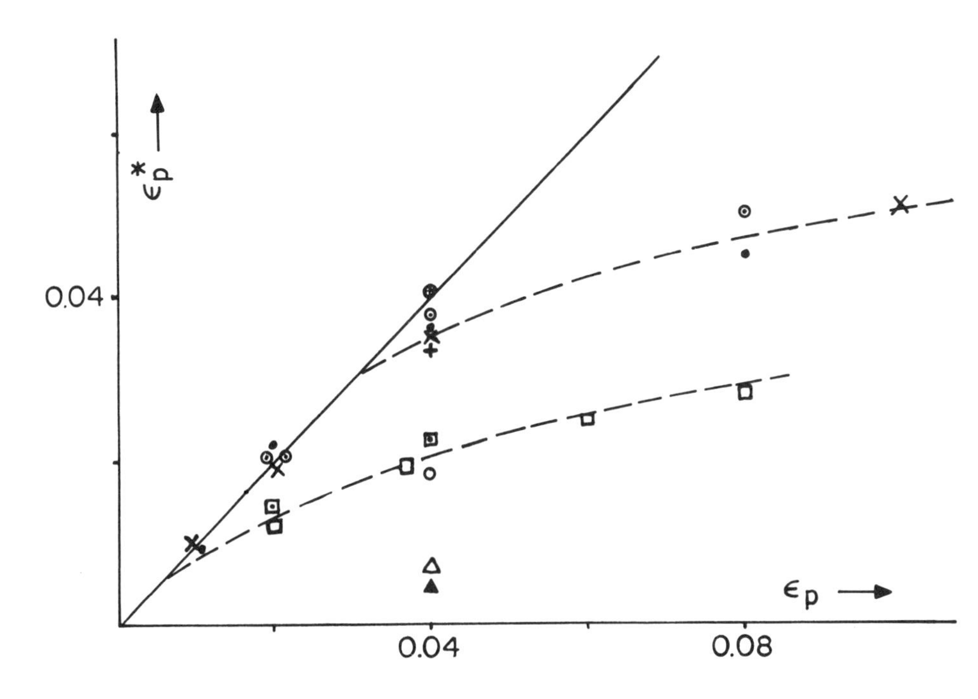

Fig. 10.10 ε_p^*, the unrelaxed strain, as a function of ε_p.

Dispersoid	Data Point for given °K 77	300	400	500	f %	$\frac{\mu^*}{\mu-\gamma(\mu^*-\mu)}$	Mean Particle Diameter Å
SiO_2 A1	⊙	⊡	◬	▲	0.6	0.8	500
SiO_2 A2	○	□			0.6	0.8	700
SiO_2 E	•				0.3	0.8	450
SiO_2 C	×				0.9	0.8	550
$A\ell_2O_3$	⊕				0.2	1.5	200
BeO	+				1.0	1.4	150

same straight line in the PMS range. The accuracy of the experiments, and uncertainties in the analysis resulting from the assumption of isotropic elasticity and from poorly known elastic constants for the oxides, together produce an uncertainty in the value of $\langle\sigma^F\rangle_M$ of about 20%. To within this possible error, the factor of two in eqn. (10.11) appears to be correct, and the assumptions of the continuum model which lead to eqn. (10.5) appear to be justified.

In the following section, we shall try to draw together the observations on dislocation structure and the mechanical behavior of dispersion-hardened metals. Our basic assumption is that the curves of fig. 10.10 accurately reflect the state of internal stress, so that they should correlate with microstructural observations when due allowance is made for the experimental constraints of electron microscopy.

10.6 MODELS AND CONSTITUTIVE EQUATIONS FOR MECHANICAL BEHAVIOR

Although it is possible to construct an equation which unifies the various types of behavior observed, it seems more reasonable to identify various well-defined regimes of particle size, strain and temperature and to describe the mechanical behavior appropriate to each regime.

10.6.1 *PMS Behavior*

It is possible to observe PMS behavior provided: (1) the particle size is small enough that new dislocations must be nucleated to provide stress relief; (ii) the strain is small enough that dislocation nucleation cannot take place; and (iii) the temperature is low enough that diffusion is inhibited and thermally activated generation of dislocations is also inhibited. These conditions are illustrated by the curves of fig. 10.10. At 77°K, the PMS range extends from zero to a strain of .03 $\pm$ .005, but at 300°K the range is greatly reduced. If the specimen contains some large particles, so that these account for an appreciable part of the volume fraction, the PMS range is suppressed. The critical particle size is estimated to be about 3000 Å diameter, and this tallies with the estimates based on electron microscopy of the particle size at which secondary relaxation becomes easy. It should be pointed out that in systems with fibrous particles, the particle geometry suppresses plastic relaxation and the PMS regime may be greatly extended. Our discussion is limited to systems with more-or-less equiaxed particles.

Experimentally, after a small initial strain, the forward work-hardening curve of the PMS is given, to within about 20%, by

$$\sigma = \sigma_o + 2\,\gamma\,\mu\,f\,\frac{\mu^*}{\mu^* - \gamma\,(\mu^* - \mu)}\,\varepsilon_p + 2.0\,\mu\,f\sqrt{\frac{b\varepsilon_p}{r_o}}\,. \tag{10.14}$$

The hardening comes from the mean stress in the matrix (eqn. 10.5) and a term called the 'source-shortening' term. This additional hardening is essentially dissipative - that is, the work done is turned directly into heat - and comes from the flexing of the dislocation as it surmounts the fluctuating stresses in the neighborhood of the particles. Theoretical estimates of this term have been given by Brown and Stobbs (1971a) and by Hart (1972); they agree with eqn. (10.4) in the dependence upon volume fraction, particle size, and strain, and the numerical agreement is in both cases within about 40%. The reader should bear in mind that theoretical estimates of this dissipative term depend upon the assumptions made about the effect of the inhomogeneous stresses, and are open to considerable doubt.

It is worth noting here that in the range of particle sizes and volume fractions observed (see the key to fig. 10.10), the forest models of Ashby (1966) and of Hirsch and Humphreys (1970) cannot be distinguished from the PMS model on the basis of the forward work-hardening curve alone. However, these models predict no Bauschinger effect and thus require substantial modification if they are to agree with experiment. The theory of the PMS, through eqn. (10.11), gives a quantitative description of the permanent softening observed in a Bauschinger test. It cannot yet give, from first principles, an understanding of the shape of the reverse stress-strain curve.

Finally, we should warn the reader that eqn. (10.14) is a simplified form of a more accurate treatment to be found in Atkinson et al. (1974). What is called the "source-shortening" term here includes the effect of a small amount of forest hardening.

10.6.2 *Temperature and Strain-rate Dependent Relaxation at $T_M/5$*

Since the work of Jones and Kelly (1968) and of Hirsch and Humphreys (1970) it has been recognized that dispersion-hardened metals with small particles display remarkable softening effects, in which about half the work-hardening introduced at 77^oK can be removed in a period ranging from a few hours (for small particles) to a few days (for larger ones) at room temperature. Large-particle systems do not show a softening; the critical particle size is one again estimated to be around

3000 Å. It seems evident that the softening process must be identified with plastic relaxation. Several pieces of evidence suggest strongly that this is the case: (i) Although the material may soften by typically 30% of the forward work-hardening increment, the flow stress in the reverse direction actually increases slightly - in other words the softening is accompanied by a marked reduction in the Bauschinger effect. This is most easily explained by the conversion of Orowan loops to some relaxed configuration, so that the forward flow stress is reduced by both the mean stress in the matrix and by the source shortening term, but increased by the dislocations resulting from relaxation; the permanent softening is then much reduced when the Bauschinger test is carried out. (ii) The Cu-Zn-$A\ell_2O_3$ system, studied by Humphreys and Hirsch, is particularly resistant to the softening, and this is the system in which Orowan loops have been seen. (iii) In the Cu-SiO_2 system, where it is most unlikely that any deformation of the particles can occur, it is also unlikely that sufficient diffusion can occur to produce softening. There is every reason to expect softening at higher temperatures due to diffusion (see below) but this occurs at temperatures of 600°K and above. In terms of homologous temperature, the low temperature relaxation occurs at $T_M/5$ or so, whereas the high-temperature relaxation occurs at $T_M/2$ - both estimates depending somewhat on particle size

Having decided that the low-temperature softening results from plastic relaxation, we must try to decide what dislocation arrangement it produces. It is observed that the softening rate is markedly affected by whether the load is on the specimen, or off it: in other words, we should distinguish dynamical softening, which can occur during a stress-strain curve, from static softening in the absence of load. The neatest interpretation of available data seems to be that static softening results in prismatic loops of primary Burgers vector (A structure) and that dynamic softening results in dislocation tangles of secondary Burgers vectors (B structure). It follows from this that the number of primary prismatic loops observed in the electron microscope is a measure of the internal stress which prevailed when the load was removed from the specimen. Figure 10.11 shows the correlation that can be achieved between the internal stress level - as measured by ε_p^* - and the observed numbers of prismatic loops - the data taken from Hirsch and Humphreys' (1970) paper. Following eqn. (10.12), it is

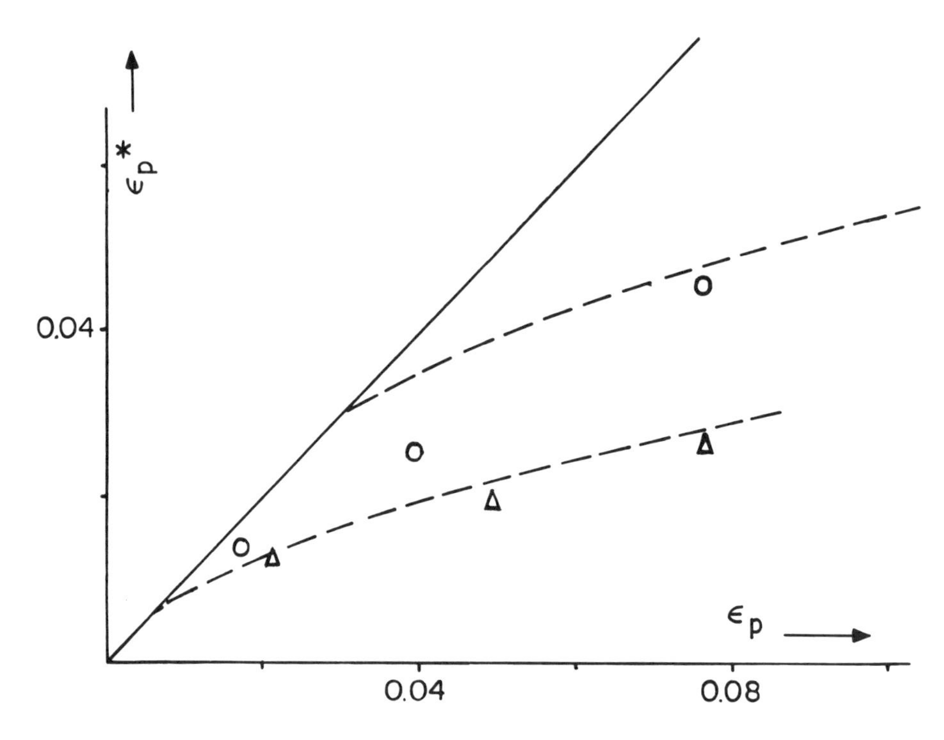

Fig. 10.11 Curves for the approximately parabolic rise of ε_p^* with ε_p, after the start of plastic relaxation, for 77°K and 300°K. The data points, △ for 300°K and ○ for 77°K, represent the values for ε_p^* which may be obtained from eqn. (10.15) using the loop count data of Hirsch and Humphreys (1970) on $Cu/A\ell_2O_3$.

assumed that

$$\varepsilon_p^* = \frac{b}{4\ r_o}\ (n + 1) \qquad (10.15)$$

where (n + 1) appears rather than n because on energetic grounds one Orowan loop at least must be left around the particle. The agreement is reasonable, considering the experimental limitations. It is worth pointing out that after a specimen has been deformed in the PMS range, and then subjected to static recovery, its flow stress should be due to the particles and the primary loop stacks alone: in this particularly simple condition the flow stress has been calculated by Foreman, Hirsch and Humphreys (1969) and shown by them to be consistent with observation.

If the dynamical softening can be described in the usual way by an activation energy U and activation volume v , the curves of fig. 10.10 allow one to calculate these quantities: they are $U = 0.9$ eV and $v = 3b^3$ (Atkinson, Brown and Stobbs 1974). These values are consistent with the observations of dislocation nucleation by pressure made by Ashby, Gelles and Tanner (1969). For the range of particle sizes studied by these authors, the nucleation of dislocations is controlled not by a nucleation barrier but by an energy criterion (Ashby and Johnson 1969, Brown and Woolhouse 1970) so the critical value of ε_p^* required to nucleate the dislocations at room temperature is less than Ashby et al.'s minimum observed value. This is clearly consistent with the curves of fig. 10.10.

If we assume that the above interpretation is correct, we may summarize the present state of affairs as follows:

1) A plastic relaxation, which we have called dynamical softening, produces a tangle of secondary dislocations when the specimen is under load. This leads to a reduction in ε_p^*, and a tentative equation describing the rate of reduction is given by Atkinson, Brown and Stobbs (1974). For deformation at 77°K, this process produces departure from PMS behavior at $\varepsilon_p \gtrsim .03$, whereas at 300°K the departure occurs at $\varepsilon_p \gtrsim .005$. At the higher temperature, relaxation leads to a characteristic strain-rate dependence of the work-hardening rate and to its reduction by a factor of about two. The reduction in ε_p^* is of course much greater: this is summarized in fig. 10.12, taken from Atkinson, Brown and Stobbs (1973).

2) Another mode of plastic relaxation, static softening, produces a stack of prismatic loops when the specimen is unloaded. This leads to a reduction in ε_p^* with a rate of reduction typically 100 times slower than that for dynamic softening. In the softened state, the loop stacks cause strengthening and this has been calculated by Foreman et al. (1969).

Throughout this section we have used the word 'softening' to describe the plastic relaxation, rather than the word 'recovery' which is more usual. We do this to avoid confusion

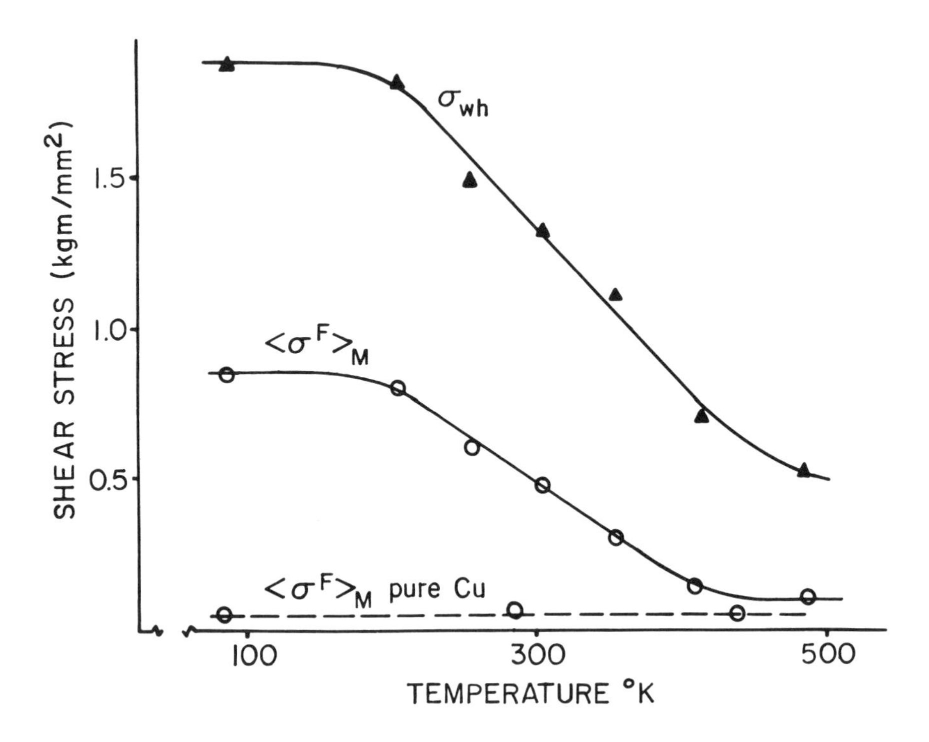

Fig. 10.12 A comparison of the mean image stress $<\sigma^F>_M$ with the total work-hardening increment, σ_{wh}, as a function of temperature. Copper 0.3%, by volume, silica specimens given a resolved shear strain of 0.08.

with the use of the phrase 'recoverable strain' in discussions of viscoelastic behavior. A recovery process implies a reduction in internal stress by return of the material to its original shape. The softening processes observed in dispersion-hardened metals are (if our interpretation is correct) accompanied by a small shape change in the direction of <u>increasing</u> deformation: lengthening of the specimen if it has previously been deformed in tension. It seems highly desirable to distinguish these two different processes.

10.6.3 *Forest Hardening and Back-stress Hardening in the Relaxed Condition*

Brown and Stobbs (1971b) on the basis of observations on the B structure proposed that the material in the tangle of dislocations is strongly forest hardened, and that this has two consequences: the material as a whole is hardened because the tangles act as hard particles and raise the flow stress; and further plastic relaxation is impeded, giving rise to a back stress which cannot be removed by plastic relaxation. Let us assume that the temperature is high enough that the stress to nucleate dislocations is negligible by comparison with the flow stress inside the tangle. Let us make the approximation that $\varepsilon_p^* << \varepsilon_p$: i.e. that the material is almost completely relaxed. Then a very simple treatment of the strength can be given, following Brown and Stobbs (1971b). Nearly every Orowan loop left around the particle will have been converted to a dislocation 'loop' inside the tangle; the total line length of dislocation per loop will be approximately $2\pi r_o$. The number of loops is related to the plastic strain by $n \approx 4\ \varepsilon_p r_o/b$. If the volume of the tangle is V_p , then the dislocation density inside the tangle is

$$\rho_{LOCAL} = \frac{C_1\ \varepsilon_p\ r_o^2}{b\ V_p} \tag{10.16}$$

where the constant C_1 should approximately equal 8π ; this is in fact an upper limit for C_1.

Electron microscopy shows that the volume of the tangle does not depend strongly on plastic strain, in the limited strain range under discussion. The volume of the tangle is given by $V_p = C_2 r_o^3$, where C_2 is a constant which electron microscopy suggests should be between 2 and 4. The volume fraction of tangle, f_p , is related to f by $f_p = 3f/4\pi r_o^3$. The elastic stress inside the inclusion is found by the stress balance:

$$\sigma^* = \mu\ \varepsilon_p^* = \alpha\ \mu\ b \sqrt{\rho_{LOCAL}} = \alpha\ \mu \sqrt{\frac{C_1\ \varepsilon_p\ b}{C_2\ r_o}} \tag{10.17}$$

where α is the forest hardening constant, thought to lie between 1/3 and 1/5. The total hardening of the material consists of a forest part caused by intersection of the glide dislocations with the tangle: this will be given by $f_p^{1/2}\ \sigma^*$ or

$$\sigma_{FOREST} = \alpha \mu b \sqrt{\frac{f b \varepsilon_p}{r_o}} \times \sqrt{\frac{3 C_1}{4\pi}} ; \qquad (10.18)$$

and in addition to the forest contribution there will be a back-stress contribution

$$\sigma_{BACK} = f \mu \varepsilon_p^* = \alpha \mu f \sqrt{\frac{C_1 \varepsilon_p b}{C_2 r_o}} \qquad (10.19)$$

The total hardening will be the sum of the forest and the back-stress contributions. It can be shown that the assumption of additivity made in this treatment is justified, and a fuller account will be published elsewhere.

It seems reasonable to identify the parabolic increase of ε_p^* observed in fig. 10.10 with the predicted parabolic increase in eqn. (10.17). At 300^oK the data of fig. 10.10 are consistent with

$$\varepsilon_p^* = \sqrt{\frac{b \varepsilon_p}{r_o}} \qquad (10.20)$$

Comparison with eqn. (10.17) gives $\alpha \sqrt{C_1/C_2} = 1$, compared to the expected value of between .3 and 1.2. It should be borne in mind that the data of fig. 10.10 are not extensive enough to establish the radius dependence given in eqn. (10.20), but the data do suggest that there is a tendency for small particles to generate a larger stress; this feature is supported more strongly by Wilson's (1965) work. It is also important to remember that the treatment given here does not apply to the nucleation stress: at 77^oK, ε_p^* will be increased by the critical strain to nucleate the zone as shown in fig. 10.10. However, it is clear that forest hardening in the tangle of secondary dislocations is capable of explaining semi-quantitatively the curves of fig. 10.10 outside the PMS range.

The total work-hardening is parabolic, and the dependence upon volume fraction is indistinguishable from the Ashby (1966) formula. His work shows that the total work-hardening is given by eqn. (10.18) with the numerical constant equal to 0.3 ± 0.1; this is slightly less than the expected value which lies between 0.5 and 0.8. Using Ashby's work, and eqn (10.20), we may write to a good approximation

$$\sigma_{TOTAL} = \sigma_o + \sigma_{FOREST} + \sigma_{BACK} = \sigma_o + (0.3 + 0.1)\mu\sqrt{\frac{f\ b\ \varepsilon_p}{r_o}} + f\ \mu\sqrt{\frac{b\ \varepsilon_p}{r_o}} \qquad (10.21)$$

Equation (10.21) is the constitutive equation for the material in the relaxed condition. It can be seen that the back-stress accounts for a small fraction of the total hardening: the ratio of back-stress to total hardening increment is given by $(3 \pm 1)\ f^{1/2}$. This ratio is broadly consistent with observation; for instance, in fig. 10.12, at about 400^oK the back stress levels off and becomes equal to about 1/6 of the total work-hardening increment; for this specimen $3f^{1/2}$ is equal to 1/5.

The reasoning of Brown and Stobbs (1971b) and the simplified version of it given here is really an extension of Ashby's forest model, taking account of the observed distribution of secondary dislocations. If the forest were uniform, then there would be no back-stress: it is because the forest is localized in precisely those regions where plastic relaxation must occur that the forest hardening is accompanied by internal stresses. These internal stresses play an important role in opening the interface between particle and matrix, so leading to fracture. They also play a role in limiting the dimensional stability of work-hardened material, for when they are relieved the specimen undergoes a shape change which may in some circumstances be of engineering significance.

A more careful treatment of the forest hardening will be given by Brown and Stobbs (1975).

10.6.4 *Diffusional Relaxation*

Stobbs (1973) has shown that the marked reduction in work-hardening at elevated temperatures can be caused by interface diffusion. In $Cu\text{-}SiO_2$, this effect becomes important, if there is no plastic relaxation, for strain rates, $\dot{\varepsilon}_p$, such that:

$$\dot{\varepsilon}_p < 60\ b^4\ \mu\ D_s/r_o^3\ kT \qquad (10.22)$$

where $D_s \approx 0.02 \exp(-E/kT) m^2 sec^{-1}$, $E = 1.71$ eV . This expression accounts quite well for data accumulated by Shewfelt (1972). If other dispersion-hardened systems show similar stress-relief mechanisms, we may assume that diffusional relaxation becomes important at temperatures above about $T_M/2$; this is also the temperature above which the flow stress starts to drop slowly because of a similar diffusional process (Shewfelt and Brown 1973).

By now it should be clear that it is the coupling of the permanent softening results with the electron microscopy information which has made it possible for us to distinguish the various classes of behavior outlined here. In Section 10.6.1 we described the behavior characterized by no relaxation, that of the PMS, while in Sections 10.6.3 and 10.6.4 we give constitutive equations for partial plastic relaxation and total diffusional relaxation respectively.

10.7 ALTERNATIVE MODELS, AND THE PUZZLE OF THE SLIP PLANE SPACING

Hirsch and Humphreys (1969) introduced a forest model in which the elastic interaction between dislocations on neighboring glide planes plays an essential role. With this model, they were able to account for parabolic hardening which in their view results from primary prismatic loops, the type A structure. Their model also predicts a slip plane spacing which is in agreement with their experiments. When the theory is modified to take account of Orowan loops around the particles (Gould et al. 1973) the theory seems able to account for the observations of the Bauschinger effect. It is striking that in the account of the work-hardening presented in this review, the slip-plane spacing plays no role; it is absent just as it is from all calculations of strength due to precipitates, short-range order, etc. On the basis of existing evidence, it seems virtually impossible to discriminate in a clear-cut way between the two models. Part of the difficulty lies in the unambiguous measurement of the slip-plane spacing. Observations of particles which show no secondary structure at all (Hirsch and Humphreys 1970) suggest that the slip plane spacing is of the same order as, or greater than, the particle radius. However, replica observations (Hirsch and Humphreys 1970b) on the same material are not consistent with this picture. It seems certain that the replica studies do not present a true picture of the internal slip plane spacing. On the other hand, the Al_2O_3 particles are coherent, and may twin or be cut by dislocations (see fig. 10.5, also see Sastry and Ramaswami 1973). At the time of writing, it does not seem possible to settle this question.

If the assumption is made that one or at most two slip planes intersect a particle, then the continuum model cannot be used, and the whole basis of the analysis presented in this review cannot hold. Empirically, using eqn. (10.10) which related the permanent softening to the back-stress calculated using the continuum model, one finds quite wide-ranging agreement with experiment. The model is simple in principle and, once one becomes accustomed to it, easy to use and unambiguous in its predictions. On the other hand, it is necessary to postulate that plastic relaxation requires the temperature-dependent stresses discussed in Section 10.6.2, and although this is consistent with known behavior at room temperature there is very little experimental corroboration for this behavior at 77°K. So the matter rests for the moment, and we refer the interested reader to the papers by Hirsch and his co-workers for the alternative account.

10.8 SUMMARY

When a dislocation passes through a dispersion hardened material it leaves 'Orowan loops' around every particle in its slip plane. Work-hardening models in these alloys concern themselves with the fate of the Orowan loops and with their effect on the strength. Both continuum and dislocation models are examined and it is concluded that the most useful model is a hybrid. This approach is made possible by a phenomenological observation of a quantitative relationship between the permanent softening, $\Delta\sigma_p$, in a Bauschinger test and the mean stress in the matrix, $\langle\sigma^F\rangle_M$, prior to backward deformation:

$$\langle\sigma^F\rangle_M = \Delta\sigma_p/2$$

It is $\langle\sigma^F\rangle_M$ which opposes plastic flow in the forward direction and aids deformation in the backward direction. It is shown that this stress is proportional to the unrelaxed applied strain and to the volume fraction of the dispersion and is quantitatively given to within 20% by the continuum model. By coupling this information with our knowledge of the dislocation microstructure as a function of the dispersion characteristics and testing conditions, we are able to come to the following conclusions:

1. Dispersion hardened alloys deform as 'Perfect Memory Solids' provided that:

a) the particle size is small enough that new dislocations must be nucleated to provide stress relief;

b) the strain is small enough that dislocation nucleation cannot take place;

c) the temperature is low enough that both diffusion and the thermally activated generation of dislocations are inhibited.

Experimentally, after a small initial strain, the forward work-hardening curve of the PMS is given to within about 20% by the sum of two 'frictional' terms, the Orowan stress and the 'source shortening' term, and one 'reversible' term, and the mean stress in the matrix (eqn. 10.14).

2. The softening process is considerably faster under load than when the load is removed. Both processes involve plastic relaxation mechanisms. In the former case we believe that stress relaxation occurs by the formation of the 'type B' secondary dislocation arrays and in the latter by the production of the 'type A' structure of aligned primary prismatic loops.

3. Both forest hardening and back stress hardening occur in the relaxed condition when the type B secondary dislocation arrays are produced at the particles. To a good approximation the total work-hardening increment is then given by the sum of forest and back stress terms (eqn. 10.21).

4. A marked reduction in work-hardening at elevated temperatures can be caused by interface diffusion. The effect becomes important under conditions indicated in eqn. (10.22).

ACKNOWLEDGEMENTS

One of us (WMS) wishes to thank the Central Electricity Generating Board for financial support.

We wish to acknowledge numerous conversations with Dr. J.D. Atkinson and his permission to publish results from a forthcoming joint publication.

REFERENCES

Ashby, M. F., (1966) Phil. Mag. 14, 1157.

Ashby, M. F., (1969) in "Physics of Strength and Plasticity", edited by A. S. Argon (Cambridge, Mass: M.I.T. Press), p. 113.

Ashby, M. F., (1970a) Phil. Mag. 21, 399.

Ashby, M. F., (1970b) in "Proc. Second Intern. Conf. on the Strength of Metals and Alloys", (Metals Park, Ohio: A.S.M.) p. 507.

Ashby, M. F., (1971) in "Strengthening Mechanisms in Crystals", edited by A. Kelly and R. B. Nicholson (Amsterdam: Elsevier) p. 137.

Ashby, M. F., Gelles, S. H. and Tanner, L. E., (1969) Phil. Mag. 19, 757.

Ashby, M. F. and Johnson, L., (1969) Phil. Mag. 20, 1009.

Atkinson, J. D., (1973) Ph.D. Dissertation, University of Cambridge, U.K.

Atkinson, J. D., Brown, L. M. and Stobbs, W. M., (1973) in "Proc. Third Intern. Conf. on the Strength of Metals and Alloys", (Cambridge: Institute of Metals) p. 36.

Atkinson, J. D., Brown, L. M. and Stobbs, W. M., (1974) Phil. Mag. 30, 1247.

Bacon, D. J., Kocks, U. F. and Scattergood, R. O., (1973) Phil. Mag. 28, 1241.

Brown, L. M., (1973) Acta Met. 21, 879.

Brown, L. M. and Ham, R. K., (1971) in "Strengthening Mechanisms in Crystals", edited by A. Kelly and R. B. Nicholson (Amsterdam: Elsevier) p. 12.

Brown, L. M. and Stobbs, W. M. (1971a) Phil. Mag. 23, 1185.

Brown, L. M. and Stobbs, W. M., (1971b) Phil. Mag. 23, 1201.

Brown, L. M. and Stobbs, W. M., (1975) to be submitted to Phil. Mag.

Brown, L. M. and Woolhouse, G. R., (1970) Phil. Mag. 21, 329.

Chapman, P. F. and Stobbs, W. M., (1969) Phil. Mag. 19, 1015.

Eshelby, J. D., (1956) Progress in Solid State Physics, Vol. 3, edited by F. Seitz and D. Turnbull (New York, London: Academic Press) p. 79.

Eshelby, J. D., (1961) in "Progress in Solid Mechanics", edited by I. N. Sneddon, and R. Hill (New York: Interscience), Vol. 2, p. 89.

Fisher, J. C., Hart, E. W. and Pry, R. H., (1953) Acta Met., 1, 336.

Foreman, A. J. E., Hirsch, P. B. and Humphreys, F. J., (1969) in "Conference on Fundamental Aspects of Dislocation Theory", edited by J. A. Simmons, R. de Wit and R. Bullough (Washington: National Bureau of Standards) p. 1083.

Gould, D., Hazzledine, P. M., Hirsch, P. B. and Humphreys, F. J., (1973) in "Proc. Third Intern. Conf. on the Strength of Metals and Alloys", (Cambridge: Institute of Metals) p. 31.

Hart, E. W., (1972) Acta Met. 20, 275.

Hirsch, P. B., (1957) J. Inst. Met. 20, 275.

Hirsch, P. B. and Humphreys, F. J., (1969) In "Physics of Strength and Plasticity", edited by A. S. Argon (Cambridge, Mass.L M.I.T. Press) p. 189.

Hirsch, P. B. and Humphreys, F. J., (1970a) Proc. R. Soc. A, 318, 45.

Hirsch, P. B. and Humphreys, F. J., (1970b) in "Proc. Second Intern. Conf. on the Strength of Metals and Alloys", (Metals Park, Ohio: A.S.M.) p. 545.

Humphreys, F. J. and Hirsch, P. B., (1970) Proc. R. Soc. A, 318, 73.

Humphreys, F. J. and Stewart, A. T., (1972) Surf. Sci. 31, 389.

Jones, R. L. and Kelly, A., (1968) in "Oxide Dispersion Strengthening", edited by G. S. Ansell, T. D. Cooper and F. V. Lenel (New York: Gordon and Breach) p. 229.

Kelly, A., (1966) "Strong Solids" (Oxford: Clarendon Press).

Lindley, T. C., Oates, G. and Richard, C. E., (1970) Acta Met. 18, 1127.

MacEwen, S. R., Hirsch, P. B. and Vitek, V., (1973) Phil. Mag. 28, 703.

Merrick, H. F., 1962, Ph.D. Dissertation, University of Cambridge [quoted by Ashby, M. F., 1971].

Mott, N. F. and Nabarro, F. R. N., (1948) in "Report on Strength of Solids (London: Physical Society), p. 1.

Orowan, E., (1959) in "Internal Stresses and Fatigue of Metals", edited by G. Rassweiler and W. Grube (Amsterdam: Elsener), p. 59.

Palmer, I. G. and Smith, G. C., (1968) in "Oxide Dispersion Strengthening", edited by G. S. Ansell, T. D. Cooper, and F. V. Lenel (New York: Gordon & Breach), p. 253.

Sastry, S. M. L. and Ramaswami, B., (1973) Phil. Mag. 28, 945.

Shewfelt, R. S. W., (1972) Ph.D. Dissertation, University of Cambridge, U.K.

Shewfelt, R. S. W. and Brown, L. M. (1973) in "Proc. Third Intern. Conf. on the Strength of Metals and Alloys", (Cambridge: Institute of Metals) p. 311.

Stobbs, W. M., (1973) Phil. Mag. 27, 1073.

Tanaka, K., (1974) Acta Met., to be published.

Tanaka, K. and Mori, T., (1970) Acta Met. 18, 931.

Tanaka, K., Narita, K. and Mori, T., (1972) Acta Met. 20, 297.

Wilson, D. V., (1965) Acta Met. 13, 807.

Wilson, D. V. and Konnan, Y. A., (1964) Acta Met. 12, 617.

Zener, C., 1948, "Elasticity and Anelasticity of Metals" (Chicago: University Press).

LIST OF SYMBOLS

σ_o	the Orowan stress
μ	(isotropic) elastic shear modulus of matrix
b	Burgers vector
N	ℓ^{-2} = no. of particles per unit area
ε_p	ε^p = symmetric plastic shear distortion
μ^*	(isotropic) elastic shear modulus of inclusion
ν	Poisson's ratio of matrix
σ^I	stress inside inclusion
γ	accommodation factor in elastic theory of inclusions
σ^c	stress in matrix caused by inclusion (constrained stress)
(r,θ,ϕ)	spherical polar coordinates
r_o	radius of inclusion
f	volume fraction of inclusion
$\langle\sigma^F\rangle_M$	mean stress in a finite matrix
η	coefficient of viscosity
$\tau_\varepsilon, \tau_\sigma$	relaxation times for standard linear solid
M_U, M_R	unrelaxed and relaxed moduli for standard linear solid
$\Delta\sigma_p$	permanent softening in Bauschinger test
ε_p^*	unrelaxed plastic strain, proportional to elastic strain in inclusion
n	number of Orowan loops
ρ	dislocation density

V_p	volume of plastic zone
f_p	volume fraction of plastic zone
C_1, C_2	numerical constants
α	forest hardening constant
σ^*	elastic stress in inclusion
D_s	surface diffusion coefficient

ABSTRACT. *The problem of deformation texture development is analyzed in terms of Taylor's model of polycrystal deformation. Within the limits of this model the active slip systems can be selected on the basis of the minimum work analysis of Taylor or of the maximum work analysis of Bishop and Hill. Numerical calculations are greatly facilitated using linear programming techniques and electronic computers. It has been shown (Chin and Mammel) that the Taylor analysis and the Bishop and Hill analysis are completely equivalent as duals in the linear programming problem. Examples of texture calculations for various slip modes associated with cubic metals and nonmetals are given. Results obtained by other workers using the original Taylor and Bishop and Hill approaches are also presented.*

11. DEVELOPMENT OF DEFORMATION TEXTURES

G.Y. Chin

11.1 INTRODUCTION

The development of crystallographic texture during plastic deformation is a direct result of lattice rotation caused by slip and/or twinning. Consequently, any model which attempts to describe deformation texture development must be based on a model of deformation of a polycrystalline aggregate. This means that the model must be capable of explaining the orientation dependence of the flow stress, the slip and/or twinning systems activated, and, of course, the lattice rotation resulting from the activation of these systems. Despite some shortcomings, the deformation model proposed by G. I. Taylor (1938) is generally judged to be both reasonably realistic and yet sufficiently simple to be treated numerically. Kocks (1970) recently reviewed the Taylor model and compared it with others. Thus a critique will not be offered here. Rather, the aim of the present chapter is to present a generalized analysis of crystal plasticity (Chin and Mammel, 1969), which is an extension of the treatments presented by Taylor (1938) and by Bishop and Hill (1951). In the generalized form, the crystal plasticity problem is viewed in terms of a linear programming problem and hence lends itself to rapid solutions using electronic computers. This analysis has been applied to the Taylor model of polycrystal plasticity to generate numerical results for the case of axisymmetric flow for several deformation modes in cubic crystals. These results will be reviewed. In addition, relevant data based on the earlier treatments of Taylor and of Bishop and Hill will also be presented.

11.2 THE TAYLOR MODEL OF POLYCRYSTALLINE PLASTICITY

The Taylor model is rather simple. There are only three basic assumptions: (1) the elastic strain is small in comparison with the plastic strain; (2) each grain deforms homogeneously and to the same strain as the aggregate; and (3) all slip systems obey the Schmid Law and harden equally, that is, the current flow stress for a given slip system depends on the sum of glide shears whether or not that system has been activated.

The first assumption is quite valid in texture calculations where the total strain is large. The second assumption is weaker. The slip pattern is more complex in the vicinity of grain boundaries, most likely due to compatibility problems arising from the discrete nature of slip bands (Armstrong et al., 1962). The thickness of this grain boundary deformation region is probably of the order of slip band spacing (Kocks, 1970). Hence while it could lead to a grain size dependence of flow strength, its effect on texture formation should be small (Thompson, 1974). In addition to complex slip near grain boundaries, deformation banding is another manifestation of inhomogeneous flow. This aspect has been discussed in some detail (Chin, 1969). Finally, for unusual textured samples, such as cold drawn wires of BCC metals, or compressed disks of FCC metals, possessing a <110> fiber texture, the grain may deform appreciably differently from the aggregate (Peck and Thomas, 1961; Hosford, 1964).

As for the third assumption, the Schmid Law appears valid for slip in FCC metals, and for BCC metals deformed at moderate temperatures to moderate strains (Christian, 1970). Under conditions of five independent shear systems generally activating in an embedded grain, twinning may also obey the Schmid Law (Chin, Hosford and Mendorf, 1969). The assumption of equal hardening among all slip systems is reasonably valid (Kocks, 1970), although quantitative texture differences can be expected when this assumption is relaxed. Examples will be given later in the paper.

11.3 GENERALIZED ANALYSIS OF CRYSTAL PLASTICITY

In this section the generalized analysis of Chin and Mammel (1969) will be summarized. If a crystal deforms by slip, the incremental plastic strain components become

$$\varepsilon_{ij} = \sum_{\ell} \gamma_{\ell} \, {}^{\ell}m_{ij} \qquad i,j = 1,2,3 \tag{11.1}$$

where γ_{ℓ} is the incremental glide shear produced by slip on the ℓ-th slip system, and ${}^{\ell}m_{ij}$ is a generalized Schmid factor:

$${}^{\ell}m_{ij} = \frac{1}{2}\left({}^{\ell}d_i \, {}^{\ell}n_j + {}^{\ell}d_j \, {}^{\ell}n_i\right) \tag{11.2}$$

In eqn. (11.2) d is the direction cosine of the slip direction and n is the direction cosine of the slip plane normal.

The resolved shear stress on the ℓ-th slip system is

$$\tau_{\ell} = \sum_{i,j} \sigma_{ij} \, {}^{\ell}m_{ij} \tag{11.3}$$

where σ_{ij} refers to the applied stress components. Assuming the Schmid Law to be valid, slip begins when τ reaches a critical value τ^*. Hence the yield criterion becomes

$$\tau_{\ell} \begin{cases} = \tau^*_{\ell} & \text{if } \gamma_{\ell} \neq 0 \text{ (active systems)} \\ \leq \tau^*_{\ell} & \text{if } \gamma_{\ell} = 0 \text{ (inactive systems)} \end{cases} \tag{11.4}$$

As discussed in the previous section, it is generally assumed that an embedded grain undergoes the same strain as the aggregate (Taylor, 1938). The problem is then one of selecting a combination of slip systems that satisfies the imposed strains (eqn. 11.1) while fulfilling the yield criterion (eqn. 11.4). Because there are six independent strain components which are reduced to five in the case of constancy of volume during slip, five independent slip systems are required. Once the active systems are selected, the incremental lattice rotation can be calculated according to the equation

$$\omega_k = \sum_{\ell} \gamma_\ell \, {}^{\ell}m'_{ij} \tag{11.5}$$

where

$${}^{\ell}m'_{ij} = \frac{1}{2}\left({}^{\ell}d_i \, {}^{\ell}n_j - {}^{\ell}d_j \, {}^{\ell}n_i\right). \tag{11.6}$$

There are two equivalent techniques of selecting the active slip systems. In the first technique, the active combination(s) of γ_ℓ's is found by minimizing the internal work $\sum \tau^*_\ell \gamma_\ell$ subject to the constraints of eqn. (11.1). This formulation was originally put forth by Taylor (1938). In the second technique, a set of applied stresses is found by maximizing the external work $\sum \sigma_{ij} \varepsilon_{ij}$ subject to the constraints of eqn. (11.4). This formulation was originated by Bishop and Hill (1951).

These two formulations are merely duals of each other in linear programming and are hence completely equivalent. For example, using the method of Lagrange multipliers, the Lagrangian function based on the minimum work approach becomes

$$F(\gamma,\lambda) = \sum_{\ell} \tau^*_\ell \gamma_\ell + \sum_{ij} \lambda_{ij} \varepsilon_{ij} - \sum_{ij} \lambda_{ij} \, {}^{\ell}m_{ij} \gamma_\ell \tag{11.7}$$

where λ_{ij} are the Lagrange multipliers. The minimum occurs at the saddle point (γ^o,λ^o) such that

$$\left.\frac{\partial F}{\partial \lambda_{ij}}\right|_o = \varepsilon_{ij} - \sum_{\ell} {}^{\ell}m_{ij} \gamma^o_\ell = 0 \tag{11.8}$$

$$\left.\frac{\partial F}{\partial \gamma_\ell}\right|_o = \tau^*_\ell - \sum_{ij} \lambda^o_{ij} \, {}^{\ell}m_{ij} \begin{cases} = 0, & \gamma > 0 \\ \geq 0, & \gamma = 0 \end{cases} \tag{11.9}$$

One can tell that eqn. (11.9) is in fact the yield criterion of eqn. (11.4) with $\lambda_{ij} \equiv \sigma^o_{ij}$ identified as the yield stresses.

By the same token, the Lagrangian function based on the maximum work approach becomes

$$G(\sigma,\mu) = \sum_{ij} \sigma_{ij}\,\varepsilon_{ij} + \sum_{\ell} \tau^{*}_{\ell}\mu_{\ell} - \sum_{ij\ell} \sigma_{ij}\,{}^{\ell}m_{ij}\,\mu_{\ell} \qquad (11.10)$$

where μ_{ℓ} are the Lagrange multipliers. The maximum occurs at (σ^{o},μ^{o}) such that

$$\left.\frac{\partial G}{\partial \sigma_{ij}}\right|_{o} = \varepsilon_{ij} - \sum_{\ell} {}^{\ell}m_{ij}\,\mu^{o}_{\ell} = 0 \qquad (11.11)$$

$$\left.\frac{\partial G}{\partial \mu_{\ell}}\right|_{o} = \tau^{*}_{\ell} - \sum_{ij} \sigma_{ij}\,{}^{\ell}m_{ij} \quad \begin{cases} = 0, & \mu > 0 \\ \geq 0, & \mu = 0 \end{cases} \qquad (11.12)$$

The equivalence of eqn. (11.11) with (11.8) and of eqn. (11.12) with (11.9) are apparent, with $\mu \equiv \gamma$ being the glide shears. Thus both analyses are completely equivalent in all respects.

A salient advantage of the linear programming analysis is that techniques have been developed to permit the attainment of a solution after evaluating only a few of the combinations of slip systems. These techniques are particularly suitable for use in conjunction with electronic computers. Computer programs were developed by W. L. Mammel for the texture problem and have been applied to a variety of deformation modes.*

---

*In his original approach, Taylor calculated the minimum internal work by evaluating the sum of shears for all possible combinations of slip systems. This approach was mathematically too tedious at the time, and remained so even with the help of modern electronic computers (Siemes, 1967; Bunge, 1970). In the Biship and Hill approach, the problem was simplified by prior searching for a list of stress states which are capable of activating five independent slip systems. The solution then consisted of finding the stress state which maximizes the external work. For {111} <110> slip applicable to FCC crystals, the list of stress states (56) is considerably shorter than the list of combinations of five independent slip systems (385). However, the problem of finding a complete list of stress states for each new slip mode is not trivial, and the list can become extremely long if several slip modes with different τ^{*} are encountered. The simplicity of calculation of maximum work is then lost. On the other hand, prior knowledge of the stress states is not necessary using the linear programming techniques.

11.4 TEXTURES RESULTING FROM AXISYMMETRIC FLOW

For axisymmetric flow the incremental strain components are:

$$\varepsilon_{yy} = \varepsilon_{zz} = - \varepsilon_{xx}/2, \ \varepsilon_{yz} = \varepsilon_{zx} = \varepsilon_{xy} = 0 \qquad (11.13)$$

where x is the axial direction of the specimen and y and z are orthogonal directions perpendicular to x . Computer results based on linear programming techniques have been obtained for {110} <111>, {112} <111>, {123} <111> slips plus {110}, {112}, {123} <111> approximate pencil glide (Chin, Mammel and Dolan, 1967), and for {111} <110> slip and {111} <112> twinning (Chin, Mammel and Dolan, 1969). Bishop (1954) has obtained the results for {111} <110> slip using the Bishop and Hill technique, while Siemes (1967) has done computer calculations for {110} <111>, {112} <111> and {123} <111> slips based on the original Taylor method. Rosenberg and Piehler (1971) calculated the lattice rotations for <111> pencil glide by applying a pattern search computer technique to a Bishop and Hill type analysis.

A typical result, for axisymmetric tension based on the {111} <110> slip mode, is shown in fig. 11.1. Each point inside the standard stereographic triangle represents the initial direction of a crystal aligned in the axial direction. The end of a ray denotes the new orientation of the x-axis after an incremental $\varepsilon_{xx} = 0.05$ based on slip in one of the allowable combinations of five slip systems. Due to the symmetry of the slip systems, several combinations of five slip systems have the same minimum internal work (or maximum external work) and are thus equally allowed. This is reflected in the several

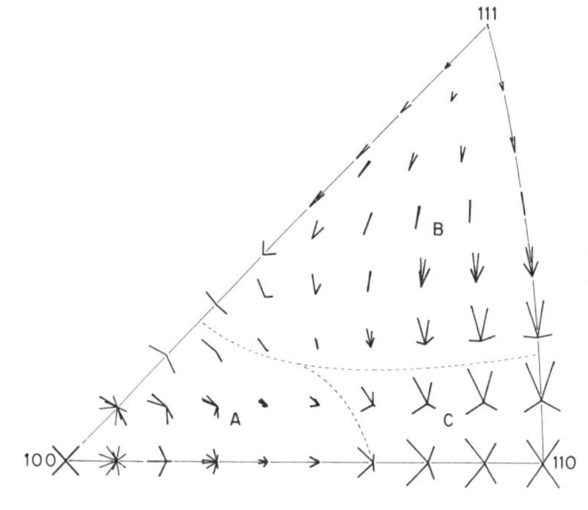

Fig. 11.1 Computer-plotted lattice rotations (ε_{xx} = 0.05) for axisymmetric tension for {111} <110> slip assuming equal hardening among all slip systems. (from Chin, Mammel and Dolan, 1967, courtesy of AIME)

rays attached to each point. Rotations in-between rays are also possible by a suitable combination of the allowed basic solutions of five slip systems.

Figure 11.1 shows three regions of axial rotation: region A where the rotation is toward <100>; region B, toward <111>; and region C, either <100> or <111>. Assuming one half of the crystals in region C rotate toward <100> and one half toward <111>, it is predicted that after axisymmetric tension, the fiber texture consists of 70% <111> and 30% <100>.

While the fiber textures of cold-drawn wires (one form of axisymmetric tension) of practically all FCC metals and alloys do contain duplex <111> + <100> components, the proportion often deviates from the 70/30 ratio. The work of English and Chin (1965), fig. 11.2, shows that the proportion of <100>

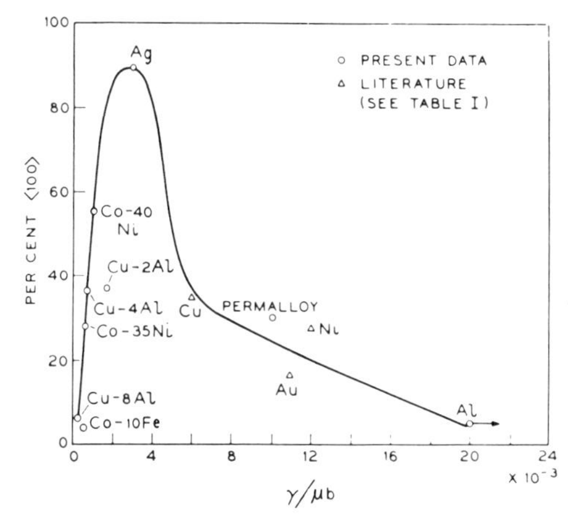

Fig. 11.2 The variation of <100> (remainder <111>) component of wire texture of FCC metals as a function of the stacking fault energy parameter $\gamma/\mu b$. (from English and Chin, 1965, courtesy of Pergamon Press)

varies with the stacking fault energy parameter $\gamma/\mu b$, (where γ is the stacking fault energy, μ the shear modulus, and b the magnitude of the Burgers vector), topping out at a value of $\gamma/\mu b \sim 3.5 \times 10^{-3}$ for Ag. The percentage of <100> is as low as $\sim$ 10 for both high and low values of $\gamma/\mu b$, and is as high as $\sim$ 90 for the peak at Ag.

The observed textures of fig. 11.2 can be partially accounted for by a modified model based on the variation of deformation behavior with $\gamma/\mu b$ (Chin, 1969). First, noting that cross-slip is favored at high $\gamma/\mu b$ and coplanar slip is favored at low $\gamma/\mu b$, the multiple solutions obtained from the equal hardening assumption were made unique by maximizing the glide shears of those systems having collinear or coplanar relationships. Both modifications depress the <100> component to $\sim$ 22%. This is in the right direction although insufficient to account for the observed ∿ 10%. Secondly, mechanical twinning on {111} <112> systems was assumed to operate along with {111} <110> slip. A Schmid law was assumed for the twinning and the minimum work calculations were applied. The results are summarized schematically in fig. 11.3. Figure 11.3a is

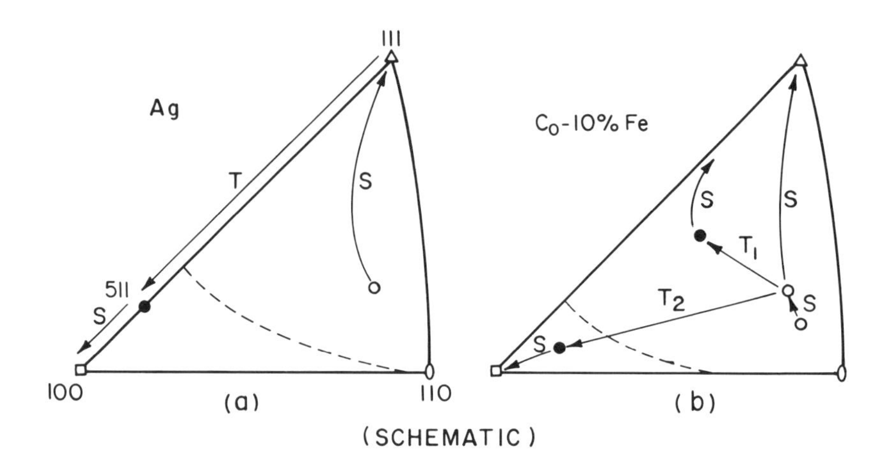

Fig. 11.3 Schematic showing trends in rotation due to slip (S) and twinning (T) in FCC metals deformed in axisymmetric tension. See text for discussion.

applicable to a material such as silver where twinning occurs after substantial prior slip. In this case, the prior slip activity rotates the axial orientation of a grain to the vicinity of [111]. When twinning occurs at this stage, the axial orientation is abruptly switched to [511], which is close to [100] and thereby accounting for the large <100> peak for Ag. On the other hand, the Co-10% Fe alloy with lower $\gamma/\mu b$ than that of Ag is known to twin very easily and yet has a small <100> fiber texture (fig. 11.2). Figure 11.3b attempts to rationalize that finding. One consequence of early twinning is that some twinning systems, e.g. T_1, rotate the axial orientation to a position away from [100]. Secondly, unless the axial orientation is very near [111] when twinning occurs, a substantial fraction of the incremental strain is accommodated by slip. Hence both these factors tend to suppress the amount of <100> in easily twinned material.

Another rationale for the drop of <100> in low $\gamma/\mu b$ materials is based on observations in Cu alloys that as $\gamma/\mu b$ is lowered, twins become finer (Venables, 1964; Peissker, 1965).

It was argued that although nucleation is easier, propagation becomes more difficult. Therefore one could expect the total volume fraction of twinned material, hence the <100> component, to decrease. This view-point was discussed by English and Chin (1965) and favored by Bunge (1971). It however cannot be applied to Co-10% Fe where copious twinning has been observed (Chin, Hosford and Mendorf, 1969).

For compression, the sense of rotation is merely the reverse of tension. Figure 11.1 shows that the compression axis tends to rotate toward <110>, with a lower rate of approach along the line joining <100>-<311>. This result is in agreement with that reported for Al, but not for 70-30 α-brass (Barrett, 1942). For the latter, there is a conspicuous absence of intensity near <100> with a corresponding rise near <111>. Semiquantitatively, this result could be explained by twinning. In contrast to the tension case, the <100> corner is favored in compression, resulting in a rotation to <221> in the vicinity of <111>.

11.5 PLANE STRAIN DEFORMATION

For plane strain deformation such as occurring during rolling the incremental strain components are

$$\varepsilon_{zz} = -\varepsilon_{xx}, \ \varepsilon_{yy} = \varepsilon_{yz} = \varepsilon_{zx} = \varepsilon_{xy} = 0 \qquad (11.14)$$

where x, y, z refers to the normal, transverse and rolling directions respectively. Computer results for FCC metals have been obtained by Kallend and Davies (1972) using the Bishop and Hill stress states, and by Bunge (1970) using Taylor's technique. Kallend and Davies included provisions for treating slip systems having cross-slip or coplanar relationships as well as for a semiquantitative treatment of twinning. Bunge, on the other hand, generalized the results to cover cases between plane strain and axisymmetric flow by varying $\varepsilon_{yy}/\varepsilon_{xx}$ between 0 and -0.5.

Figure 11.4 shows Kallend and Davies' comparison between theory and experiment for Cu and α-brass in the form of {111} pole figures. They started with an initially random array of {111} poles and followed their movement in small increments of strain. The solutions were not significantly different between cases of favoring cross-slip and coplanar slip, a conclusion also reached by Dillamore and Katoh (1971). The array shown in fig. 11.4 was obtained after about 80% reduction in thickness. Superimposed on this array of {111} poles are equi-intensity contours on a sample rolled to this reduction. Figure 11.4a is for Cu and 11.4b is for α-brass, where twinning was added to slip in the analysis. The agreement is quite good in both cases.

Figure 11.5 shows Bunge's calculations of the so-called skeleton lines of stable orientations where the lattice rota-

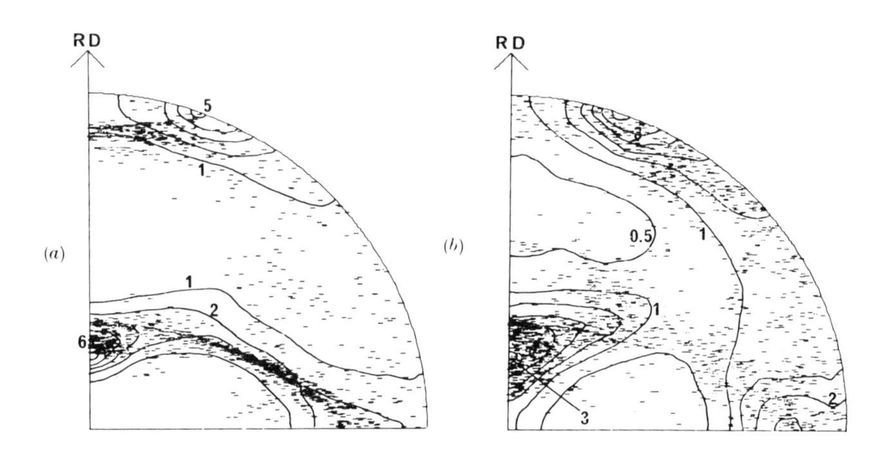

Fig. 11.4 Comparison of {111} pole figures between theory (array of points) and experiment (equi-intensity contours) for Cu(a) and brass (b), for a deformation of 80% reduction by rolling. (from Kallend and Davies, 1972, courtesy of Taylor & Francis)

tion is zero. One major component is a spread around {4, 4, 11} <11, 11, 8> (near {112} <111>). There are also some minor components. For comparison, fig. 11.6a gives the observed skeleton lines in rolled Cu which agrees well with calculation. Figure 11.6b is a similar figure for steel, a material with BCC structure. It is practically identical to that of Cu if the rolling and normal directions are reversed. The primary slip mode in BCC metals is {110} <111>, which is the same as {111} <110> except for an interchange of slip plane and direction. This interchange does not alter the generalized Schmid factor. Hence all the solutions for {111} <110> slip apply to {110} <111> slip, except that the sense of lattice rotation is reversed (see eqns. 11.5, and 11.6). This reversal accounts for the interchange of the normal and rolling directions between figs. 11.6a and 11.6b. The same reasoning accounts for the observed <110> tension texture and <111> + <100> compression texture in BCC metals as obtainable from their FCC counterparts.

Other slip modes are known to occur in BCC metals. These include {112} <111> and {123} <111> slips as well as <111> pencil glide. The results for axisymmetric flow (Chin, Mammel and Dolan, 1967; Rosenberg and Piehler, 1971), however, indicate only minor variations are expected, with a tendency for a somewhat greater ratio of <111> to <100> in the compression texture. For plane strain deformation, the results of calcula-

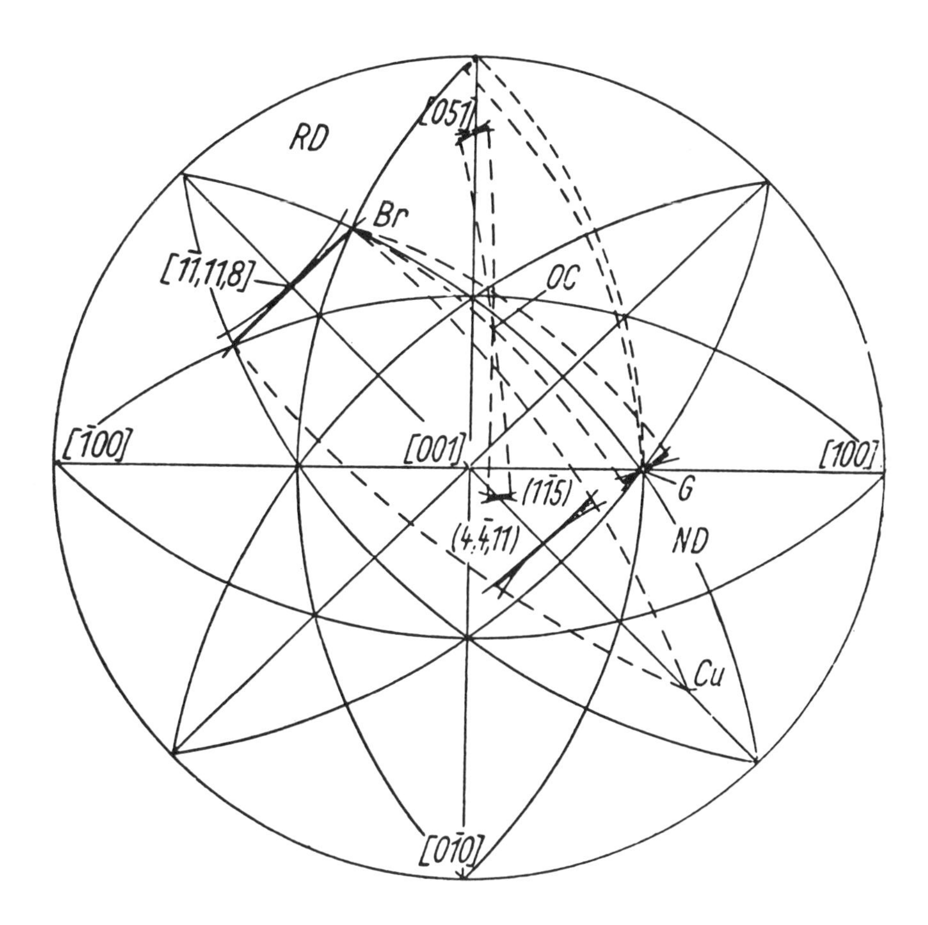

Fig. 11.5 The skeleton lines of predicted stable orientations for FCC metals deformed in plane strain. See text for details. (from Bunge, 1970, courtesy of Akademie Verlag)

tions for <111> pencil glide (Dillamore and Katoh, 1971) are also quite similar to those for {110} <111> slip (Bunge, 1970).

11.6 IONIC MATERIALS OF CUBIC STRUCTURE

Ionic crystals with the rock salt structure generally exhibit primary {110} <110> slip at low temperature. At elevated temperature additional slip on {100} <110> and {111} <110> systems are observed. Those with the fluorite structure show the opposite trend, the {100} <110> slip mode being predominant at

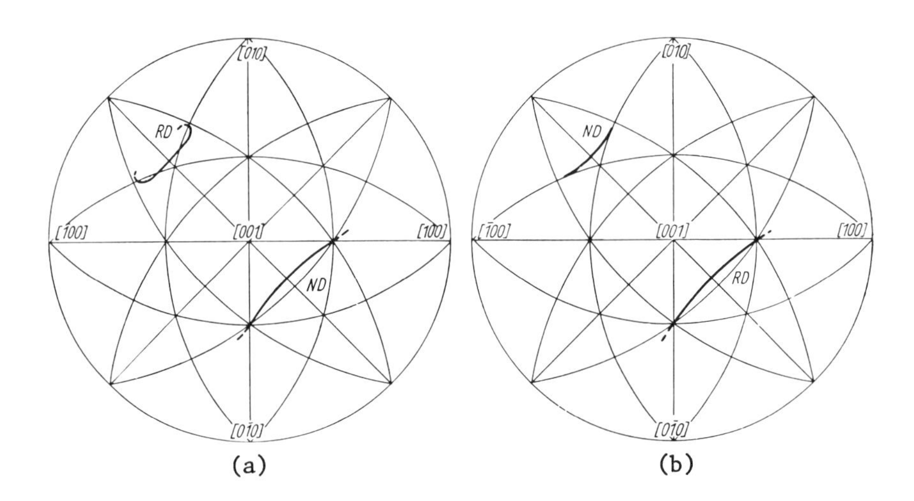

Fig. 11.6 The skeleton lines of texture observed in 90% cold-rolled Cu(a) and 70% cold-rolled steel (b). (from Bunge, 1970, courtesy of Akademie Verlag)
Note similarity of the two figures with the interchange of RD and ND.

low temperature and the {110} <110> mode appearing as the additional mode at elevated temperatures.

A detailed general analysis of the crystal plasticity of ionic crystals, including texture predictions, was given recently (Chin, 1973, 1974; Chin and Mammel, 1973, 1974). For general plasticity which requires the activation of five independent slip systems, one can either activate {111} <110> slip alone, or one of the combinations involving {111} + {100}, {111} + {110}, {111} + {100} + {110} and {100} + {110}, all with <110> as slip direction. The {100} and {110} slip modes cannot be activated alone because they respectively possess three and two independent slip systems only, which are below the required value of five. Which combination of slip modes will be activated is solely dependent on the relative value of the critical resolved shear stress (CRSS) for slip among the three slip modes. This has been worked out.

As far as the texture predictions are concerned, there is surprisingly little difference among the various slip mode combinations, for the case of axisymmetric flow. Figure 11.7 shows the calculated lattice rotations for axisymmetric tension, allowing slip to occur on {111}, {100} and {110} planes with the same value of CRSS. The results are quite similar to that of fig. 11.1. It would be of interest to see if the plane

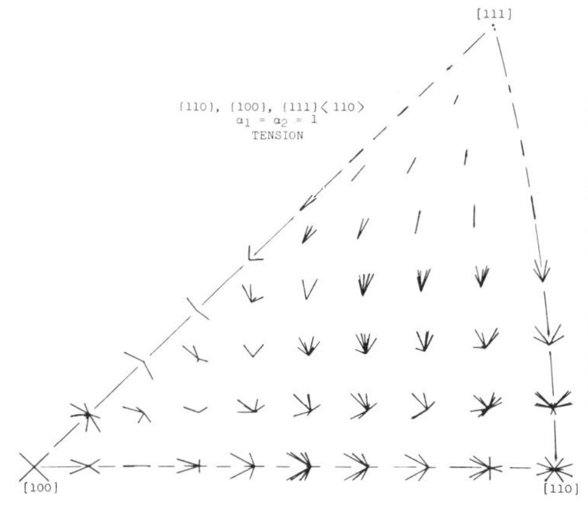

Fig. 11.7 Computer-plotted lattice rotations (ε_{xx} = 0.05) for axisymmetric tension for {111} <110> + {110} <110> slip assuming equal values of CRSS. (from Chin and Mammel, 1974, courtesy of AIME). Compared with fig. 11.1.

strain calculations are also similar.

The general analysis also treated the case of approximate <110> pencil glide by dividing the zone of slip planes containing a <110> slip direction into a number (up to 12) of equally spaced planes. A Bishop and Hill type analysis of true <110> pencil glide (continuous distribution of slip planes) was also worked out.

11.7 HEXAGONAL CLOSE-PACKED METALS

The crystal plasticity of HCP metals is complicated by the applearance of many slip and twinning modes. The most common slip mode is basal slip, (0001) $\langle 2\bar{1}\bar{1}0\rangle$, while prism $\{01\bar{1}0\}$ $\langle 2\bar{1}\bar{1}0\rangle$ and pyramidal $\{01\bar{1}1\}$ $\langle 2\bar{1}\bar{1}0\rangle$ slip modes have also been observed. The prism slip mode is favored over basal for metals of low c/a ratio, such as Zr and Ti. Another form of pyramidal slip, $\{11\bar{2}2\}$ $\langle\bar{1}\bar{1}23\rangle$ has been reported for zinc and cadmium. The most common twinning system is $\{10\bar{1}2\}$ $\langle\bar{1}011\rangle$, while additionally $\{10\bar{1}1\}$ $\langle 10\bar{1}\bar{2}\rangle$ and $\{10\bar{1}3\}$ $\langle 30\bar{3}\bar{2}\rangle$ have been cited for Mg and $\{11\bar{2}1\}$ $\langle\bar{1}\bar{1}26\rangle$ and $\{11\bar{2}2\}$ $\langle 11\bar{2}\bar{3}\rangle$ have been observed in Ti and Zr. Additional slip and twinning modes have been reported, but the above list is sufficient to point out the difficulty of a generalized treatment for hcp metals. As a result, some simplifications have been worked out although much work remains to be done.

One simplification concerned an analysis of mixed slip among (0001) basal, $\{01\bar{1}0\}$ prism and $\{01\bar{1}1\}$ pyramidal modes (Chin and Mammel, 1970). These slip modes together provide four independent slip systems (extension or compression in c-axis direction is not possible). Depending on the relative values of the CRSS, the four systems can be made up of (1) pyramidal slip only, (2) pyramidal plus prism, (3) pyramidal plus basal, (4) pyramidal, basal plus prism and (5) basal plus prism. The various regions of allowed slip modes can be presented in a deformation mode diagram as shown in fig. 11.8. Hence once

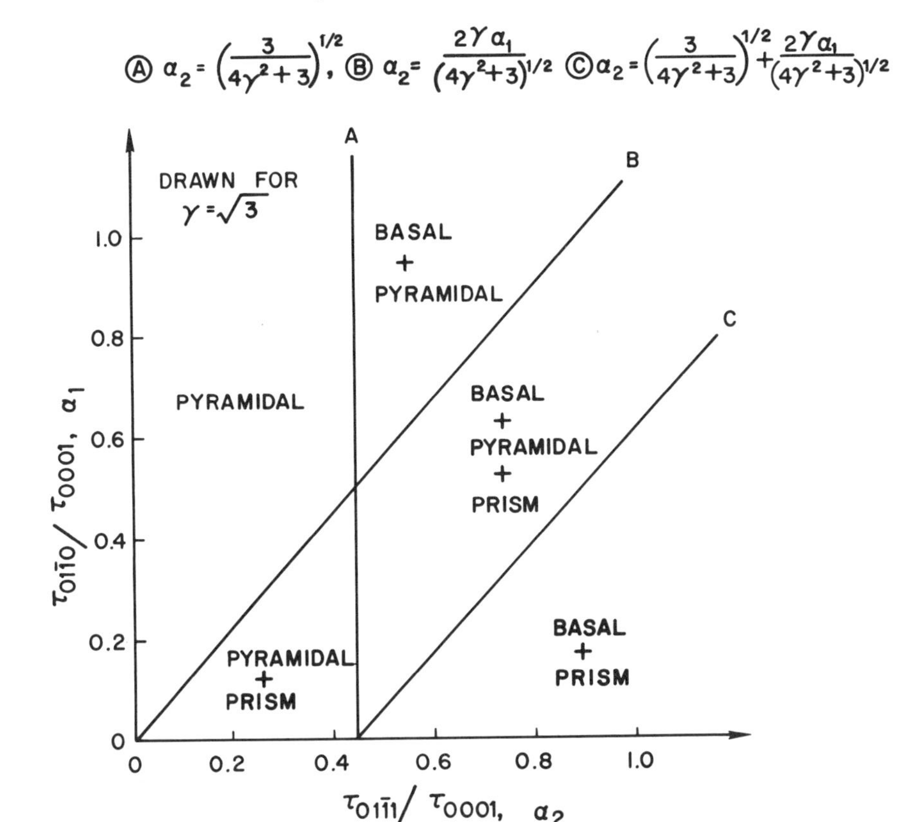

Fig. 11.8 Deformation mode diagram showing various regions of permissible activity of slip modes in HCP metals. Equations of the lines are marked on top of figure, which was drawn for a value of $\gamma = \sqrt{3}$. ($\gamma \equiv c/a$). (from Chin and Mammel, 1970, courtesy of AIME)

the values of the CRSS are known, an analysis can be focussed in one region alone.

Thornburg and Piehler (1974) used a Bishop and Hill type approach and obtained a list of 336 stress states for the mixed basal, prism and pyramidal slip modes and identified these stress states in the various regions of fig. 11.8. In this way the number of stress states for a given location in fig. 11.8 is no more than 120 and can be as low as 42. In addition, they provide for c-axis extension by adding $\{11\bar{2}1\}\langle 11\bar{2}6\rangle$ or $\{10\bar{1}2\}\langle 10\bar{1}1\rangle$ to the mixed slip modes, and for c-axis compression by adding on the $\{11\bar{2}2\}\langle 11\bar{2}\bar{3}\rangle$ twinning mode. This was done specifically for those stress states applicable to Ti. The procedure consisted of substituting the four stress values of a given stress state to the resolved shear stress equations for all the twinning systems of one twinning mode. The fifth stress value can then be solved for each twinning system. The twinning system with the lowest value of the fifth stress is then selected as companion to the slip systems of that stress state.

Thornburg and Piehler (1973) used the stress states to analyze the texture development in Ti and Ti-Al alloys. Some measure of success was gained in comparison with experiment. The discrete lattice rotation brought on by twinning, however, was not treated and further progress could be made along this line. For a general analysis of HCP metals the linear programming approach should prove to be more versatile.

Dillamore, et al. (1972) also pointed out the difficulty of a general approach in HCP metals where twinning plays an important role. In discussing texture development of plane strain magnesium extrusions, they considered the qualitative trends of lattice rotation developed by various slip and twinning systems consistent with plane strain deformation.

11.8 SUMMARY AND CONCLUDING REMARKS

In this chapter, the analysis of deformation texture development starts with a generalized analysis of crystal plasticity (Chin and Mammel, 1969). This analysis combines the minimum work analysis of Taylor (1938) and the maximum work analysis of Bishop and Hill (1951) into a linear programming problem in which they are duals of each other. The generalized analysis is then applied to Taylor's model of polycrystal deformation. The theoretical results adequately accounts for the observed tension and compression textures of FCC metals if mechanical twinning and differential hardening among octahedral slip systems are taken into account. Twinning is judged to have the greater effect on texture. Similar conclusions are reached by Kallend and Davies (1972) in their computer simulation of FCC rolling textures using the Bishop and Hill approach. For BCC metals, the change of slip mode from $\{110\}\langle 111\rangle$ to $\langle 111\rangle$ pencil glide is not expected to change the texture significantly (Dillamore and Katoh, 1971). Although some progress has been made in analyzing HCP metals (Chin and Mammel, 1970; Thornburg and Piehler, 1974), much more work is needed in the future. A better treatment of twinning is also required for all metals.

REFERENCES

Armstrong, R., Codd, I., Douthwaite, R. M., and Petch, N. J., (1962) Phil. Mag., 7, 45.

Barrett, C. S., (1942) Structure of Metals (N.Y.L McGraw-Hill), p. 386.

Bishop, J. F. W., (1954) J. Mech. Phys. Solids, 3, 130.

Bishop, J. F. W., and Hill, R., (1951) Phil. Mag., 42, 414, 1298.

Bunge, H. J., (1970) Kristall u. Technik, 5, 145.

Bunge, H. J., (1971) Kristall u. Technik, 6, 677.

Chin, G. Y., (1969) Textures in Research and Practice, edited by J. Grewen, and G. Wassermann (Berlin: Springer-Verlag), p.51.

Chin, G. Y., (1973) Met. Trans., 4, 329.

Chin, G. Y., (1974) Met. Trans. 5, 331.

Chin, G. Y., and Mammel, W. L., (1969) Trans. AIME, 245, 1211.

Chin, G. Y., and Mammel, W. L., (1970) Met. Trans., 1, 357.

Chin, G. Y., and Mammel, W. L., (1973) Met. Trans., 4, 335.

Chin, G. Y., and Mammel, W. L., (1974) Met. Trans., 5, 325.

Chin, G. Y., Hosford, W. F., and Mendorf, D. R., (1969) Proc. Roy. Soc. (London), A309, 433.

Chin, G. Y., Mammel, W. L., and Dolan, M. T., (1967) Trans. AIME, 239, 1854.

Chin, G. Y., Mammel, W. L., and Dolan, M. T., (1969) Trans. AIME, 245, 383.

Christian, J. W., (1970) Proc. 2nd Int. Conf. on Strength of Metals and Alloys, Vol. 1 (Metals Park, Ohio: ASM), p. 31.

Dillamore, I. L., and Katoh, H., (1971) BISRA Rept. No. MG/39#71; Quantitative Analysis of Textures (Proc. of Int. Seminar, Cracow, 1971), p. 315.

Dillamore, I. L., Hadden, P., and Stratford, D. J., (1972) Texture, 1, 17.

English, A. T., and Chin, G. Y., (1965) Acta Met., 13, 1013.

Hosford, W. F., (1964) Trans. AIME, 230, 12.

Kallend, J. S., and Davies, G. J., (1972) Phil. Mag., 25, 471.

Kocks, V. F., (1970) Met. Trans., 1, 1121.

Peck, J. F., and Thomas, D. A., (1961) Trans. AIME, 221, 1240.

Peissker, E., (1965) Z. Metallkde, 56, 155.

Rosenberg, J. M., and Piehler, H. R., (1971) Met. Trans., 2, 257.

Siemes, H., (1967) Z. Metallkde, 58, 228.

Taylor, G. I., (1938) J. Inst. Metals, 62, 307; S. Timoshenko Anniversary Volume (N.Y.: MacMillan), p. 218.

Thompson, A. W., (1974) Met. Trans., 5, 39.

Thornburg, D. R., and Piehler, H. R., (1973) Titanium Science and Technology, Vol. 2, edited by R. I. Jaffee, and H. M. Burte (N.Y.: Pergamon Press), p. 1187.

Thornburg, D. R., and Piehler, H. R., (1974) Met. Trans., 5, in press.

Venables, J., (1964) J. Phys. Chem. Solids, 25, 685, 693.

ABSTRACT. *Based on the results of earlier calculations of the interaction of dislocations and dislocation dipoles a set of integro-differential equations is derived, which governs the statistics of the production of dislocations and dipoles. Well established empirical relations allow the calculation of the plastic strain amplitude from the dislocation density. The results explain the oscillatory behavior of the strain amplitude in stress controlled tests ("strain bursts") as well as the monotonic cyclic hardening in strain controlled tests.*

12. MODELLING OF CHANGES OF DISLOCATION STRUCTURE IN CYCLICALLY DEFORMED CRYSTALS*

P. Neumann

12.1 INTRODUCTION

Cyclic deformation tests can be carried out in many different ways. One important parameter is the plastic strain amplitude $\hat{\gamma}_{p\ell}$. $2\hat{\gamma}_{p\ell}$ is the maximal amount of uninterrupted unidirectional deformation in such a test. Therefore it must be expected that the structure and properties produced by cyclic deformation are the more related to those produced by unidirectional deformation the larger the plastic strain amplitude $\hat{\gamma}_{p\ell}$ is. We are interested, however, in the effects due to the cyclic nature of the deformation. These features will become dominant only if there are as many reversals of the direction of deformation as possible per total amount of strain,

*Work performed at Argonne National Laboratory, Argonne, Illinois, under the auspices of the U. S. Atomic Energy Commission.

i.e. if the plastic strain amplitude is as small as possible. The amplitude cannot be chosen arbitrarily small, because cyclic hardening up to high cycle fatigue stresses must be obtained within a reasonable number of cycles. Because of these reasons we shall confine ourselves to experiments with average plastic strain amplitudes of the order of 10^{-3} to 10^{-4}.

Cyclic hardening under such conditions produces a hardened state which is quite different from that produced by monotonic deformation: It has less internal stresses and is less stable. We shall discuss both points in more detail:

The reduction of long range internal stresses after cyclic deformation was first observed with the help of X-ray diffraction (Kemsley 1958/59) and nowadays more directly by observation of the radius of curvature of dislocations which were pinned under stress (Kralik and Mughrabi 1973, Grosskreutz and Mughrabi 1974). Electron microscopy (Basinski et al. 1969, Hancock and Grosskreutz 1969) shows very clearly, that dislocations of opposite sign are thoroughly mixed so that their long range stresses cancel and that dipole stresses are left at most. These dipoles are clustered in walls of high dislocation density. This indicates that perhaps even the dipole stresses are cancelled. Altogether the experiments show that shaking of dislocations by cyclic deformations with small strain amplitude puts the dislocations in a very low energy configuration.

This low energy configuration is, however, not very stable in the following sense: If a cyclically hardened crystal is monotonically deformed, it shows a long low work-hardening range (Broom and Ham 1959, Basinski et al. 1969) and the deformation occurs on extremely long, straight and sharp slip lines. This indicates, that the dislocation motion is hindered only up to a certain applied stress level. As soon as this is exceeded, gross dislocation motion is possible without further increase of the applied stress. Furthermore the long and sharp slip lines indicate, that deformation in strained regions is even easier than in others. Basinski et al. (1969) pointed out very clearly that it is surprising to get long and straight slip lines starting from a dislocation structure with walls of very high dislocation density. He concluded from these facts that this dislocation structure in itself must be metastable.

An even more spectacular manifestation of this lack of stability is the strain bursts which can be observed during stress controlled cyclic tests with slowly increasing stress amplitude (Neumann 1967, 1968, 1970). Electron microscopic pictures of underlying dislocation structure (pinned under stress) show again patches or bundles of edge dislocation dipoles in a material which is almost free of long range internal stresses (Kralik and Mughrabi 1973, Grosskreutz and Mughrabi 1974, cf. Chapter 7). Because of the very small plastic strain amplitude between the bursts (can be of the order of 10^{-4}) these dipoles must be stable against the applied stress. If the stress amplitude is, however, continually increased, the widest dipoles will soon become unstable and dissociate into free dislocations which will interact with the remaining dipoles. Therefore the following events will be determined by the interaction of free dislocations and dislocation dipoles. If this interaction is such that free dislocations aid the applied stress to break up the dipoles, this would lead to a chain reaction producing an avalanche of free dislocations and thus a strain burst. Such effects could account for most of the observations discussed above.

In the following the most important experimental facts about the strain bursts will be reported, followed by a short summary of the interactions between dislocations and dipoles, and finally an attempt to model the changes in dislocation and dipole densities will be described.

12.2 EXPERIMENTAL FACTS ABOUT STRAIN BURSTS

Most of the results reported here can be found in a previous paper by the author (Neumann 1968). On the other hand, the low temperature results shown in figs. 12.1 to 12.3 are new. They were obtained with the same mechanical equipment as the old results but the values of the plastic strain amplitude were recorded digitally after every third half-cycle. The large data sets obtained in this way were evaluated and plotted with the help of a computer.

When FCC or HCP metal single crystals of single slip orientation are cyclically deformed in a stress controlled machine (e.g. an electrodynamic vibrator) under slowly increasing stress amplitude, the plastic deformation happens almost exclusively in the form of regularly spaced strain bursts (cf. fig. 12.1). The last but one burst in fig. 12.1 is shown in full detail in fig. 12.2. Single cycles can be resolved. It was demonstrated experimentally that the slip processes which make up a single burst are evenly distributed all over the gauge length. Most likely slip activity starts at some place and spreads within the duration of the burst all over the specimen.

Every burst has a half width of about 50 cycles (cf. fig. 12.2). The spacing of the bursts is governed by the following law: If $\hat{\sigma}_i$ is the applied stress at which the i-th burst is

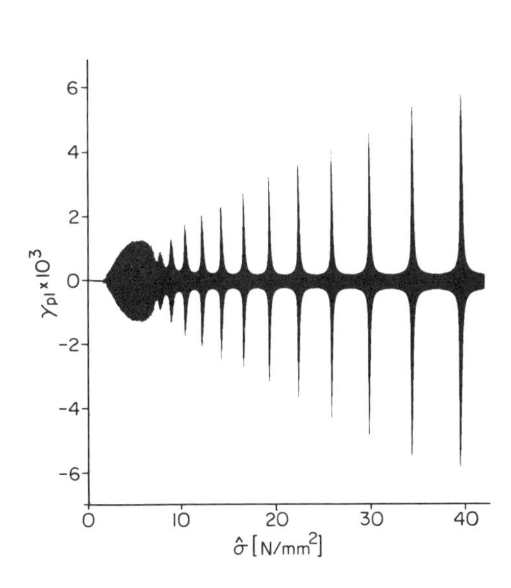

Fig. 12.1 Strain bursts in a middle-oriented copper single crystal at 78°K and 50 cps. The plastic strain $\gamma_{p\ell}$ is plotted versus the shear stress amplitude , which is increasing linearly with time at a rate of 7.1 10^{-3} N/(mm^2 cycle). Therefore the horizontal axis can also be regarded as a time axis. Since one cycle corresponds to 7.1 10^{-3} N/mm^2 stress amplitude increase only, single cycles cannot be resolved in this figure; the steps visible on the sides of the bursts are artefacts due to data reduction. (Note: 1 N/mm^2 = 10^2 g/mm^2)

observed, then

$$\frac{\Delta \hat{\sigma}_i}{\hat{\sigma}_i} = \frac{\hat{\sigma}_{i+1} - \hat{\sigma}_i}{\hat{\sigma}_i} = 0.14 \pm 0.01 \qquad (12.1)$$

i.e. the relative stress amplitude increase between neighboring bursts is constant. In order to resolve the bursts, their spacing must be larger than twice their halfwidth or

$$\Delta \hat{\sigma}_i > \frac{d\hat{\sigma}}{dN} \cdot 100 \text{ cycles}$$

from eqn. (12.1) we have

$$0.14 = \frac{\Delta \hat{\sigma}_i}{\hat{\sigma}_i} > \frac{d\hat{\sigma}}{\hat{\sigma}\, dN} 100 \text{ cycles} \quad \text{or} \quad \frac{d\hat{\sigma}}{\hat{\sigma}\, dN} < 1.4 \cdot 10^{-3}/\text{cycle} \qquad (12.2)$$

In fig. 12.1 with $d\hat{\sigma}/dN = 7.1 \cdot 10^{-3}$ N/(mm^2 cy) this condition

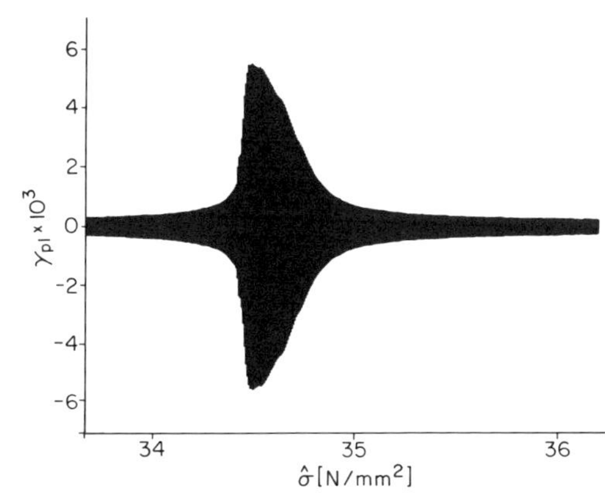

Fig. 12.2 The last but one burst out of fig. 12.1, shown with stretched horizontal axis and without data reduction. Single cycles can be resolved. The halfwidth of the burst is 44 cycles.

is fulfilled only for $\hat{\sigma} > 7.1 \cdot 10^{-3}/1.4 \cdot 10^{-3}$ N/mm$^2 \approx 5$ N/mm^2. This is the reason why we get overlapping bursts below this limit. From eqn. (12.2) it follows that an exponential stress amplitude increase with constant $d\hat{\sigma}/\hat{\sigma}\, dN$ would be more suitable to resolve bursts over a wide stress range. In fig. 12.3 this ideal case was approximated by selecting three different

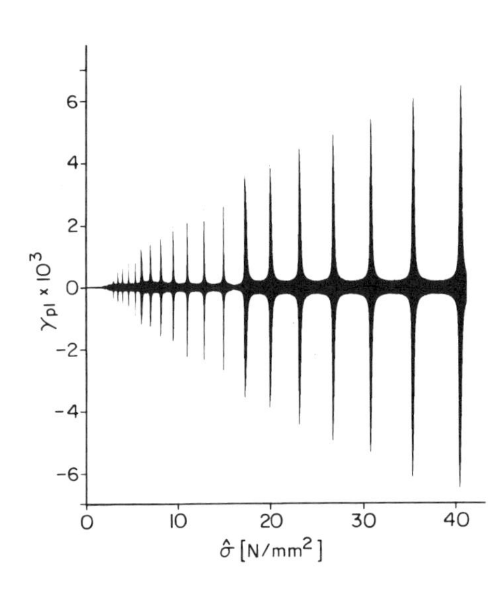

Fig. 12.3 The same as fig. 12.1 but with three different stress rates (0.9, 2.7, 7 x 10^{-3} N/(mm^2 cycle)) in three different stress ranges (0,5.5, 16, 52 N/mm^2) such that 1.6 10^{-4} < $d\hat{\sigma}/\hat{\sigma}\, dN$ < 5 10^{-4} for 1.9 N/mm^2 < $\hat{\sigma}$ < 42 N/mm^2

stress rates in three different stress ranges. In this way bursts could be obtained over more than a decade in stress. If the stress amplitude is increased too slowly ($d\hat{\sigma}/\hat{\sigma}\, dN$ < 10^{-4}/cycle) then the bursts are irregularly spaced.

The bursts are more regular at low temperatures than at room temperature. They were observed at frequencies down to 0.1 cps. Then, however, the maximal strain amplitude in a burst is only about twice the minimal strain amplitude between bursts. These results were obtained at room temperature. Corresponding tests at 78°K were not carried out.

Bursts were observed in all FCC or HCP metals which were tried (Neumann and Neumann 1970). In the BCC metals Iron and Niobium, bursts could not be detected. In alloys bursts exist but look very different (Desvaux 1970). Prestraining into stage II greatly reduces the size of the bursts, whereas dispersion hardening by particles like SiO_2 in copper (Hashimoto and Haasen 1974) are not detrimental for the appearance of bursts. In single crystals oriented for multiple slip as well as in polycrystals no bursts were found.

12.3 THE INTERACTION BETWEEN DISLOCATIONS AND DISLOCATION DIPOLES

In the optimal configuration the interaction between a dislocation and a dipole can be as strong as the interaction with another dislocation (Chen et al. 1964). The question remains, however, under which conditions these configurations are reached if a dislocation approaches a dipole. In a previous paper by the author this problem was treated in detail (Neumann 1971). Figure 12.4 shows the general configuration and defines

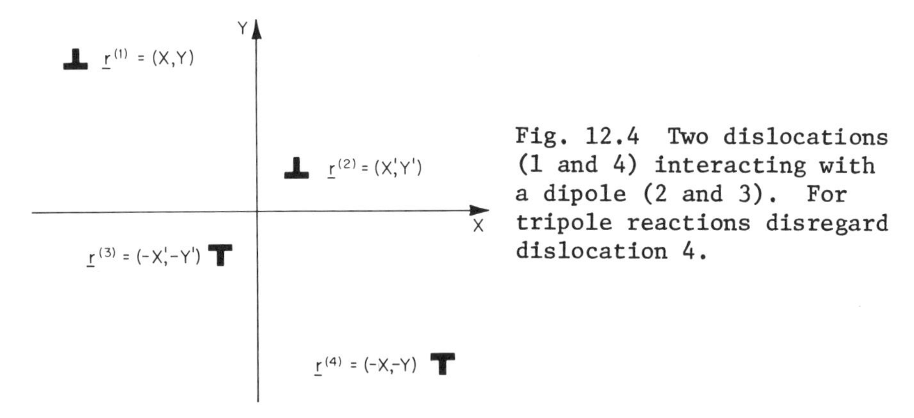

Fig. 12.4 Two dislocations (1 and 4) interacting with a dipole (2 and 3). For tripole reactions disregard dislocation 4.

the coordinates of the dislocations involved. For the discussion of the tripole reaction disregard dislocation 4. The y-coordinate of the approaching dislocation 1 as well as the width of the dipole are statistically distributed for different individual reactions but are constant during every reaction. Thus any possible configuration during an individual reaction can be specified by only two independent parameters X and X'.

In any macroscopic crystal such elementary reactions will always happen in a very short time during which the applied stress can be considered constant. Therefore we will calculate the reactions for constant applied stress only. The mechanical forces due to the applied stress and the dislocation stresses define a potential energy of the whole system, which can be a function of X and X' only. This function can be easily calculated and is plotted in three dimensions in fig. 12.5. Both X and X' are taken in units of the stress dependent parameter Y_c $(\hat{\sigma})$. $Y_c(\hat{\sigma})$ is half the critical dipole

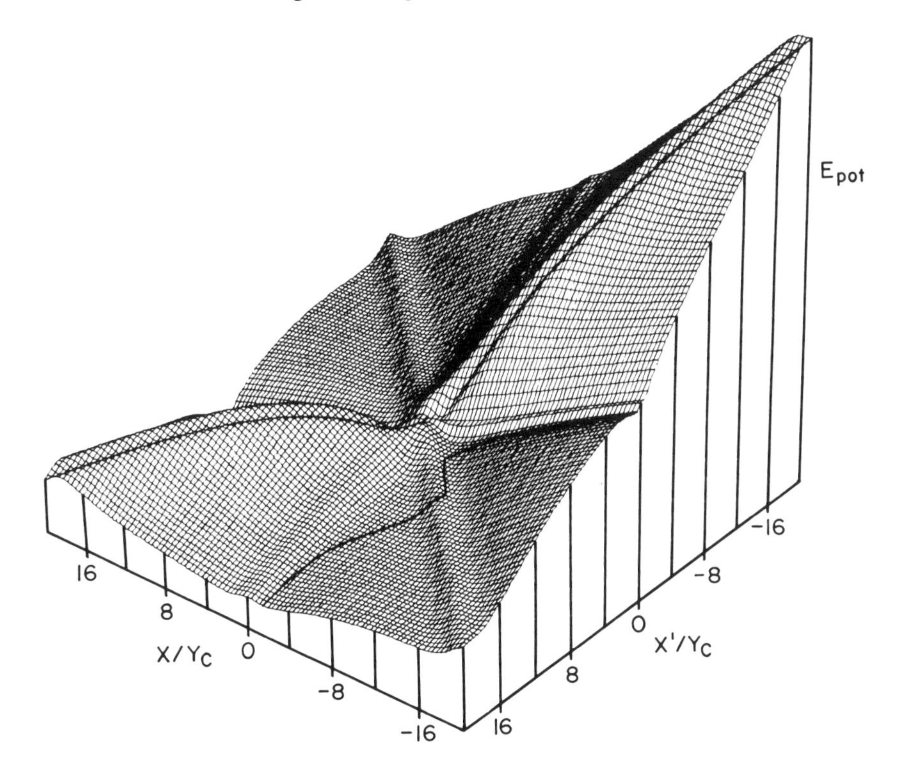

Fig. 12.5 Potential energy surface of a symmetrical quadrupole for $Y = 1.8\ Y_c$, $Y' = 0.4\ Y_c$. The path of the system during the reaction, starting from $X = -20\ Y_c$, $X' = Y'$, is drawn on the energy surface.

width, i.e. half of that dipole width above which a dipole is dissociated by the applied stress $\hat{\sigma}$. We have from the stress field of edge dislocations

$$Y_c = \frac{b\mu}{16\pi(1 - \nu)} \frac{1}{\hat{\sigma}} \tag{12.3}$$

Normalization of all lengths by Y_c makes the results valid for all values of the applied stress due to the principle of similitude (Kuhlmann-Wilsdorf, 1970).

The whole reaction is a set of successive configurations which can be represented as a path in the X,X'-plane. If we assume that the velocity of every dislocation is proportional to the total force exerted on it by the applied stress and the interaction stresses, it is easy to show (Neumann 1971) that the path of the reaction in the X,X'-plane is always parallel to the gradient of the potential energy. In fig. 12.5 such a path for a quadrupole reaction is shown. The reaction starts out with large negative X-values and a dipole (X' = Y'). Forced by the applied stress which tilts the whole energy surface towards large X and X' values, the free dislocations 1 and 4 approach the dipole (X becomes less negative). After some interactions X' becomes large (the dipole becomes dissociated). Beyond the range of fig. 12.5 X will also start increasing again. Thus after the reaction there are four free dislocations. The important point is that minima do exist in the energy surface. They correspond to stable quadrupole equilibria but they are not reached in the course of the reaction and therefore the dipole is dissociated instead of the approaching dislocations being trapped by the dipole.

Many reactions with this kind of starting configuration were calculated with one (tripole case) or two approaching dislocations (quadrupole case). The result of the reaction depends on the Y and Y' values (cf. fig. 12.4). Therefore the results of these calculations could be summarized in four phase diagrams (Neumann 1971) one of which is repeated here in fig. 12.6. Again all length parameters are taken in units of Y_c in order to make the figure valid for all applied stresses. Obviously Y' must be smaller than Y_c because otherwise the dipole would be dissociated by the applied stress before the reaction has started. This limits the range of the horizontal axis in fig. 12.6 from zero to unity. The vertical Y'-axis is not limited but for large Y values the approaching dislocation simply passes the dipole. For Y' values just below Y_c the dipole is almost always dissociated. For Y values close to

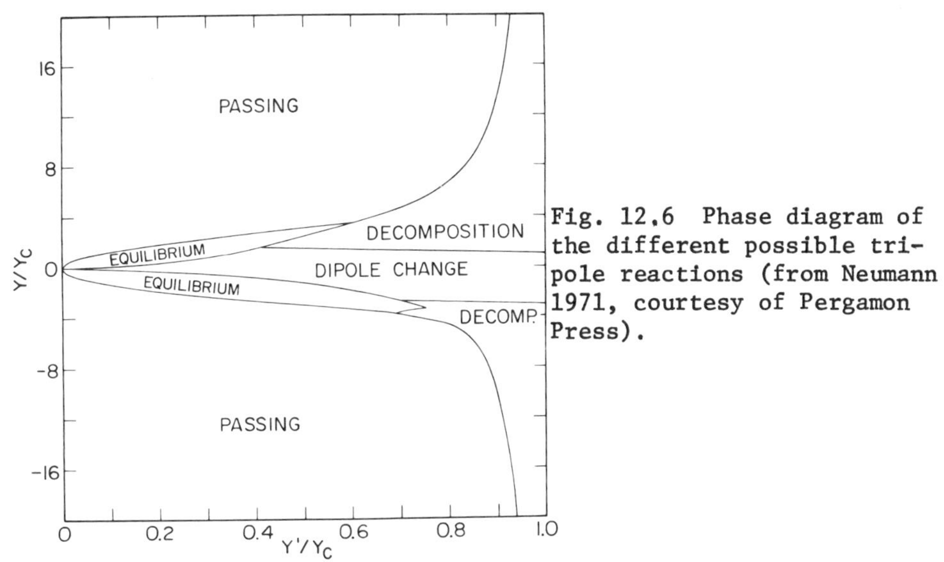

Fig. 12.6 Phase diagram of the different possible tripole reactions (from Neumann 1971, courtesy of Pergamon Press).

Y_c stable equilibrium configurations are reached or the dislocations involved re-arrange to form a new dipole during the reaction (dipole change). The most important point is that decomposition of the dipole is possible in the range $Y_c/2 < Y' < Y_c$ with the help of a dislocation. If two dislocations interact with the dipole, the decomposition field becomes considerably larger. Similar results were obtained, if more dipoles are involved (Hazzledine 1971).

12.4 THE STATISTICS OF THE REACTIONS BETWEEN DISLOCATIONS AND DISLOCATION DIPOLES

If we consider reactions in a real crystal we have an ensemble made up out of individual reactions with Y and Y' as statistical variables: Let $\tilde{D}(Y', N)$ be the distribution function of dipole widths, i.e. $\tilde{D}(Y', N)\, dY'$ is the number of dipoles with half widths between Y' and $Y' + dY'$ per unit area after N cycles. As far as Y is concerned, it is reasonable to assume that all values within a certain range are equally probable. We take as such a range that of fig. 12.6 ($|Y| \leq 20\, Y_c$).

We abbreviate

$$w = \frac{Y'}{Y_c} = Y' \alpha\hat{\sigma} \qquad \text{with} \qquad \alpha = \frac{16\pi(1 - \nu)}{b\mu} \tag{12.4}$$

Further we denote by $h_p(w)$ the total height of the passing fields of fig. 12.6 at $Y'/Y_c = w$. Correspondingly we define $h_d(w)$, $h_e(w)$, $h_c(w)$ for the decomposition, equilibrium and dipole change fields. From these definitions we have for all w

$$h_p(w) + h_d(w) + h_e(w) + h_c(w) = 40 \tag{12.5}$$

Since the probability for all Y values with $|Y| \leq 20\ Y_c$ was assumed to be the same, $h_p(w)/40$ is the probability to get passing in a reaction of a free dislocation with a dipole with normalized half width w. The symbols h_d, h_e, h_c have the corresponding meanings. Therefore we can get immediately from fig. 12.6 the conditional probabilities for getting a specific reaction if it is known that a reaction has taken place. From this it is easy to obtain the expectation values of the change of the number of dipoles per reaction: In a tripole reaction, passing and dipole change (one free dislocation is left over) leave the number of dipoles and free dislocations unchanged. Equilibrium reactions produce one tripole and consume one free dislocation whereas decomposition destroys one dipole and produces two new free dislocations. Now we make an important simplification by counting the tripole as one and a half dipole (on the average) having a width which is equal to the width of the dipole before the reaction. We can justify this simplification only by pointing out that the binding forces of a stable tripole - if it is once formed - are as large as those of a dipole (Cheng et al. 1964). In spite of that, this simplification will lead to severe errors but they do not obscure the key features of the situation - as the results will show.

The expectation value $p_t(w)$ for the change of the number of dipoles of width w per reaction of one dislocation with a dipole of width w therefore is

$$p_t(w) = \frac{1}{40}\left(1.5\ h_e(w) - h_d(w)\right) \tag{12.6}$$

Correspondingly we obtain for the quadrupole reaction

$$p_q(w) = \frac{1}{40}\left(h_c(w) + h_e(w) - h_d(w)\right) \tag{12.7}$$

Thus $p_t(w)$ and $p_q(w)$ can be obtained from the phase-diagrams (Neumann 1971). If we have in mind the severe simplifications which we introduce by considering tripole and symmetrical quadrupole reactions only, it seems adequate if we approximate these complicated functions in the following way by elementary functions:

$$\left.\begin{aligned} p_t(w) &= 0.2\left(1 - e^{7(w - 0.6)}\right) \\ p_q(w) &= 0.2\left(1 - e^{7(w - 0.5)}\right) \end{aligned}\right\} \begin{aligned} &\text{as long as } p_{t,q}(w) > -1, \\ &\text{otherwise } p_{t,q}(w) = -1. \end{aligned} \tag{12.8}$$

Let f be the conditional probability that a dipole interacts with just one dislocation (tripole reaction). Since we assume that only tripole and quadrupole reactions take place, the probability for quadrupole reactions must be $1 - f$. Thus

$$f\, p_t(w) + (1 - f)\, p_q(w) \tag{12.9}$$

is the expectation value per reaction for the change of the number of dipoles of width w if both kinds of reactions can occur. The functional dependence of f on the number of free dislocations will be discussed later on.

Finally, we consider the expectation value of the dipole production during one cycle of deformation. For this we have to make an assumption about the number of reactions per cycle. From the knowledge of the structure (Kralik and Mughrabi 1973) it seems to be reasonable to assume, that every free dislocation makes about one reaction per cycle. Let $\tilde{d}\,(N)$ be the number of free dislocations per unit area after N cycles. Then $\tilde{d}$ reactions will take place. The probability that a free dislocation will react with a dipole of width $Y' \pm dY'/2$ is

$$\frac{2\,\tilde{D}\,(Y',\,N)\,dY'}{\rho}$$

where ρ is the total dislocation density. Therefore we finally obtain

$$2\int_0^{Y_c} \tilde{D}\,(Y',N)\,dY' + \tilde{d}\,(N) = \rho \tag{12.10}$$

and

$$\frac{\partial \tilde{D}\ (Y',N)}{\partial N} = \frac{2\ \tilde{D}\ (Y',N)}{\rho}\ \tilde{d}\ (N)\ \ (f(N)\ p_t(w) + (1 - f(N))\,p_q(w)) \tag{12.11}$$

If we measure all densities in units of the total dislocation density by defining

$$\bar{D}(Y',N) = \frac{2\ \tilde{D}\ (Y',N)}{\rho} \quad ; \quad d\ (N) = \frac{\tilde{d}\ (N)}{\rho} \tag{12.12}$$

we can rewrite eqns. (12.10) and (12.11) as

$$2\int_0^{Y_c} \bar{D}\ (Y',N)\ dY' + d\ (N) = 1 \tag{12.13}$$

$$\left(\frac{\partial \bar{D}}{\partial N}\right)_{Y'} = 2\ \bar{D}\ d\left((f\ p_t + (1 - f)\ p_q\right) \tag{12.14}$$

These are the central equations which govern the statistics of the reactions. If an initial distribution $\bar{D}\ (Y',N_o)$ is given, $d(N)$ and $\bar{D}(Y',N)$ can be calculated from the coupled integro-differential equations (12.13) and (12.14).

Because all functions are given in terms of w and N we first express also $\bar{D}$ in terms of these variables by defining

$$D(w,N)\ dw = \bar{D}(Y',N)\ dY' \text{ at constant } N \tag{12.15}$$

It is important to distinguish carefully between the variables w and Y' as soon as we consider different cycles because from eqn. (12.4)

$$w = \alpha\, Y'\ \hat{\sigma}\ , \tag{12.16}$$

and $\hat{\sigma}$ usually depends on N .

Therefore eqn. (12.15) can be rewritten as

$$\bar{D} = D\left(\frac{\partial w}{\partial Y'}\right)_N = \alpha\ D\ \hat{\sigma}(N) \tag{12.17}$$

From eqn. (12.14) we have

$$2\,\bar{D}d\left(f\,p_t + (1-f)\,p_q\right) = \alpha\left(\frac{\partial(D\hat{\sigma})}{\partial N}\right)_{Y'} = \alpha\left[\left(\frac{\partial D}{\partial N}\right)_{Y'}\hat{\sigma} + D\hat{\sigma}'\right]$$

$$= \alpha\left[\left(\left(\frac{\partial D}{\partial N}\right)_w + \left(\frac{\partial D}{\partial w}\right)_N\left(\frac{\partial w}{\partial N}\right)_{Y'}\right)\hat{\sigma} + D\hat{\sigma}'\right] = \alpha\left[\left(\frac{\partial D}{\partial N}\right)_w\hat{\sigma} + \left(\frac{\partial D}{\partial w}\right)_N\right.$$

$$\left. Y'\alpha\,\hat{\sigma}'\hat{\sigma} + D\hat{\sigma}'\right] = \alpha\,\hat{\sigma}\left[\left(\frac{\partial D}{\partial N}\right)_w + \frac{\hat{\sigma}'}{\hat{\sigma}}\left(\left(\frac{\partial D}{\partial w}\right)_N w + D\right)\right]$$

yielding

$$\left(\frac{\partial D}{\partial N}\right)_w = -\frac{\hat{\sigma}'}{\hat{\sigma}}\left(\left(\frac{\partial D}{\partial w}\right)_N w + D\right) + 2\,Dd\left(f\,p_t + (1-f)\,p_q\right) \qquad (12.18)$$

eqn. (12.13) simply gives

$$2\int_0^1 D(w,N)\,dw + d(N) = 1 \qquad (12.19)$$

In this form the equations contain only dimensionless quantities within fixed ranges and are suitable for numerical solution. The first term in eqn. (12.18) describes the change in the distribution of the normalized width w by the change of the cycle dependent normalization factor Y_c, whereas the second term describes the changes due to real reactions.

A program was written which calculated stepwise the development of the distribution $D(w,N)$ for successive N's starting from a linear distribution

$$D(w,N_o) = 1 - (1 + d\,(N_o))\,w \qquad (12.20)$$

which fulfills eqn. (12.19).

The applied stress amplitude was increased exponentially and f was set to zero (only quadrupole reactions are allowed). The resulting distribution function is shown in a three-dimensional plot in fig. 12.7. After some damped oscillations a stationary distribution is obtained. A similar result is obtained for $f = 1$ (only tripole reactions allowed). The distribution function is, however, larger at $w > 0.5$ and smaller at $w < 0.5$. This reflects the fact that tripole

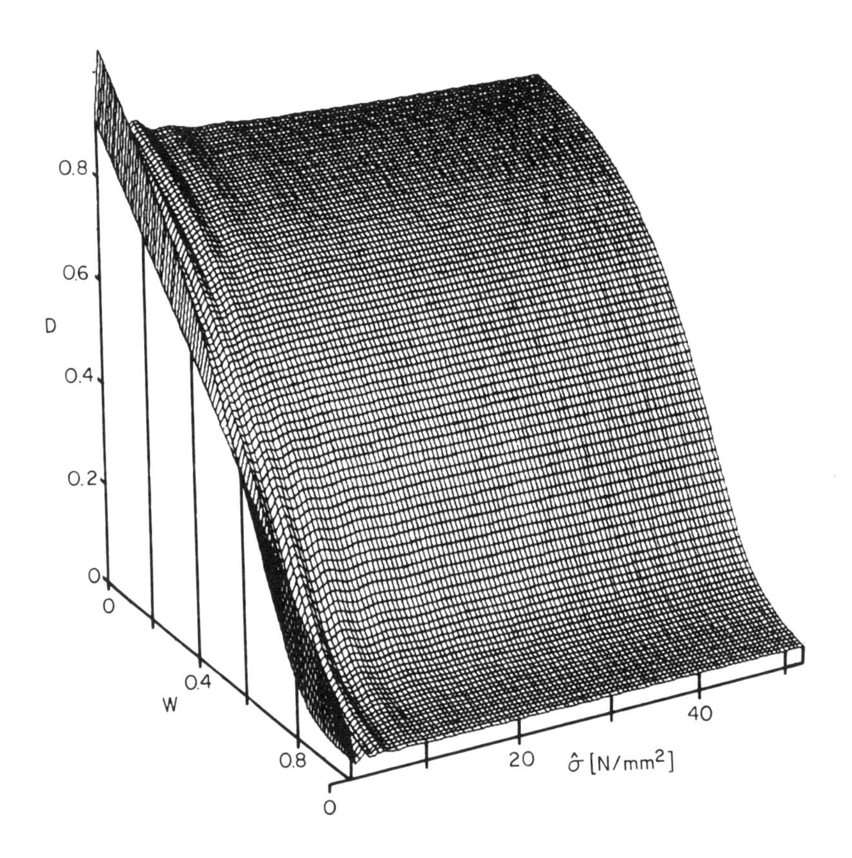

Fig. 12.7 Solution of eqns. (12.18), (12.19) for f = 0 (only quadrupole reactions are allowed) and $\hat{\sigma}'/\hat{\sigma} = 6.10^{-3}$.

reactions are less effective in destroying dipoles at $w > 0.5$. In both cases d is of the order of some per cent.

If f decreases with d , i.e. if the quadrupole reaction becomes more likely with increasing number of free dislocations then a chain reaction is possible in either direction: an increase in d favors the quadrupole reaction which is more effective in destroying dipoles, i.e. which produces more dislocations. But the opposite is true also: a decrease in d favors the tripole reaction which produces less free dislocations. Therefore the distribution function should oscillate between the two stationary distributions belonging to the tripole and quadrupole reactions alone. Figure 12.8 shows the

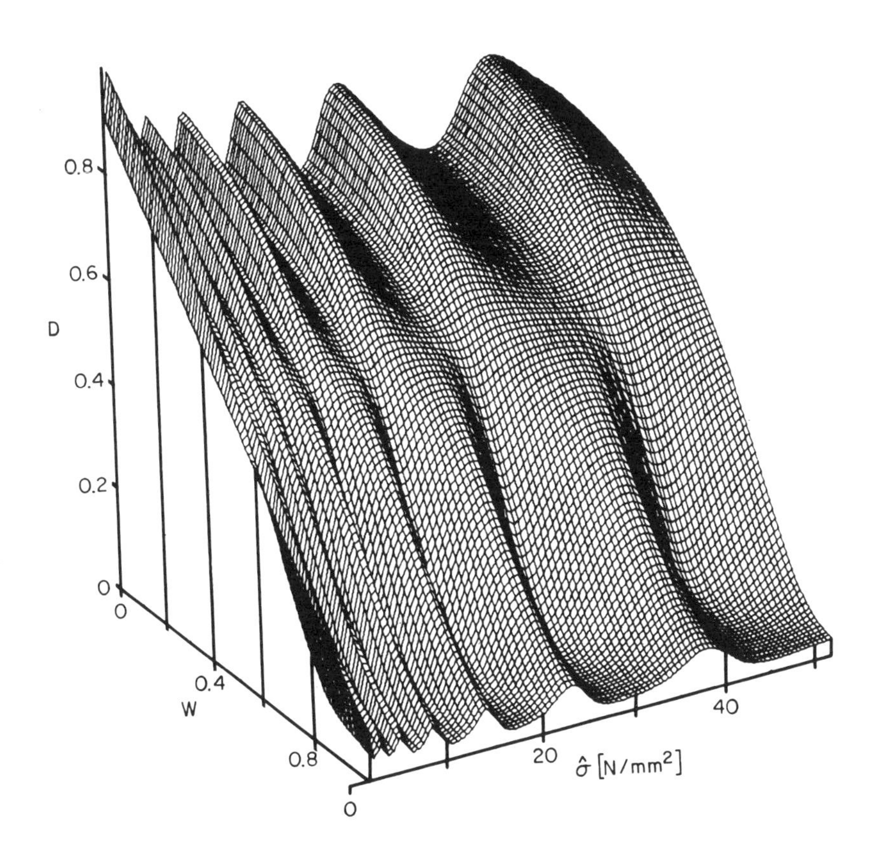

Fig. 12.8 As fig. 12.7 but now with f = exp (-50.d). Thus both tripole and quadrupole reactions are allowed. The distribution function oscillates between those corresponding to tripole or quadrupole reactions alone (cf. fig. 12.7).

distribution resulting from

$$f = \exp(-50\, d)\ , \tag{12.21}$$

which indeed behaves as discussed above. The dependence of f on d, given by eqn. (12.21), was selected on the following grounds: f has to be always less than one, for $d = 0$ it has to be 1, (only tripole reactions) and for $d = 0.05$ the quadrupole reaction has to be dominant already, i.e. $f \approx 1$ (exp (-50 x 0.05) = 0.082). From a statistical point of view the last condition is certainly not a very realistic one since the probability to find two dislocations near a dipole is definitely smaller than the probability to find only one dislocation reacting with a dipole if the number of dislocations is only ten per cent of that of the dipoles. In a real crystal, however, the possibility of chain reactions amplifies fluctuations in the densities of dislocations and dipoles and thus produces strong inhomogeneities which spread through the crystal. In a completely homogeneous theory like ours we have to incorporate these effects in a very integral way by exaggerating the dependence of f on d.

In addition to the distribution function $D(w,N)$ the solution of eqn. (12.18) and (12.19) yields the number of free dislocations $d(N)$ as a function of cycles. By using well established empirical relations the plastic strain amplitude $\hat{\gamma}_{p\ell}(N)$ can be calculated from $d(N)$: Any plastic strain γ is connected with the density of mobile dislocations ρ_m, the mean free path of a single mobile dislocation ℓ, and the Burgers vector b by the relation

$$\gamma = \rho_m b\ell . \tag{12.22}$$

In our case γ has to be set equal to $2\hat{\gamma}_{p\ell}$, the maximal unidirectional strain during one cycle. ρ_m is obviously equal to $d\cdot\rho$. For ℓ we take the cell diameter for which a relation of the form

$$\ell = \frac{\beta}{\hat{\sigma}}$$

is well established (e.g. Staker and Holt 1971). In cyclic deformation ℓ is 1.5 μ at a stress amplitude of 30 N/mm^2 (Woods 1973) giving a value of $\beta = 0.05$ N/mm. This yields from eqn. (12.22)

$$2\hat{\gamma}_{p\ell} = d\rho b\beta \frac{1}{\hat{\sigma}(N)} \tag{12.23}$$

Another well established empirical relation connects the total dislocation density $\rho(N)$ with the yield stress, which in our tests is equal to the applied stress amplitude $\hat{\sigma}$ within 10 per cent:

$$\rho = \delta\hat{\sigma}^2$$

According to Grosskreutz and Mughrabi (1974) ρ is of the order of $2.10^{10}/cm^2$ at a stress amplitude of 30 N/mm^2 giving a value of $\delta = 2.10^5 \; mm^2/N$. Using this in eqn. (12.23) gives

$$\hat{\gamma}_{p\ell}\,(N) \;=\; d(N)\,\hat{\sigma}(N)\;10^{-3}\;\frac{mm^2}{N} \tag{12.24}$$

in fig. 12.9 $\hat{\gamma}_{p\ell}$ is plotted according to eqn. (12.24). Strain bursts do indeed occur. A comparison with fig. 12.3 shows that all parameters are of the right order of magnitude:

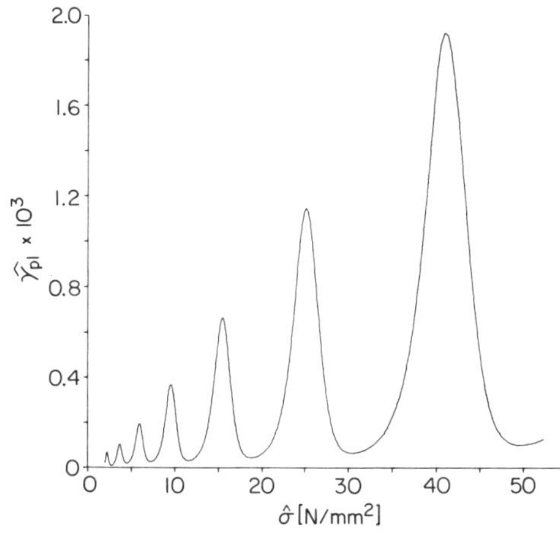

Fig. 12.9 Plastic strain amplitude $\hat{\gamma}_{p\ell}$ as a function of the applied stress amplitude $\hat{\sigma}$ according to eqn. (12.24) using the solution of eqns. (12.18), (12.19) for the same parameters as in fig. 12.8.

The height of the bursts is about one third of the measured height; the calculated spacing is about three times as large as in reality.

12.5 DISCUSSION

The agreement between the calculated bursts in fig. 12.9 and the measured bursts in fig. 12.3 is astonishingly good in the light of the severe simplifications which were done in the course of deriving the central equations (12.18) and (12.19):

a. The whole theory is strictly homogeneous and does not allow for inhomogeneities which are inevitable as soon as instabilities occur.

b. Only straight dislocations were considered. Although electron microscopic studies of Kralik and Mughrabi (1973) show that under the conditions necessary for bursts, the internal structure is free of internal stresses and contains only small amounts of secondary dislocations, pinning and bowing out of dislocations certainly will occur (with strictly straight dislocations no multiplication is possible).

c. Only reactions of the dislocations with dipoles were considered. This means, if a dislocation is running into a patch of dipoles we considered only the interaction with the dipole next to the dislocation. This simplification can be justified to some extent by the very short range of the stress fields of dipole clusters (even dipole fields are cancelled).

From the results of the calculations it can be concluded, however, that the main features of the situation are well described, e.g. at the first glance one might expect, that in a strain controlled test one should get oscillations in the stress amplitude. To the knowledge of the author such oscillations have not yet been found. Instead, linear hardening is observed if the strain amplitude is kept constant (Basinski et al. 1969, Hancock and Grosskreutz 1969). The very same equations (12.18) and (12.19), which describe the oscillatory behavior under stress control give also monotonic solutions for $\hat{\sigma}$ (N) if $\hat{\gamma}_{p\ell}$ (N) is prescribed: Numerical solutions of eqns. (12.18) and (12.19) under the boundary conditions of constant plastic strain amplitude give linear hardening with errors of the order of 2 per cent. For an exponentially increasing strain amplitude the corresponding result (exponentially increasing stress amplitude) can be even proven without numerical methods: If we try as a solution

$$\hat{\gamma}_{p\ell}(N) = \gamma_o \, \ell^{\chi N} \; ; \; \hat{\sigma}(N) = \sigma_o \, \ell^{\chi N} \; ; \; D = D(w) \qquad (12.25)$$

equation (12.24) yields $d = \text{const.}$ Then eqns. (12.18) and (12.19) reduce to

$$0 = -\chi \, (D'w + D) + 2 \, D \, d \, (f \, p_t + (1 - f) \, p_q)$$

$$2\int_0^1 D(w) \, dw + d = 1 \qquad (12.26)$$

All quantities in these equations do not depend on N. Therefore eqn. (12.26) has a solution $D(w)$ which is independent of N , i.e., eqn. (12.25) is a solution of eqns. (12.18), (12.19) if for $D(w)$ the existing solution of eqn. (12.26) is used.

It should be noted that eqn. (12.25) contains the same

stress increase which was used as a boundary condition for the numerical solution for the burst of fig. 12.9. Thus the oscillating solution shown in fig. 12.8 is obviously not the only solution which is compatible with the given $\hat{\sigma}$ (N) . Equation (12.25) is another one which is not stable, however, if no further conditions exist for $\hat{\gamma}_{p\ell}$ (N) . If $\hat{\gamma}_{p\ell}$ (N) is controlled to be non-oscillating it definitely is compatible with a monotonically increasing $\hat{\sigma}$ (N) . In reality non-oscillating behavior of $\hat{\gamma}_{p\ell}$ (N) may also be obtained by oscillating solutions in different parts of the crystal which are out of phase and add up to the smooth function required by the control. Based on the present knowledge it cannot be decided whether this really is what happens; in any case the eqns (12.18) and (12.19) represent a homogeneous model, which renders solutions for both kinds of tests.

12.6 CONCLUSIONS

a. Two integro-differential equations were derived which govern the statistics of dipole formation and dissociation under cyclic deformation.

b. Solutions were found numerically, which can explain the phenomenon of strain bursts during stress controlled tests.

c. The same equations have also solutions which describe linear hardening in strain controlled tests with constant plastic strain amplitude.

d. Depending on the boundary conditions oscillating or monotonic behavior of $\hat{\gamma}_{p\ell}$ (N) are both compatible with the same stress history $\hat{\sigma}$ (N)

ACKNOWLEDGEMENT

The author is grateful to U. F. Kocks and R. O. Scattergood for many stimulating discussions during the course of this work.

REFERENCES

Basinski, S. J., Basinski, Z. S., and Howie, A., (1969) Phil. Mag, 19, 899.

Broom, T., and Ham, R. K., (1959) Proc. Roy. Soc., (London) A251, 186.

Chen, H. S., Gilman, J. J., and Head, A. K., (1964) J. Appl. Phys., 35, 2502.

Desvaux, M. P. E., (1970) Z. Metallk., 61, 206.

Grosskreutz, J. C. and Mughrabi, H., (1974) in "Constitutive Equations in Plasticity", edited by A. S. Argon, (Cambridge, Mass: M.I.T. Press) p. 251.

Hancock, J. R., and Grosskreutz, J. C., (1969) Acta Met., 17, 77.

Hashimoto, O., and Haasen, P., (1974) Z. Metallk., 65, 178.

Hazzledine, P. M., (1971) Scripta Met., 5, 847

Kemsley, D. S., (1958/59) J. Inst. Met., 87, 10.

Kralik, G., and Mughrabi, H., (1973) in Proc. III. Int. Conf. on the Strength of Metals and Alloys, (Cambridge: Institute of Metals) p. 410.

Kuhlmann-Wilsdorf, D., (1970) Met. Trans., 1, 3173.

Neumann, P., (1967) Z. Metallk., 58, 780; (1968) Z. Metallk., 59, 927; (1971) Acta Met., 19, 1233.

Neumann, R., and Neumann, P., (1970) Scripta Met., 4, 645.

Staker, M. R., and Holt, D. L., (1971) Naval Research Technical Report No. 14-67-A204-39, NR 31-741-3.

Woods, P. J., (1973) Phil. Mag., 28, 155

ABSTRACT. *In Sections 1 and 2, a macroscopic approach is presented insofar as no structural details are considered, but rather represented by an overall-parameter ξ. A straightforward general method to obtain strain rate and strain at any moment for given initial conditions is formulated as an iterative process. In Section 3, this general framework is filled with a set of model equations for random dislocation distributions. In Section 4, the treatment is extended to cellular dislocation arrangements. Essential assumptions are that dislocation generation is a reaction within the cell volume, controlled by the effective stress. Dislocation annihilation, in contrast, is a reaction at the cell wall, controlled by the internal stress. The cell diameter is found to adjust to changes in stress, decreasing inversely with increasing σ. Mechanisms for this cell growth/cell refinement are discussed.*

13. MODELLING OF CHANGES OF STRAIN RATE AND DISLOCATION STRUCTURE DURING HIGH TEMPERATURE CREEP

B. Ilschner

13.1 INTRODUCTION

This chapter focuses on the dislocation structure during high temperature deformation. Experiments show that this structure is not only characterized by a more or less homogeneous dislocation network, but - in addition - by regular arrays which may be described as cell walls, (see also Chapter 8). For an example for a non-metallic material - fig. 13.1.

A further result of microstructural investigations is that cell walls as well as dislocation arrays within the cells are not permanent elements of the structure. Rather, they are subject to growth and decay, to knitting and rearranging processes. In particular, cell walls may absorb or release individual dislocations. This variable substructure will be described in the following by two variables: the dislocation density ρ (not including those dislocations which are integrated into cell walls) and the average cell diameter, ℓ_c. The latter may be replaced, for convenience, by the cell wall

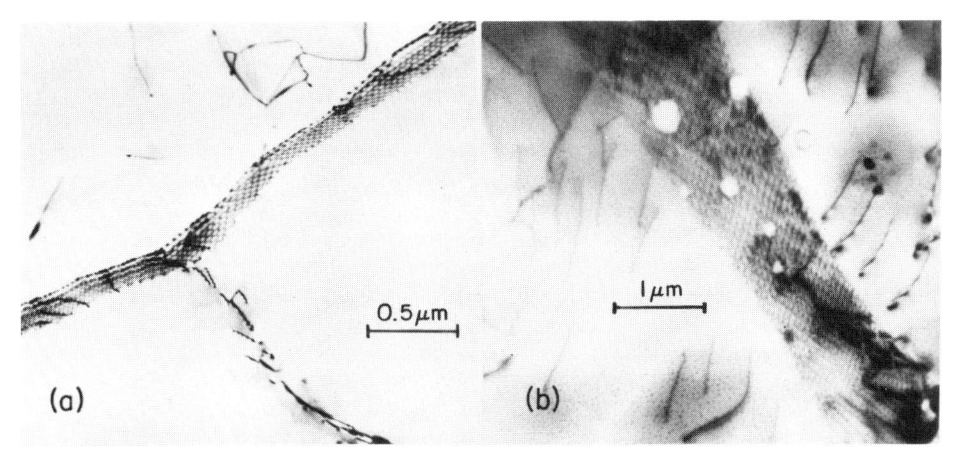

Fig. 13.1 Dislocation cell walls in hot deformed
a) magnesium oxide single crystal, deformed 40% at 1970 K and 85 MN/m^2, (from Hüther, 1973)
b) fayalite (ferrous silicate) hot pressed to a strain of 20% at 1120 K and 50 MN/m^2, (from Streb, 1974).

area per unit volume, $S_c \cong 1/\ell_c$

At the beginning of service life - or of a test - a set of structural data ρ_o and ℓ_{co} exists, which is inherited from the prior thermal and mechanical history of the sample under consideration. Starting from here, thermally activated dislocation motion leads to a continuous change of the substructure. An important experimental result related to this structural change is that ρ and ℓ approach equilibrium values, ρ_e and ℓ_{ce}, if only σ and T are maintained at constant values for a sufficient period of time, equivalent to a sufficient amount of strain (say, 15 pct. in a single-phase metallic alloy). If σ and T are cycled between different values, corresponding equilibrium values of ρ and ℓ are approached in a reproducible manner, irrespective of prior history. They are functions of stress and temperature only, as long as no other structural parameters of the alloy (e.g. matrix composition, particle distribution) change simultaneously.

$$\rho_e = \rho_e(\sigma,T) \; ; \; \ell_e = \ell_{ce}(\sigma,T) \; . \qquad (13.1)$$

Neglecting structural changes other than those of the dislocation and cell wall array, it is an obvious assumption that the macroscopic strain rate, $\dot{\varepsilon}$, will depend only on the following three parameters: (1) the stress, providing the driving force for dislocation motion; (2) the temperature, controlling the thermally activated motion velocities; and (3) the dislocation structure, providing the deformation carriers on a microscopic scale.

Combining ρ and ℓ_c in the symbol ξ, this statement reads

$$\dot{\varepsilon} = \dot{\varepsilon}(\sigma, T, \xi) . \tag{13.2}$$

The structural equilibrium states, as mentioned, are characterized by constant ξ for constant σ, T. All variables in eqn. (13.2) being constant, then, the strain rate will also assume a constant value, $\dot{\varepsilon}_e$. This regime, in which the material deforms under constant stress at constant rate, is termed steady-state. Each steady-state, as discussed above, is approached from the preceding steady-state (or from another initial state) through a transient period; primary creep is an example for this.

Modelling constitutive equations may be understood as a process to construct either state equations or rate equations, or both. While state equations interrelate state variables under equilibrium conditions, rate equations endeavour to control the time-dependent change of a system. Clearly, both types of relation are within the scope of this chapter. We aim at providing a procedure to forecast the transition from one state of dynamic equilibrium in high-temperature deformation to another, following changes in load or temperature, and accompanied by reproducible structural changes.

In doing so, there should be no doubt about the fact that, in most technical applications, steady-state is never achieved.

13.2 "MACROSCOPIC" APPROACH

For a first step, microscopic features will be symbolized by ξ and not be discussed in any detail. Equation (13.2) will serve as a general basic equation. One will notice that it does not explicitly contain the strain, ε. That is, the treatment rests on the assumption that prior strain has an effect on the strain rate at a given moment only:

a) by changing the cross sectional area of the sample and, hence, changing σ at constant load in a creep-rupture test;

b) by making the sample stay for a long time at elevated temperature, thereby allowing secondary precipitation, Ostwald-ripening, and other diffusion-controlled transformations to proceed:

c) by changing the dislocation substructure, so that $\xi = \xi(\varepsilon)$.

It is the state variable ξ and not the "shape variable" ε which controls the process. The engineer, on the other hand, has to be interested in the rate of shape change, $\dot{\varepsilon}$. - The sequence of events envisaged may now be represented as indicated below:

initial state →	transient state →	steady-state
ξ_o	$\xi_o \rightarrow \xi_e$	$\xi_e = \text{const}$
$\dot{\varepsilon}_o$	$\dot{\varepsilon}_o \rightarrow \dot{\varepsilon}_e$	$\dot{\varepsilon}_e = \text{const}$

For practical purposes, we are interested in the total strain after a given time. During this time, a load program, $\sigma(t)$, and a heating program, $T(t)$, is performed. $\varepsilon(t)$ may be obtained by integration of eqn. (13.2):

$$\varepsilon(t) = \varepsilon_o + \int_0^t \dot{\varepsilon}[\sigma(t), T(t), \xi(\varepsilon(t))]\, dt \;. \qquad (13.3)$$

Equation (13.3) includes within the integral transient as well as steady-state periods. If at $t = 0$ an initial structure, ξ_o, is inherited, then for any $t > 0$ for $\sigma, T = \text{const}$

$$\dot{\varepsilon}(\varepsilon) = \dot{\varepsilon}\,(\sigma, T, \xi_o) + \int_0^{\varepsilon} \left(\frac{\partial \dot{\varepsilon}}{\partial \xi} \cdot \frac{d\xi}{d\varepsilon} \right) d\varepsilon \;. \qquad (13.4a)$$

At constant T and σ, we expect an equilibrium state, $\xi \rightarrow \xi_e$. This implies that $d\xi/d\varepsilon \rightarrow 0$. For convenience, we may define a transient strain ε_t by saying that for $\varepsilon > \varepsilon_t$, $d\xi/d\varepsilon = 0$.

Then,

$$\dot{\varepsilon}_{\varepsilon > \varepsilon_t} \equiv \dot{\varepsilon}_e = \dot{\varepsilon}_o + \int_0^{\varepsilon_t} \left(\frac{\partial \dot{\varepsilon}}{\partial \xi} \cdot \frac{d\xi}{d\varepsilon}\right) d\varepsilon = \text{const} \qquad (13.4b)$$

where $d\xi$ may be split into contributions due to "hardening" and due to "recovery", i.e.:

$$d\xi = \underset{harden.}{(\partial\xi/\partial\varepsilon)\, d\varepsilon} + \underset{recov.}{(\partial\xi/\partial t)\, dt} \,. \qquad (13.5)$$

From eqns. (13.4) and (13.5), we obtain (again for σ, T = const)

$$\dot{\varepsilon}(\varepsilon) = \dot{\varepsilon}\,(\sigma, T, \xi_o) + \int_0^{\varepsilon} \frac{\partial \dot{\varepsilon}}{\partial \xi} \left[\left(\frac{\partial \xi}{\partial \varepsilon}\right) + \left(\frac{\partial \xi}{\partial t}\right) \dot{\varepsilon}^{-1}\right] d\varepsilon \,. \qquad (13.6)$$

Under steady-state conditions, the expression in brackets becomes zero; this holds in particular for $\dot{\varepsilon} = \dot{\varepsilon}_e$ at $\xi = \xi_e$, though this may be approached only asymptotically.

The practical procedure to follow in view of these equations appears to be an iterative one:

a) Have a data set: $\dot{\varepsilon}, \varepsilon; \sigma, T; D(T), \eta(T) ; \langle \xi_i \rangle$

b) Evaluate, for each of the ξ_i, the change during the next strain increment, $(d\xi_i/d\varepsilon)\, \Delta\varepsilon$

c) Provide data on the structural sensitivity of the strain rate, $\partial\dot{\varepsilon}/\partial\xi_i$

d) Obtain the next data set, in particular $\dot{\varepsilon}(\varepsilon + \Delta\varepsilon)$;

e) Repeat the procedure as in (b), (c) ...

For step (d), a differential form of eqn. (13.4b) is appropriate:

$$\Delta\dot{\varepsilon} = (\partial\dot{\varepsilon}/\partial\xi)(d\xi/d\varepsilon)\, \Delta\varepsilon \,. \qquad (13.7)$$

For practical application, the main difficulty rests in the abundance and complexity of unit processes involving dislocations, point defects, cell walls, grain boundaries, voids, cracks, particle sizes, particle shapes, etc., all contributing to $\dot{\varepsilon}$. It is at this point that the general approach usually gets stuck, at the present state of knowledge. Very promising and also successful calculations have been performed, however, and are presented in Chapter 14 in this book. - Often more specialized and, therefore, more simplified approaches are required.

13.3 MODELLING THE CHANGE OF HOMOGENEOUS DISLOCATION ARRAYS

A "control circuit" as in fig. 13.2 may be used to discuss

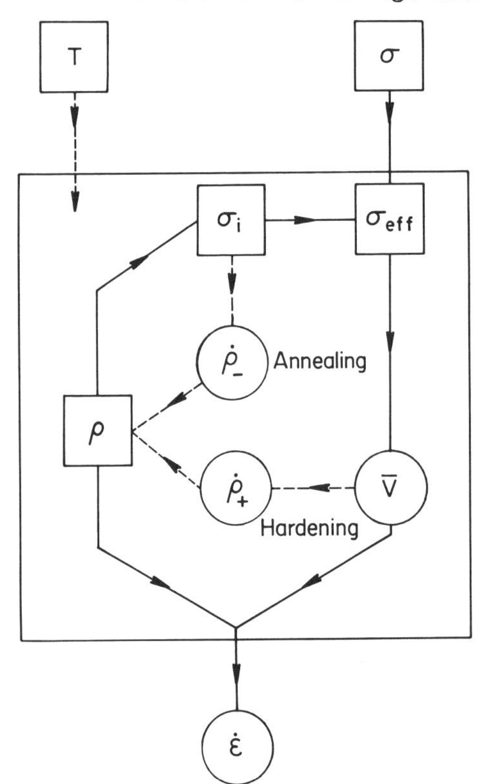

Fig. 13.2 Control circuit for systems with homogeneous dislocation distribution of density ρ .

this situation. The basic assumption is that only random distributions of dislocations are relevant for deformation behavior. Cell structures and, hence, quantities like ℓ_c or S_c are not taken into account.

In fig. 13.2, stress σ and temperature T (the latter one with the purpose to maintain the necessary mobility of defects) are "input" quantities into the solid crystalline system, which is symbolized by a box. $\dot{\varepsilon}$ is - according to eqn. (13.6) - the expected "output". In fact, the dislocation density ρ is the third "input" value, in the sense that ρ at any moment t is "inherited" from the foregoing moment $(t - \Delta t)$, or else from the total prior history of the material.

With ρ_o being defined as earlier, we may determine the internal stress level σ_{io}, due to the dislocation network:

$$\sigma_{io} = \alpha\mu b\sqrt{\rho_o} \tag{13.8}$$

Hence, the local driving stress for dislocations will be the sum of σ_{io} and of the applied stress σ. However, for macroscopic plastic deformation to occur, it is necessary that the (averaged) minima of this sum (or of the appropriate force vector sum) are larger than certain threshold values due to obstacles. The applied (external) stress must, in other words, be able to lift a dislocation across an obstacle even if the local internal stress acts in the opposite direction. This leads, roughly, to an effective stress

$$\sigma_{eff/o} = \sigma - |\sigma_{io}| = \sigma - \alpha\mu b\sqrt{\rho_o} . \tag{13.9}$$

Knowledge of σ_{eff} for a given structure and given applied stress enables us to calculate the mean thermal drift velocity of dislocations, $\bar{v}$. This velocity $\bar{v}$ is naturally linked with D, the diffusion coefficient, as well as with some factor expressing the thermal activation. Without going into the details here, we might propose

$$\bar{v}_o = \text{const } D \rho_o^{1/2} \sinh(-bA\sigma_{eff/o}/kT) , \tag{13.10}$$

where A is an activation area. Equation (13.10) allows one to calculate the deformation rate

$$\dot{\varepsilon}_o = b \rho_o \bar{v}_o . \tag{13.11}$$

It will be remembered that the subscript zero refers to the beginning time interval in the iteration process of Section 13.2. Equation (13.11), multiplied by Δt_1, yields

$$\varepsilon_1 = \varepsilon(t + \Delta t_1) .$$

The <u>change</u> in the dislocation structure will now be calculated as a change in dislocation density. There are "hardening" processes with the tendency to increase ρ :

$$\dot{\rho}_{+/o} = + \rho_o \bar{v}_o / L \; . \tag{13.12}$$

The last expression may be proposed for hardening by dislocation multiplication, L being a characteristic length. On the other hand,

$$\dot{\rho}_{-/o} = + \text{const}\; D \rho_o^{\;2} \tag{13.13}$$

characterizes dislocation annihilation processes as a second order reaction, diffusion controlled by the rate of mutual approach. The occurrence of D in eqn. (13.13) as well as in eqn. (12.10) for $\bar{v}_o$ indicates that the temperature dependence of $\dot{\rho}_+$ and $\dot{\rho}_-$ is essentially the same. Obviously, ρ_1 - the dislocation density after the time interval Δt_1 - results from a balance of eqn. (13.12) and eqn. (13.13):

$$\rho_1 = \rho_o + (\dot{\rho}_{+/o} - \dot{\rho}_{-/o})\, \Delta t_1 \; . \tag{13.14}$$

In principle, this sequence of equations - (13.8) through (13.14) - make the general rules of the preceding section applicable. Naturally, the presentations selected here for $\bar{v}$, $\dot{\rho}_+$ and $\dot{\rho}_-$ are open to criticism and exhibit some ambiguity in the choice of parameters. For practical purposes, this must be overcome by fitting computer calculated curves to a limited number of test values of $\varepsilon(t)$ and - if possible - $\rho(t)$.

Transient phenomena in load change experiments may easily be explained (at least in a qualitative way) by these equations. In view of the observed cell formation, however, some further refinement is required.

13.4 MODELLING THE CHANGE OF CELL STRUCTURES

In contrast to Section 13.3, but in accordance with Section 13.1, we will base the following treatment on the assumption that a network of cell walls spreads throughout the material in addition to the dispersed dislocation lines on slip planes and elsewhere within the cells. If a material has been, say, cold rolled, swaged and recrystallized prior to high temperature deformation, or has just been prepared by casting or zone melting, it will not contain the type of cell structure envisaged here, but rather a random population of dislocations.

What we expect in this case (and, in fact, observe experimentally) is that from the onset of high temperature plastic deformation, the dislocation density increases until a certain supersaturation occurs with respect to the formation (or "precipitation") of cell walls. Holt (1970) has proposed to apply the concept of spinodal decomposition of supersaturated solid solutions to this problem, which is to some degree analogous. "Nucleation" of the first cell wall segments within the crystal results in rapid acquisition of further individual dislocations, thereby reducing the ρ level built-up before: this is illustrated in fig. 13.3. The absorption capacity of the wall for segments is due to the lower strain energy contents of segments in a wall, as compared with an equal length of individual dislocation line away from the wall.

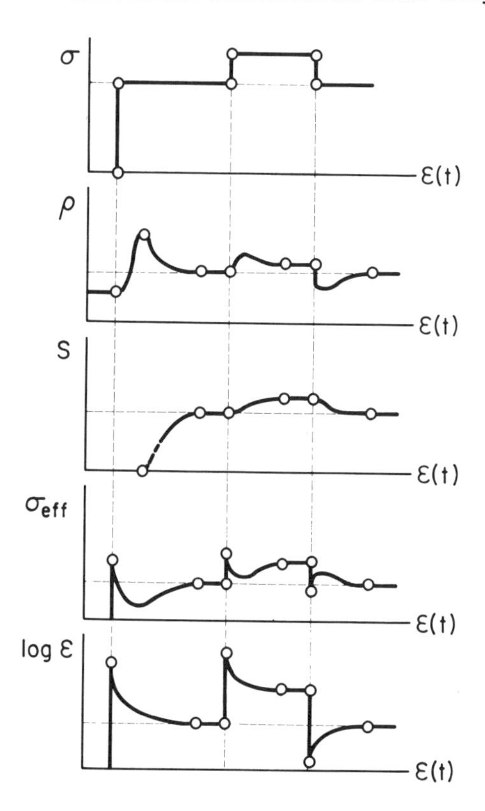

Fig. 13.3 Change of dislocation density and cell wall area at beginning of a test and after a load change.

The kinetics of these reactions are determined by: a) the ability to "knit" networks, to integrate and to release single lines into or from the network and by, b) the high mobility of dislocations within the cells.

The occurrence of cell wall networks has to be considered together with another experimental fact: There is a functional relationship between average cell size, ℓ_c, and applied stress (for data, see Ilschner 1973, p. 135). This relation may in an approximative way be represented by

$$\ell_c \sim \sigma^{-1} \; ; \; S_c \sim \sigma \qquad (13.15)$$

(more exactly, ℓ_c is proportional to σ_i^{-1}, which results in a somewhat lower power exponent than 1.). As with ρ, stress-dependent values of ℓ_c are found in a reproducible manner also in load-cycling experiments, if the system is allowed some time for the transient. These observations lead to the conclusion that cell walls are dynamic equilibrium structures. This means that some reaction mechanisms must be operative which enables the system to attain equilibrium configurations corresponding to the given stress (or strain rate).

The following assumptions will be made, see also Blum (1972):

a) Dislocation motion is limited by cell walls - cell walls are considered to be impermeable;

b) Dislocation generation occurs mainly by dislocation interactions within the cells ("volume reaction");

c) Dislocation annihilation occurs, in contrast, mainly by reaction with the cell walls ("boundary reaction");

d) Driving force for dislocation generation is the effective stress;

e) Driving force for dislocation annihilation is the internal stress.

With these assumptions in mind, we proceed to devise model equations for $\dot{\rho}_+$ and $\dot{\rho}_-$, appropriate to the cell wall concept. In the following, generation and annihilation rates will be given per average cell (and not per unit volume). With the cell diameter ℓ_c,

$$\dot{\rho}_+ = \ell_c^3 \cdot k_1 \cdot \rho \cdot \sigma_{eff} = \ell_c^3 k_1 \rho (\sigma - \sigma_i) \qquad (13.16)$$

Making use of eqn. (13.8), this may also be re-written:

$$\dot{\rho}_+ = \ell_c^{\,3} \cdot k_1' \cdot \sigma_i^{\,2} (\sigma - \sigma_i) \quad (13.17)$$

where $k_1' = k_1/(\cdot\alpha\mu b)^2$

For the annihilation reaction, we assume a diffusion controlled process due to the long range interaction between cell walls and dislocation line segments. We may contend that this interaction fades out for dislocations at distances from the wall larger than the average inter-dislocation spacing $\bar{\ell}_d$. Applying then the same reasoning for stress-induced motion of dislocations as usual, we find

$$\dot{\rho}_- = \ell_c^{\,2} (\text{const } D\sigma_i/\bar{\ell}_d) = \ell_c^{\,2} (\text{const } D\sigma_i \, \rho^{1/2})$$

$$\dot{\rho}_- = \ell_c^{\,2} k_2 \sigma_i^{\,2} = \ell_c^{\,3} S_c k_2 \sigma_i^{\,2}$$

$$k_2 = \text{const } D/\alpha\mu b \quad (13.18)$$

Under steady-state conditions, i.e., $\dot{\rho}_+ = \dot{\rho}_-$, eqn. (13.16) together with (13.18) yields

$$p \equiv \sigma_i/\sigma = [1 + (k_2 \sigma_i/k_1 \rho) S_c]^{-1} \quad (13.19)$$

In view of eqn. (13.8), this is equivalent to

$$p \equiv \sigma_i/\sigma = [1 + (k_2/k_1' \sigma_i) S_c]^{-1} \quad (13.20)$$

Equation (13.20) includes the following <u>limiting</u> cases:

a) $k_2 S_c >> k_1' \sigma_i$; it follows

$p = \sigma_i/\sigma << 1$ equivalent to $\sigma_{eff} > \sigma_i$

Hence, the deformation rate will be controlled by the generation and transportation processes within the cell (<u>transportation control</u>).

b) $k_2 S_c << k_1' \sigma_i$; here follows

$p = \sigma_i/\sigma \approx 1$ equivalent to $\sigma_{eff} << \sigma_i$

Hence, nearly all of the applied stress is used up for "disposal" of dislocations at cell walls (<u>annihilation control</u>).

Experimentally observed values $0,3 < p < 0,7$ indicate a reasonable balance between the two limiting cases. Steady-state creep in a cellular structure would, then, be neither "volume" nor "boundary controlled", but represent a mixed case. One is thus lead to the conclusion that the system prefers to adjust itself at intermediate p-levels by rearranging the cell wall area, S_c, in such a way that neither an overload nor under-utilization with respect to annihilation rates per unit cell wall area is encountered.

If this is indeed a valid assumption, i.e., if under steady-state conditions for not too large variations in applied stress

$$\sigma_i/\sigma \sim \text{const},$$

then eqn. (13.20) implies a functional dependence of the average cell size on stress. With $k_2' = k_2/(1 - p)$, eqn. (13.20) results for $p = \text{const}$ in

$$1/\ell_c = S_c(\sigma) = (k_1'/k_2')\sigma . \tag{13.21}$$

This is indeed the observed form. The cell wall area per unit volume increases linearly with stress, but depends only to a minor degree, if at all, on temperature. This is so since the temperature dependence of both k_1' and k_2' is primarily given by that of the diffusion coefficient, D(T). It thus cancels.

While eqns. (13.19) and (13.20) have been derived for steady-state conditions, we are left with the need to discuss what happens in the event of a load change, or stress change.

Let us first consider an increase in stress by $\Delta\sigma$. Immediately after the stress change, ρ and, hence, also σ_i are unchanged, while σ_{eff} has increased by $\Delta\sigma$. Thus, $\dot{\varepsilon}$ increases instantaneously.

On the other hand, the increase in σ_{eff} accelerates immediately the "breeding rate", $\dot{\rho}_+$, thus increasing the dislocation density. This, in turn, will gradually reduce σ_{eff} and slow down both the deformation and the breeding rate. Simultaneously, the annihilation rate at the cell walls, being still present in their prior, "inherited" configuration, is accelerated such as to keep up again with the breeding rate. However, closer examination shows that a new dynamic equilibrium at a higher (ρ,σ) level could not be reached by this means - the breeding rate would continue to exceed the annihilation rate.

There appear to be three ways to establish a new steady-state after a certain transient period:

a) Increase $p = \sigma_i/\sigma$ instead of holding it at a constant value. This would help to reduce the ratio $\dot{\rho}_+/\dot{\rho}_-$ and eventually bring it back to 1 (unity). However, experiment shows that nature does not accept this bypass - if it changes at all, p instead decreases slightly with increasing σ. We have also briefly discussed above why this is the "natural" response.

b) Consider the contribution of volume annihilation processes, such as we have discussed in Section 13.3, eqn. (13.13). There appears to be no reason to assume that this mechanism be completely absent in the presence of a cell structure, so that eqn. (13.18) should actually be replaced by the sum of eqns. (13.13) and (13.18). Volume annihilation processes then would predominate at high dislocation densities and thus indeed help to solve our present problem. This is an essential point, even though it is somewhat adverse to our basic assumption (c) above. We feel, however, that the evidence by transmission electron microscopy does suggest a predominant contribution of cell wall reactions - cell volume annihilation serving probably more as an emergency valve for dislocation overflow following drastic load increases.

c) Adjust the cell wall web, i.e., change the cell size distribution, thereby increasing S_c. This would, according to eqn. (13.21), increase the annihilation rate even at constant p and without participation of volume reactions. Thus, a dynamic equilibrium corresponding to steady-state could be reestablished.

The latter mechanism is considered here as basic for the structural adjustment to stress change. To save space, we omit the analogous discussion of the consequences of a stress decrease. It should, however, be indicated how the change in cell wall area might be envisaged to proceed in reality.

If the cell structure were a regular honeycomb array, it would indeed be extremely difficult to adjust its parameters to the applied stress at any period of time. It is, however, far more realistic to treat the cellular arrangement as being one of fairly random nature with a statistical distribution of (local) ℓ_c values. S_c, in a sense, averages over all individual cell wall areas within a large volume. The distribution has a mean value, $\bar{\ell}_c$. A shift in $\bar{\ell}_c$ is obtained either by subdivision of large, or by coalescence of small individual cells.

These processes are required for either formation or dissolution of cell wall area. It is not too difficult to visualize how a free edge of a cell wall is either growing by addition of single dislocations, or being dismantled by release of network dislocations. It might be a reasonable guess that the linear growth rate (positive or negative) of a wall section with length ℓ_w is given by

$$d\ell_w/dt \cong \overline{(\dot{\rho}_+ - \dot{\rho}_-)} \qquad (13.22)$$

The bar over the right-hand side denotes an average over the dislocation balance within a local area (essentially one cell). If generation outweighs annihilation, there will be a tendency for the free wall element to grow edgewise, and vice versa.

If, however, free edges are not available in sufficient number, because all of them have been growing until they formed some link with other cell walls, or have been shrinking until they have disappeared - then eqn. (13.22) is inapplicable. What is now necessary is nucleation of new edges within the otherwise edge-free cellular network. This might be brought about by one of two processes:

a) A hole is formed in a cell wall (like when thinning a metal foil for electron microscopy). We expect that the probability for this to occur increases with decreasing size or length (ℓ_w) of a wall element, since the effect of the triple joints is then more pronounced. Moreover, the hole nucleation probability P should increase as $\dot{\rho}_- > \dot{\rho}_+$:

$$P(\text{hole}) = (k_5/\ell_w)(\dot{\rho}_- - \dot{\rho}_+) \qquad (13.23)$$

Once the hole is formed, its edges will further retreat according to eqn. (13.22), or it will heal, if $\dot{\rho}_+ > \dot{\rho}_-$. In the former case, a whole wall element eventually disappears, forming a larger cell from two smaller ones, see fig. 13.4 at "A".

b) A branch is formed in the middle of a cell wall by knitting new dislocations not into, but onto it, see fig. 13.4 at "B". We expect that the probability for this to happen increases with increasing size of a wall element, since three closely spaced triple joints would be energetically unfavorable. Moreover, the branch nucleation probability P should increase as $\dot{\rho}_+ > \dot{\rho}_-$:

$$P(\text{branch}) = k_6\, \ell_w\, (\dot{\rho}_+ - \dot{\rho}_-) \qquad (13.24)$$

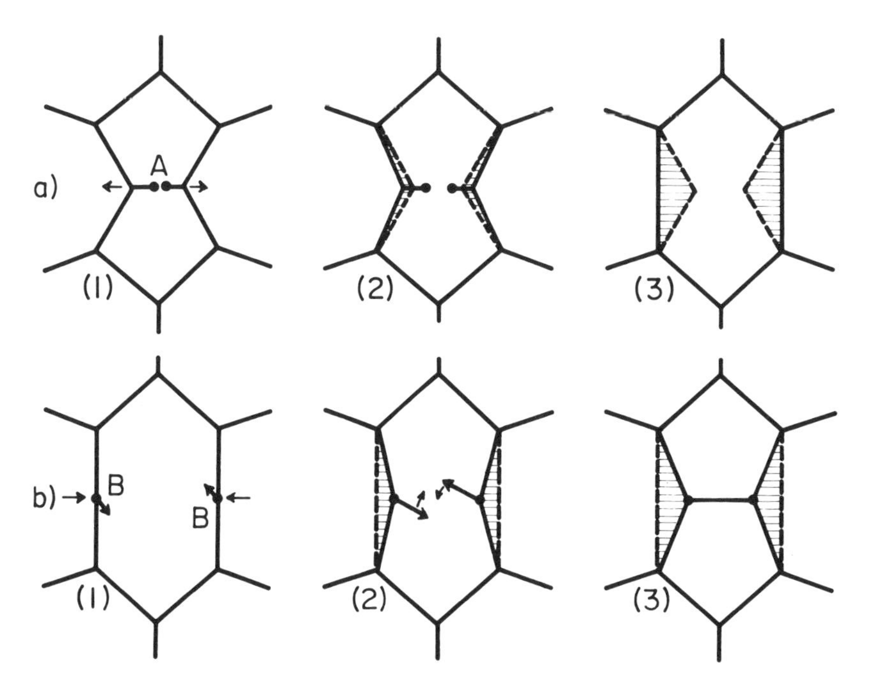

Fig. 13.4 Formation of "holes" (at A) and of "branches" (at B) as a mechanism to change cell size distributions.

Once the branch is formed, its edge will further proceed according to eqn. (13.22), or it will dissolve again, if $\dot{\rho}_- > \dot{\rho}_+$. In the former case, eventually a whole wall element is formed, extending throughout a large cell and subdividing it either by joining another growing edge, or by getting attached to the opposite wall of the cell.

13.5 CONCLUDING REMARKS

This chapter which is only a report of work in progress provides a general outline and framework rather than a manual for handling specific technical creep problems. In due course, however, it will be possible to simulate the high temperature behavior of real materials by model systems such as the one described. This will require use of numerical data processing methods, and should give an opportunity to evaluate the sensitivity of the model with respect to variations in the basic assumptions and in the input data. This will probably lead to refinements as well as corrections. Only then, an attempt to fit experimental data to calculated theoretical creep curves (with the goal of extrapolation) seems to be justified. The more expanded treatment given in Chapter 9 by Zarka as well as the successful and realistic approach of Gittus in Chapter 14 should be compared and, perhaps, linked to the line of thinking presented here.

The intention of this chapter is to demonstrate how equations may be modelled which describe, in a semi-quantitative way, the interrelation between variations in load-temperature programs, deformation rates and structural changes, respectively. As for the latter, both homogeneous dislocation distributions and cellular arrangements are considered, with a pronounced bias towards the statistical methods of kinetics of chemical reactions.

ACKNOWLEDGEMENT

I would like to thank Dr. W. Blum and Dr. B. Reppich for providing relevant experimental results and valuable discussions.

REFERENCES

Blum, W., (1972) Z. Metallkunde, 63, 757.

Holt, D.L., (1970) J. Appl. Phys., 41, 3197.

Hüther, W., (1973) Diploma Thesis, University of Erlangen, see also Reppich, B., and Hüther, W., (1974) Phil. Mag., in press.

Ilschner, B., (1973) Hochtemperatur-Plastizität, (Berlin: Springer).

Streb, G., (1974) Ph.D. Thesis, University of Erlengen.

ABSTRACT. *If stress and temperature change markedly from point to point and during the service lifetime of an engineering structure then the prediction of deformation-behavior usually requires the use of a computer model. In such cases a complicated constitutive equation may be needed, and if it is to be a reasonable means of interpolating and extrapolating the data which it seeks to represent, then it must be based on a sound physical model of the microstructural processes involved in deformation. Such a model has been developed in the course of studies of the behavior of the metal cladding tubes of fast breeder and thermal nuclear reactors. The present chapter describes and summarizes the results of such a model. The model is of quite general utility and its application is not limited to nuclear fuel materials. In the model two sources of dislocation creep are analyzed, that due to thermal climb, and that due to the absorption of irradiation-induced point defects on edge dislocations. In both cases the model is based on detailed descriptions of physical mechanisms, and is general enough to predict behavior under multiaxial stress including both transient and steady state behavior, as well as delayed elastic processes and Bauschinger effects. When data for actual materials is used the model requires no adjustments in parameters. Finally, the multiaxial creep model is successfully applied to predict the behavior of a reactor fuel element.*

14. MICROSTRUCTURE BASED MODELLING OF CONSTITUTIVE BEHAVIOR FOR ENGINEERING APPLICATIONS

J.H. Gittus

14.1 INTRODUCTION

Nuclear fuel elements, concrete pressure vessels, superheater-header tubes and parts of gas turbine engines are amongst the engineering structures whose design involves knowledge of the deformation properties of the material of construction. In the simplest cases, it may be adequate to know some simple parameter such as the stress-rupture life or the stress needed to give 0.1% creep strain in 10,000 hours at the mean service temperature. A creep-rupture parameter or the equation relating the steady state creep-rate to the stress and temperature is all that is required to fill this requirement. However, if the stress and temperature change markedly during the service life-time of the structure, if the stress is a function of position within the structure and if there are significant variations in temperature from one point to another, then the use of a computer model (possibly of the finite element type) will probably be called for and in such cases the constitutive equation needs to be more sophisticated.

Suppose then that we are given the full history of stresses and temperatures to which a solid has been subjected and wish to deduce its current state of strain. To be adequate, the equation which enables such a calculation to be performed (the mechanical equation of state) must cater for the effects of stress-reduction ("stress-dip"), stress-removal ("recovery"), stress-reversal ("creep-fatigue"), constant extension (i.e. stress-relaxation), and temperature-changes. Moreover, it must predict not only the volume-conserving component of strain but also the strain due to void and crack formation and the viscoelastic volume dilation. Finally, it must take account of the environment and related effects such as the adsorption of environmental impurities on the grain-boundaries in the case of metals or the migration of water in concrete. None of the mechanical equations of state which have so far been advanced for various materials comes near to filling all these requirements. Nevertheless the incentive to develop improved approximations is a very real one in the contemporary world. The mechanical equation of state (MEOS) must of course relate to the full strain and stress-tensor and it must predict not only the steady-state creep-contribution (if any) but also the transient creep which follows every change in the stress-system. In this chapter we indicate the pitfalls and the progress that is being made to circumvent them and some of the resultant equations that have been developed for metals. The emphasis is on dislocation creep and the separate effects of temperature and of bombardment with energetic particles (in a nuclear reactor) are considered.

14.2 STEADY STATE THERMAL CREEP BY DISLOCATION MOVEMENT

Our procedure will be to commence with an analysis of the simplest aspect of dislocation creep behavior - the steady state - and then introduce one by one the various complicating features of the transient, viscoelastic, Bauschinger and related effects.

Ashby (1972), has remarked (following Mukherjee, et al., 1969, and Weertman, 1973) in his recent paper on deformation maps, that the steady-state rate of creep ($\dot{\varepsilon}_s$) due to the diffusion-controlled movement of crystal dislocations, obeys the constitutive relation:

$$\dot{\varepsilon}_s = A \frac{D_v \mu b}{kT} \left(\frac{\sigma}{\mu}\right)^n \tag{14.1}$$

where A and n are materials constants, D_v is the bulk

self-diffusion coefficient, μ the shear modulus, b magnitude of the Burgers vector, k Boltzmann's constant and T the absolute temperature.

A contemporary review (Gittus, 1974) shows that there are half a dozen theories of steady-state creep none of which can really be said to have gained wide acceptance. Amongst the better known are those of Weertman (1955, 1957, 1968) and Nabarro (1959). In the course of this review it became evident that the following (comparatively simple) explanation of the steady-state creep process, although it relies only on long established equations for the recovery and hardening of the structure, had not received attention (Gittus, 1974). Thus Bailey (1926) and Orowan (1946) formulated the following equation for the balance between work-hardening and recovery-processes for a material undergoing steady-state creep:

$$d\sigma = \left(\frac{\partial\sigma}{\partial\varepsilon}\right) d\varepsilon + \left(\frac{\partial\sigma}{\partial t}\right) dt \tag{14.2}$$

Taylor (1934) was the first to develop a theory of work-hardening, based on the long-range stress-fields of dislocations. Evans (1973) has argued that Taylor's theory is applicable to a material undergoing creep. His equation for the work-hardening coefficient is

$$\left(\frac{\partial\sigma}{\partial\varepsilon}\right) = \frac{\mu}{2\pi} \tag{14.3}$$

He remarks that work-hardening coefficients measured by changing the stress during a creep-test (characteristically above $T_m/2$) are of the order of magnitude predicted by eqn. (14.3) and are larger by a factor of about 50 than those observed during rapid tensile tests on metal crystals below half their absolute temperature of melting (T_m).

Turning to recovery: Friedel (1964, his eqn. 8.32, page 239) shows that the rate at which a dislocation-network will coarsen due to jog-controlled climb is, theoretically:

$$\frac{\partial r_s}{\partial t} = \frac{D_v\ \mu b^3 c_j}{kT} \cdot \frac{1}{r_s} \tag{14.4}$$

where r_s is the network-spacing and c_j is the jog-concentration given by (cf. Friedel, 1964, p. 23)

$$c_j = \exp\left(-\chi\mu b^3/kT\right) \tag{14.5}$$

where $0.2 > \chi \gtrsim \frac{1}{8\pi}$

Coarsening, reduces the flow stress of the network, as:

$$\frac{\partial\sigma}{\partial r_s} = -\frac{\mu b}{2\pi r_s^{\,2}} \tag{14.6}$$

where the flow stress is defined as $\sigma = \mu b/2\pi r_s$ (cf. Cahn, 1965).
By combining eqns. (14.4) and (14.6) and substituting for r_s:

$$\frac{\partial\sigma}{\partial t} = -\frac{4\,D_v\,b\,c_j\,\pi^2\,\sigma^3}{\mu\,kT} \tag{14.7}$$

The network spacing will become constant when the rate of network-refinement due to straining equals the rate of coarsening due to climb, and at a constant external stress, a constant rate of creep will occur. From eqn. (14.2) for that condition (i.e. for $d\sigma/dt = 0$):

$$\dot{\varepsilon}_s = -\frac{(\partial\sigma/\partial t)}{(\partial\sigma/\partial\varepsilon)} \tag{14.8}$$

and so by substituting eqns. (14.7) and (14.3) into eqn. (14.8):

$$\dot{\varepsilon}_{st} = (8\pi^3\,c_j)\,\frac{D_v\,\mu b}{kT}\left(\frac{\sigma_t}{\mu}\right)^3 \tag{14.9}$$

$$R_\varepsilon = \frac{\dot{\varepsilon}_{st}}{\dot{\varepsilon}_s} = \frac{8\pi^3\,c_j}{A}\left(\frac{\sigma_t}{\mu}\right)^{(3-n)} \tag{14.10}$$

and from eqn. (14.1):

$$R_\sigma = R_\varepsilon^{\,1/n} \tag{14.11}$$

Table 14.1 shows the results obtained by evaluating eqns. (14.10) and (14.11).

Table 14.1 Comparison of Actual and Theoretical Creep Strengths and Creep Rates*

Element	R_σ, ratio of creep strengths for:		R_ε, ratio of creep rates for:	
	$\sigma_t = 10^{-3}\mu$	$\sigma_t = 10^{-4}\mu$	$\sigma_t = 10^{-3}\mu$	$\sigma_t = 10^{-4}\mu$
Silver	1.9	5.1	29.0	5.79
Copper	1.2	2.0	2.56	161.5
Gold	1.0	2.7	0.80	243
Nickel	0.8	1.7	0.29	11.7
Aluminum	0.4	0.9	0.02	0.5
Lead	1.3	2.5	2.88	45.6
Gamma Iron	0.4	1.2	0.004	2.25
Cadmium	1.0	2.0	0.87	17.35
Zinc	0.7	2.2	0.09	113
Alpha Thallium	1.1	3.0	1.56	311
Molybdenum	0.4	0.7	0.014	0.28
Tantalum	0.5	0.9	0.043	0.68
Tungsten	0.2	0.6	9.4×10^{-5}	0.06
Beta Thallium	0.6	1.8	0.053	33.44
Alpha Iron	0.4	1.4	1.51×10^{-3}	12.0
Delta Iron	0.4	1.4	1.05×10^{-3}	8.34
Germanium	0.02	0.04	1.13×10^{-7}	1.79×10^{-6}
M_gO	0.004	0.005	9.17×10^{-9}	1.82×10^{-8}
Stoichiometric UO_2	0.6	1.2	0.068	2.15

*(from Gittus, 1974, courtesy of Pergamon Press)

The entry $R_\sigma = 1.9$ for silver in column 1 of Table 14.1 means that at a stress of $1.9 \times 10^{-3}\mu$ the real material (eqn. 14.1 with Ashby's values of A and n) creeps at the same rate as the "theoretical material" (eqn. 14.9) does at a stress of $1.00 \times 10^{-3}\mu$. The second column in the table deals with stresses an order of magnitude lower. The entry $R_\varepsilon = 0.8$ in column 3 of Table 14.1 for gold means that at a stress of $10^{-3}\mu$ the theoretical creep-rate is 0.8 of the actual creep-rate.

In columns 1 and 2 of Table 14.1, there are a number of cases where the fit is acceptable, for example beta thallium, stoichiometric UO_2, nickel and cadmium. It would be over-optimistic, of course, to hope that a single theoretical creep relationship would represent such a disparate group of materials in their entirety.

A temperature of Tm/2 was chosen for the calculation of c_j, the jog density with $\chi = 1/8\pi$. If $\chi > 1/(8\pi)$ then the comparison is appropriate to some other temperature above $T_m/2$ (cf. eqn. 14.5).

In eqn. (14.9) we have used the symbol σ_t to signify that it is the theoretical value of the stress which the equation is concerned with. Which stress? Tacitly the theoretical analysis assumed that it was the applied stress and that the only source of internal stress is the back stress due to the forces which the network of dislocations exert on one another. This is of course the assumption made in the Taylor theory of work-hardening. It is reasonable for a material in which the only other stress opposing dislocation glide is a Peierls stress of negligible magnitude. But if there are hardening precipitates or solutes present then these will impose an additional resistance to the movements of dislocations. If the effect upon the climb-rate is negligible then the behavior during dislocation creep may still be represented by eqn. (14.9) but with σ_t set equal to the difference between the external stress and the internal stress due to the hardening precipitate or solute. In a number of cases it is possible to calculate the latter - for example in a dispersion hardened alloy at high stress it may be equal to the Orowan stress - so that the creep equation is still free from arbitrary constants. The implications of these improvements are dealt with elsewhere (Gittus 1975).

14.3 GENERALIZATION TO THE TRANSIENT CASE

Consider, as before, a dislocation-network in which the network-spacing is a function of the density of dislocations present. Figure 14.1 shows the sequence of events when the speci-

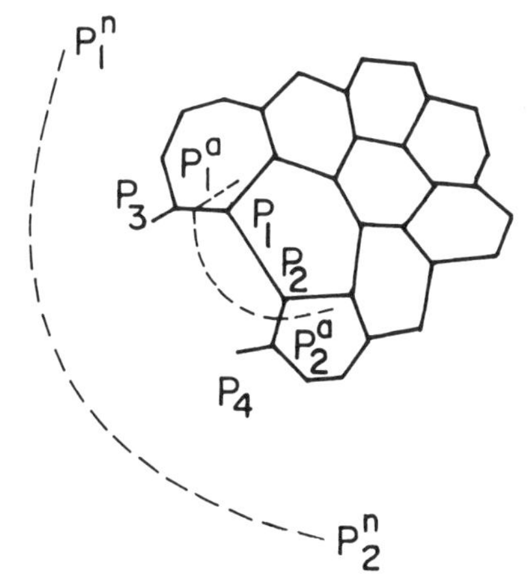

Fig. 14.1 Unit of plastic strain as it might occur in the 3-dimensional dislocation network.

men containing the network is stressed (McLean, 1970). The link P_1P_2 is the first to move in this area of the network; P_1 passes through node P_3 and P_2 through through P_4 etc. Two positions in the links movement are shown dashed at $P_1^a\ P_2^a$ and $P_1^n\ P_2^n$, where it is finally arrested by neighboring parts of the three-dimensional network. As the moving dislocation increases in length, so the local dislocation-density increases and with it the local internal stress opposing further glide until eventually the latter becomes equal to the applied stress and it is this that halts the further glide of the dislocation-link. In steady-state creep the dislocation links gliding at mean velocity v, move an average distance L_s before being halted by these internal-stress peaks where $v = \dot{\varepsilon}/(b\rho)$. And so the frequency with which mobile dislocations are halted is $v/L_s = x\dot{\varepsilon}_s/(b\sqrt{\rho})$ where x expresses the proportionality between the network-spacing and L_s. According to Lagneborg (1972), L_s is several times the network-spacing. Electron-microscope data on dislocation-densities taken together with creep data for the same specimens indicate that, in fact, $v/L_s \sim 10\ \dot{\varepsilon}_s/(b\sqrt{\rho_o})$ where ρ_o is the asymptote towards which the mobile dislocation-density moves as the stress causing creep is reduced towards zero.

If the sinks which halt glide are sub-boundaries then we can calculate the resulting change in sub-boundary angle, $\Delta\theta_s$.

Thus if the density of dislocations is ρ and the sub-grain size is d_s then the number of dislocations threading a plane through a (cube-shaped) sub-grain is $(d_s)^2\rho$. Let these all go into the sub-boundaries. In a plane section, the length of sub-boundary which must be added to outline the cross-section of one new sub-grain is $2d_s$. So the density of dislocations in the sub-boundary (ρ_e) is $[(d_s)^2\rho]/2d_s$, i.e. $\rho\, d_s/2$. But $\Delta\theta_s = \rho_e b = b\rho\, d_s/2$ if all of the dislocations have the same sign. Moreover, on the Taylor-hardening model, $L_s \sim d_s$ and $\varepsilon = b\rho\, L_s = b\rho\, d_s$. But $\Delta\theta_s = b\rho\, d_s/2$. So:

$$\Delta\theta_s = \varepsilon/2 \tag{14.12}$$

In one practical case discussed by Orlova, et al. (1972), the process leading to an increase in sub-boundary angle has only been about 5% efficient, compared with the (maximum) effect predicted by eqn. (14.12). This may be partly due to the fact that many of the dislocations which enter the sub- boundaries cancel one another's effects upon θ_s ("redundant dislocations"). A theoretical equation similar to (14.12) was derived by McLean (1962) but with $\Delta\theta_s = 2\varepsilon/3$, a result which he later confirmed by experiments in which the sub-boundary angle was measured by X-ray diffraction techniques on specimens which had undergone various amounts of creep. However, Orlova et al. (1971) report that their group's work on alpha-iron gave a smaller value and they refer to supporting work on aluminum and upon an aluminum magnesium alloy.

If the theoretical value of the constant (= 1/2) in eqn. (14.12) was found to hold good (as has been the case in some experiments) then the sub-boundaries would be perfect polygon walls at all stages of the deformation process and would exert no significant force on within-sub-grain dislocations at a distance greater than $1/\rho_e = h_s$ (= $2/(\rho\, d_s)$ from the boundaries. We can relate d_s to ρ via Holt's (1970) analogy between the spinodal decomposition of a supersaturated solid solution and the behavior of a statistical distribution of parallel dislocations. It is energetically favorable for the distribution to become less homogeneous with a non-uniformly modulated dislocation-density. The wavelength of the density-modulation is the sub-boundary spacing and Holt's theory indicates that $d_s = K'/\sqrt{\rho}$ where K' is of order 10. That is to say, the sub-boundary-separation is proportional to the mean distance which would separate adjacent dislocations if they were homogeneously spread through the specimen. So,

$$h_s = 2/(\rho\ d_s) \quad , \text{ i.e.,}$$

$$h_s = 2/K' \sqrt{\rho}) \qquad (14.13)$$

According to eqn. (14.13), with $K \sim 10$, dislocations must come within 1/5th of their mean "homogeneous" separation $(1/\sqrt{\rho})$ of a sub-boundary before they experience a force ("internal stress") due to its presence. This is a limiting case and in circumstances where the sub-boundary angle changes more slowly than eqn. (14.12) predicts, L_s will be greater than the value given by eqn. (14.13) and at the upper limit $L_s \sim 1/\sqrt{\rho}$.

We shall assume that this discussion always provides a reasonable approximation to the frequency with which dislocations are immobilized, during primary as well as secondary creep. So at an instant when there are ρ mobile dislocations, the rate of immobilization will be

$$-i\rho = \frac{d\rho}{dt} = -\lambda\dot{\varepsilon}_s\rho \qquad (14.14)$$

$$\text{where } \lambda \approx \frac{10}{b\sqrt{\rho_o}} \qquad (14.15)$$

If the operation of dislocation sources can continue to supply dislocations even in the absence of network recovery, then the mobile dislocation density will eventually reach an asymptote, ρ_∞ say, and so from eqns. (14.14) and (14.15) with this restriction:

$$\dot{\rho}_A = -\lambda\dot{\varepsilon}_s(\rho - \rho_\infty) \qquad (14.16)$$

where $\dot{\rho}_A$ is the rate at which the mobile dislocation density is changing due to immobilization at stress-peaks and source-operation.

Consider now the recovery-process. As before we consider first very low external stresses. Here the coarsening of the network will approximate the rate given by Friedel's equation. The network is densest where a dislocation-link has come to rest: indeed it was this very increase in local dislocation-density, contributed by the advent of the link, which raised the local internal stress to a level at which it halted the further glide of that link. If the extra density contributed by the link could be eliminated by local coarsening then the internal stress would once again be lowered to a level at which another link in this region can glide. The mean rate of network coarsening at $\sigma = 0$ is

$$D_v G b^3 c_j \rho_o^2 / 2kT$$

if ρ_i is the immobile dislocation-density ($\rho_i << \rho_o$ for low σ) then for <u>this</u> component of the total:

$$\dot{\rho}_i \approx \gamma \rho_i \approx \gamma(\rho_o - \rho) \tag{14.17}$$

$$\text{where} \quad \gamma = D_v G b^3 c_j \rho_o / 2kT \tag{14.18}$$

For every immobilized link whose length (and contribution to an internal stress peak) is eliminated by this recovery-process, another neighboring link can glide. And so the total rate of change of mobile dislocation-density is given by $\dot{\rho}_A + \dot{\rho}_i$ i.e. from eqns. (14.17) and (14.18)

$$\dot{\bar{\rho}} = - \lambda \dot{\varepsilon}_s (\bar{\rho} - \bar{\rho}_\infty) + \gamma(1 - \bar{\rho}) \tag{14.19}$$

here we have made the substitutions

$$\bar{\rho} = \rho/\rho_o : \quad \bar{\rho}_\infty = \rho_\infty/\rho_o \tag{14.20}$$

14.3.1 *Remobilization by Stress Reversal*

A link that has become immobilized is subject to an internal stress which opposes the external stress and prevents further glide. If the external stress is removed, then the internal stresses relax and the dislocation-recoil produces a viscoelastic strain. Removing or reversing the stress will cause the link to move back along its glide-path. So if the stress is reversed the $(\rho_o - \rho)$ immobilized links will be remobilized: they will, however, become immobilized once more at internal stress-peaks which oppose glide in this new direction. The rate of this immobilization will be $i(\rho_o - \rho)$. Now each of these freshly-immobilized dislocations will be remobilized yet again if the stress is once more reversed (i.e. restored to its original direction), and so $i(\rho_o - \rho)$ measures the rate at which the mobile dislocation-density in tension is being increased by a period in compression. During that period the network coarsening-process will continue to operate and will itself increase the tension-mobile dislocation-density at a rate $\gamma(\rho_o - \rho)$. So during compression the tensile-mobile dislocation-density changes at a total rate

$$\dot{\bar{\rho}} = (i + \gamma)(1 - \bar{\rho}) \quad , \quad \text{i.e.,}$$

$$\dot{\bar{\rho}} = (\lambda\dot{\varepsilon}_{sc} + \gamma)(1 - \bar{\rho}) \tag{14.21}$$

where $\dot{\varepsilon}_{sc}$ = the steady-state creep-rate under the prevailing compressive-stress.

We can write equations identical with (14.20) and (14.21) for the rate of change, during compression and tension, of the compression-mobile dislocation-density, $\bar{\rho}_c$.

14.3.2 *The Recoverable Component of Strain*

The local internal stress which halts a gliding dislocation is generated, according to the Taylor-hardening theory, by the force exerted on it by neighboring dislocations. That same force causes these neighboring dislocations to recoil and if the immobilized segment is removed, e.g. by reversing the stress, or eliminated by network-coarsening, then the strain due to recoil will be recovered (cf. Salama and Roberts, 1970). This strain is of order $1/2\ br_s$ per immobilized dislocation and its sense depends on whether that dislocation was immobilized by a tensile stress or a compressive one. So the total recoverable (recoil) strain is

$$\xi = \ell_a(1 - \bar{\rho}) - \ell_a(1 - \bar{\rho}_c) \quad , \quad \text{i.e.,}$$

$$\xi = \ell_a(\bar{\rho}_c - \bar{\rho}) \tag{14.22}$$

$$\text{where} \quad \ell_a = \frac{1}{2}\rho_o br_s = \frac{b}{2}\sqrt{\rho_o} \tag{14.23}$$

14.3.3 *The Total Creep Strain Rate*

The total creep-strain rate (= $\dot{\varepsilon}$ say) is the sum of the strain-rate due to the ρ mobile dislocations moving at mean velocity v (i.e. $\dot{\varepsilon}$) and that due to recoil of the immobilized dislocations (i.e. $\dot{\xi}$):

$$\dot{e} = \dot{\varepsilon} + \dot{\xi} \qquad (14.24)$$

Where $\dot{\xi}$ is defined by differentiating eqn. (14.22) and using eqns. (14.19) and (14.20) to define $\dot{\bar{\rho}}$ (and their counterparts to define $\dot{\bar{\rho}}_c$). To define $\dot{\varepsilon}$ we shall assume that the mobile dislocations are sufficiently remote from the internal stress-peaks that halt their motion for the stress exerted on them by those peaks to be negligible. This is reasonable, since we have already made a separate estimate (ξ) of the strain contributed by movements <u>within</u> the stress-field of an internal stress-peak. So $v = f(\sigma)$ and therefore

$$\frac{\dot{\varepsilon}}{\bar{\rho}} = \frac{\dot{\varepsilon}_s}{\bar{\rho}_L} \qquad (14.25)$$

Where $\bar{\rho}_L$ is the value of the mobile dislocation-density and ε_s the value of the creep-rate as $t \to \infty$ under the prevailing stress and temperature.

14.3.4 *Equations for the Creep Strain*

We have now defined the creep-rate in terms of the parameters ε_s, ρ_o, b, D_v, G, c_j, k, T. By integrating the differential equations presented above for a sequence of m time periods, during each of which the stress and temperature are constant, the following equations for creep-strain are obtained:

Let $\dot{\varepsilon}_{Tsm}$ imply $|\dot{\varepsilon}_{Tsm}|$, the steady-state creep-rate during the m th period of tension.

Let $\dot{\varepsilon}_{csm}$ imply $|\dot{\varepsilon}_{csm}|$ the steady state creep-rate during the m th period of compression.

If $\sigma > 0$, $\dot{\varepsilon}_{cs} = 0$ (tension)

If $\sigma \leq 0$, $\dot{\varepsilon}_{Ts} = 0$ (compression)

Let $\mu_{Tm} = \lambda \dot{\varepsilon}_{Tsm} + \gamma_m$

where γ_m is the value of γ (eqn. 14.18) during the m th period.

Then the limiting value of the mobile dislocation-density, which would be attained at $t_m \to \infty$, is (eqn. 14.19, with $\dot{\bar{\rho}} = 0$):

$$\bar{\rho}_{TLm} = (\lambda \dot{\varepsilon}_{Tsm} \bar{\rho}_\infty + \gamma_m)/\mu_{Tm} \tag{14.26}$$

Integrating eqn. (14.19) for the m th period, duration t_m:

$$\bar{\rho}_{Ti(m+1)} = (\bar{\rho}_{Tim} - \bar{\rho}_{TLm}) \exp - \mu_{Tm} t_m + \bar{\rho}_{TLm} \tag{14.27}$$

Here $\rho_{Ti(m+1)}$ is the value of the tension-mobile dislocation-density (called $\bar{\rho}$ in the differential equations but now given the subscript T to emphasize that it is tension-mobility that we are computing) at the end of period m - i.e. at the beginning of period (m+1).

Evidently $1/\mu_{Tm}$ is the time-constant for the change in mobile dislocation-density.

Integrating eqn. (14.25), using eqn. (14.27) to provide a value for the mobile dislocation-density at times from $t_m = 0$ to $t_m = t_m$ we obtain, for the tensile creep-strain during the m th period:

$$\varepsilon_{Tm} = \dot{\varepsilon}_{Tsm} \left[\frac{\bar{\rho}_{Tim} - \bar{\rho}_{Ti(m+1)}}{\mu_{Tm} \bar{\rho}_{TLm}} + t_m \right] . \tag{14.28}$$

Integrating eqn. (14.21):

$$\bar{\rho}_{Ti(m+1)} = (\bar{\rho}_{Tim} - 1) \exp - \mu_{cm} t_m + 1 . \tag{14.29}$$

This tells us how a period of compressive straining ($\mu_{cm} = \lambda \dot{\varepsilon}_{csm} + \gamma_m$) tends to restore tension-mobility to some

of the immobilized dislocations.
Similarly:

$$\bar{\rho}_{ci(m+1)} = (\bar{\rho}_{cim} - 1) \exp - \mu_{Tm} t_m + 1 \ , \tag{14.30}$$

$$\bar{\rho}_{ci(m+1)} = (\bar{\rho}_{cim} - \bar{\rho}_{cLm}) \exp - \mu_{cm} t_m + \bar{\rho}_{cLm} \ , \tag{14.31}$$

and

$$\varepsilon_{cm} = \dot{\varepsilon}_{csm} \left[\frac{\bar{\rho}_{cim} - \bar{\rho}_{ci(m+1)}}{\mu_{cm} \bar{\rho}_{cLm}} + t_m \right] . \tag{14.32}$$

Finally,

$$e = \varepsilon + \xi \ . \tag{14.33}$$

Equation (14.33) with ε defined by successive applications of eqns. (14.28) and (14.32) and with ξ defined by eqn. (14.22), is the required constitutive equation. It defines the behavior of a specimen subjected to <u>any arbitrary sequence of temperatures</u>. A very simple numerical example which illustrates the use of these equations is given by Gittus (1971), whilst a FORTRAN computer program (similar to that given further below) that solves the above equations and furnishes the multiaxial creep-curves for a given stress-temperature-time history has been described by Gittus (1973a).

14.3.5 *Theoretical and Actual Values of* λ γ ℓ_a *and* $\bar{\rho}_\infty$

In this section we shall compare the theoretical values of the creep-constants with those measured for a 20 Cr - 25 Ni - Nb stainless steel. To solve eqn. (14.18) for this steel we use the following data:

$$D_v = D_o \exp(-Q/RT)$$

$D_o = 0.2 \times 10^{-4}\ m^2/sec$, $Q_d = 65{,}000$ cal/mol

$\mu = 5 \times 10^{10}\ N/m^2$, $b = 2.58 \times 10^{-10} m$

$\rho_o = 10^{14}\ m^{-2}$, $k = 1.38 \times 10^{-23}\ J/^oC$,

$T = 1023^oK$

With c_j given by eqn. (14.5) this leads to the expression:

$$\gamma = \exp\left(E' - \frac{Q}{RT}\right) \tag{14.34}$$

where $E' = 19.5$ and $Q = Q_d = 65{,}000$ cal/mole. So that

$$\gamma = 4.7 \times 10^{-6} \text{ at } 1023^{\circ}\text{C}$$

For comparison, (cf. Gittus, 1973a) the best fit to creep-data for this material was provided by $E' = 18.2$ and $Q = 60{,}000$ cal/mole at 1023°C to 1073°K : i.e. $\gamma = 14.7 \times 10^{-6}$ at 1023°K.

From eqn. (14.23):

$$\ell_a = \frac{b}{2}\sqrt{\rho_o}$$

$$= 1.29 \times 10^{-10} \times 10^{7} = 1.29 \times 10^{-3} \tag{14.35}$$

The best fit to this data was provided by $\ell_a = 10^{-3}$. Finally, from eqn. (14.15):

$$\lambda \sim \frac{10}{b\sqrt{\rho_o}} = \frac{5}{\ell_a} = 3876 \tag{14.36}$$

and the best fit to the creep data was given by $\lambda = 8460$.

In view of the uncertainties surrounding c_j , the agreement between theory and experiment is quite satisfactory and strongly supports the view that this model of creep-behavior is adequate.

The theory would lead us to expect that $1 \gg \rho_\infty \to 0$ since at constant stress and in the absence of recovery the back stresses due to the internal stress-peaks should stop the dislocation-sources from operating (i.e. the Taylor-hardening model). The best fit to the data was given by $\bar{\rho}_\infty = 0.113$ (cf. Gittus, 1973a) so here again the agreement is reasonable.

14.3.6 *Comparison with Data*

Values of λ, γ, ℓ_a and $\bar{\rho}_\infty$ for experimental data were obtained by Gittus (1971, 1973a) by fitting the creep-equation to data from creep-tests on 20 Cr - 25 Ni - Nb stainless-steel. In some of the creep tests the stress was maintained constant but in others it was reduced, removed or reversed. Figure 14.2 shows the effects of cyclically reversing the

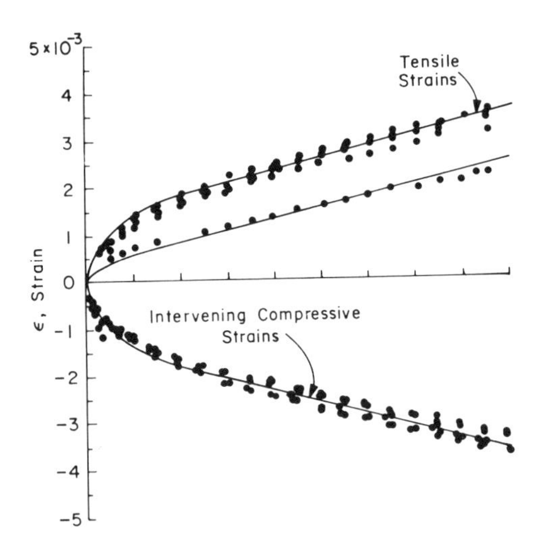

Fig. 14.2 Stress-reversed creep test on 20 Cr - 25 Ni - Nb stainless steel at 750°C. Sections have been superimposed to permit comparison.

stress during a creep test: an equilibrium state is predicted to be approached asymptotically (Gittus, 1971). At equilibrium, reversing the stress should theoretically regenerate primary creep and produce a creep-curve that is the mirror-image of that due to the preceding period under a stress of opposite sign. The data of fig. 14.2 support this prediction.

If the stress cycles are not symmetrical, so that (for

example) each period under tension is shorter than the intervening periods in compression then an equilibrium is still expected on theoretical grounds. Figure 14.3 shows that it is,

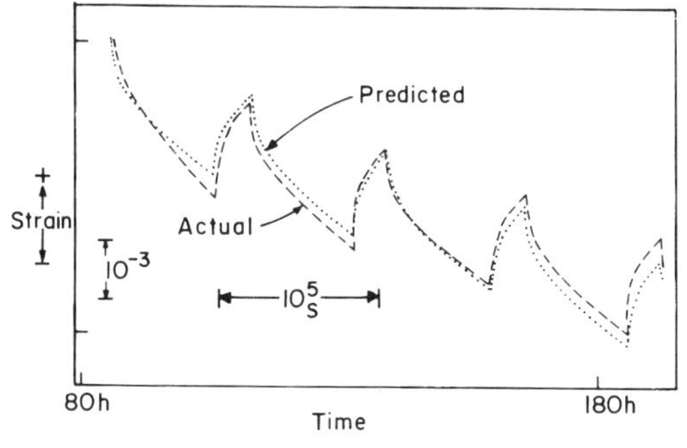

Fig. 14.3 Part of the elastic-plus-plastic strain-time curve for a stress-cycling creep experiment ... = predicted --- = actual results.

indeed, attained in the case of the stainless-steel.

Figure 14.4 shows the effect which, according to the creep-

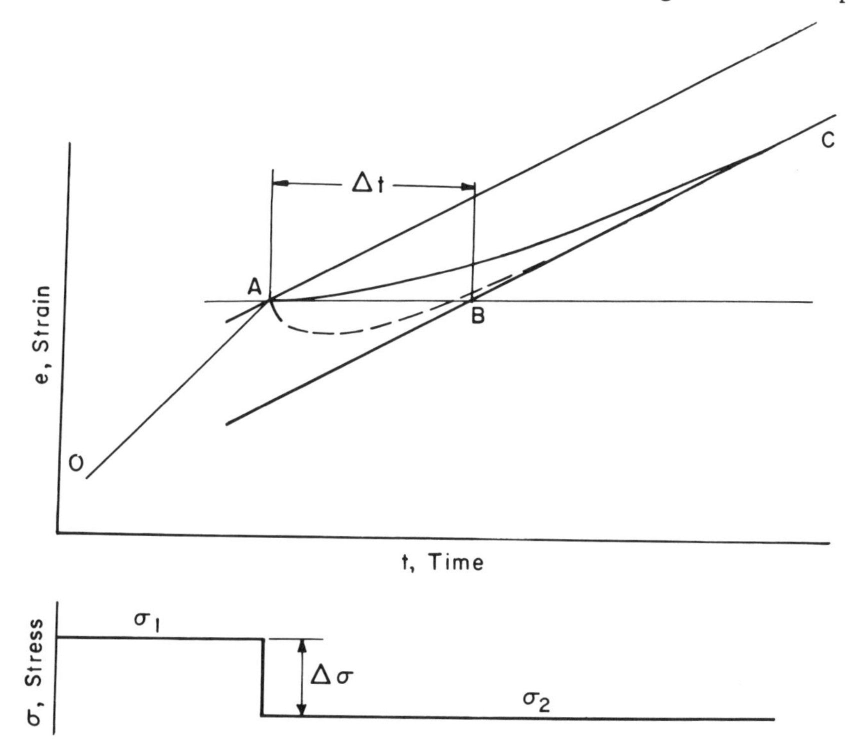

Fig. 14.4 Stress-dip behavior predicted by theoretical creep-equations (schematic).

equations, should be produced in a stress-dip experiment. Reducing the stress by the critical amount $\sigma_1 \rightarrow \sigma_2$ causes creep to cease, instantaneously ($\dot{e} = 0$) at point A . If the stress is reduced by more than the critical amount then creep reverses for a period (broken line ABC). Figure 14.4 shows the geometric construction used by Mitra and McLean (1966, 1967) to measure a "recovery-time" Δt . These authors found, as the theory predicts, that $\Delta\sigma/\Delta t$ constant as $\Delta\sigma \rightarrow 0$. Figure 14.5 shows the theoretical variation of $\Delta\sigma/(\mu\cdot\Delta t)$ with

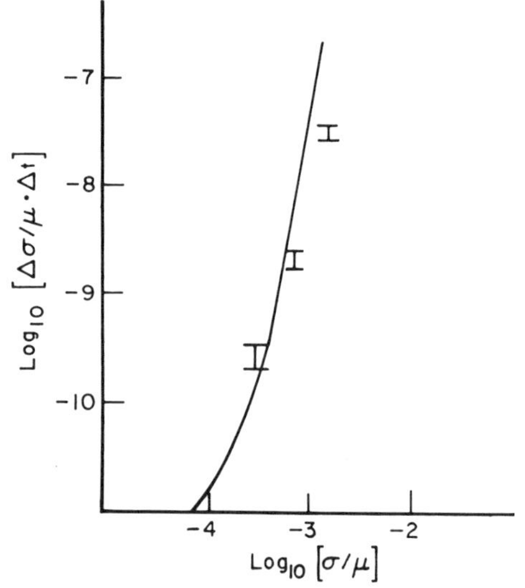

Fig. 14.5 Stress-dip results for 20 Cr - 25 Ni - Nb stainless-steel at 750°C and a line predicted by the theoretical creep equation.

(σ/μ) together with experimental data for the 20 Cr - 25 Ni - Nb steel.

To generate the $\Delta\sigma/\Delta t$ data it is not necessary to reduce the stress by the critical amount which causes creep to cease. If, however, such critical stress-reductions are made then data of the kind shown in fig. 14.6 are obtained. Here the prediction arrived at by solving the theoretical creep-equation for 20 Cr - 25 Ni - Nb stainless steel is shown as a line. The bars are for that steel: good agreement is seen to exist between prediction and data. Also included on this figure are data-points for Aℓ - 5% Mg (from Weertman, 1968) and pure copper (from Lloyd and Embury, 1970): theoretically at certain σ_1/μ values it should be possible to choose a temperature for any metal at which its (σ_1/σ_2) vs. (σ_1/μ) graph will resemble the continuous line of fig. 14.6 These data for Aℓ - 5% Mg

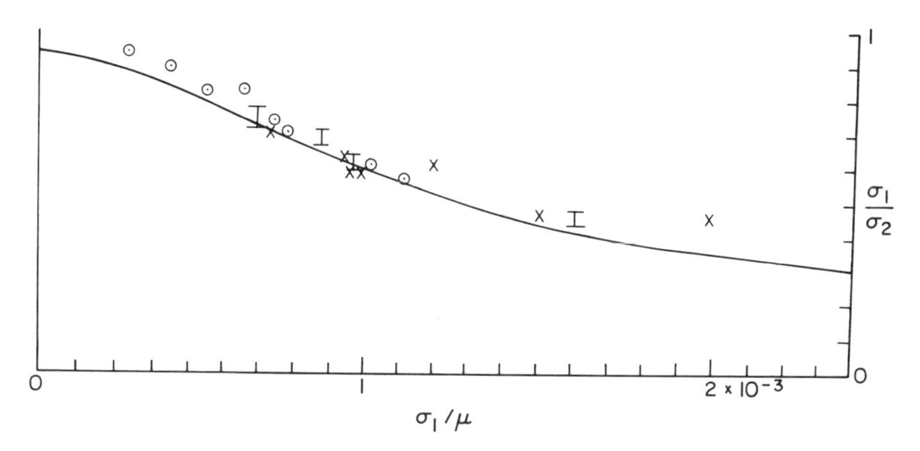

Fig. 14.6 Stress-dip results for 20 Cr - 25 Ni - Nb stainless-steel at 750°C (I), for A - 5% Mg at 350°C (x) and for pure copper at 530°C (0). Theoretical line for the stainless steel.

and copper are found to support this deduction.

14.4 ANTICIPATED FEATURES OF THE MULTIAXIAL EQUATIONS

So far we have limited the discussion to the case of a material subjected to a uniaxial stress system: this has enabled us to describe the model of dislocation creep in a particularly straightforward manner. The complications which arise when the analysis is generalized to the full stress and strain tensors will next be examined from a general standpoint.

For the rigid-plastic material we have the relationships

$$d\varepsilon_{ij} = \sigma_{ij}' \, d\lambda \tag{14.37}$$

and

$$\sigma'_{ij}\,\sigma'_{ij} = (2/3)\,\sigma_y^{\,2} = 2k^2 \tag{14.38}$$

where k^2 is defined in eqn. (14.39), and λ is a scalar factor of proportionality.

$$2J_2' = \sigma_{ij}'\,\sigma_{ij}' = (\sigma_1')^2 + (\sigma_2')^2 + (\sigma_3')^2 = 2k^2 \tag{14.39}$$

Consider now the time-independent idealization for which a definite elastic range (interior of the current yield surface) exists at each stage of the deformation history (cf. Rice, 1970). The essential content of the theory of time-independent plasticity is then expressed by the maximum plastic work inequality:

$$(\sigma_{ij} - \sigma_{ij}^{o})\, d\varepsilon_{ij} \geq 0 \qquad (14.40)$$

Here $d\varepsilon_{ij}$ is a plastic strain increment under a stress $\underline{\sigma}$ on the current yield surface and $\underline{\sigma}^{o}$ is any other stress-state lying either within or on the yield surface. This inequality leads to:

1. Normality of the plastic strain increments to the yield surface at smooth points and

2. The requirement for convex yield surfaces.

Inequality (14.40) is the key element expressing the validity of certain limit theorems associated with the non-strain-hardening idealization, uniqueness, and the variational theorems. Other and seemingly more fundamental postulates have been introduced as a proposed basis for inequality (14.40). For example, it has been postulated that any additional set of stresses must do non-negative work by virtue of the strains which they produce. Again, a cycle of straining from an arbitrary deformed state should, it has been specified, do non-negative work. Rice (1970) remarks that whilst such postulates may seem to be rather general thermodynamic principles, they are in reality nothing more than reasonable classifications of behavior for classes of materials (for metals in particular). For example, in the case of polymers and other materials where the plastic resistance is significantly pressure dependent such formalisms are found to be inadequate.

We wish to consider the counterpart of inequality (14.40) for the time-dependent case, of course. How does the current creep-rate depend on the current stress? Rice (1970, see also Chapter 2) has considered this and has arrived at:

$$(\sigma_{ij}^{A} - \sigma_{ij}^{B})\, \dot{\varepsilon}_{ij}\, (\underline{\sigma}^{A}) \geq 0 \qquad (14.41)$$

In eqn. (14.41), $\underline{\sigma}^{B}$ is a stress-state chosen to lie on or within the surface $\Omega(\underline{\sigma})$ = constant, passing through the point

$\underline{\sigma}^A$. Here $\Omega(\underline{\sigma})$ is a potential function of stress which has a value for each slipped state and whose derivatives with stress yield the components of the creep-strain-rate tensor. Evidently $\Omega(\underline{\sigma}^B) \leq \Omega(\underline{\sigma}^A)$.

Equation (14.41) is identical in form to the maximum plastic work inequality of the time independent case (eqn. 14.40) and so it implies both normality and convexity (1 and 2 above).

Hill (1950) has shown that because the macroscopic yield surface is a polyhedron with an enormous number of faces, continuum slip can lead to the formation of corners (or vertices) on the yield surface. Each face is the plane in stress space representing the critical shear stress of a single slip system in a single grain of the polycrystal. Since faces corresponding to all active slip systems must share the current stress point during plastic deformation, their envelope results in a vertex on the yield surface at the loading point. This behavior cannot be included in approximate descriptions of the yield surface which depend on some small number of parameters such as an equivalent strain and a set of rest-stresses. However, the time-dependent case is simpler. It can be argued that a surface of constant flow potential cannot contain a corner (cf. Rice, 1970) because constant Ω surfaces have continuously varying normals. The important exception is a point at which the strain rate is, for a given point in stress-space, equal to zero: continuity does not require the flow potential surface to have a unique normal at that point.

We now turn our attention to the first component of the multiaxial deformation by dislocation creep: the viscoelastic or recoverable strain, which is completely recoverable, in time, if the stress is removed.

14.4.1 *The Recoverable Strain Diagram*

We shall represent the potential magnitude of the recoverable strain in any direction in deviatoric stress-space by the distance, from the origin, of points lying on a surface. For an isotropic specimen in which $\sigma_{ij}' = 0$, $t \to \infty$, this surface will evidently be a spherical shell with its center at the origin. A point S can be located having the principal external deviatoric stresses (σ_1', σ_2', σ_3') as co-ordinates. It is noted that $(OS)^2 = 2J_2'$; J_2' is the second deviatoric external stress tensor invariant and that S lies on the Mises circle (cf. von Mises, 1913). In fig. 14.7 let OS cut the spherical shell at S_1 and S_2 .

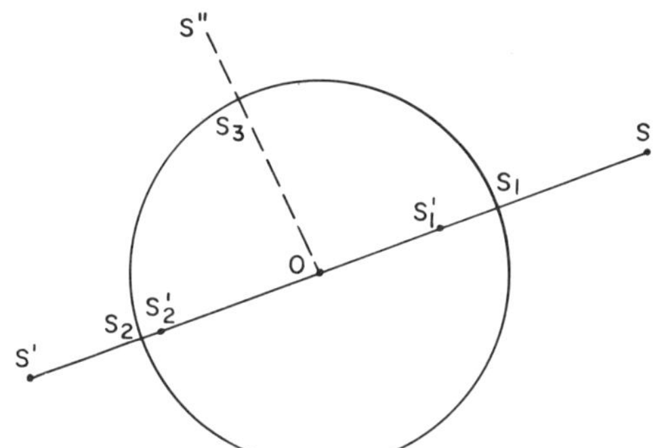

Fig. 14.7 Section through recoverable strain diagram.

When external stress-tensor σ_{ij} has been operating for a finite period of time it will have produced recoverable strain. Position a point S_1' on OS such that $S_1 S_1'$ is proportionate to this component of recoverable strain.

Suppose that the principal stresses are now reversed, producing stress-state S'. This would cause S_1' to move back towards S_1 and simultaneously other dislocations will begin to move and the recoverable strain which is so produced can be represented by the distance S_2S_2'.

The state of anelastic strain comprises, then, the sum of two opposing components (i.e. $S_1 S_1' - S_2 S_2'$ or $OS_2' - OS_1'$). A strain increment of this magnitude but of opposite sign would be produced if following the stress-history described above (i.e. $0 \to S \to S'$) the stresses were all set to zero ($\sigma_{ij} = 0$) and the specimen maintained at elevated temperature for $t \to \infty$, to enable all of the dislocations to recoil.

In the fully annealed, unstresses specimen, then, the principal recoverable strains are all equal (and equal to zero) and are given by

$$\xi_1 = \bar{x}_2 - \bar{x}_1 \quad \text{(etc)} \tag{14.42}$$

where $\bar{x}_2$ and $\bar{x}_1$ are the moduli of the x co-ordinates of the points at which the x axis (i.e. the axis of one of the principal deviatoric external stresses) cuts the spherical shell.

If we can determine the surface into which the spherical-shell has been transformed by the action of a finite stress system (such as S) then the intercepts made by this new surface should, via eqn. (14.42) define the new values of the three principal recoverable strains.

A priori, all that we know of the new surface is that points S_1' and S_2' lie upon it. The simplest assumption which can be made, therefore, is that the surface has remained spherical and symmetrical about OS .

It will now be shown that this assumption produces predicted values of the recoverable strains that are consistent with (a) the Levy-Mises relationships and (b) the conservation of volume. First, a relevant theorem will be proved.

Theorem: If the center of a sphere lies on the plane $x + y + z = 0$, then the algebraic sum of the co-ordinates of its intercepts with orthogonal axes is zero.

The equation of the sphere, radius r and center $C(x_c, y_c, z_c)$ is

$$(x - x_c)^2 + (y - y_c)^2 + (z - z_c)^2 = r^2 \qquad (14.43)$$

Let its center be a distance R from the origin (0) or co-ordinates and let L, M and N be the direction cosines of OC. Then the equation of OC is

$$\frac{x_c}{L} = \frac{y_c}{M} = \frac{z_c}{N} = R \qquad (14.44)$$

The sphere intersects the x axis at

$$x = y = z = 0 \qquad (14.45)$$

Substituting eqns. (14.44) and (14.45) into (14.43) we obtain a quadratic equation in x whose roots (x_2 and x_1) are the co-ordinates of the intercepts which the sphere makes with the x axis:

$$x^2 - 2x\,LR + (R^2 - r^2) = 0 \qquad (14.46)$$

Solving

$$\left.\begin{aligned} x_1 + x_2 &= 2LR \\ y_1 + y_2 &= 2MR \\ z_1 + z_2 &= 2NR \end{aligned}\right\} \qquad (14.47)$$

But $x_c + y_c + z_c = 0$ and so (eqn. 14.44)

$L + M + N = 0$. Hence from eqn. (14.47):

$$x_1 + x_2 + y_1 + y_2 + z_1 + z_2 = 0 , \tag{14.48}$$

and the theorem is proved.

Also, as $x_1 + x_2 = \bar{x}_1 - \bar{x}_2$ (14.49)

$$\therefore \quad (\bar{x}_1 - \bar{x}_2) + (\bar{y}_1 - \bar{y}_2) + (\bar{z}_1 - \bar{z}_2) = 0 \tag{14.50}$$

14.4.2 *The Levy-Mises Equations, and Volume-Conservation Criteria for the Equilibrium Case*

Let a stress σ_{ij} act for $t \to \infty$ producing a limiting strain

$$\xi_L = (S_1 S_1 - S_2 S_2) = 2R \tag{14.51}$$

By our argument about its symmetry, the center of the sphere now lies on OS , a line in the plane $\sigma_1' + \sigma_2' + \sigma_3' = 0$.

From eqn.s (14.47), (14.49) and (14.51)

$$\xi_i = \bar{x}_1 - \bar{x}_2 = x_1 + x_2 = 2LR = L\xi_L$$

$$\xi_2 = M\xi_L \quad \text{(by similar argument)} \tag{14.52}$$

$$\xi_3 = N\xi_L$$

Now $L = \sigma_1'/\sqrt{(2J_2')}$ etc. (14.53)

From eqns. (14.52) and (14.53):

$$\frac{\xi_1}{\sigma_1'} = \frac{\xi_2}{\sigma_2'} = \frac{\xi_3}{\sigma_3'} = \frac{\xi_L}{\sqrt{(2J_2')}} \tag{14.54}$$

Equations (14.54) are the Levy-Mises relationships.

From eqns. (14.50) and (14.52):

$$\xi_1 + \xi_2 + \xi_3 = 0 \qquad (14.55)$$

Equation (14.55) shows that volume is conserved during straining.

Evidently then the hypothesis that the shape of the recoverable strain surface is invariably a sphere whose center, at equilibrium, lies on OS , leads to predicted principal recoverable strains that satisfy the Levy-Mises and volume-conservation criteria.

14.4.3 *The Approach to Equilibrium*

We have discussed, for a stress state S , the equilibrium state to which the spherical shell moves as $t \to \infty$, arguing on the basis of symmetry that its center must lie on OS .

If the prior stress-path (for example the stress-path of fig. 14.7 $0 \to S \to S'$) does not affect the position and radius of an equilibrium sphere, then an equilibrium center for the sphere will always lie on OS , where S represents a stress-state that has prevailed for $t \to \infty$.

If we change the stress state from S' to S'' (fig. 14.7) then the center of the sphere will move from C_1 , its present position (not necessarily an equilibrium position, for the time spent at S' may have been finite) towards C_2 , the equilibrium position which would be attained if the new stress-state, S'' , prevailed for $t \to \infty$.

The theorem shows that if the center C of the sphere during its movement from C_1 towards C_2 , remains in the plane $\sigma_1' + \sigma_2' + \sigma_3' = 0$ (a plane containing OC_1 and OC_2) then volume will be conserved at all times during its transit (i.e. eqn. (14.55) will apply).

Consider now a specimen whose recoverable strain diagram is that shown in fig. 14.8, a sphere whose center C has been displaced a distance $R(= OC = 2\xi)$ from the origin of coordinates 0 , by the prior operation of a stress-system. The internal stress (σ_I) due to dislocation curvature is proportionate to the strain due to dislocation-curvature (eqns. 14.31, 14.32). The recoverable creep-rate will be a function of the net stress, $\sigma_n = (\sigma - \sigma_I)$, i.e. the difference berween the Peach-Koehler stress and the internal stress due to dislocation curvature.

Then the locus of new stress-states (S) that produce the same effective recoverable creep-rate will (fig. 14.8)

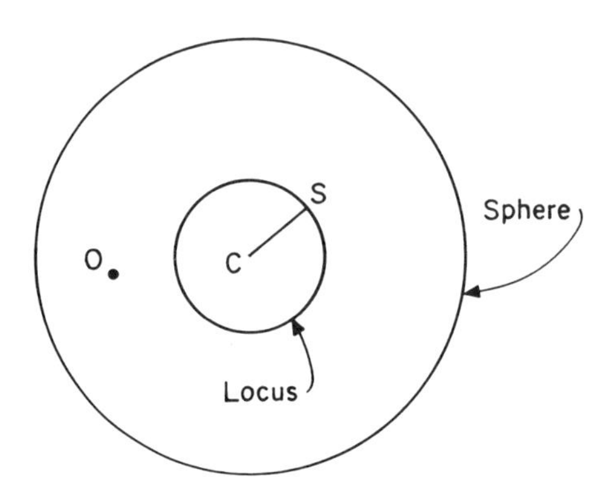

Fig. 14.8 Recoverable-strain sphere and the locus of equivalent stress-states in the plane $J_1' = 0$.

be a circle lying in the plane $J_1' = 0$, concentric with the sphere and of radius $CS = \sqrt{(2J_2 (\sigma_{ni}'))}$ where $J_2(\sigma_{ni}')$ is the second invariant of the deviatoric net stress tensor:

$$i = 1, 2, 3 : \sigma_{ni}' = \sigma_i' - \sigma_{Ii}' \qquad (14.56)$$

$$\text{where} \quad \sigma_{Ii}' = \frac{2}{3} E_m' \xi_i \qquad (14.57)$$

E_m' is the equivalent for recoverable strain, of Young's modulus and we have made use in eqn. (14.57) of the fact that we are dealing with the non-dilational component of the strains.

Hence

$$\dot{\xi}_i = \sigma_{ni}' \dot{\lambda} \qquad (14.58)$$

$$\text{where} \quad \dot{\lambda} = \frac{f\ [J_2\ (\sigma_{ni}'),\ T]}{\sqrt{(2J_2\ (\sigma_{ni}'))}} \qquad (14.59)$$

The function f links the effective recoverable strain-rate with net stress and temperature. It can be determined experimentally by measuring the instantaneous creep-rate in uniaxial tension for a fully-annealed material ($\sigma_I = 0$) at $t = 0$. In such cases $\dot{\xi}_1 \sqrt{3} = f\ [\sigma_1/\sqrt{3},\ T]$, a relationship which enables f to be evaluated by curve-fitting techniques.

Because $\dot{\lambda}$ depends, in the general case, upon ξ_1, ξ_2 and ξ_3, eqn. (14.58) has to be solved numerically.

Under a constant stress (σ_{ij}) the path of the center of the sphere (e.g. CS in fig. 14.8) is the straight-line joining its initial position (at the instant when the stress changed) to its new equilibrium position (attained as $t \to \infty$ under the new stress state). This is the general class of behavior first termed "kinematic hardening" by Prager (see for example Drucker, 1958). It leads to inequality (14.41).

14.4.4 *Dilation, E′ and ν′*

Recoverable creep strains, like elastic strains, are recoverable if the stress is removed. In both cases this recovery-process occurs because of the tendency for atoms, that had been displaced from their equilibrium positions, to return to those positions when the perturbing external stress is removed.

In the case of elastic straining, all of the atoms in the body behave in this manner whilst in the case of recoverable creep straining it is the atoms adjacent to dislocations that are involved.

This similarity makes it conceivable that there could be a dilational strain associated with recoverable creep, just as there is in the case of elasticity.

Now the elastic equations relating principal elastic-strains, ε_{e1}, ε_{e2}, ε_{e3} with stress, Young's modulus (E_m) and Poisson's ratio (ν) are

$$\begin{bmatrix} \varepsilon_{e1} \\ \varepsilon_{e2} \\ \varepsilon_{e3} \end{bmatrix} = \frac{1}{E_m} \begin{bmatrix} 1 & -\nu & -\nu \\ -\nu & 1 & -\nu \\ -\nu & -\nu & 1 \end{bmatrix} \begin{bmatrix} \sigma_1 \\ \sigma_2 \\ \sigma_3 \end{bmatrix} \tag{14.60}$$

whilst the recoverable creep, strain-stress equations are

$$\begin{bmatrix} \xi_1 \\ \xi_2 \\ \xi_3 \end{bmatrix} = \frac{1}{E'_m} \begin{bmatrix} 1 & -\nu' & -\nu' \\ -\nu' & 1 & -\nu' \\ -\nu' & -\nu' & 1 \end{bmatrix} \begin{bmatrix} \sigma_1 \\ \sigma_2 \\ \sigma_3 \end{bmatrix} \tag{14.61}$$

where, from eqn. (14.54),

$$\nu' = 1/2$$

and

$$E'_m = 3\sqrt{(J_2/2)}/\xi_L \qquad (14.62)$$

The analogy discussed above, between recoverable creep and elasticity, would lead us to expect some similarity between ν and ν'. In particular if ν' (as ν) is less than 0.5 then dilational strains will occur, superimposed upon the conservative-strains with which this section has so far been concerned.

There is experimental evidence to justify this. Thus Levy (1967) subjected specimens of an aluminum alloy (DTD 5070A) to tensile-creep ($\sigma_2 = \sigma_3 = 0$) for various periods, then removed the stress ($\sigma_{ij} = 0$) and measured the recoverable strains which ensued (ξ_1 and ξ_2), as a function of time, t. As $t \to 1000$ h, $\dot{\xi}_{1,2} \to 0$. According to the present work, therefore, $-\xi_1$ and $-\xi_2$ should have been the contributions to the creep made by the recoverable straining that had occurred during the prior period under tension. So

$$-\xi_1 = \sigma_1/E'_m$$

$$-\xi_2 = -\nu'\sigma_1/E_m$$

i.e. $\nu' = -\xi_2/\xi_1$

Levy (1967) found that $-\xi_2/\xi_1 = 0.31$ which is not only less than 0.5 (confirming that dilational recoverable creep strains occurred) but was actually identical with the value of Poisson's ratio, i.e. for their work:

$$\nu' = \nu \qquad (14.63)$$

The same relationship holds for other materials, including concrete.

In the dislocation network of Mott (1952), and Friedel (1953), the theoretical relationship between E_m and E'_m is

$$E'_m = 2E_m \qquad (14.64)$$

We expect this equation also to be obeyed at temperatures and stresses where significant dislocation-climb can occur. Now E_m for Levy's alloy would have been approximately 0.6×10^5 MN/m^2 and, in good agreement with the above equation, his creep-data leads to values of E'_m which approach 1.2×10^5 MN/m^2 at the highest temperatures and stresses used by him. We shall return to eqns. (14.63) and (14.64) when we deal with the viscoelastic component of irradiation creep.

14.4.5 *The Recoverable Strain due to Recoil of Obstacle Dislocations*

If the internal stress, which tends to return the dislocations to their zero-stress rest positions when the external stress is removed, is mainly due to the forces exerted by the dislocations on one another, as in the Taylor (1934) hardening theory,

$$S_1S_1' = \ell_a (1 - \bar{\rho}) \qquad (14.65)$$

where $\bar{\rho}$ is the mobile dislocation density for an external stress "vector" S . Evidently:

$$OS_1' = \bar{\rho} \qquad (14.66)$$

Similarly, the mobile dislocation density would be $OS_2' = \bar{\rho}_c$ if we suddenly rotated the stress vector to S' -- i.e., in the uniaxial case, changed the stress from tensile to compressive. We now have a method of reading off the value of the mobile dislocation density for <u>any</u> arbitrary direction of the applied stress vector. For example, if it changes from OS to OS'' (fig. 14.7) then the length of the intercept which it makes (OS_3) is the new value of the mobile dislocation density. Of course OS_3 would only be the value of $\bar{\rho}$ at the <u>instant</u> after the vector was changed from (say) OS to OS''; immediately after the change the sphere will start to change its position and radius (in the way which we have already described) and the intercept will therefore change, too, portraying the change in mobile dislocation density which determines the dislocation creep rate. So we have:

$$S_2S_2' = \ell_a (1 - \bar{\rho}_c) \tag{14.67}$$

and at the limit, when a given stress-vector has been acting for infinite time, the delayed elastic strain will move to:

$$\xi_L = \ell_a (1 - \bar{\rho}_L) \tag{14.68}$$

where $\bar{\rho}_L$ is the limiting value of the mobile dislocation density under the prevailing conditions of stress and temperature.

We use eqn. (14.68) with eqns. (14.61) and (14.62) to calculate the delayed elastic strain tensor for this case.

We shall not develop the use of the intercept OS_3 as a means of calculating the non-recoverable component of the creep-rate here. Instead we use an approximation (cf. Gittus, 1972a) which has three advantages:

1. It does not _rely_ on the Levy-Mises equations to partition the strain between the components of the strain-tensor. Instead it actually _leads_ to those relationships.

2. It clarifies the roles of latent hardening and latent softening.

3. It produces equations for the creep-strain which do not require to be integrated by numerical methods.

The disadvantage is that the potential functions of stress (constant Ω surfaces) now can have steps in them and this we have seen is unlikely for the creep case. However, the inaccuracy so introduced, particularly for the case in which the stress is low, is often tolerable and can be offset (in the practical case of a finite difference model of some creeping structure) by the greater detail in the model itself, which becomes possible through the integrable creep-expressions.

The details of the development of this approximation have already been fully described by Gittus (1972a), and so, it will suffice to summarize the main features: first the effects of slip on intersecting systems upon the immobilization of dislocations and their remobilization by stress reversal are considered. Then the form of a function linking the velocity of the dislocations on a specific shear system with the shear stress causing their motion is arrived at. Finally, the strain-time equation is written down in terms of the mobile dislocation density in the "forward" direction: simultaneously the value of the "reverse" dislocation density is calculated (i.e. the density when would determine the creep rate if the stress vector were suddenly rotated 180°). The new value of the forward dislocation density which would apply following a change in the stress-system (i.e. the length OS in fig. 14.7) is approximated by the following equations:

Let the delta function (δ) be defined as follows, for $n = 1, 2, 3$:

If signum $\dot{\varepsilon}_{ns(n+1)}$ = signum $\dot{\varepsilon}_{nsm}$ then, $\delta_n = 1$.

If signum $\dot{\varepsilon}_{ns(m+1)} \neq$ signum $\dot{\varepsilon}_{nsm}$ then, $\delta_n = 0$.

$$\text{Let } \delta = (\delta_1 + \delta_2 + \delta_3)/3 , \tag{14.69}$$

Here m, (m+1) are time intervals during which the stresses and temperature remain constant. Then,

$$\bar{\rho}_{Fi(m+1)} = \bar{\rho}_{Fem} \cdot \delta + \bar{\rho}_{Rem} (1-\delta) , \text{ and} \tag{14.70}$$

$$\bar{\rho}_{Ri(m+1)} = \bar{\rho}_{Rem} \cdot \delta + \bar{\rho}_{Fem} (1-\delta) \tag{14.71}$$

Equations (14.70) and (14.71) reduce to those previously derived when the stress is uniaxial. In eqns. (14.70) and (14.71) $\bar{\rho}_{Fi(m+1)}$ is the fractional value of the mobile dislocation density in the forward direction at the commencement of period (m+1), and $\bar{\rho}_{Rem}$ is the fractional value of the mobile dislocation density in the reverse direction at the end of period m .

14.4.6 *Algorithm (MEOS) that Solves the Multiaxial Equations*

An Algorithm (MEOS) written in the high level computer language FORTRAN which solves the multiaxial equations and so calculates the sum of the recoverable and the non-recoverable strains is given in Appendix I as Table A1. The values of the input parameters that best fit the experimental creep-data for 20 Cr - 25 Ni - Nb stainless steel are given also in Appendix A1 as Table A2 which also gives for comparison the theoretical values calculated from the above equations. An example of the output of the computer program is shown in fig. 14.9 where the results predicted for arbitrary sequences of stresses and tem-

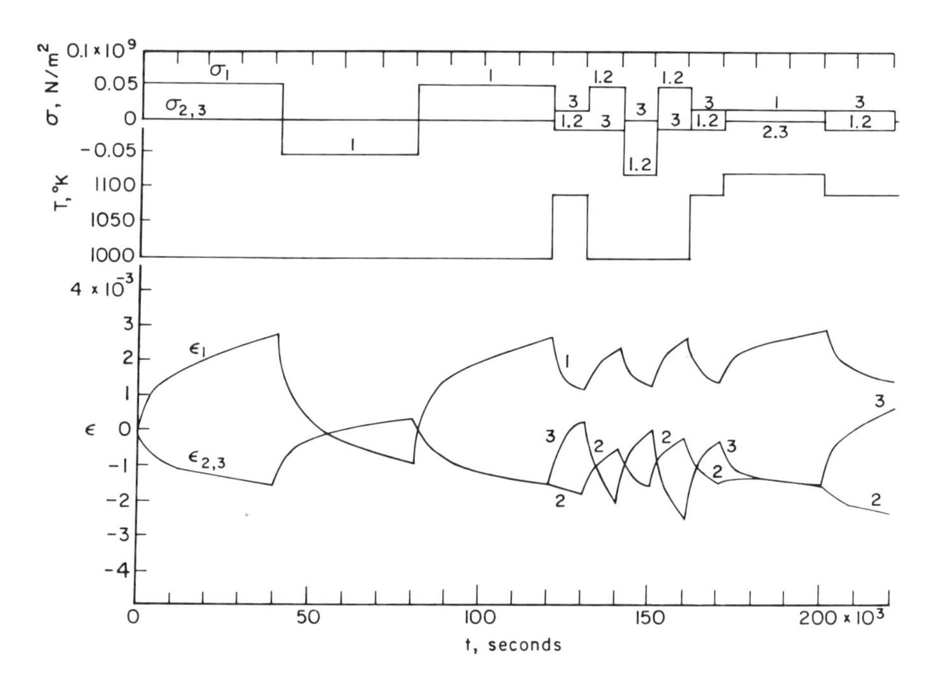

Fig. 14.9 Stress (σ), temperature (T) and creep strain (e) versus time (t).

peratures are illustrated. For this computation, experimental values of the parameters (Table A2) were used giving good agreement with the results of variable stress (including stress-reversal) experiments.

14.5 IRRADIATION INDUCED STEADY STATE CREEP

There are almost as many theories of the effects of irradiation with energetic particles upon the creep-rate as there are data points (cf. Gittus, 1974a). However, recently Gittus et al. (1972) have proposed a unifying idea that may reduce the number of possible mechanisms. The new model stems from an earlier analysis of the creep under irradiation of a specimen containing a network of dislocations which climbed by the acquisition of irradiation-induced point defects to positions in the structure where they could bow out under the influence of the external stress. It was then realized that the climbing dislocations would bow out under the influence of the <u>internal</u> stresses which they exert on one another when, adventitiously, they climb into positions of close proximity. Thus the flux of irradiation-induced point defects is due to the bombardment of the specimen with energetic particles and is independent of internal or external stress. So the internal stresses cannot produce concentration gradients of (for example) interstitials. So there can be no tendency for the irradiation-induced climb to polygonize the dislocations. This is in strong contrast to the thermal climb process (at temperatures above $T_m/2$, typically) where the vacancy concentration is dependent on the local value of the stress and the dislocations tend to climb into positions where the stresses which they exert on one another are zero: polygonized sub-boundaries tend to result.

Under irradiation then, the climb due to the absorption of irradiation-induced interstitials is virtually inexorable: in fact a stress about an order of magnitude higher than the macroscopic yield stress would be needed to boil point defects off the dislocations at a rate sufficiently great to stop the climb all together. So the dislocations can climb so close to one another that the stresses which they exert on their neighbors equals their glide resistance which makes them glide until the climb process has separated them by a sufficient distance to once more reduce the interaction stress below their glide resistance. The climb and the associated glide strains will be randomly directed and it is the perturbation of the glide-steps by an external stress which gives rise to the phenomenon or irradiation creep. The external stress cannot, as have already explained, significantly perturb the irradiation induced climb component of the dislocation movements.

Part of the glide comprises the bowing out of the dislocations between their pinning points. This component is recoverable and so the strain due to the tendency for an external stress to bias the bowing will be recovered if the external stress is removed. When climb has raised the stress to the level at which dislocations can break free from their pinning points, they start to glide freely and the strain produced by this free glide is not recoverable. Accordingly, when the external stress is removed, the steady-state strain (that had occurred) will not be recovered: this results from the biassing effect of the external stress upon the strain due to the free glide of the dislocations.

The original climb-glide model of irradiation creep whose generalization has enabled the new model to be arrived at, led to an equation for the creep rate which can be written in the form:

$$\frac{\dot{\varepsilon}}{\sigma} = \frac{\dot{S}}{Y_p} \cdot \frac{\sqrt{2}}{\pi} \tag{14.72}$$

where Y_p is the yield point of a material containing a dislocation network. It is essentially the same as the Roberts-Cottrell equation for the creep of alpha uranium except that instead of $\dot{\varepsilon}_g$ (the single crystal growth rate) we now have $\dot{S}$ (the isotropic void swelling rate). If this similarity is something more than a coincidence it could imply that the equation can be used to predict the irradiation creep rate of materials whose yield strength is determined not only by dislocations but by precipitates (such as the gamma-prime precipitate in Nimonic PE16) as well.

Gittus et al. (1972) have recently established an experimental technique which makes it possible to measure the dislocation-flux produced when a thin film of a metal is bombarded with electrons. Figure 14.10 shows experimental results obtained, using this new technique for 99.999% pure

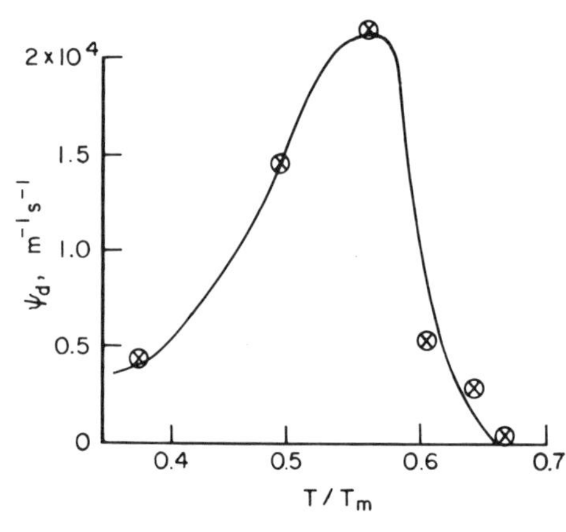

Fig. 14.10 Dislocation-flux measurements for copper versus homologous temperature (T/Tm) for $\emptyset = 1.5 \times 10^{-2} s^{-1}$.

copper bombarded with 1 MeV electrons at temperatures between 250°C and 635°C in an EM7 microscope. Similar results have been obtained with nickel, aluminum and stainless steel. The dislocation-flux (dimensions $L^{-1}T^{-1}$) was calculated from measurements of the frequency with which the projected images of dislocations intersected a random point on the foil-surface. This frequency did not depend upon the direction of dislocation-movement and so the dislocation flux ψ_d, was random ($\psi_d = 2/(\tau \cdot th)$ where τ = time interval between successive intersections of dislocations with a fixed point on the foil surface and th is the foil thickness). The internal strain rates produced by the dislocation movements are self-cancelling and of magnitude ($\dot{\varepsilon}_i$) given by Orowan's equation which with the present symbols becomes:

$$\dot{\varepsilon}_i = b\ \psi_d$$

where b is the magnitude of the Burgers vector.

The magnitude of the internal stresses (I_s) associated with this random dislocation-flux has been measured experimentally (Buckley and Manthorpe, 1972) for some pure metals and an aluminum-silicon alloy. Some of the results obtained by them for aluminum are plotted in fig. 14.11. Here the local in-

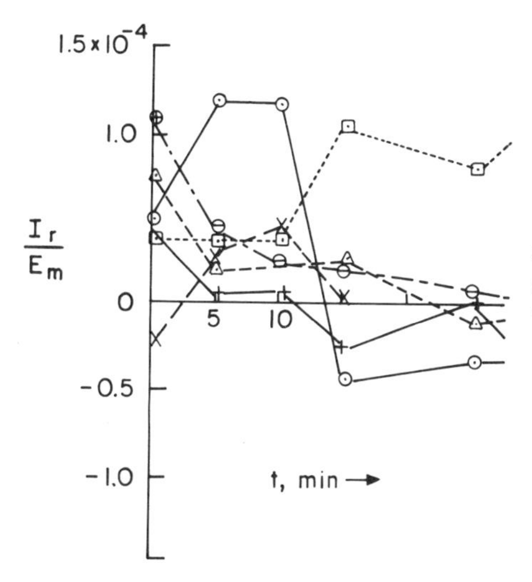

Fig. 14.11 Local shear stress divided by Young's modulus based on measurements of dislocation-bow at various instants.

ternal stress (expressed as a fraction of Young's modulus, E_m) has been calculated from the ratio of the Burgers vector to the radius of a bowed-out dislocation. The local internal stress (I_r) fluctuates between limits of about $10^{-4}\ E_m$ and, instantaneously, different sites have values lying at random within this range. As in the dislocation-flux experiments the specimen was undergoing bombardment by 1 MeV.

These experiments demonstrate that when cubic metals are bombarded with energetic particles, conflicting internal strain-rates can be produced together with internal stresses which are in the same order of magnitude as the yield-stress. Anderson and Bishop (1962) effectively showed that Cottrell's equation (cf. eqn. 14.72) was, theoretically, applicable for both steady and fluctuating internal stresses. Subsequently

Cottrell (1964) pointed out that this type of creep is not limited to the case where the self-stresses are produced by thermal strains, but that any other source of spontaneous plasticity (e.g. neutron irradiation growth in uranium; mechanical vibrations; electro-magnetically induced oscillation) could also produce it, and that the Newtonian flow of a true liquid might be interpreted similarly, since small regions deform spontaneously by thermal motion and the applied stress perturbs this motion only slightly. In the present case the small regions that deform spontaneously do so by irradiation-induced motion and the applied stress, σ_{ij}, per perturbing that motion slightly, produces a creep rate $\dot{\varepsilon}_{ij}$ given by

$$\frac{\dot{\varepsilon}_{ij}}{\sigma'_{ij}} = \frac{1.5\, b\psi_d}{I_s} \tag{14.73}$$

Equation (14.73) is the form of Cottrell's equation appropriate to the situation under discussion: in it σ'_{ij} is the deviatoric external stress tensor and $\dot{\varepsilon}_{ij}$ the creep-rate tensor (cf. Gittus, 1973b).

If all of the dislocations had the same Burgers vector, then their climb during bombardment would cause the specimen to elongate (even with $\sigma'_{ij} = 0$) in the direction of their Burgers vector. More generally, if an above-random number of dislocations have the same Burgers vector then their climb will cause the specimen to change shape (at rate $\dot{g}_{\alpha\beta}$, say) at $\sigma_{ij}' = 0$.

The rate of change of shape referred to the external-stress axes then has components

$$\dot{g}_{ij} = \ell_{i\alpha}\, \ell_{j\beta}\, \dot{g}_{\alpha\beta}$$

where ℓ_{12} is the cosine of the positive angle turned by the positive arm of axis X_2 (one of the principal axes of the zero-stress shape-change) in rotating about the original to the positive arm of X'_1 (one of the external-stress axes). α and β are repeated suffixes, implying summation.

The deformation produced, if bombardment reduces the specific-gravity of the material (at rate $\dot{S}$), is isotropic and of magnitude:

$$\dot{S}_{ij} = \delta_{ij}\, \dot{S}/3$$

Combining these equations

$$\dot{\varepsilon}_{ij} = (1.5\, \sigma_{ij}{}'\, b\, \psi_d / I_s) + (\ell_{i\alpha}\, \ell_{j\beta}\, \dot{g}_{\alpha\beta}) + (\delta_{ij}\, \dot{S}/3) \tag{14.74}$$

Equation (14.74) is the general equation for the nine components of the deformation-rate tensor of a material undergoing bombardment by energetic particles subjected to deviatoric stress-tensor $\sigma_{ij}{}' \ll I_s$.

Figure 14.11 exemplifies the experimentally-measured magnitude of I_s: we shall next calculate its theoretical magnitude for various situations, commencing with the case in which the dominant events are the climb and bow of network-dislocations. This is the mechanism which was analysed in the original theory of "I"-creep.

14.5.1 *Theoretical Value of I for "I" Creep*

Irradiation-induced interstitials are preferentially attracted to dislocations so that the latter tend to receive more interstitials than they do vacancies. The unpaired vacancies (equal in number to the excess interstitials that cause dislocation climb) go to sinks such as three-dimensional clusters, whose characteristic is that they do not preferentially attract a particular class of point defect.

The strain contributed by a dislocation-segment of length ℓ , subjected to an internal stress I_1 and external stress σ is

$$\ell^3\, (\sigma + I_1)/6\mu$$

where μ is the shear modulus.

The total strain, due to bowing of the N segments which comprise the network, is therefore:

$$\Delta\varepsilon = \sum_{(r)} \ell^3\, (\sigma + I_r)/6\mu$$

Now $\Delta\varepsilon$ is an increment of <u>creep</u>-strain and therefore equals zero if there is no external stress. So,

$\sum_{(r)} I_r \ell^3/6\mu = 0$, Combining the last two equations:

$$\Delta\varepsilon = (N\ell^3/6)(\sigma/\mu)$$

Hence the equation for I-creep can be rederived, taking account now of the internal stresses as well as the external ones:

$$\dot{\varepsilon} = \psi_d\ (\sigma'/\mu)\ \sqrt{2/\rho} \qquad (14.75)$$

Equation (14.75) is derived from a detailed consideration of the deformation mechanism in a simple material hardened only by network-dislocations: it is therefore the version of the general equation (eqn. 14.74) appropriate to such a (special) material. Comparing the two equations, the condition for them to be equivalent is:

$$I_s = 1.5\ \mu b\ \sqrt{(\rho/2)}\ ,$$

i.e. I_s is of the order of the yield-point of a material hardened solely by a dislocation-network. Hence

$$I_s/E_m = 3\ b\ \sqrt{\rho/2}/4(1 + \nu)$$

For a density of network-dislocations typical of that in pure metal ($3 \times 10^{12}\ m^{-2}$) and with Burgers vector = 2.5×10^{-10}m this last equation gives $I_s/E_m = 10^{-4}$ which should be compared to the experimentally-measured values plotted in fig. 14.11. Note that if the dislocation density increases with (ϕt) then we expect the irradiation creep rate to fall - i.e. $\varepsilon \propto (\rho)^{-1/2} \propto (\phi t)^{-1/2}$. This is precisely what Buckley and Manthorpe (1972) found in their experiments on copper. Now to inhibit the nucleation of loops, the experiments which produce the data of fig. 14.11 were conducted at low displacement-flux/diffusion-rate ratios. We can, however, calculate a value of internal stress for the case in which loop-formation is not inhibited:

14.5.2 *Theoretical Value of I_s during Loop Formation*

During the early stages of particle-bombardment in, for example nickel and stainless steel, isolated, climbing interstitial dislocation-loops are formed. At higher doses these loops merge to form (or extend) the dislocation-network. The loops may offer additional mechanisms for internal-stress relief. Thus Brailsford and Bullough (1972) have proposed that loops will nucleate preferentially upon crystallographic planes that are normal to a tensile stress. A similar argument had earlier been advanced to explain the anisotropic irradiation-growth of alpha-uranium single crystals where the tensile-stress was an internal stress generated by the anisotropic thermal contraction of material in a displacement spike.

It could be held therefore that the internal stresses produced by random climb of dislocations will be relieved by the nucleation of loops upon preferred crystallographic planes. Then,

$$I_s = \frac{kT}{\Omega n_i} \sinh^{-1}\left[\frac{9\rho_d}{\rho_d^L}\right]$$

where Ω is the atomic volume, n_i the number of interstitials in the loop at nucleation, ρ_d the density of network dislocations and ρ_d^L the density of loop-dislocations.

Solving for 500^oK and the values given by Brailsford and Bullough (1972) ($\Omega = 1.17 \times 10^{-29}$ m^3: $n = 10$) for stainless steel we obtain the values presented in Table 14.2. A tenfold variation of ρ_d/ρ_d^L (bracketing Brailsford and Bullough's value of 1.0) only produces a two-fold change in the calculated value of I_s .

Table 14.2 Theoretical Values of Internal Stress (I_s) for Stainless Steel: Biassed Loop Formation Mechanism

ρ_a/ρ_d^L	0.5	1	2	5
I_s (MNm^{-2})	160	200	240	310
I_s/E_m	10×10^{-4}	13×10^{-4}	15×10^{-4}	20×10^{-4}

15.5.3 *Thermal Creep and a Theoretical Values of I_s*

The value of I_s may be independent of $b\psi_d$ if internal straining involves preferred loop-nucleation sites or network-yielding. If, on the other hand, to relieve the climb-induced strains the dislocations move by one of the mechanisms that operates during thermal creep, then

$$I_s = f\,(b\psi_d)$$

Internal straining by such mechanisms is quite likely: thus (cf. Nabarro, 1948, Hesketh, 1969) although internal stresses cannot perturb the flux to dislocations of irradiation-induced point defects, the flux of thermal-vacancies can be perturbed by stress during bombardment. The precise mechanism may not be important providing that we have relevant data relating tensile thermal creep (at rate $b\psi_d$) with tensile stress (i.e. I_s) at given temperature, T.

Thus thermal creep-data for pure nickel have been obtained by Norman and Duran (1970) and can be used to form an estimate of I_s/E_m. Let $\phi = 10^{-6}$ dpa/s and let $b\psi_d = 10^{-3}\phi$. At 650^oC the form of the previous equation deduced from the thermal creep of pure nickel indicates that $I_s/E_m = 10^{-4}$. The ratio is not strongly temperature-dependent and at 650^oC it is approximately 1/4 the 350^oC-value. The concept that it is thermal creep that relieves the internal stress was of course central to both Blackburn's (1960) extension of the Roberts-Cottrell theory and Hesketh's (1968) application of that theory to zirconium. In those cases thermal creep relieved the stresses produced by randomly-directed climb of gorups of identical dislocation (i.e. radiation-growth of grains) whereas in the cubic metals to which we have now extended Cottrell's theory, thermal creep relieves the stresses produces by the randomly directed climb of individual dislocations.

The theoretical magnitude of the internal stress, I_s, for the three deformation-mechanisms which we have discussed are collected together in Table 14.3. To arrive at estimates of the theoretical creep-rate some means of predicting the dislocation-flux, ψ_d, is needed, too.

Table 14.3 Values of the Internal Stress Expressed as a Fraction of Young's Modulus

I_s/E_m	Source of value
10^{-4}	Calculated from dislocation bow in high voltage electron microscopy experiments on Aluminum
13×10^{-4}	Theoretical value for stainless steel (loop mechanism)
10^{-4}	Theoretical value for pure nickel (thermal creep mechanism)
$\leq 22 \times 10^{-4}$	Deduced from stainless steel creep data of Foster et al. [1972]
$\geq 23 \times 10^{-4}$	Deduced from stainless steel creep data of Walters et al. [1972]
27×10^{-4}	From mean irradiated yield strength of stainless steel

14.5.4 *Theoretical Relationship between ψ_d, $\dot{S}$ and $\dot{g}$*

There must be an unpaired vacancy for every excess interstitial that condenses on a dislocation. The former can go to a three-dimensional vacancy cluster or to some other sink such as a vacancy-loop, grain-boundary, sub-boundary, dislocation-dipole, precipitate-matrix interface or free surface. Hence

$$\psi_d = \dot{S}/b + \psi_o$$

where in unit of time $b\psi_o$ is the fractional volume of unpaired vacancies that goes to sinks other than three-dimensional clusters.

Now ψ_d is the random component of the dislocation-flux and if $g_{\alpha\beta} \neq 0$ there will also be a non-random flux of dislocations whose climb causes the specimen to change shape even at zero external stress. It is suggested, by analogy to the above analyses of the non-random dislocation flux during thermal creep that the second invariant of tensor $\dot{g}$ (that is $J_2(g)$) determines the magnitude (ψ_g) of the non-random dislocation flux, i.e.

$$J_2(\dot{g}) \;\alpha\; b\psi_g \;=\; \frac{\sqrt{2}}{3}\,[(\dot{g}_1 - \dot{g}_2)^2 + (\dot{g}_1 + \dot{g}_3)^2 + (\dot{g}_2 - \dot{g}_3)^2]$$

where $g_{1,2,3}$ are the <u>principal</u> strains which the irradiation-induced shape-change would produce at zero external stress ($\sigma_{ij} = 0$). The total dislocation-flux $(\psi_d + \psi_g) = (\dot{S}/b + \psi_o)$ and hence:-

$$\psi_d \;=\; (\dot{S}/b) + \psi_o - \psi_g$$

$\dot{S}$ can either be obtained by experiment or deduced from a theory such as that of Bullough, et al. (1970) and ψ_g has been defined.

14.5.5 *Comparison with Experiments*

Equation (14.74) when employed together with data such as those exemplified in figs. 14.10 and 14.11 predicts a creep rate of the order 1 (ed/dpa)* in reasonable agreement with the results given by actual creep experiments. We shall next compare its predictions with data obtained by two groups of workers for the creep of stainless steel.

Foster, et al. (1972) analysed data obtained by machining slits in the stainless steel cladding tubes from irradiated fuel pins and measuring the extent to which these slits opened or closed as a result of the macroscopic unbalanced internal stresses that were present. From these measurements they deduce

$$\dot{\varepsilon} \;=\; B\,E\,\phi + D\,\sigma\,\dot{S}$$

---

*elastic deflection per displacement per atom.

where $\bar{E}$ was the neutron energy (MeV), and B and D were constants.

This, with $D = 1/I_s$, is just the algebraic form which we would expect from the present work if some of the unpaired vacancies go to three-dimensional clusters and some to other sinks causing the specific gravity to decrease at rate $\dot{S}$. The term $B\bar{E}\phi$ was negligible for type 316 steel at swelling rates of a few per cent per year. This result provided the first experimental verification of the writer's prediction (cf. Gittus, 1972b) that the creep-rate would be highest when the irradiation induced vacancies are precipitating as three-dimensional clusters.

Foster et al., found that $D \geq 2 \times 10^{-5}$ psi^{-1}. So $I_s \leq 5 \times 10^4$ psi (350 MNm^{-2}), which is similar in magnitude to the theoretical values in Table 14.2 for stainless steel.

Another set of data, in this case for the circumferential creep (rate - $\dot{\varepsilon}_\theta$) of internally pressurized stainless steel tubes bombarded with fast neutrons, was that of Walters, et al. (1972). Their results indicate that: $b\psi_g \sim 2.5 \times 10^{-11}$ s^{-1}; $\dot{\varepsilon}_\theta = 1.4 \times 10^{-12}\ \sigma_\theta$; and $\dot{S} = 3.6 \times 10^{-10}$ s^{-1}.

Using these data in the theoretical relationship between ψ_d , $\dot{S}$ and $\dot{g}$ derived above:

$$b\psi_d \geq 3.6 \times 10^{-10} - 0.25 \times 10^{-10}$$

$$\geq 3.35 \times 10^{-10} \text{ , and so}$$

$$I_s\ (= 1.5\ b\psi_d\ \sigma_\theta{}'/\dot{\varepsilon}_\theta) \geq 370 \text{ MNm}^{-2}$$

The stainless steel creep data from these two sources are seen, therefore, to be in good agreement, indicating that $I_s/E_m \sim 22 \times 10^{-4}$ (Table 14.3). For comparison the data of Hunter, et al., (1972) suggest that the average value of the yield-strength during the tube creep trials of Walters et al. (1972) will have been $\sim 27 \times 10^{-4}\ E_m$.

14.5.6 *Creep of Uranium*

The orthorhombic symmetry in an alpha-uranium single crystal impels most of the irradiation-induced interstitials to condense upon one family of lattice planes and the vacancies upon another. As a result of this phenomenon dislocations lying on these two families of planes climb and under irradiation a single alpha-uraniun crystal elongates at rate $\dot{\varepsilon}_g = b\psi_g = \dot{g}$. So we do not expect a single alpha-uranium crystal to exhibit irradiation-creep: all of the dislocation-clumb movements are co-operative ($\psi_o = \psi$ and $\dot{S} = 0$ in the equation relating ψ_d, $\dot{S}$ and $\dot{g}$) and none are conflicting. The internal stress will, at equilibrium, be no higher than the external stress and so $\varepsilon = 0$, $\dot{e} = \dot{g}$.

Consider now an alpha-uranium polycrystal, devoid of texture. The individual crystals are prevented from elongating by their neighbors and so their growths have to be accommodated by plastic deformation, i.e. $b\psi_g = 0$ (the polycrystal does not change shape), $\dot{S} = 0$ and $b\psi_o = g$, so:

$$\dot{\varepsilon}/\sigma = \dot{g}/I_s .$$

This, with I_s = the yield stress, is the equation for the creep of an alpha-uranium polycrystal proposed by Cottrell and known to provide a good representation of experimental data. We have derived it here from eqn. (14.74) to illustrate the applicability of that relationship not only to cubic metals but also to those which (like alpha-uranium, zirconium and possibly graphite) owe their irradiation creep to the co-operative climb of the families of like dislocations and dislocation loops which is a source of single crystal irradiation growth.

14.6 IRRADIATION INDUCED TRANSIENT CREEP

The biased bowing of the dislocations is the source of the transient or primary component of irradiation creep.

If the external stress is removed the biased bow will disappear as their line-tension causes the dislocations to straighten. A small strain will be so produced opposite in sign to the preceding creep strain. By analogy with the equations of linear elasticity, this viscoelastic response may be expected to follow an analogue of Hookes law:

$$t \to \infty,\ \xi_{1L} \to (1/E_m')(\sigma_1 - \nu' (\sigma_2 + \sigma_3)) . \qquad (14.76)$$

Here we have considered the limiting asymptote ξ_{1L} of the viscoelastic strain due to the biasing effect upon dislocation bowing which the external stresses $\sigma_{1,2,3}$ produce as time, t, approaches infinity: E'_m is the analogue of Young's modulus and ν' is the analogue of Poisonn's ratio. The analysis of Mott (1952) and Friedel (1953) of the bowing of dislocations preicts that:

$$E'_m = 2E_m \tag{14.77}$$

As with elastic strains, viscoelastic strains are recoverable if the stress is removed. This similarity makes it conceivable, as we discussed above for thermal creep, that there could be a dilational strain associated with irradiation-induced viscoelasticity, just as there is in the case of elasticity. The simplest case would be one in which the two are equal, i.e.:

$$\nu' = \nu \tag{14.78}$$

Turning to the kinetics of the viscoelastic straining process: bowing proceeds at a rate which depends on the difference between the external stress and the opposing internal stress due to dislocation line-tension. If the bowing-rate obeys first order kinetics, like the steady-state irradiation creep rate, then the strain, is given by the equations of linear viscoelasticity (cf. Axelrad, 1970) and in particular:

$$d\xi_n/d\Phi = \Delta\xi_n/\Delta\Phi \tag{14.79}$$

In eqn. (14.79) Φ is the dose of irradiation (displacements per atom), $\Delta\xi_n = (\xi_n - \xi_{nL})$, and $\Delta\Phi$ is a materials parameter (the "e-folding dose").

Now at low stresses much of the primary thermal creep of metals is recoverable viscoelastic and obeys equations algebraically identical with eqn. (14.79), except of course, that Φ is now a time-temperature parameter since it is the thermal displacement of atoms, not their displacement by bombardment, which then controls the creep-rate. This encourages us to examine the view that most of the primary irradiation creep strain may also be described by eqn. (14.79).

14.6.1 *Comparison with Experiments*

Integrating eqn. (14.79) we obtain:

$$\xi_n = (\xi_{no} - \xi_{nL}) \exp(-\phi/\Delta\Phi) + \xi_{nL} \quad (14.80)$$

where $\xi_n = \xi_{no}$ when $\Phi = 0$. McElroy, et al., (1970) examined the data which Hesketh (1963) had obtained for the irradiation creep of iron, high carbon steel, stainless steel and zirconium after fast neutron doses of 2×10^{19} and 3×10^{20} n/cm^2. They found, empirically, that eqn. (14.80) gave a good fit. In fact, Hesketh (1967) had himself reported that eqn. (14.80) fits his data for the irradiation-creep of the annealed austenitic stainless-steel alloy EN 58B which had been subjected to irradiation creep at 430°C in the reactor HERALD.

Lewthwaite and Proctor (1973) found that their data for the primary irradiation-creep of Ti, Zr, stainless-steels types 316 (CW), 316 (A) and FV 548 (CW): Nimonic PE16 (FHT, ST), Nimonic 80A and Ni all conformed to an empirical equation algebraically identical with eqn. (14.80). From their data we have calculated E_m/E_m' and $\Delta\Phi$: the results, given in Table 14.4, shows that E_m/E_m' varies between 0.2 and 3.4, which may be compared with the theoretical value of 0.5 for a Mott-Friedel network.

Table 14.4 Parameter Values Computed from the Data of Lewthwaite and Proctor (1973).

Material	E_m/E_m'	Φ, n/cm^2*
Ti	1.7	2.6×10^{20}
Zr	1.5	2.4×10^{20}
316 (CW)	0.7	0.6×10^{20}
316 (A)	0.20	-
FV 548 (CW)	0.5	0.3×10^{20}
PE16 (FHT, ST)	0.2	0.3×10^{20}
Ni	3.4	2.5×10^{20}

*10^{21} n/cm^2 produce, very approximately, 1 dpa. The exact relation depends on the position within the core of the reactor and upon the reactor. 1 dpa = one displacement per atom.

Gilbert, et al., (1972) found that eqn. (14.80) gave a very precise fit to their high accuracy primary irradiation creep data for type 316 (CW) stainless steel. Harkness and Yaggee (1972) produced irradiation-creep by cyclotron-bombardment and their data are also reasonable well fitted by eqn. (14.80).

All of the above experiments were done at either constant external stress or, in a few cases, constant external strain. They do not, therefore, test the recoverable nature of irradiation-induced primary creep. Such recovery is to be expected because if the external stress is removed the dislocations will try to straighten out and the strain which their biased-bow had produced will eventually all be recovered. Equations (14.76) and (14.80) predict the occurrence, magnitude and kinetics of such a recovery-process. Confirmation stems from the work of Gilbert and Blackburn (1970). During irradiation creep, in their experiment, the stress was suddenly reduced from 240 MN/m^2 to 138 MN/m^2: the response was that predicted from eqn.s (14.76) and (14.80). Part of the prior primary creep strain which, too, had obeyed eqn. (14.80) was recovered. The amount recovered was, as predicted by eqn. (14.76), a fraction of the original primary creep strain (i.e., (240 - 138) $\div$ 240). The value of $\Delta\Phi$ which fitted the initial primary creep curve was indistinguishable from the value of $\Delta\Phi$ which fitted the subsequent creep recovery curve as should be the case, according to eqn. (14.80).

If $\nu' < 0.5$, then during irradiation creep under simple tension there will be a volume-increase but not during creep under pure shear. This is what Gilbert and Straalsund found (1970) when they analyzed data for the irradiation-induced creep-relaxation of stainless steel in either tension or torsion. Most of the strain in these experiments can be shown to have been due to primary-creep.

14.7 PRACTICAL APPLICATION OF THE CONSTITUTIVE EQUATIONS

We have now described and compared with experimental data, equations for the transient and steady state components of the thermal and irradiation creep of crystalline solids in general, and metals such as stainless steel in particular. The practical purpose of this work is the design and appraisal of engineering structures and in this section we briefly describe the success which has attended the use of the constitutive equations in a computer model of a nuclear fuel element destined for service in the British AGR reactors (cf. Gittus, 1973c).

When Beryllium was abandoned as a cladding material for AGR fuel-pins, on economic grounds and not because the remaining technical problems were believed insuperable, attention was turned to a 20 Cr-25 Ni-niobium stabilized stainless steel which had been under development as a back-up material. Early work, which has since been confirmed, had indicated that oxidation-resistance in CO_2, compatibility with UO_2 and fission products, and fabricability were all favorable. But calculations suggested that although it should perform satisfactorily under the driver-charge conditions in the Windscale Advanced Gas-cooled Reactor (WAGR), the more onerous conditions in the Board's power reactors might cause mechanical failures. The program designed to avoid this problem received an impetus when failures began to occur, more or less as predicted, in early WAGR experiments designed to approximate the values of the parameters of the power reactor. That program has now led to the satisfactory design of fuel pin and clad variant described below.

14.7.1 *Endurance of the Early WAGR Fuel-Pins*

The WAGR driver charge consisted of the conservatively designed Mk II fuel pin (0.51 m long with cans of 0.38 mm thick stainless steel containing solid UO_2 pellets 10.2 mm in diameter) which operated at nominal can surface temperatures of up to 650°C. Some of these pins have undergone burnups of over 31,000 MWd/te but only two have failed, (MWd/te = megawatt-day/tonne of uranium).

For an advanced gas cooled reactor to be competitive, however, it was necessary to reduce fuel manufacturing costs and attain higher gas outlet temperatures. The Mk III and Mk IV pins were aimed at these requirements and had larger solid pellets (14.5 mm diameter): About 10,000 were irradiated in WAGR with can temperatures of up to 850°C but they tend to fail after a sudden increase in rating. Thus fig. 14.12, adapted from Evans, et al. (1973) shows how large values of a rating parameter $Y(\varepsilon)$ ($\sim$ magnitude of the sudden increase in rating plus value of the subsequent mean rating) can produce a high proportion of failed fuel pins.

14.7.2 *The Cause and Proposed Cure of the Early Failures*

Metallographic examination of failed Mk III and Mk IV fuel pins revealed splits in the clad, adjacent to radial cracks in the pellets and usually on the hottest side of pins that had had a cross-pin temperature gradient. Fracture initiation was intergranular and may in some cases have been assisted by cor-

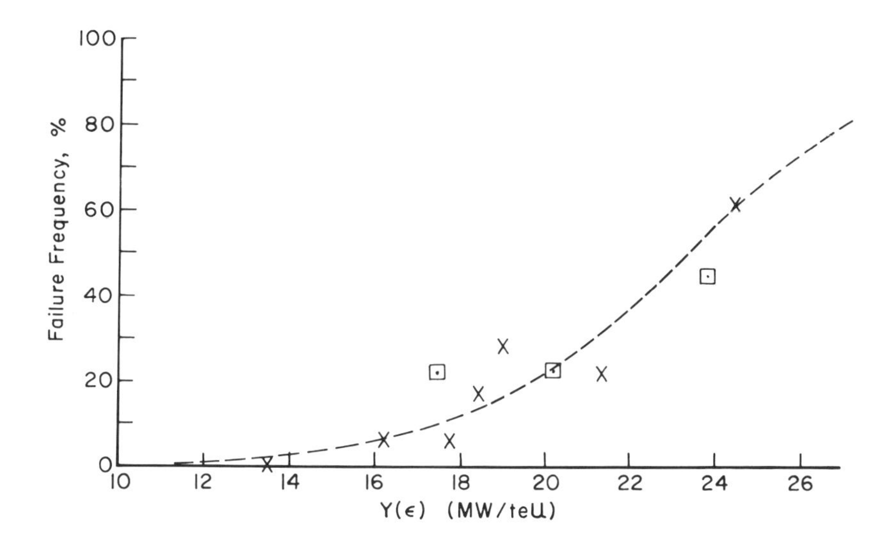

Fig. 14.12 Failure-frequency of Mk III and Mk IV fuel-pins versus a rating-parameter.

rosion of the clad bore by impurities from the UO_2 but 10% reduction in area occurred and failure seemed to be largely due to mechanical interaction between pellet and clad.

A theoretical analysis (cf. Gittus, 1972c) showed that radial cracks which form due to the expansion of the centers of pellets during an upward power increase (see fig. 14.13)

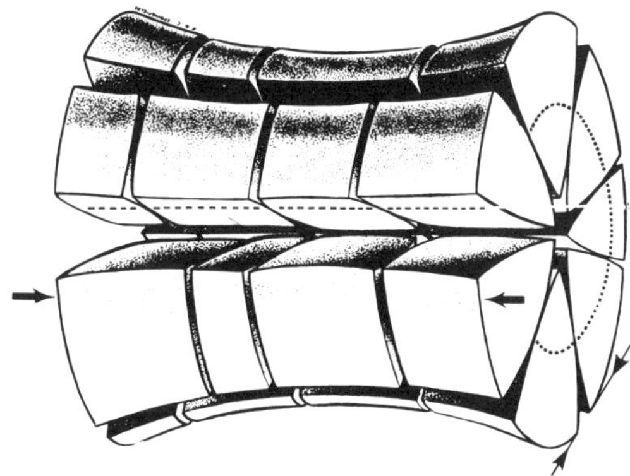

Fig. 14.13 Theoretical shape of a UO_2 pellet on-power in a can (not shown).

should indeed propagate through the surrounding cladding if the pellet was in contact and the cladding sufficiently brittle.

The magnitude of the strains produced by expansion of the

center of the pellet would be reduced, it was believed, if that center was removed and a hollow pellet adopted: there was already a strong incentive to do this since fission gas pressured would also be reduced to more acceptable levels.

The tolerance of the clad for strains imposed by the expansion of the pellet would be increased if its ductility could be enhanced: laboratory experiments had revealed very low hoop-ductilities in cans from irradiated Mk III fuel pins.

Finally, if the clad had a high creep strength then coolant pressure should not collapse it into contact with the pellet during brief periods at low rating: pellet expansion during a subsequent power increase might then be partly accommodated by the gap between pellet and can. As increased ductility is often only attainable at the expense of creep resistance it was recognized that the calculation techniques that had confirmed mechanical interaction as a plausible source of failure would have to be refined if they were to predict the advantage of raising the ductility and lowering the creep strength of the clad. The fuel pin computer model SEER was therefore developed (cf. Gittus, et al. 1970): its input comprised not only the physical and mechanical properties of the original and improved clad but also, of course, those of the pellet.

14.7.3 *Pellet-Clad Interaction*

If the pellet were welded to the clad, then pellet-cracks would always propagate at least a short distance into the bore of the clad during an upward power excursion. As discussed by Gittus (1972c) it is the existence of frictional sliding at the pellet-clad interface that ameliorates this situation, and if the coefficient of friction (μ_F) were zero then the clad would be subjected to much lower peak strains. The actual value of μ_F is about 0.5. It has been deduced from classical sliding experiments and confirmed by calculation from the strains produced in an actual cladding tube when a pellet is expanded by inflating a central tube.

14.7.4 *Some Predictions of the SEER Model*

The SEER computer model was fed with data on the properties of UO_2, UO_2-clad interface and the constitutive equations of the various clad variants. It was used to predict how the clad variants should have behaved in WAGR experiments (Table 14.5) and how they would behave in a power reactor.

The kind of agreement between prediction and observation

Table 14.5 Comparison of Predicted and Observed Wall Thinning in a Number of WAGR Experiments

	Fuel Pellets	Clad route	Clad thickness mm	Local wall thinning predicted by SEER %	Observed local wall thinning %
Cycling Experiments	Solid	Flash annealed	0.25	16	20
in the	* Hollow	C	0.25	26	28
WAGR loops	* Hollow	C	0.25	20	16
Experiments	Solid	A	0.38	6.	7.7
in the	Solid	A	0.25	1	8.0
WAGR core	Solid	A	0.38	4	< 4
	Solid	B	0.25	5	6.5

(* Two failure positions on the same fuel pin)

exemplified in Table 14.5 has been obtained without making any arbitrary changes in the SEER model or its input package (i.e. without any adjustment of parameters). This gives confidence in the SEER prediction that variant C will survive satisfactorily in a power-reactor.

From this work has emerged the hollow-pelleted fuel-pin, clad in ductile (route C) 20-25-Nb stainless steel which has been manufactured by BNFL for service in the Board's AGRs.

In WAGR 200 channels containing 7000 hollow pellet pins, using types A, B and C clad, have been irradiated at ratings up to 28 MW/te and nominal can surface temperatures up to 850°C. Of these, 120 channels had the selected Route C clad. Burnups extending up to 26,000 MWd/te have been achieved (cf. Evans, et al., 1973).

The experimentally-observed failure probability due to up-rating is low for this type of fuel and indeed it has not been possible to show statistically at the 95% confidence level, that up-rating has led to any failures, compared with the original fuel (cf. fig. 14.12). By radial shuffling, from low rating to high rating positions, four failures have been produced, but three of these were in fuel pins having type B cladding. The one failure which has occurred in a CAGR type pin (hollow pellet + type C clad) was produced by a very large effective up-rate of 12 MW/teU.

A similar improvement is emerging from current WAGR loop experiments where the rating is cycled a few times a day (cf. Evans, et al., 1973). Between 50 and 150 such cycles were sufficient to fail the original fuel whereas in the most recent experiment hollow-Route C fuel pins are surviving after 740 cycles in each of which the power was changed from 100% to 75%.

14.8 CONCLUSION

The development of constitutive equations for the dislocation creep of crystalline materials in general (and metals in particular) has been described. The effects of temperature and of bombardment with energetic particles have been dealt with and the resulting equations have been compared to the results of laboratory and in-reactor creep experiments. In some of these the response of the model equations to changes in the stress tensor have been verified - a critical test of their adequacy.

The constitutive equations were developed against a practical need, the design and evaluation of engineering structures. An example of the successful application of the particular equations described in this chapter is given: the use of a computer model to predict the creep in service of the cladding-tubes which isolate nuclear fuel from the coolant in a reactor.

REFERENCES

Anderson, R. G., and Bishop, J. F. W., (1962) in "Inst. Metals Symposium on Uranium and Graphite", (London: Institute Metals), paper 3, p. 1.

Ashby, M. F., (1972) Acta Met., 20, 887.

Axelrad, D. R., (1970) Advances in Molecular Relaxation Processes, 2, 41.

Blackburn, W. S., (1960) Phil. Mag., 6, 503.

Brailsford, A. D., and Bullough, R., (1972) in "Proceedings of European Conference on Irradiation Embrittlement and Creep in Fuel Cladding and Core Components" (London: British Nuclear Energy Society), paper 29, p. 1.

Buckley, S. N., and Manthorpe, S. A., (1972) in "Proceedings of European Conference on Irradiation Embrittlement and Creep in Fuel Cladding and Core Components" (London: British Nuclear Energy Society), paper 27, p. 1.

Bullough, R., Eyre, B. L., and Perrin, R. C., (1970) Nucl. Appl. Tech., 9, 346.

Cahn, R. W., (Ed.) (1965) 'Physical Metallurgy', North-Holland. (See esp. pp. 756 and 757).

Cottrell, A. H., (1964) "The mechanical properties of matter" (New York: Wiley).

Drucker, D. C., (1960) in "Structural Mechanics", edited by J. N. and N. J. Holf (London: Pergamon Press), p. 407.

Evans, D. M., Gallie, R. R., and Thorpe, G., (1973) "Windscale Advanced Gas Cooled Reactor Fuel Pin Operating Experience". Paper to the October, 1973 BNES Meeting (London: The British Nuclear Energy Society), to be published.

Evans, H. E., and Williams, K. R., (1973) Phil. Mag., 28, 227.

Foster, J. P., Wolfer, W. G., Biancheria, A., and Boltax, A., (1972) in "Proceedings of European Conference on Irradiation Embrittlement and Creep in Fuel Cladding and Core Components (London: The British Nuclear Energy Society), paper 26, p. 1.

Friedel, J., (1953) Phil. Mag., 44, 444.

Friedel, J., (1964) 'Dislocations' (Pergamon Press).

Gilbert, E. R., and Blackburn, L. D., (1970) 'Irradiation-induced creep in austenitic stainless steel' WHAN-FR-30, USA: Hanford Laboratory, Washington.

Gilbert, E. R., and Straalsund, J. L., (1970), Nuclear Engineering and Design, 12, 421.

Gilbert, E. R., Kaulitz, D. C., Holmes, J. J., and Claudson, T. T., (1972) HEDL-SA 469, USA: Hanford Laboratory, Washington.

Gittus, J. H., (1971) Phil. Mag., 24, 1423.

Gittus, J. H., (1972a) Phil. Mag., 25, 1233.

Gittus, J. H., (1972b) Phil. Mag., 25, 345.

Gittus, J. H., (1972c) Nuclear Engineering and Design, 18, 69.

Gittus, J. H., (1973a) in "Proceedings of Int. Conf. on Creep and Fatigue in Elevated Temperature Applications", Philadelphia and Sheffield, (London: Inst. Mech. Eng.) paper No. C155/73, p. 155.1.

Gittus, J. H., (1973b) Phil. Mag., 28, 261.

Gittus, J. H., (1973c) "The influence of materials properties on the design of AGR fuel", paper No. 24, Int. Conf. Phys. Met. Reactor Fuel Elements, CEGB, Berkeley Nuclear Labs., Gloucestershire, England.

Gittus, J. H., (1974) Acta Met., 22, 789; (1974a) "Creep, Viscoelasticity and Creep-Rupture in Solids" (London: Elsevier); (1975) "Theoretical equation for steady-state dislocation-creep in a material having a threshold stress", Proc. Roy. Soc. (forthcoming).

Gittus, J. H., Howl, D. A., and Hughes, H., (1970) Nuc. Appl. Tech., 9, 40.

Gittus, J. H., Makin, J., and Anderson, R. G., (1972) in "Proceedings of European Conference on Irradiation Embrittlement and Creep in Fuel Cladding and Core Components" (London: British Nuclear Energy Society), paper 32, p. 1.

Harkness, S. D., and Yaggee, F. L., (1972) in "Proceedings of European Conference on Irradiation Embrittlement and Creep in Fuel Cladding and Core Components" (London: British Nuclear Fuel Society), paper 28, p. 1.

Hesketh, R. V., (1963) Phil. Mag., 8, 1321.

Hesketh, R. V., (1967) BNL 50083, USA: Brookhaven National Laboratory.

Hesketh, R. V., (1968) J. Nuclear Materials, 26, 77.

Hesketh, R. V., (1969) J. Nuclear Materials, 29, 217.

Hill, R., (1950) "The Mathematical Theory of Plasticity" (Oxford: Clarendon Press).

Holt, D. L., (1970) J. Appl. Phys., 8, 3197.

Hunter, C. W., Fish, R. L., and Holmes, J. J., (1972) Trans. American Nuclear Society, 15, 254.

Lagneborg, R., (1972) Metal Science Journal, 6, 127.

Levy, R., (1967) "The Mathematical Theory of Plasticity" (Oxford: Clarendon Press), (see especially equation 5 on page 16).

Lewthwaite, G. W., and Proctor, K. J., (1973) Journal of Nuclear Materials, 46, 9.

Lloyd, D. J., and Embury, J. D., (1970) Metal Science Journal, 4, 6.

McElroy, W. N., Dahl, R. E., and Gilbert, E. T., (1970) Nuclear Engineering and Design, 14, 319.

McLean, D., (1962) Met. Rev., 7, 481; (1970) "The Mechanical Behavior of Metals". Paper given at a conference of the Japan Society of Metals.

von Mises, R., (1913) Gottingen Nachrichten, Math.-Phys., Klasse, 528.

Mitra, S. K., and McLean, D., (1966) Proc. Roy. Soc. (London), 4, 295, 288.

Mitra, S. K., and McLean, D., (1967) Metal Science Journal, 1, 192.

Mott, N. F., (1952) Phil. Mag., 43, 1151.

Mukherjee, A. K., Bird, J. E., and Dorn, J. E., (1969) Trans. ASM, 62, 155.

Nabarro, F. R. N., (1948) in "Proceedings of Bristol Conference on Strength of Solids" (London: The Physical Society), p. 75.

Nabarro, F. R. N., (1959) in "The Application of Modern Physics to the Earth and Planetary Interiors", edited by S. K. Runcorn, (London: Wiley Interscience) p. 251.

Norman, E. C., and Duran, S. A., (1970) Acta Met., 10, 723.

Orlova, S., Pahutova, M., and Cadek, J., (1971) Phil. Mag., 25, 865.

Orlova, A., Tobolova, Z., and Cadek, J., (1972) "Internal Stress and Dislocation Structure of Aluminum in High-temperature Creep" (Personal communication).

Orowan, E., (1946-47) West of Scotl. Iron and Steel Inst., 54, 45.

Rice, J. R., (1970) J. Appl. Mech., 37, 728.

Salama, K., and Roberts, J. M., (1970) Scripta Met., 4, 749.

Taylor, G. I., (1934) Proc. Roy. Soc., (London) A145, 362.

Walters, L. C., Walker, C. M., and Pugacz, M. A., (1972) J. Nuclear Materials, 43, 133.

Weertman, J., (1955) J. Appl. Phys., 21, 1213.

Weertman, J., (1957) J. Appl. Phys., 28, 362.

Weertman, J., (1968) Trans, ASM Quarterly, 61, March, 681-694.

Weertman, J., (1973) Trans. AIME, 227, 1475.

14A.1 APPENDIX ON ALGORITHMS

Table A1 Algorithm MEOS that Solves the Multiaxial Equations for Anelastic Plus Plastic Strains

Algorithm
REAL LASTE(3),LASTEDOTS(3) DIMENSION SFAC(3),ANT",ALAN" DIMENSION EDOTS(3),SIG",EPS",IDELTA" TOTIM=IM=∅
INPUT,AN,EA,EPRIME,Q,LAMBDA,ROINF,ROFJ,RORI,AL
REAL MU
STAR:IM=IM+1
INPUT,SIG(1),SIG(2),SIG(3),T,TIME
IF(TIME)DONE,ALONG,ALONG
ALONG:SIGV=∅.3333*1.414*SQRT(SIG(1)-SIG(2))**2+(SIG(1) + -SIG(3))**2+(SIG(2)-SIG(3))**2)
GAMMA=EXP(EPRIME-Q/2/T)
EDOTVS=SIGV**AN*EXP(EA-Q/2/T)
IDEL=∅
DO ONE,N=1,3 SFAC(N)=(SIG(N)-(SIG(1)+SIG(2)+SIG(3))/3)/SIGV
EDOTS(N)=(SIG(N)-(SIG(1)+SIG(2)+SIG(3))/3)*EDOTVS/SIGV
ONE: MU=LAMBDA*EDOTVS+GAMMA
ROL=(LAMBDA*EDOTVS*ROINF+GAMMA)/MU ALIM=AL*(1-ROL)
IF(IM-2)D,E,E

Table A1 (continued)

	Algorithm
	E:IDEL=Ø DO C,N=1,3 J-SIGN(1,EDOTS(N))-SIGN(1,LASTEDOTS(N)) IF(J)A,B,A A:IDELTA(N)=Ø;GOTO C B:IDELTA(N)=1 C:IDEL=IDEL+IDELTA(N) DEL=IDEL;DEL=DEL/3
	ROFI=ROFE*DEL+RORE*(1-DEL)
	RORI=RORE*DEL+ROFE*(1-DEL)
	D:DO THREE,M=1,1Ø TIM=TIME*M/1Ø
 (1)	DO TWO,N=1,3 ANT(N)=(ALAN(N)-ALIM*SFAC(N))*EXP(-MU*TIM)+ALIM*SFAC(N) FAC=(ROFI/ROL=1)*(1-EXP(-MU*TIM))/MU+TIM TWO:EPS(N)=EDOTS(N)*FAC+LASTE(N)+ANT(N) TOTIM=TOTIM+TIME/1Ø PRINT,TOTIM,EPS(1),EPS(2),EPS(3)
	THREE: DO FOUR,N=1,3 ALAN(N)=ANT(N) LASTEDOTS(N)=EDOTS(N)
	FOUR:LASTE(N)=EDOTS(N)*FAC+LASTE(N)
	ROFE=(ROFI-ROL)*EXP(-MU*TIME)+ROL
	RORE=(RORI-1)*EXP(-MU*TIME)+1
	GOTO STAR DONE:

NOTE (1). This statement calculates the viscoelastic strain.

Table A2 MEOS Parameters for Stainless Steel. (Input for the Algorithm)

Parameter	n	E_a	E'	Q	λ	$\bar{\rho}_\infty$	$\bar{\rho}_{Fi1}$	$\bar{\rho}_{Ri1}$	ℓ_a
Algorithmic name	AN	EA	EPRIME	Q	LAMBDA	ROINF	ROFI	RORI	AL
Experimental value	2.4	20.1	18.2	6×10^4	8460	0.113	1.0	1.0	0.001
Theoretical value	3.0	18 to 20	19.5	6.5×10^4	3876	$\rightarrow 0$	1.0	1.0	.0013

LIST OF SYMBOLS

$\dot{\varepsilon}_s$	steady-state creep-rate
t	time
A	materials constant
n	materials constant
D_v	bulk self-diffusion-coefficient
μ	shear modulus
b	Burgers vector
k	Boltzmanns constant
T	temperature, deg Kelvin
T_m	temperature of melting, ^oK
r_s	dislocation-spacing
c_j	jog concentration
σ	stress
v	mean glide-velocity of a dislocation
L_s	mean free glide-path of a dislocation-network link
$\Delta\theta_s$	change in sub-boundary angle
d_s	sub-grain size
ε	strain (non-recoverable)
h_s	distance between adjacent sub-boundary dislocations
K'	"Holt's" constant (of order 10)
i	rate of immobilization of dislocations
λ	materials constant
ρ	dislocation-density

ρ_o	maximum mobile dislocation density
ρ_∞	minimum mobile dislocation density
$\bar{\rho}$	ρ/ρ_o
$\bar{\rho}$	ρ_∞/ρ_o
γ	rate of climb-induced remobilization of dislocations
$\bar{\rho}_c$	fractional number of compression-mobile dislocations
ξ	recoverable component of creep-strain
ℓ_a	$(b/2)\sqrt{\rho_o}$
e	$= \varepsilon + \xi$ = total creep-strain
$\bar{\rho}_L$	limit of ρ as $t \to \infty$ at constant stress + temperature
Q	activation energy
$J'_{1,2,3}$	deviatoric tensor invariants
ε_e	elastic strain
E_m	Young's modulus
ν	Poisson's ratio
E'_m, ν'	viscoelastic constants
Y_p	yield point
$\dot{S}$	swelling-rate due to void-formation
ψ_d	dislocation-flux (random)
I_s	internal stress
$\ell_{i\alpha}$	direction cosine
$\dot{g}$	rate of anisotropic growth
$\bar{E}$	neutron energy

ϕ	neutron flux
Φ	neutron dose
$\Delta\Phi$	e-folding dose

ABSTRACT. *The LIFE computer code has been developed to predict fuel-element performance in sodium-cooled fast-breeder reactors. The code attempts to model the complex thermal, mechanical, and nuclear behavior of fuel and cladding under fast-neutron irradiation in a physically meaningful way. The models developed for the code include fuel swelling, hot pressing and creep, fuel cracking and healing, cladding swelling and creep, fuel-cladding mechanical and chemical interactions, and the effects of sodium on the cladding. The models are described and directions of current work are indicated.*

15. THE FUEL ELEMENT LIFE CODE, AN ULTIMATE APPLICATION OF CONSTITUTIVE EQUATIONS TO A HIGH TECHNOLOGY PROBLEM(*)

R.W. Weeks, V.Z. Jankus, and R.B. Poeppel

15.1 INTRODUCTION

The core of a large, fast-breeder nuclear reactor will contain tens of thousands of fuel elements. The safety, reliability, and economics of the reactor system will depend to a great extent on our understanding of in-reactor fuel-element behavior. The LIFE computer code (Jankus 1973, Jankus and Weeks 1970 and 1972, Weeks, Jankus, Katsuragawa and Lambert 1971) represents the integration of a great many detailed fuel and cladding materials studies in an attempt to predict the thermal, nuclear, and mechanical behavior of fuel elements irradiated in sodium-cooled fast-breeder reactors. The code was originally developed at Argonne National Laboratory (ANL) and is currently being refined by the U. S. commercial nuclear industry and ANL in a coordinated effort under the auspices of the U. S. Atomic

*Work performed under the auspices of the U. S. Atomic Energy Commission.

Energy Commission. In addition to providing a test for our understanding of fuel-element behavior, the code allows a means with which to define the most critical experiments to be run in-reactor, to focus materials research on key areas related to improved performance, and to estimate a priori the effects of changes in design and operating variables.

The analytical procedures used in the code have been discussed elsewhere (Jankus 1973, Jankus and Weeks 1970 and 1972) and will be mentioned here only in their relation to modeling the various materials phenomena in a manner that is physically meaningful yet analytically tractable within a reasonable amount of computer time. Perhaps the most important feature of the analytical procedure is that it is designed to follow fluctuations in reactor operating conditions. This is important in predicting fuel-element behavior since most of the materials properties are strongly temperature dependent, and thermal stresses generated during reactor cycling play an important role in the deformation of the element.

Fuel elements for liquid-metal fast-breeder reactors (LMFBRs) basically consist of a column of solid cylindrical fuel pellets loaded into a long hollow tube of metal cladding. A plenum is provided at the top of the fuel column to hold gases released during fissioning of the fuel. The cladding tube is 0.25 in. in diameter, and 3 ft. in length, with a 15-mil wall thickness. Initially, a diametral gap of 1-2 mils exists between the fuel and cladding. The element is filled with helium and sealed to prevent radioactive fission products from contaminating the coolant. For near-term LMFBR cores in the U. S. A., the fuel will be mixed oxide, $(U,Pu)O_2$, and the cladding will be cold-worked Type 316 austenitic stainless steel.

When a fuel element is irradiated by fast neutrons, several phenomena occur. The poor thermal conductivity of the oxide fuel results in temperature gradients on the order of $3000^{o}C/cm$. This, in turn, causes fuel cracking due to thermal stresses, and restructuring of the fuel as a result of sintering and porosity migration. As irradiation proceeds, the fuel swells because of the accumulation of solid and gaseous fission products, and the cracks heal. The swelling of the fuel is modified by hot pressing, the restraint of the cladding, and fission-gas release. The fuel relieves deviatoric stresses through several creep mechanisms, some of which are induced or enhanced by the fissioning process. The cladding also creeps and swells, although the swelling in this case is due to the creation and unbalanced elimination of point defects following fast-neutron collisions with the cladding. While these mechanical and nuclear phenomena are occurring, the fuel-element chemistry is also changing. Vaporization and condensation within the fabricated porosity can cause porosity migration and radial variations of density and U/Pu ratio, both of which can significantly affect the temperature distribution. The stoichiometry of the fuel also changes as metal atoms are transformed, and the excess oxygen reacts with fission products, which can migrate from the fuel to the cladding inner surface to produce a corrosive attack of the cladding wall. Additionally, the hot sodium coolant acts as a nonmetallic mass-transfer agent, normally leaching carbon and nitrogen from the cladding and depositing them in cooler parts of the reactor system. These last two chemical effects must be considered in the formulation of cladding-failure criteria.

From the foregoing considerations, it is clear that the behavior of a fuel element can be quite complex and highly sensitive to its irradiation conditions. However, because of the great expense of fuel-element irradiations, the years normally required to carry them out, and the necessity to extrapolate the data since the test environment is not prototypical, a strictly empirical assessment of fast-reactor fuel-element behavior is severely limited at best. The LIFE computer code was formulated to supplement and help guide the experimental program, and to provide the designer with an objective means of predicting fuel-element performance.

15.2 THERMAL ANALYSIS

15.2.1 *Temperature Distribution*

The temperature and temperature gradient distributions are established for each time step in the LIFE analysis using temperature and structure-dependent conductivities, and as fine a radial mesh as desired. The process is repeated for up to ten axial nodes to account for axial variations in power, flux, and coolant temperature.

The conductance across the fuel-cladding interface is one of the most difficult parameters to establish, since it is a function of temperature and gas composition or interface pressure, depending on whether the gap is open or closed. A formulation used for the gap conductance (h_{fc}) is given by

$$\frac{1}{h_{fc}} = \frac{1}{h_{gap}} + \frac{1}{h_o(1 + p_{fc}/P_o)} , \tag{15.1}$$

where

h_{gap} = $k_{gas}/\Delta x$,

k_{gas} = the conductivity of the gas in the fuel-cladding gap when a gap exists,

Δx = the width of the gap,

h_o, P_o = constants used to fit the data (see fig. 15.1),

P_{fc} = the interface pressure between the fuel and cladding.

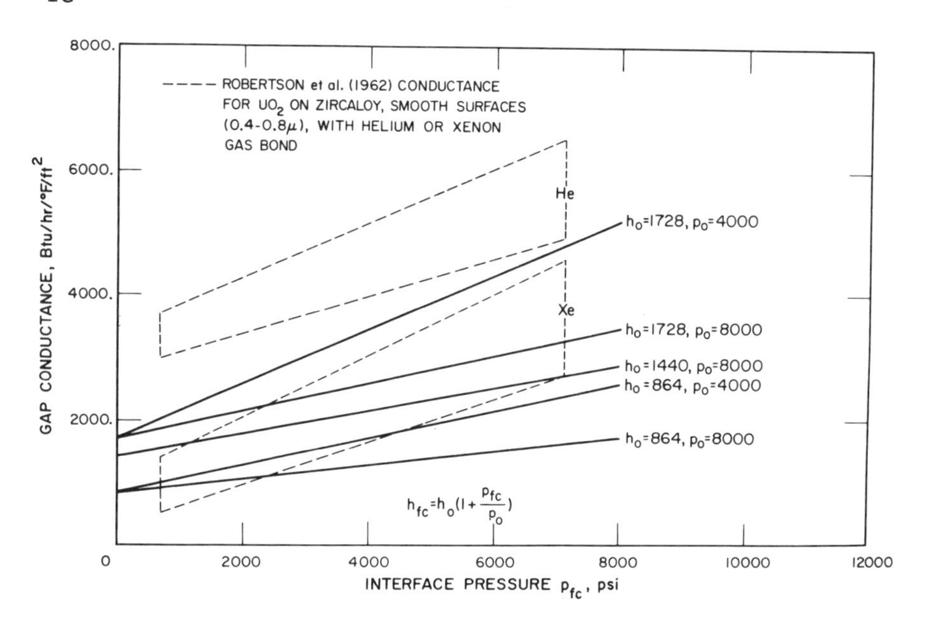

Fig. 15.1 Various linear fits to the data of Robertson, Ross, Notley and MacEwan (1962).

A parametric study of the influence of the gap conductance on fuel behavior (Jankus and Weeks 1972) indicated gross changes in the predicted fuel restructuring and diametral expansion of the fuel element, depending on the gap conductance used in the analysis. For this reason, indirect measures from post-irradiation analyses of short-term fuel-element irradiations are also used to help establish reasonable conductance values.

15.2.2 *Fuel Restructuring*

A radial cross section of an element after irradiation in the U. S. Experimental Breeder Reactor (EBR-II) is shown in fig. 15.2. The idealization of the restructured cross section used

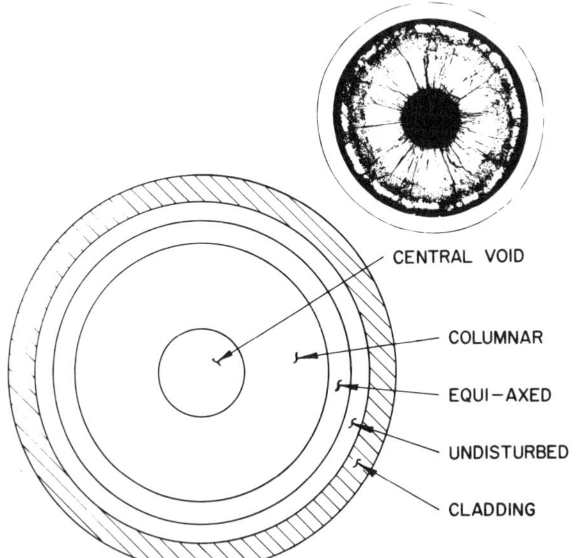

Fig. 15.2 Radial cross section of a fuel element after irradiation.

in the LIFE code is also shown. The central void was produced by the densification and outward migration of the fuel in response to the temperature gradient. The columnar-grain region is believed to be formed through the migration of large fabricated pores up the temperature gradient by an evaporation-condensation mechanism (fig. 15.3). In the LIFE code, the rate of restructuring has been modeled by computing the velocity of a migrating pore at the boundary of the region. The pore velocity in oxide fuel is derived as the product of the pore mobility by evaporation-condensation times the thermal gradient driving force (Weeks, Scattergood, and Pati 1970), or

$$v_{pore} = C_1 T^{-3/2} [\exp(-C_2/T)] dT/dr , \tag{15.2}$$

where C_1 and C_2 involve properties of the fuel, and dT/dr is the temperature gradient. A similar treatment is used for the formation of the equiaxed region, although an alternate sintering model would perhaps be more appropriate. An extensive review of pore migration in temperature gradients has

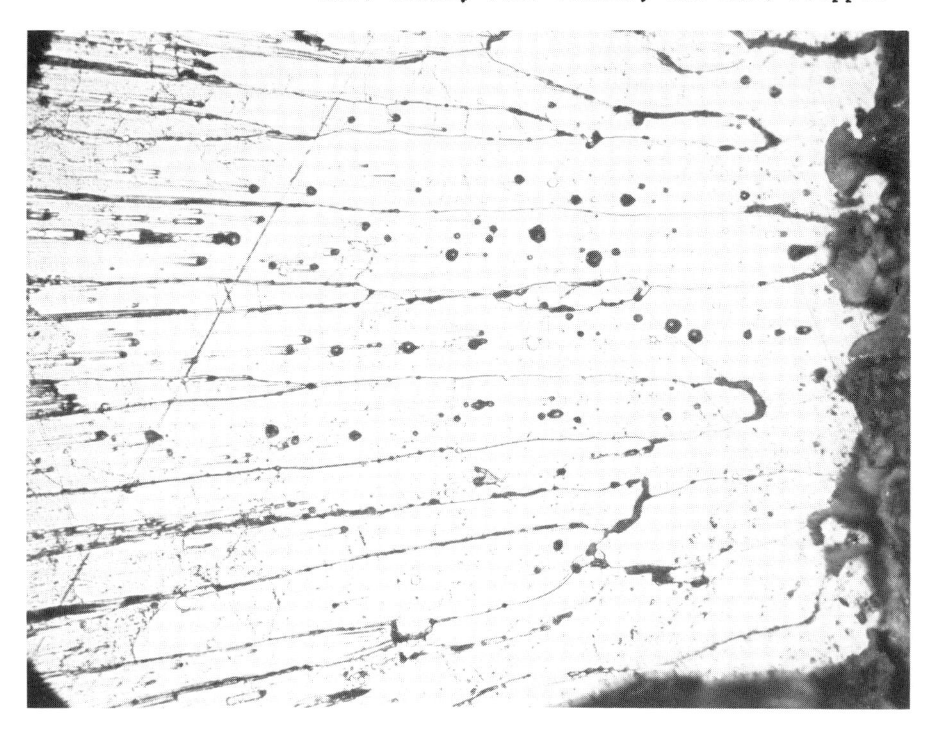

Fig. 15.3 Large pores migrating up the temperature gradient in the columnar-grain region to the central void. (Courtesy of L. A. Neimark, Argonne Nat. Lab.)

been given by Nichols (1969). It should be noted that not all the materials properties required by the constants in eqn. (15.2) are well known, most notable Q^*, the surface heat of transport. For this reason, the restructuring models are calibrated against short-term irradiation results. For a ten-day irradiation in EBR-II, typical LIFE-code calculations (Jankus and Weeks 1972) for the temperature history and restructuring of an element are shown in fig. 15.4. The temperature variations correspond with reactor power variations during the irradiation. The peak temperatures of each region decrease somewhat with time as a result of fuel densification and narrowing of the initial fuel-cladding gap.

15.3 MECHANICAL ANALYSIS

15.3.1 *General Constitutive Relations*

The LIFE code employs an iterative generalized plane-strain model in the mechanical analysis of the fuel element, using

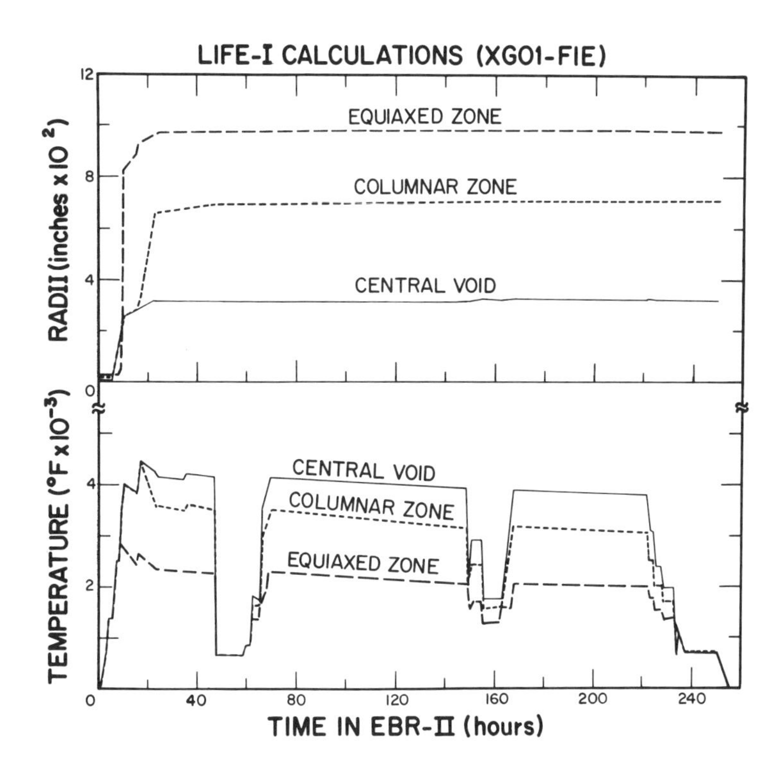

Fig. 15.4 Predicted restructuring and zone-boundary temperatures for a ten-day irradiation of fuel element XG01-F1E in the EBR-II reactor.

concentric cylinders to account for radial variations in properties and operating conditions. The code solves the problem several times with boundary conditions appropriate for various positions along the length of the element to account for axial changes in flux, power, and coolant temperature (see figs. 15.5 and 15.6). The mechanics have been discussed in detail elsewhere (Jankus 1973, Jankus and Weeks 1970 and 1972) but it may be noted that the constitutive equations for any radial region are formally simple. The total radial strain is

$$\varepsilon_r^T = E^{-1}[\sigma_r - \nu(\sigma_\theta + \sigma_z)] + \alpha T + \varepsilon_r^p + \Delta\varepsilon_r^p + \varepsilon_r^s + \Delta\varepsilon_r^s , \tag{15.3}$$

and similarly for ε_θ^T and ε_z^T, where

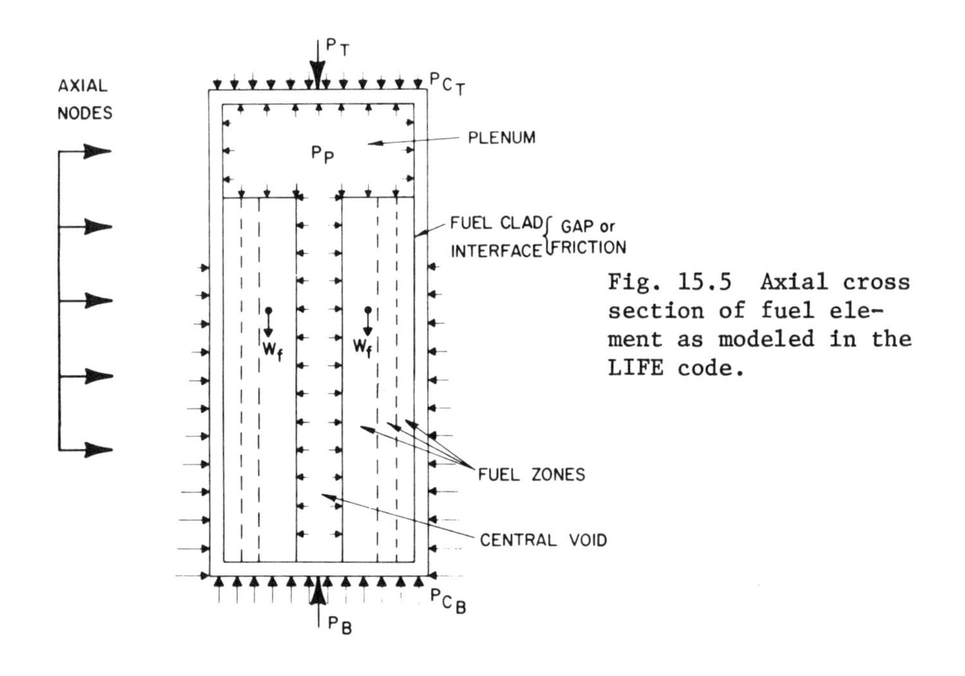

Fig. 15.5 Axial cross section of fuel element as modeled in the LIFE code.

$\varepsilon^T_{r,\theta,z}$ = components of total strain at time $t + \Delta t$,

$\sigma_{r,\theta,z}$ = components of stress at time $t + \Delta t$,

$\varepsilon^p_{r,\theta,z}$ = components of the creep strain accumulated up to time t.

$\Delta\varepsilon^p_{r,\theta,z}$ = components of creep strain that occur during the time interval Δt,

$\varepsilon^s_{r,\theta,z}$ = components of swelling strain accumulated up to time t,

$\Delta\varepsilon^s_{r,\theta,z}$ = components of swelling strain that occur during time interval Δt,

E, ν = elastic modulus and Poisson's ratio,

αT = $\int \alpha dT$ the linear thermal expansion from room temperature.

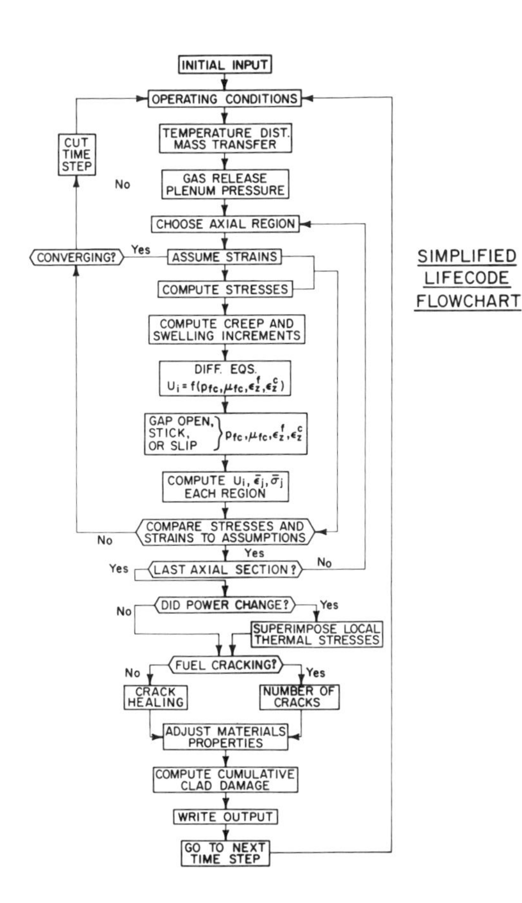

Fig. 15.6 Calculational flow-chart for the LIFE code.

The Prandtl-Reuss flow laws are used to determine the creep-strain components from uniaxial creep-strain data, and swelling phenomena are assumed isotropic.

15.3.2 *Fuel Treatment*

a) Creep Deformations. The creep deformation of the fuel is based on both in-reactor and out-of-reactor data for $(U,Pu)O_2$ (Roberts, Routbort, Voglewede and Solomon 1973) and includes a low-temperature term linear in stress, the usual thermal creep with a higher stress exponent, and two fission-enhanced creep terms, one temperature dependent and the other athermal. Although each is physically meaningful, the combination of the four terms yields a good representation of the steady-state creep behavior of $(U,Pu)O_2$ observed in- and out-of-reactor.

A deformation map for the fuel (after Ashby 1972) has been constructed by Roberts, Routbort, Voglewede and Solomon (1973) and is shown in fig. 15.7. The analytical formulation in the

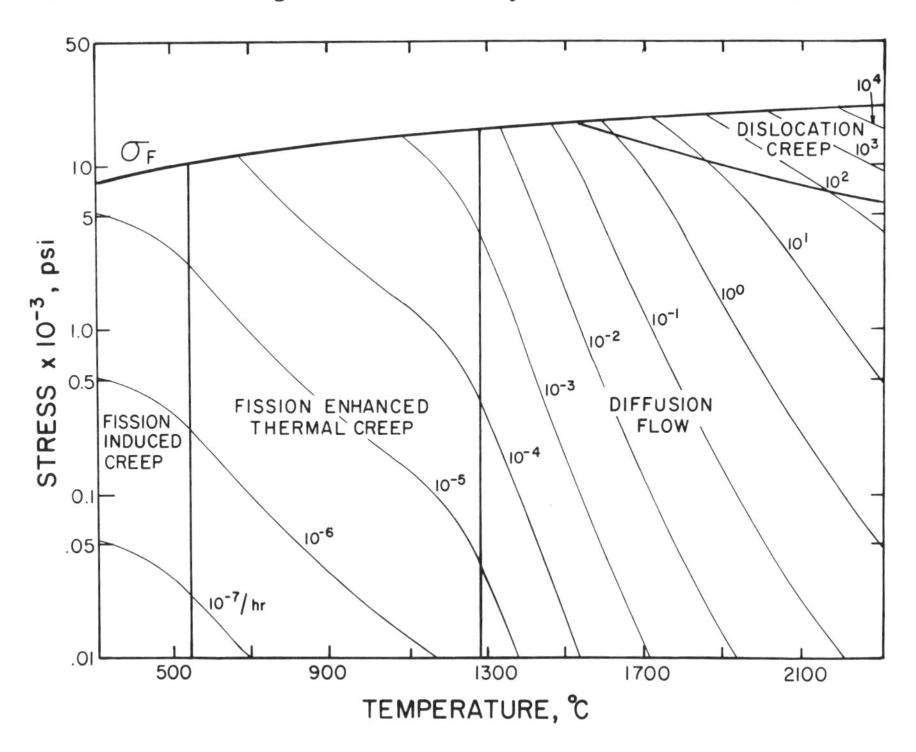

Fig. 15.7 Deformation map of 4-μm 95% TD UO_2-20 to 25 wt% PuO_2 subjected to fission rates of 5 x 10^{13} fissions/cc-sec (Roberts, Routbort, Voglewede and Solomon 1973, courtesy of Argonne Nat. Lab.)

code is sufficiently general to include primary creep, and work is in progress in this area.

b) <u>Fuel Swelling</u>. The fuel swells as a result of thermal expansion and the accumulation of solid and gaseous fission products. The treatment of thermal expansion and solid fission-product expansion are straightforward. With the assumption of incompressibility of the solid fission products, the fission rate in each region is simply multiplied by the net solid volume per fission. The gaseous fission products, however, are both compressible and highly mobile.

To compute fuel swelling due to gaseous fission products, the amount of gas generated, accumulated, and released from each region of the fuel during any time step must first be computed. Furthermore, the swelling depends not only on the amount of fission gas present, but on its form, which can range from individual incompressible gas atoms or small bubbles, to large, highly compressible bubbles.

Gas release can take place through several mechanisms including the migration of bubbles up the temperature gradient (Nichols 1969), the linkup of bubbles trapped on grain boundaries (Michels, Poeppel and Neimark 1970), or by fuel cracking during power transients. The code, at present, does not model each of these mechanisms, but rather adopts the following simpler, more macroscopic approach to gas release and fuel swelling.

The fission gas generated is computed from the fission rate. The gas release rate from the fuel is taken as an exponential function of the temperature of the region, and is proportional to the amount of gas remaining in the region. The swelling due to the fission gas has been treated with a simple gas-law approach.

$$\left(\frac{\Delta V}{V}\right)_i^{gas} = \frac{n_i R\bar{T}_i}{V_i(P_i + P_i')} + \left(A\,\frac{n_i}{V_i}\right), \tag{15.4}$$

where the first term on the right is the compressible portion, the last term is the incompressible portion, and

ΔV_i = change in volume of the ith region,

V_i = original volume of the ith region,

n_i = number of gas moles remaining in the region,

$\bar{T}_i$ = average temperature of the gas, assumed equal to the average temperature of the region,

P_i = $-(\sigma_r^i + \sigma_\theta^i + \sigma_z^i)/3$,

$\sigma_{r,\theta,z}^i$ = the average stresses in the ith region,

P_i' = a constant similar to the usual surface-tension

correction term, but referring to the entire ith region,

R = the universal gas constant,

A = a constant.

This simplified approach presently requires calibration against integral fuel-element irradiations, although, in principle, all the data required can be obtained from independent experiments.

A much more detailed and mechanistic approach to fission-gas behavior and fuel swelling, called the GRASS subroutine, has been developed by Poeppel (1971). A transient version of the subroutine (called HORSE) has been used for the analysis of fuel behavior in accident situations (Diaz and Poeppel 1973). GRASS is designed for compatability with the LIFE code. The fuel treatment includes the production, migration, coalescence, and release of fission-gas-bubble distributions, as well as fission-gas resolution (Pati, Dapht, O'Boyle and Patrician 1973) and the interaction of bubbles with structural defects (Weeks, Scattergood and Pati 1970). The use of the GRASS model, in addition to providing a more accurate prediction of fuel swelling and gas release, will allow comparison of the predicted bubble size and spatial distributions with those observed in postirradiation analyses (Poeppel, Makenas and Michels 1971).

c) Fuel Hot Pressing. Hot pressing of fuel is the closing down of fuel porosity under the influence of heat and hydrostatic pressure. The porosity is generated in the fuel fabrication process and also by cracking of the fuel during power transients. A diffusional model of hot pressing originally developed by Rossi and Fulrath (1965) for alumina compacts proved to underestimate the phenomena in oxide fuels as observed in out-of-reactor tests (Roberts, Routbort, Voglewede and Solomon 1973). An attempt to relate the fuel hot-pressing rate to the fuel creep rate has been made by Routbort in the study just mentioned, following MacKenzie and Shuttleworth's (1949) original treatment of hot pressing and sintering and has resulted in a much better fit to the experimental data. The resultant equation is of the form

$$\dot{\varepsilon}_{hp} = \frac{P}{1-P} \dot{\varepsilon}_{ssc} , \qquad (15.5)$$

where $\dot{\varepsilon}_{hp}$ is the volumetric strain rate due to hot pressing,

P is the volume fraction porosity, and $\dot{\varepsilon}_{ssc}$ is the steady-state creep rate. A parametric study to explore the influence of fuel hot pressing on the resultant interface pressure between the fuel and the cladding for various fuel creep rates indicated a tradeoff between fuel creep and hot pressing. A "soft" fuel with a high creep rate (i.e. less fuel self-restraint) tends to generate more interface pressure with the cladding, whereas a high hot-pressing rate tends to decrease the interface pressure. This suggests that a low-density fuel might exert nearly as much pressure on the cladding as a high-density fuel. A low-density fuel has been thought necessary to decrease fuel-element deformations, but a high-density fuel is best for breeding and thermal performance; therefore, these results have important implications for fuel-element design.

d) Fuel Cracking and Healing. The radial cracks observed in the fuel after irradiation are caused by the thermal stresses generated during reactor power changes. Substantial evidence is available to show that the cracks also heal during steady-state periods of reactor operation. It is difficult to account for fuel cracking and healing rigorously, since the LIFE code formulation is only approximately two-dimensional, and isotropy and radial symmetry have been assumed in the mechanical analysis. Nevertheless an excellent approximate representation for fuel cracking, consistent with the LIFE-code assumptions, was developed (Jankus and Weeks 1972). Basically, the effects of fuel cracking are averaged over the fuel by creating an equivalent isotropic material in which elastic properties reflect the number of cracks that have occurred. An outline of the mathematical procedure is given elsewhere (Jankus and Weeks 1972), but it is important to note that the elastic properties are not adjusted arbitrarily. The constants after each crack occurs are consistently related to the original, and after all cracks have healed, the original elastic properties of the fuel are recovered. Crack healing occurs when the stresses in a region become compressive, the power level is steady, and the temperature is above a minimum level. Experimental studies to define fuel-cracking and crack-healing parameters have been, and are being, carried out at Argonne by Roberts and Wrona (1973). An example of LIFE-code calculations involving fuel cracking during a reactor startup is shown in fig. 15.8.

15.3.3 *Cladding Treatment*

a) Thermoelastic and Creep Deformations. The thermoelastic deformation of the cladding is computed in a straightforward

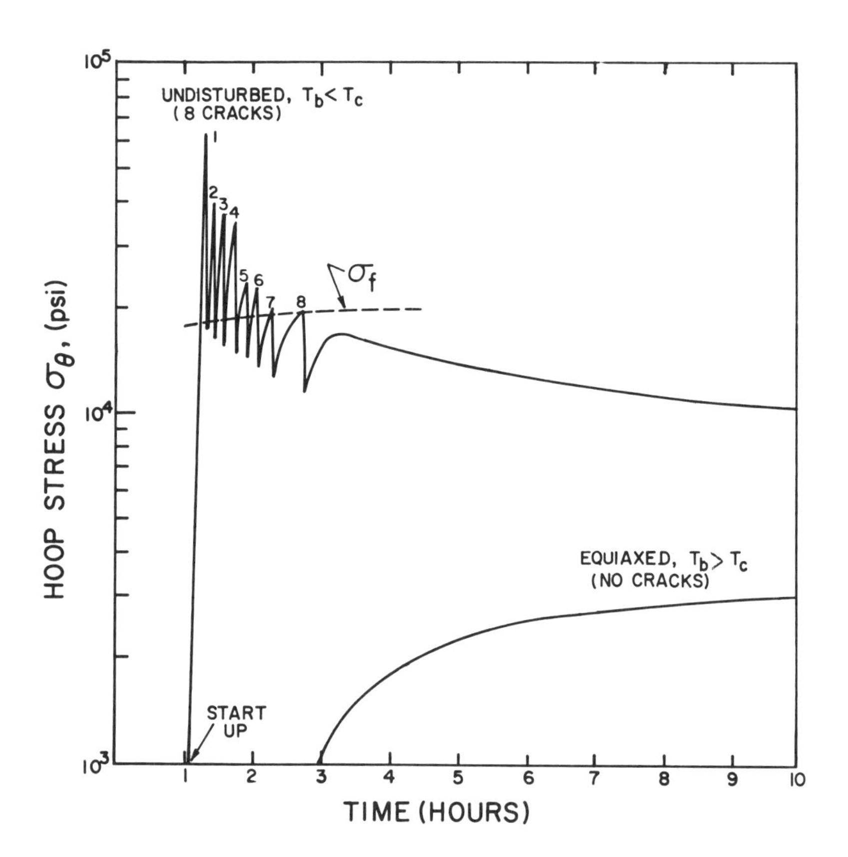

Fig. 15.8 Example of fuel-cracking predictions during start-up of an irradiation. Note that cracking stops when the stresses in a region fall below the rupture stress, σ_f, or when the temperature is sufficiently high to permit rapid stress relaxation through creep.

manner with the coefficient of thermal expansion taken as a linear function of temperature. A flux enhancement of the creep rate occurs, however, and the creep law is taken as

$$\Delta\varepsilon_p = [A\sigma_e^m \exp(-Q/kT) + B\sigma_e^n \phi]\Delta t \ , \tag{15.6}$$

where

$\Delta\varepsilon_p$ = equivalent creep-strain increment,

σ_e = $\frac{1}{\sqrt{2}}[(\sigma_r - \sigma_\theta)^2 + (\sigma_r - \sigma_z)^2 + (\sigma_\theta - \sigma_z)^2]^{1/2}$, the equivalent uniaxial stress,

Q = average activation energy for thermal creep,

k = Boltzmann's constant,

T = absolute temperature,

ϕ = neutron flux,

Δt = time interval,

A,B,m,n, = experimentally determined constants.

A more fundamental treatment of the flux enhancement is also available, based on the enhanced vacancy formation rate during irradiation (Harkness, Tesk and Li 1969). The creep law for the cladding, as all other constitutive relations discussed here, are updated and revised as further experimental data become available.

b) Cladding Swelling. The cladding swells as a result of the preferential attraction of interstitial defects over vacancies to sinks, with the resultant growth of voids. The point defects are generated by the neutron collisions with the cladding lattice structure. An example of voids in irradiated cladding is shown in fig. 15.9. A theoretical model of this phenomena by Harkness and Li (1971) has been included in the LIFE code, along with several empirical swelling relations developed strictly on the basis of the available data. The empirical relations undergo periodic revision, but one such expression used for 20% cold-worked Type 316 stainless steel was given by (Sutherland 1971)

$$\Delta V/V[\%] = 9 \times 10^{-35} (\phi t)^{1.5}(4.028 - 3.712 \times 10^{-2T} \cdot T + 1.0145 \times 10^{-4} T^2 - 7.879 \times 10^{-8} T^3) \,, \qquad (15.7)$$

where $\Delta V/V$ is the volumetric expansion, ϕt is the fast fluence, and T is in ^{o}C.

c) Fission-Product Attack of the Cladding. The kinetics of cladding attack by volatile fission products such as cesium and

EBR-II IRRADIATED 304 STAINLESS STEEL

T_{irr} = 417°C Fluence = 9.5 x 10^{22} n/cm^2
void volume fraction = 10%
void concentration = 1.5 x 10^{16} cm^{-3}
mean void diameter = 200 A

Fig. 15.9 TEM picture of voids in irradiated Type 304 stainless steel. (Courtesy P.R. Okamoto, Argonne Nat. Lab.)

tellurium (fig. 15.10) have been studied and duplicated in out-of-reactor experiments (Maiya and Busch 1973). The severity of attack is strongly dependent on the oxygen potential, and low oxygen-to-metal (O/M) ratio fuels and oxygen getters have been used in an attempt to eliminate the problem. The effects of the attack on cladding mechanical properties have also been evaluated in out-of-reactor tests (Rosa, Maiya and Weeks 1974). It was determined that the effects could be accounted for analytically by a reduction in the effective cladding thickness. The kinetics data can be combined with a knowledge of O/M changes with fuel burnup to predict the extent of the attack, and this, coupled with the mechanical-properties data, can be used as part of a cladding-failure criterion.

d) Sodium Effects on the Cladding. A phenomenon that has only recently been understood and is still not widely appreciated is the nonmetallic mass transfer that takes place in sodium-cooled reactor systems (Natesan, Kassner and Li 1972-73). Basically, carbon and nitrogen are diffused out of the higher temperature portions of the system, such as the reactor core, and are deposited in the cooler parts, such as the intermediate heat exchanger. This transfer is highly dependent on the temperature of the system and the amount of carbon and nitrogen in the steel and in the sodium.

Whereas the effects of carbon or nitrogen loss on the mechanical properties of cladding are currently being investigated (Weeks, Kassner and Weins 1973), the possible influence of this loss on cladding swelling and creep rates may only be inferred

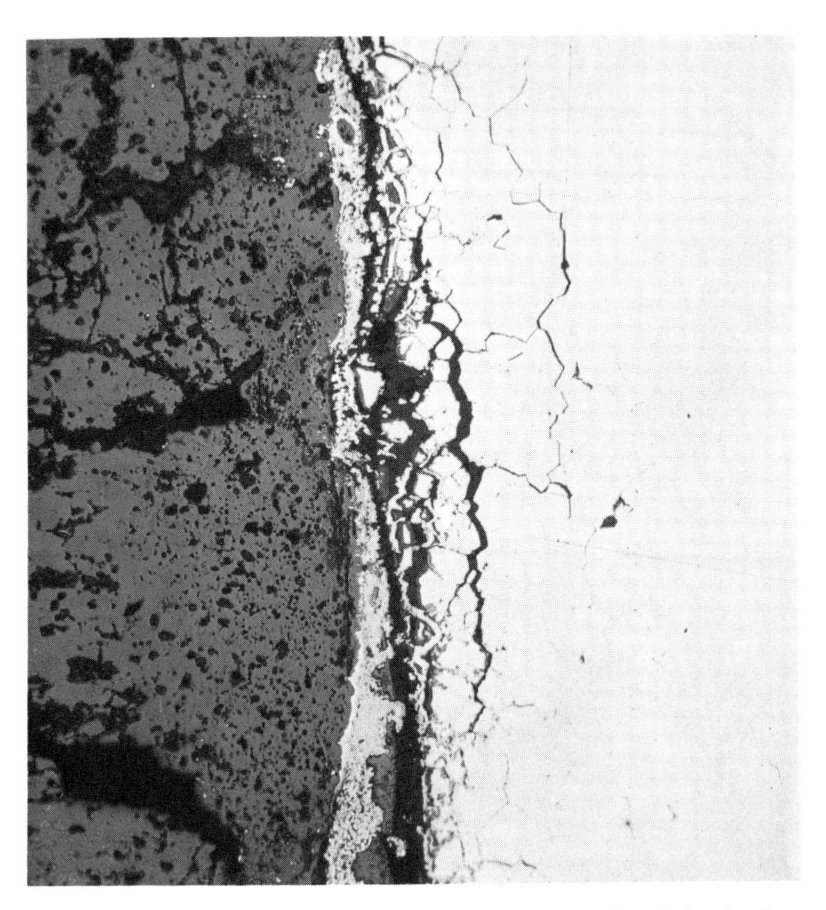

Fig. 15.10 Example of intergranular attack of fuel-element cladding by fission products. (Courtesy of L.A. Neimark, Argonne Nat. Lab.)

as a result of the lack of controlled data. The effects on the creep-rupture life of the cladding, however, have been shown to be severe and possibly design limiting (Natesan, Kassner and Li 1972-73). A computer program to predict the carbon loss or gain in sodium-system components has recently been developed by Snyder, Natesan and Kassner (1973). A typical carbon level-time-temperature profile map from the program is shown in fig. 15.11. By combining this program with the LIFE code and the data on the mechanical properties of decarburized cladding, the effects of sodium on the constitutive behavior of the cladding and on cladding-failure criteria may be assessed.

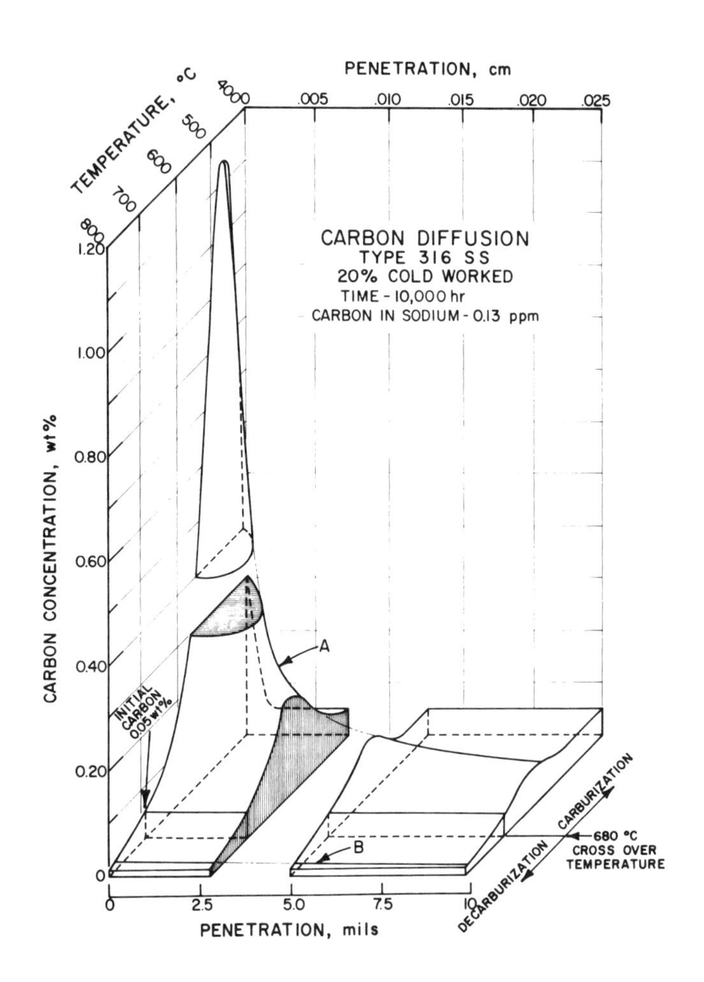

Fig. 15.11 Effect of temperature between 400 and 800°C on carburization-decarburization behavior of 20% cold-worked Type 316 stainless steel after 10,000-hr exposure to sodium containing 0.13 ppm carbon (from Snyder, Natesan and Kassner, 1973, courtesy of Argonne Nat. Lab.)

15.4 SUMMARY

The LIFE-code representation of fast-reactor fuel-element behavior represents a concerted attempt to predict the behavior of a complex materials system in a mechanistic way. The nature of the system is such, furthermore, that less ambitious representations usually yield results of limited value.

In developing the LIFE representation, the conventional view of constitutive relations has been expanded to include a great variety of materials behavior, including fuel swelling, hot pressing and creep, fuel cracking and healing, cladding swelling and creep, fuel-cladding mechanical and chemical interactions, and the effects of sodium on the cladding. This development has been aided, of course, by the relatively simple geometry of the problem. Much work remains to be done on the model, especially in the areas of plutonium migration in the fuel, the incorporation of primary creep, the continued development of the GRASS swelling model, and the inclusion of fission-product attack and sodium effects in a cladding-failure criterion. But the work to date has provided the designer with a powerful tool, the potential of which, is only beginning to be realized.

ACKNOWLEDGEMENTS

The authors wish to thank their fellow workers at Argonne National Laboratory for their invaluable help in this effort. In addition, Dr. Paul Shewmon of the National Science Foundation and Dr. Che-Yu Li of Cornell University provided much helpful advice during the course of the work. This work was performed under the auspices of the U. S. Atomic Energy Commision.

REFERENCES

Ashby, M. F., (1972) Acta Met., 20, 887.

Diaz, J.W., and Poeppel, R. B., (1972) USAEC Report, Argonne National Laboratory, ANL-7992.

Harkness, S. D., and Li, Che-Yu (1971) Trans. AIME, 2, 1457.

Harkness, S. D., Tesk, J. A., and Li, Che-Yu, (1969) Trans. Am. Nucl. Soc., 12, 523.

Jankus, V. Z., (1973) Proc. 2nd Intl. Conf. Struc. Mech. in Reactor Tech., Berlin.

Jankus, V. Z., and Weeks, R. W., (1970) USAEC Report, Argonne National Laboratory, ANL-7736.

Jankus, V. Z., and Weeks, R. W., (1972) Nucl. Eng. and Design, 18, 83.

MacKenzie, J. K., and Shuttleworth, R., (1949) Proc. Phys. Soc. (London), 62B, 833.

Maiya, P. S. and Busch, D. E., (1973) Met. Trans., 4, 663.

Michels, L. C., Poeppel, R. B., and Neimark, L. A., (1970) Trans. Am. Nucl. Soc., 13, 601.

Natesan, K., Kassner, T. F., and Li, Che-Yu, (1972-73) Reactor Tech., 15, 244.

Nichols, F. A., (1969) J. Nucl. Mater., 30, 143.

Pati, S. R., Dapht, M. J., O'Boyle, D. R., and Patrician, T. J., (1973) USAEC Report, Argonne National Laboratory, ANL-8003.

Poeppel, R. B., (1971) Proc. Fast Reactor Fuel Element Tech., New Orleans, 227.

Poeppel, R. B., Makenas, B. J., and Michels, L. C., (1971) Conf. Abstrs., Joint Full Mtg. (TMS-AIME, ASM, MPIF, ASNT), Detroit, 140.

Roberts, J. T. A., Routbort, J. L., Voglewede, J. C., and Solomon, A. A., (1973) USAEC Report, Argonne National Laboratory, ANL-8028.

Roberts, J. T. A., and Wrona, B. J., (1973) J. Am. Cer. Soc., 56, 297.

Robertson, J. A. L., Ross, A. M., Notley, M. J. F., and MacEwan, J. R., (1962) J. Nucl. Mater., 7, 225.

Rosa, F., Maiya, P. S., and Weeks, R. W., (1974) Proc. Spring Mtg. Metals Prop. Council, Miami.

Rossi, R. C., and Fulrath, R. M., (1965) J. Am. Ceram. Soc., 48, 558.

Snyder, R. B., Natesan, K., and Kassner, T. F., (1973) USAEC Report, Argonne National Laboratory, ANL-8015.

Sutherland, W. H., (1971) Hanford Engineering Development Laboratory, private communication.

Weeks, R. W., Jankus, V. Z., Katsuragawa, M., and Lambert, J. D. B., (1971) Proc. Fast Reactor Fuel Element Tech., New Orleans, 227.

Weeks, R. W., Kassner, T. F., and Weins, J. J., (1973) Proc. Intl. Conf. Creep Fatigue in Elevated Temp. Appl., Philadelphia.

Weeks, R. W., Scattergood, R. O., and Pati, S. R., (1970) J. Nucl. Mater., 36, 223.

LIST OF SYMBOLS

A	molar volume of fission gas deemed incompressible (15.4)
A	an empirical constant for creep law (15.6)
B	an empirical constant for creep law (15.6)
C_1	multiplication constant in expression for pore velocity (15.2)
C_2	temperature in expression for pore velocity (15.2)
E	Young's modulus (15.3)
h_{fc}	conductance of the gap between fuel and cladding
h_{gap}	conductance of gas in gap
h_0	pressure-less contact conductance of gap (15.1)
k	Boltzmann's constant (15.6)
k_{gas}	thermal conductivity of gas in fuel-cladding gap (15.1)
m	an exponent in creep law (15.6)
n	an exponent in creep law (15.6)
n_i	number of fission gas moles in i-th region (15.4)
P	volume fraction porosity (15.5)
P_i	average hydrostatic pressure in fuel in the i-th region (15.4)
P'_i	average supplementary pressure due to bubble surface tension in the i-th region
P_{fc}	interface pressure between fuel and cladding (15.1)
P_0	characteristic pressure in contact conductance (15.1)
Q	average activation energy for thermal creep (15.6)
r	radius in cylindrical coordinates

R — the universal gas constant

T — absolute temperature

$\bar{T}_i$ — average temperature in the i-th region (15.4)

V_i — original volume of the i-th region (15.4)

v_{pore} — pore velocity (15.2)

α — average coefficient of thermal expansion, so that total linear expansion is αT (15.3)

Δt — time interval (15.6)

Δx — width of the gap (15.1)

$\Delta\varepsilon_p$ — equivalent creep-strain increment (15.6)

$\Delta\varepsilon^p_{r,\theta,z}$ — creep strain increments occurring during time interval Δt in the radial, hoop, axial directions

$\Delta\varepsilon^s_{r,\theta,z}$ — swelling strain increments during time interval Δt in the radial, hoop, axial directions

$\varepsilon^p_{r,\theta,z}$ — radial, hoop or axial components of creep strain accumulated at time t

$\varepsilon^s_{r,\theta,z}$ — components of swelling strain at time t in radial, hoop, or axial directions

$\varepsilon^T_{r,\theta,z}$ — components of total strain in radial, circumferential, or axial directions at time $(t + \Delta t)$

$\dot{\varepsilon}_{hp}$ — volumetric strain rate due to hot pressing (15.5)

$\dot{\varepsilon}_{ssc}$ — steady state creep rate (15.5)

ν — Poisson's ratio (15.3)

σ_e — equivalent uniaxial stress (15.6)

$\sigma_{r,\theta,z}$ — components of stress at time $(t + \Delta t)$

$\sigma^i_{r,\theta,z}$ — average radial, hoop or axial stresses in the i-th region (15.4)

ϕ — neutron flux

Underlined numbers refer to references at the ends of chapters.